Quantum Computing in Solid State Systems

Quantum Computing in Solid State Systems

With 170 Figures

Edited by:

B. Ruggiero

Istituto di Cibernetica "E. Caianiello" CNR
Italy

P. Delsing

Chalmers University of Technology
Sweden

C. Granata

Istituto di Cibernetica "E. Caianiello" CNR
Italy

Y. Pashkin

RIKEN/NEC
Tsukuba, Japan

P. Silvestrini

Seconda Università di Napoli
Italy

Under auspices of Regione Campania

Springer

Library of Congress Control Number: 2005927931

ISBN-10: 0-387-26332-2
ISBN-13: 978-0387-26332-8

Printed on acid-free paper.

Printed in the United States of America. (SPI/HAM)

9 8 7 6 5 4 3 2 1

springer.com

Preface

The volume reports some fundamental aspects of quantum physics, enhancing the connection between the quantum behavior of macroscopic systems and information theory.

The aim of the volume is to report on the recent theoretical and experimental results on the macroscopic quantum coherence of mesoscopic systems, as well as on solid state realization of qubits and quantum gates. Particular attention has been given to coherence effects in Josephson devices. Other solid state systems, including quantum dots, optical, ions and spin devices, exhibiting macroscopic quantum coherence, have also been discussed.

For the applied aspect we have tried to collect discussions relevant to practical implementation of the quantum computing and information processing devices and in particular observations of quantum behavior in several solid state systems. On the theoretical side, the complementary expertise of the contributors provides models of the various structures in connection with the problem of minimizing decoherence.

Our previous volumes on this field have been ennobled by the first observations of Macroscopic Quantum Coherence in mesoscopic systems, two decades after Leggett's proposal to experimentally test the quantum behavior of macroscopic systems using Josephson devices. The current volume proposes, among many stimulating results, several mesoscopic systems exhibiting a quantum two-level system behavior controlled by external signals to work as qubits, as well as the first realization of coupled qubits to work as conditional gates.

The volume is one of results of the scientific spreading of MQC2 Association on 'Macroscopic Quantum Coherence and Computing" in collaboration with Città della Scienza and the Istituto Italiano per gli Studi Filosofici, under the auspices of the Italian Society of Physics (SIF) and Assessorato Ricerca Scientifica of Regione Campania. We are indebted to V. Corato, S. Rombetto and R. Russo for scientific assistance.

Berardo Ruggiero
Paolo Silvestrini

Contents

List of Corresponding Authors

K.Arutyunov	University of Jyvaskyla, Finland konstantin.arutyunov@phys.jyu.fi
P. Bertet	Delft University of Technology, The Netherlands bertet@qt.tn.tudelft.nl
S. Chung	Western Michigan University, USA chung@wmich.edu
V.Corato	SUN and IC "E. Caianiello" CNR, Italy v.corato@cib.na.cnr.it
S.Corlevi	Royal Institute of Technology, Stockholm, Sweden corlevi@fy.chalmers.se
J.E.Elzerman	University of Technology, Delft, The Netherlands elzerman@qt2.tn.tudelft.nl
D. Esteve	CEA-Saclay, France esteve@drecam.saclay.cea.fr
M. Everitt	University of Sussex, UK M.J.Everitt@sussex.ac.uk
R. Fazio	Scuola Normale Superiore, Italy fazio@sns.it
T. Fujisawa	NTT Atsugi, Japan fujisawa@will.brl.ntt.co.jp
H.S. Goan	University of New South Wales, Australia goan@physics.uq.edu.au
N. Grønbech- Jensen	University of California at Davis, USA ngjensen@ucdavis.edu
T.Hakioglu	Bilkent University, Turkey hakioglu@fen.bilkent.edu.tr
M.D. Kim	Seoul National University, Korea mdkim@phya.snu.ac.kr
S.Komiyama	University of Tokyo, Japan skomiyama@thz.c.u-tokyo.ac.jp
A. Konstadopoulou	University of Bradford, UK akonstad@bradford.ac.uk

E. Pazy	University of the Negev, Israel epazy@bgumail.bgu.ac.il
T. Roscilde	University of Southern California, USA roscilde@usc.edu
R. Schäfer	Forschungszentrum Karlsruhe, Germany Roland.Schaefer@ifp.fzk.de
F. Sciarrino	Università di Roma "La Sapienza", Italy fabio.sciarrino@uniroma1.it
I. Siddiqi	Yale University, USA irfan.siddiqi@yale.edu
M. Sillanpää	Helsinki University of Technology, Finland masillan@cc.hut.fi
R. Simmonds	NIST, USA simmonds@boulder.nist.gov
J. Sjöstrand	Stockholm University, Sweden js@physto.se
P. Sodano	Università di Perugia, Italy sodano@pg.infn.it
R. M. Stevenson	Toshiba Research Europe Ltd., UK mark.stevenson@crl.toshiba.co.uk
P.B. Stiffell	University of Sussex, UK P.B.Stiffell@sussex.ac.uk
V. Tognetti	Università di Firenze, Italy tognetti@fi.infn.it
A.V. Ustinov	Erlangen University, Germany ustinov@physik.uni-erlangen.de
A. Vourdas	University of Bradford, UK A.Vourdas@bradford.ac.uk
T. Yamamoto	NEC/RIKEN, Tsukuba, Japan yamamoto@frl.cl.nec.co.jp
S.P. Yukon	Air Force Research Laboratory, MA, USA stanford.yukon@hanscom.af.mil
M. Wallquist	Chalmers University of Technology, Sweden mwallquist@fy.chalmers.se
A. Zorin	PTB, Brannschweig, Germany Alexander.Zorin@ptb.de
P. Cappellaro	MIT Cambridge, USA pcappell@mit.edu
G. Campagnano	Delft, The Netherlands g.campagnano@tnw.tudelft.nl
F. Chiarello	INFN – CNR,Roma,Italy Chiarello@ifn.cnr.it
B. Militello	Università di Palermo, Italy bdmilitello@fisica.unipa.it
A.Naddeo	Università di Napoli "Federico II", Italy naddeo@na.infn.it
M.Scala	Università di Palermo, Italy scala@unipa.it

1

Decoherence of a Josephson Quantum Bit during its Free Evolution: The Quantronium

G. ITHIER[1], E. COLLIN[1], P. JOYEZ[1], P.J. MEESON[1], D. VION[1], D. ESTEVE[1], F. CHIARELLO[2], A. SHNIRMAN[3], Y. MAKHLIN[4], AND G. SCHÖN[3]

[1] Quantronics Group, Service de Physique de l'Etat Condensé, Direction des Sciences de la Matière, CEA-Saclay, 91191 Gif-sur-Yvette, France
[2] Istituto di Fotonica e Nanotecnologia, CNR, Via Cineto Romano, 42 00156 Roma, Italy
[3] Institut für Theoretische Festkörperphysik, Universität Karlsruhe, 76128 Karlsruhe, Germany
[4] Landau Institute For Theoretical Physics, Kosygin st. 2, 119334 Moscow, Russia

Abstract

Relaxation and coherence times of the quantronium, a Josephson Quantum Bit (qubit), as a function of its operating point have been measured. Several coherence times are obtained from complementary techniques such as resonance linewidth, Ramsey fringes, detuning pulse, or spin-echo measurements. We find we can explain their variations by a simple model involving modified l/f charge and phase noise spectral densities.

1.1 Introduction

Superconducting circuits based on Josephson junctions can behave quantum mechanically with a coherence time long enough to perform simple manipulations of their quantum state.[1–3] These circuits are potential candidates for implementing quantum bits. Nevertheless, decoherence due to the coupling between the qubit and its environment still severely hinders using these circuits for the development of a quantum processor.[4] The quantitative characterization and understanding of decoherence processes is thus a central issue. In this work, we present experiments that characterize decoherence in a particular Josephson qubit, the quantronium,[2] for which atomic physics-like and NMR-like manipulation has been demonstrated.[5]

In the next section, the experimental setup and its noise sources are described, and the relevant decoherence rates is introduced. In Section 3, characterization of decoherence at different operating points are explained, and the experimental results within a simple model for the noise sources are discussed.

1.2 The Quantronium Circuit and its Decoherence Sources

1.2.1 *Principle*

The quantronium circuit (Fig. 1.1) is derived from the Cooper pair box.[6–8] It consists of a superconducting loop interrupted by two adjacent small Josephson junctions with Josephson energies $E_J(1 \pm d)/2$, with d an asymmetry coefficient made as small as possible, and by a larger Josephson junction ($E_{J0} \approx 15E_J$) for readout. The island between the small junctions, with total capacitance C_Σ and charging energy $E_C = (2e)^2/2C_\Sigma$, is biased by a voltage source U through a gate capacitance C_g. The eigenstates of this system are determined by the dimensionless gate charge $N_g = C_g U/2e$, and by the superconducting phase $\delta = \gamma + \varphi$ across the two small junctions, where γ is the phase across the large junction and $\varphi = \phi/\varphi_0$, with ϕ the external flux through the loop and $\varphi_0 = \hbar/2e$. The two lowest energy states $|0\rangle$ and $|1\rangle$ form a qubit, whose transition frequency υ_{01} depends on the bias point $P = (\delta, N_g)$. At the optimal working point $P_0 = 01(\delta = 0, N_g = 1/2)$, υ_{01} is stationary, which makes the quantronium insensitive to noise at first order.[7,8] For readout,[2,8] a trapezoidal readout pulse $I_b(t)$ with a peak value slightly below $I_0 = E_{J0}/\varphi_0$ is applied so that the switching of the large junction to a finite voltage state is

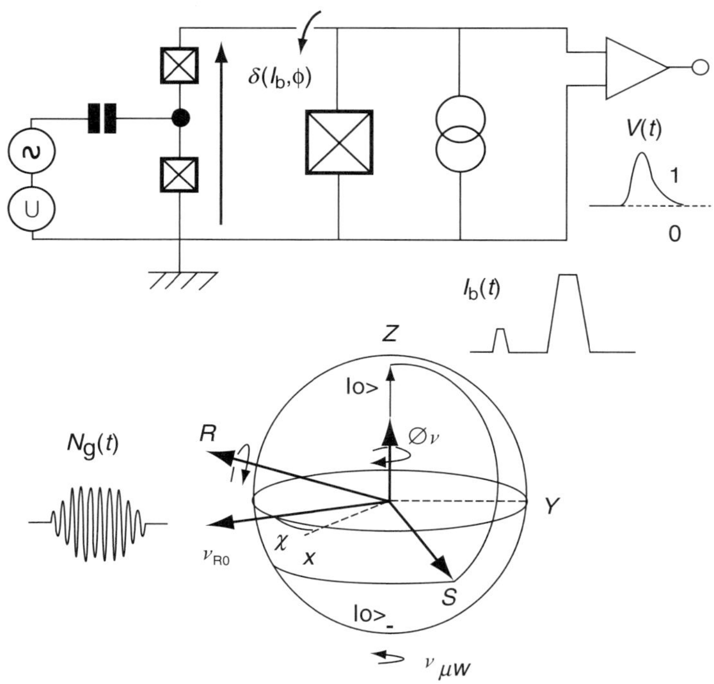

FIGURE 1.1. Top: Circuit diagram of the quantronium qubit. The control parameters are $N_g = C_g U/2e$ and the phase δ, which is determined by the flux ϕ imposed through the loop and by the bias-current I_b. The switching to finite voltage of the larger readout junction in response to an $I_b(t)$ pulse enables discrimination of the qubit states. Bottom: Bloch sphere representation in the rotating frame. Gate microwave pulses induce rotations of the effective spin S around the R axis, whereas adiabatic $N_g(t)$ or $I_b(t)$ pulses induce rotations around Z.

induced with a large probability p_1 for state $|1\rangle$ and with a small probability p_0 for state $|0\rangle$. The manipulation of the qubit state is achieved by applying $N_g(t)$ and $I_b(t)$ pulses. When a nearly resonant microwave modulation $\Delta N_g \cos(2\pi \upsilon_{\mu w} t + \chi)$ is applied to the gate, the Hamiltonian $\hat{h}$ described in a frame rotating at the microwave frequency, is that of a spin 1/2 in an effective magnetic field $\vec{H}$:

$$\hat{h} = -\vec{H}.\vec{\hat{\sigma}}/2,\ \vec{H} = h\Delta\upsilon\vec{z} + h\upsilon_{R0}[\vec{x}\cos\chi + \vec{y}\sin\chi] \tag{1.1}$$

Here, $\Delta\upsilon = \upsilon_{01} - \upsilon_{\mu w}$ is the detuning, and $\upsilon_{R0} = 2E_C\Delta N_g\langle 1|\hat{N}|0\rangle/h$ the Rabi frequency. Rabi precession of the spin takes place around an axis with polar angles $\theta = \pi/2$ -arctan $(\Delta\upsilon/\upsilon_{R0})$ and χ. Free evolution corresponds to a rotation of the spin around Z at frequency $-\Delta\upsilon$. Rotations around Z can thus be performed by changing υ_{01} using adiabatic N_g or I_b detuning pulses.[5]

1.2.2 Experimental implementation

The sample used in this work was fabricated using standard e-beam lithography and double angle shadow evaporation of aluminum. The readout junction was also connected to a parallel on-chip coplanar capacitor $C_j \approx 0.6$pF, in order to lower its plasma frequency. Separate gates with capacitances 40 and 80 aF were used for the DC and microwave N_g signals, respectively. The sample was mounted in a copper shielding box thermally anchored to the mixing chamber of a dilution refrigerator with base temperature 15 mK. The impedance of the microwave gate line as seen from the qubit was defined by a 50Ω attenuator placed at 600 mK. That of the DC gate line was defined below 100 MHz by a 1kΩ resistor at 4 K, and its real part was measured to be close to 80Ω in the 6–17 GHz range explored by the qubit frequency. The bias resistor of the readout junction, $R_b = 4.1$kΩ, was placed at the lowest temperature. Both the current bias line and the voltage measurement lines were shunted above a few 100 MHz by two surface mounted 100Ω–47pF RC shunts. The microwave gate pulses used to manipulate the qubit were generated by mixing continuous microwaves with 1 ns rise time trapezoidal pulses of variable duration τ (defined as the time between 50% of the rise and 50% of the fall). The switching probability p was measured over 25000–60000 events with a 10–60 kHz repetition rate. The relevant parameters $E_J = 0.87k_BK$, $E_{C=0.66k_BK}$, $\nu_{01}(P_0) = 16.41$GHz, $d = 3$–4%, $I_0 = 427$nA were measured as reported in.[9] The readout sensitivity was optimized by using 100 ns readout pulses ending at $\delta_M \approx 130^{deg}$. The fidelity, *i.e.*, the largest achieved value of $p_1 - p_0$, was $\eta \approx 0.3 - 0.4$, which is greater than in our previous work,[2] but nevertheless much smaller than the expected value $\eta \sim 0.95$. This loss of visibility is attributed to spurious relaxation of the qubit during the adiabatic ramp of the readout pulse. In this regard, we note that the signal loss after 1 microwave π pulse is approximately the same that after three adjacent π pulses.

1.2.3 Decoherence sources

The quantronium is subject to decoherence due to its interactions with uncontrolled degrees of freedom inducing noise in $\lambda = N_g$ or $\lambda = \delta/2$ (we neglect here noise on E_J, which is smaller). Decoherence is described here in terms of relaxation of the qubit energy into a quantum system on one hand, and in terms of random dephasing between the two qubit states due to adiabatic variations of ω_{01}, on the other hand. Each process is conveniently described by a quantum spectral density $S_\lambda(\omega) = 1/(2\pi)\int dt\langle\delta\lambda(0)\delta\lambda(t)\rangle\exp(-i\omega t)$ that quantifies, at positive ω, the ability of the source to absorb one energy quantum $\hbar\omega$. Classical spectral densities

$Sc_\lambda(\omega) = 1/2[S_\lambda(-\omega) + S_\lambda(+\omega)]$ when $k_BT \gg \hbar|\omega|$ are also used. Fig. 1.2 presents the main noise sources that have been identified, the actual circuit being represented as an effective equivalent circuit for decoherence.

The gate series impedance Z_g gives a spectral density[7,8]

$$S^G_{N_g}(\omega) = \kappa_g^2 \frac{\hbar^2\omega}{E_C^2}\left[1 + \coth\left(\frac{\hbar\omega}{2k_BT}\right)\right]\frac{\mathrm{Re}\left[Z_g(\omega)\right]}{R_k} \tag{1.2}$$

with $\kappa_g = C_g/C_\Sigma$ and $R_k = h/e^2$. Using the parameters mentioned previously, we find

$$Sc^G_{N_g}(|\omega| < 2\pi.10\mathrm{MHz}) = (30 \times 10^{-9})^2/(\mathrm{rad/s}), \tag{1.3}$$

$$S^G_{N_g}(6\mathrm{GHz} < \omega < 17\mathrm{GHz}) = \left(1 - 3 \times 10^{-9}\right)^2/(\mathrm{rad/s}). \tag{1.4}$$

The admittance Y_R in parallel with the readout junction gives[7,8]

$$S^R_{\delta/2\pi}(\omega) = \frac{\mathrm{Re}[Y_R(\omega)]}{(\varphi_0\omega|Y_\delta|)^2}\frac{\omega}{2\pi}\left[1 + \coth\left(\frac{\omega}{2k_BT}\right)\right] \tag{1.5}$$

where Y_δ is the parallel combination of Y_R and of the readout junction inductance L_J. Using the actual parameters of the sample, we find

$$Sc^R_{\delta/2\pi}(|\omega| < 2\pi.10\mathrm{MHz}) = (2 \times 10^{-9})^2/(\mathrm{rad/s}), \tag{1.6}$$

$$S^R_{\delta/2\pi}(6\mathrm{GHz} < \omega < 17\,\mathrm{GHz}) = \left(80 - 20 \times 10^{-9}\right)^2/(\mathrm{rad/s}). \tag{1.7}$$

In addition, the noise δI_b of the Arbitrary Waveform Generator (AWG) used for the readout pulse has been measured to be white up to 200 MHz, and corresponds to a spectral density

$$Sc^{AWG}_{\delta/2\pi}(|\omega| < 2\pi.200\mathrm{MHz}) = (15 \times 10^{-9}/\cos\gamma)^2/(\mathrm{rad/s}). \tag{1.8}$$

Besides, as any other Coulomb blockade device, the quantronium suffers from a Background Charge Noise (BCN) due to charged two-level fluctuators. The corresponding spectral density is of the $1/f$ type at low frequency,

$$Sc^{\mathrm{BCN}}_{N_g}(|\omega| < 2\pi.100\mathrm{kHz}) \sim A_g/|\omega|, \tag{1.9}$$

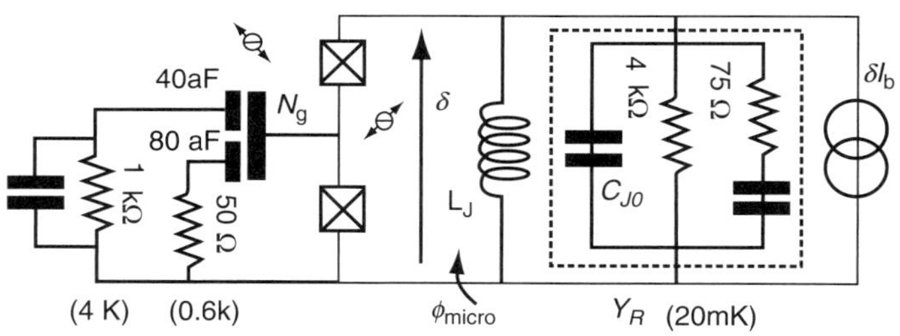

FIGURE 1.2. Equivalent circuit diagram for decoherence. The 50 Ω and 1 kΩ resistances represent the real part of the series gate impedance at the qubit frequency and at low frequency, respectively. Inductance L_J represents the readout junction whereas the 75 Ω and 4 kΩ resistances are parallel admittances at the qubit and at low frequency, respectively. Microscopic charged two-level fluctuators are displayed as double arrows and flux fluctuators are encapsulated in ϕ_{micro}.

and has an amplitude A_g commonly found in the range $[10^{-6}, 10^{-7}]$ for the parameters of our experiment. Finally, as any other SQUID type device, the quantronium should experience a phase $1/f$ noise:[10]

$$Sc^{\text{micro}}_{\delta/2\pi}(|\omega| < 2\pi.1\text{kHz}) \sim A_\delta/|\omega| \; A_\delta \approx 10^{-10}. \tag{1.10}$$

1.2.4 *Theoretical relaxation and decoherence rates*

We consider here the case when, after its initial preparation, the effective spin precesses freely under the influence of the field $\vec{H}(\lambda)$ including its classical and quantum fluctuations. One distinguishes two time scales, the relaxation time T_1 for the decay of the longitudinal Z component of the spin density matrix, and the decay time T_2 of the transverse part, which is the coherence time. In addition, we also consider the case of spin-echoes, *i.e.*, when a π pulse is applied to the qubit in the middle of a period of free evolution in order to eliminate the effect of low frequency noise.[5] T_2 is in that case replaced by T_E.

Applying first the Fermi golden rule, one finds the relaxation rate[8]

$$T_1^{-1} = \pi D_{\lambda,\perp}^2 S_\lambda(\omega_{01})/2 \tag{1.11}$$

where $D_{\lambda,\perp}$ is the transversal component of $\vec{D}\lambda = \;^{-1}\partial\vec{H}/\partial\lambda$. From the sample parameters we compute

$$D_{\delta/2\pi,\perp}(\delta = 0\text{or } N_g = 1/2) = 380d\left[1 + 6.0\left(\frac{\delta}{2\pi}\right)^2\right]10^9 \text{ rad/s}, \tag{1.12}$$

$$D_{N_g,\perp}(\delta = 0 \text{ or } N_g = 1/2) = 193 \times 10^9 \text{ rad/s}. \tag{1.13}$$

Then we compute[5,8] the average factors $f_{z,R}(t) = \langle\exp[i\varphi(t)]\rangle$ and $f_{z,E}(t) = \langle\exp[i\varphi_2(t/2) - i\varphi_1(t/2)]\rangle$ involving the phases $\varphi(t)$ or $\varphi_2(t/2) - \varphi_1(t/2)$ accumulated during a period t of free evolution without and with a π pulse at half time, respectively. Assuming that the noise $\delta\lambda$ is Gaussian and assuming a non vanishing linear coupling to the noise, $\partial\omega_{01}/\partial\lambda = D_{\lambda,z} \neq 0$, a simple semi-classical calculation gives[7,8]

$$f_{z,R}(t) = \exp\left[-\frac{t^2}{2}D_{\lambda,z}^2 \int_{-\infty}^{\infty} d\omega S_\lambda(\omega)\,\text{sin c}^2\left(\frac{\omega t}{2}\right)\right], \tag{1.14}$$

$$f_{z,E}(t) = \exp\left[-\frac{t^2}{2}D_{\lambda,z}^2 \int_{-\infty}^{\infty} d\omega S_\lambda(\omega)\sin^2\left(\frac{\omega t}{4}\right)\text{sin c}^2\left(\frac{\omega t}{4}\right)\right]. \tag{1.15}$$

Taking now relaxation into account, the transverse polarization decays actually as $g(t) = f_z(t)$ $\exp[-t/2T_1]$. We define T_2 and T_E by $g(T_{2,E}) = 1/e$. It is interesting to notice that when the noise power S_λ is smooth near $\omega \approx 0$, $f_{z,R}(t) = \exp(-t/T_\varphi)$ with $T_\varphi^{-1} = \pi D_{\lambda,z}^2 S_\lambda(0)$. From the sample parameters we compute

$$D_{\delta/2\pi,z}(\delta = 0 \text{ or } N_g = 1/2) = -850\frac{\delta}{2\pi}10^9 \text{ rad/s}, \tag{1.16}$$

$$D_{N_g,z}(\delta = 0 \text{ or} N_g = 1/2) = +290(N_g - 1/2)10^9 \text{ rad/s}. \tag{1.17}$$

At P_0, the two $D_{\lambda,z}$'s vanish and a second order calculation of the dephasing involving $\partial^2\omega_{01}/\partial\lambda^2$ is required.[11,12] The sample parameters lead to $\partial^2\omega_{01}/\partial(\delta/2\pi)^2 = -850 \times 10^9$ rad/s and $\partial^2\omega_{01}/\partial(N_g)^2 = +290 \times 10^9$ rad/s.

1.3 Experimental Characterization of Decoherence During Free Evolution

1.3.1 *Relaxation Time (T_1) measurement*

Relaxation of the longitudinal polarization is measured from the decay of the switching probability p after a π pulse. Fits by an exponential decay[5] lead to the relaxation times T_1, which vary with P as shown in Fig. 1.3: T_1 is about 0.5μs in the vicinity of P_0 and show rapid variations by a factor up to 4 away from P_0 in the phase direction. In the parameter range explored, the matrix element $D_{N_g,\perp}$ is approximately constant, whereas $D_{\delta/2\pi,\perp}$ varies smoothly with δ by a factor of only 2. Consequently, the measured variation of T_1 reflects the variation with frequency of the density of environmental modes coupled to the qubit. Can the relaxation be fully attributed to the biasing and measuring circuit? To answer this question, we compute T_1 at P_0, from Eq. (1.11) and from the noise spectra Eqs. (1.4)–(1.7) due to Z_g and Y_R. We obtain values of about 2μs and 3–6 μs, respectively. The combined effect of Z_g and Y_R gives thus $T_1 \approx 1 - 1.5\mu$s, which is 2–3 times longer than the measured value. The error with which we can estimate impedances above 14 GHz could be as high as a factor 2. We thus conclude that the circuit contribution to relaxation is significant and that the contribution of microscopic environmental modes is undetermined.

1.3.2 *Coherence time (T_2 and T_E) measurements*

T_2 was determined using three different methods. The first one is the Ramsey fringe method.[2,5] It consists in measuring directly the temporal decay of the average transverse polarization of the spin using two microwave $\pi/2$ pulses (detuned by a few tens of MHz) separated by Δt. The switching probability at the end of the sequence oscillates with Δt at the frequency $\Delta\upsilon$, with an amplitude decaying as $[1 + g_R(\Delta t)]/2$ up to small corrections due to detuning. The second method uses an adiabatic detuning N_g or δ pulse in the middle of a Ramsey sequence performed at the optimal point P_0 [5]. This detuning pulse moves the system temporarily during a time Δt_1 from P_0 to a point P where T_2 is to be measured. The measured signal oscillates with Δt_1 at the new detuning frequency $\Delta\upsilon(P)$, with amplitude that decays with a characteristic time $T_2(P)$. The usefulness of the method is that the parameters of the Ramsey pulses can be optimized and kept constant for all points P. These first two methods are of course applicable only to points where $T_2 > 1/\Delta\upsilon(P)$. When T_2 is too short, it is more convenient to operate in the frequency domain, by recording the line shape of the resonance line at low microwave power (desaturated line). In this case the lineshape is the Fourier transform of the Ramsey signal. One has $T_2 = \alpha/(\pi\,\Delta\upsilon_{\mu w})$ where $\Delta\upsilon_{\mu w}$ is the full width at half maximum and α is a coefficient that depends on the exact shape of the line (1 for Lorentzian, 0.8 for Gaussian). In practice, desaturating the line leads to a rather low signal to noise ratio and it is difficult to discriminate between the two shapes. Near P_0, the Ramsey decays look close to exponential and we take $\alpha = 1$. Anyway, our error bars on T_2 are larger than the difference between the two α's. Spin-echo-like experiments were performed by inserting a microwave π pulse in the middle of a Ramsey sequence.[5,8] It was possible experimentally to vary the sequence duration Δt while keeping the π pulse precisely in

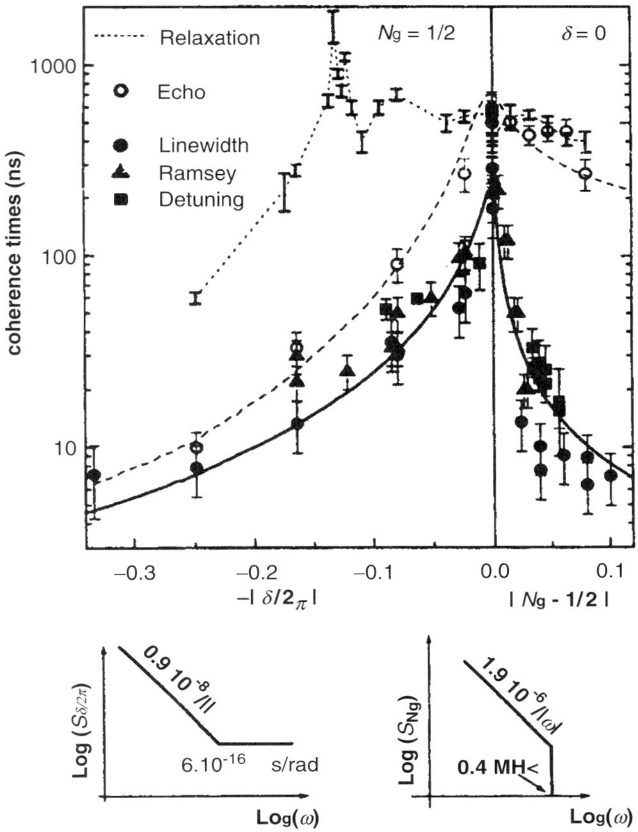

FIGURE 1.3. Summary of all the characteristic times, T_1 (dotted line), T_2 (solid symbols), and T_E (open circles), as a function of the biasing point. For T_2, the legend indicates the measurement method. The vertical line separating the two top frames corresponds to the optimal point P_0. The dashed and solid lines are fits of the T_2 and T_E's, assuming that N_g and δ noises have the modified $1/f$ spectral densities shown in the bottom. The fitting parameters are also indicated.

the middle of the sequence. Using this protocol, one maps directly the echo maximal amplitude as a function of Δt.[5] Dephasing makes this amplitude decrease as $[1 + g_E(\Delta t)]/2$. Examples of the experimental signals recorded with all methods described above are found in,[2,5] and extensive results will be published elsewhere. We summarise in Fig. 1.3 all T_1, T_2 and T_E values obtained at different P. This plot forms the main result of this work.

1.3.3 Discussion

Figure 1.3 shows first that decoherence of this quantronium is limited at all P by pure dephasing rather than by relaxation. Nevertheless $T_E(P_0)$ approaches the limit $2T_1$. Next, we observe that Ramsey decays "detuning pulses," and lineshapes measurements all lead to T_2's in reasonable agreement with each other. The concept of optimal point P_0, where T_2's and T_E's are maximum, is also justified. The echoes are seen to be much more efficient away from P_0 in the charge direction than away from P_0 in the phase direction. This result shows that a large part of the charge noise is at frequency lower than $T_E^{-1} \approx 1\,\text{MHz}$. Another important result is that the T_2's are too short for the decoherence to be explained by the dissipative elements of the circuit. Injecting spectral densities Eqs. (1.3), (1.6) and (1.8) in Eqs. (1.14)–(1.17) leads to theoretical $T_2(P_0)$ values of a few tenth of second for the gate line, of 160 μs for the readout impedance and of 7 μs for the white noise from the AWG, respectively. The main sources of noise are thus of microscopic nature. Given that their spectral densities are known to be essentially of the $1/f$ type, we tried to fit the data with pure $1/f$ spectra, using Eqs. (1.14)–(1.17). The fits were good for the T_2's. However we computed $T_E/T_2 \approx 4$ everywhere (as expected from Ref. [8]), which is too large in the phase direction and too small in the charge direction, when compared to the data. Consequently, we are led to introduce purely empirically a high frequency cut-off in the charge noise and a white contribution to the phase noise, as shown in Fig. 1.3. The corresponding best fits (shown on Fig. 1.3) are now in better agreement with the data. Note that the second order contribution of the noises, which dominate very close to P_0, were roughly estimated assuming Gaussian N_g^2 or δ_g^2 noises. More accurate estimations also lay within our experimental error bars and confirm that the main contribution to decoherence at P_0 originates from the microscopic charge noise. The fitted amplitude $A_g = 1.6 \times 10^{-6}$ of the charge noise is in the expected range whereas that for the phase noise, $A_\delta = 0.9 \times 10^{-8}$, is 100 times larger than expected from literature. The cut-off found at 0.4 MHz was not expected and it would be interesting to check this result with a fast electrometer experiment. Finally, the amplitude 6×10^{-16}/(rad/s) of the white phase noise corresponds to the contribution of the bias current generator.

1.4 Conclusion

We have characterized decoherence in the quantronium using techniques adapted from NMR and atomic physics. We have found that whereas relaxation might be limited by the circuit, microscopic noise sources are responsible for decoherence. We have validated the concept of optimal operating point P_0 and shown that for $E_J/E_C \sim 1$, the main source of decoherence at P_0 is charge noise. We have analysed the data using a simple model for the interaction of the qubit with its environment, and for the noise sources. A more complete analysis including decoherence during driven evolution will be published later.

Acknowledgments. The essential technical contributions of P.F. Orfila, P. Senat and J.C. Tack, and the financial support from the European Commission (SQUBIT2) and of the "Dynasty Foundation" are gratefully acknowledged.

References

1. Y. Nakamura, Yu. A. Pashkin, J. S. Tsai, *Nature* **398**, 786 (1999).
2. D. Vion *et al.*, *Science* **296**, 886 (2002).
3. I. Chiorescu *et al.*, *Science* **299**, 1869 (2003).
4. M. A. Nielsen and I. L. Chuang, *Quantum Computation and Quantum Information* (Cambridge University Press, Cambridge, 2000).
5. E. Collin *et al.*, *Phys. Rev. Lett.* **93**, 157005 (2004).

6. V. Bouchiat *et al.*, *Phys. Scr.* **T76**, 165 (1998).
7. A. Cottet, PhD thesis, UniversitÉParis VI, (2002); *www-drecam.cea.fr/ drecam/spec/Pres/Quantro/*
8. D. Vion, in *Quantum Entanglement and Information Processing*, D. Estève, J.M. Raimond, and J. Dalibard (eds), (Elsevier, 2004), and refs therein.
9. G. Ithier *et al.*, *Phys. Rev. Lett.*
10. F. C. Wellstood, C. Urbina and J. Clarke, *Appl. Phys. Lett.* **50**, (1987).
11. Y. Makhlin and A. Shnirman, *Phys. Rev. Lett.* **92**, 178301 (2004).
12. E. Paladino *et al.*, cond-mat/0407484 (2004).

2

Conditional gate operation in superconducting charge qubits

T. YAMAMOTO[1,2,3], YU. A. PASHKIN[2], O. ASTAFIEV[2], Y. NAKAMURA[1,2,3], AND J. S. TSAI[1,2,3]

[1] NEC Fundamental and Environmental Research Laboratories, Tsukuba, Ibaraki 305-8501, Japan
[2] The Institute of Physical and Chemical Research (RIKEN), Wako, Saitama 351-0198, Japan
[3] CREST, Japan Science and Technology Agency (JST), Kawaguchi, Saitama 332-0012, Japan

Abstract

A variety of Josephson-junction-based qubits have recently been implemented with remarkable progress in coherence time and read-out scheme. These developments, together with its potential scalability, have renewed the position of this solid-state device as a strong candidate as a building block for the quantum computer. On the other hand, coupling multiple qubits to construct a logic gate is another important step toward the realization of quantum computer. In this paper, we present our experiments on conditional gate operations using coupled superconducting charge qubits.

2.1 Introduction

Low-capacitance Josephson junction offers a way to build an artificial two-level system, which can be used as a quantum bit (qubit) for the quantum information processing.[1] The potential of this system as a qubit has been already confirmed by many kinds of Josephson-junction-based qubits implemented so far.[2–6] Recently, researchers have been trying to improve the *quality* of the single qubit by, for example, implementing new read-out scheme with high efficiency and low back action[7–9] or by importing the technique of nuclear magnetic resonance to efficiently control the quantum state.[10] Needless to say, decoherence in superconducting qubits have been also one of the most important issues.[11–15]

Besides all this, coupling multiple qubits is another important issue, because to realize the universal gate in quantum computation, two-qubit conditional gate is required on top of the single-qubit rotation gate. The first experiment on coupled superconducting qubits has been reported by our group using capacitively-coupled Cooper-pair boxes.[16] By applying pulse voltage on the gate, two qubits are brought to the resonance simultaneously, which induces the beating

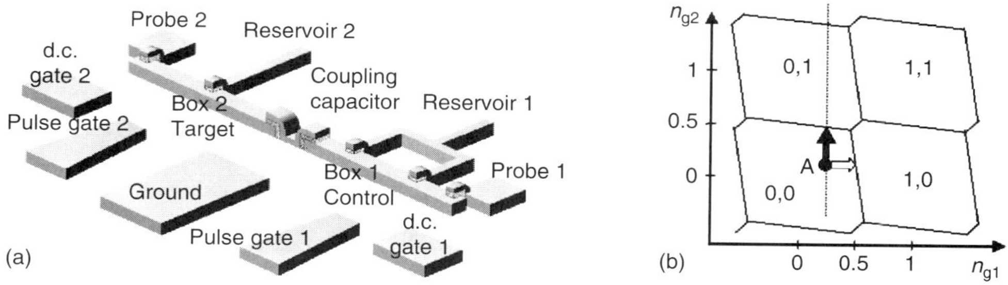

FIGURE 2.1. (a) Schematic layout of two capacitively coupled qubits. The coupling capacitor is realized by using an extra island overlapping each of the boxes. The characteristic energies of this sample estimated from the d.c. current-voltage measurements are $E_{c1} = 580$ μeV, $E_{c2} = 671$ μeV and $E_m = 95$ μeV. From the pulse measurements, E_{J1} is found to be 45 μeV at a maximum and E_{J2} to be 41 μeV. The superconducting energy gap is 209 μeV. (b) Ground-state charging diagram of the coupled charge qubit system. Point A is the operation point and the black and white arrows from Point A represent the pulse voltage for the conditional gate operation and input preparation, respectively.

signal in the output current. After this, two-qubit experiments in other types of superconducting qubits have also been reported.[17–19] However, two-qubit conditional gate had not yet been realized.

Recently, we have successfully demonstrated the conditional gate operation using the device similar to that used in Ref. [16]. By utilizing the difference of the degeneracy condition between two pairs of the charge states, namely, |00>, |01> and |10>, |11>, we can flip the state of the target qubit only when the control qubit is in the |0> state. The main results of the experiment have been published in our previous study.[20] In this chapter, the experimental setup and procedure for that experiment is described in more details.

2.2 Experimental Details

2.2.1 *Device Structure*

Figure 2.1(a) shows the schematic device structure of coupled superconducting charge qubits. The device was fabricated by electron-beam lithography and three-angle evaporation of Al on a SiN_x insulating layer above a gold ground plane on the oxidized Si substrate. Two Cooper-pair boxes are capacitively coupled by an on-chip capacitor. The right qubit has Superconducting Quantum Interference Device (SQUID) geometry so that we can control the Josephson coupling of the box to the reservoir by an external magnetic field. In the experiment, we use this qubit as the control qubit and the left one as the target qubit, although this structural asymmetry is not essential for the logic operation. Both qubits have independent pulse gates and they enable us to address each qubit individually. We measure the pulse-induced currents through probes 1 and 2, which are biased at ~ 650 μV to enable Josephson-Quasiparticle (JQP) cycle. These currents are proportional to the probability of the respective qubit having one extra Cooper pair.

2.2.2 Operation Scheme

In the two-qubit charge basis |00>, |10>, |01> and |11>, the Hamiltonian of the coupled charge qubit system is given as

$$H = \sum_{n1,n2=0,1} E_{n1n2}|n1, n2 >< n1, n2|\frac{E_{J1}}{2} \sum_{n2=0,1} (|0 >< 1| + |1 >< 0|) \otimes |n2 >< n2|$$
$$\frac{E_J2}{2} \sum_{n1=0,1} |n1 >< n1| \otimes (|0 >< 1| + |1 >< 0|), \qquad (2.1)$$

where E_{J1} (E_{J2}) is the Josephson coupling energy of the first (second) box to the reservoir, $E_{n1n2} = E_{c1}(n_{g1} - n_1)^2 + E_{c2}(n_{g2} - n_2)^2 + E_m(n_{g1} - n_1)(n_{g2} - n_2)$ is the total electrostatic energy of the system (n_1, n_2 = 0, 1 is the number of excess Cooper pairs in the first and second boxes, and $n_{g1,2}$ are the gate-induced charges on the corresponding qubit divided by $2e$). $E_{c1(2)} = 4e^2C_{\Sigma 2(1)}/2(C_{\Sigma 1}C_{\Sigma 2} - C_m^2)$ are the effective Cooper-pair charging energies ($C_{\Sigma 1(2)}$ are the sum of all capacitances connected to the corresponding island including the coupling capacitance C_m between the two boxes). Finally, $E_m = 4e^2C_m/(C_{\Sigma 1}C_{\Sigma 2} - C_m^2)$ is the coupling energy.

Figure 2.1(b) shows the ground-state charging diagram of the coupled charge qubit system.[21] The charge states shown in the figure denote the ground states in each hexagonal cell in the absence of Josephson coupling. At boundaries, charging energies of two (or three at triple point) neighbouring states degenerate. Pulsed voltage applied to Pulse gates 1 and 2 move the system non-adiabatically along n_{g1} and n_{g2} axes, respectively. In our previous experiment to demonstrate quantum beating,[16] we first set the operation point (point A) sufficiently far away from the co-resonant point (n_{g1}, n_{g2})=(0.5,0.5) and brought the system non-adiabatically to the co-resonant point, where four eigenenergies become close to each other and let the system evolve freely. For the conditional gate, however, we fix n_{g1} as a constant value and operate along the dashed line in the figure.

Figure 2.2(a) shows the energy bands of the present system along the dashed line in Fig. 2.1(b). Here, four energy bands can be regarded as two pairs of nearly independent single-qubit energy bands. For the lower two bands, the first qubit (control qubit) is always in the |0> state, while for the higher two bands, the first qubit is in the |1> state. Importantly, the charging energies of each of the two-level systems degenerate at different n_{g2}. This difference originates from the electrostatic coupling between the qubits and can be utilized for the conditional gate operation.

For the conditional gate operation, we apply the voltage pulse to Pulse gate 2 so that it brings the system to the degeneracy point for lower two bands (Fig. 2.2(a)). Suppose we start from the |00 > state. Application of this pulse induces the oscillation between |00> and |01> states with maximum amplitude, as schematically shown by the Bloch sphere in Fig. 2.2(b). By properly timing the length of the pulse, the oscillation can be stopped when the system is in the |01> state. By the same pulse, we obtain |00> state from the input state of |01>. On the other hand, when the initial state is |10> or |11>, the pulse induces oscillation between |10> and |11> states. However, because the system is not brought to the degeneracy point for |10> and |11> states, the oscillation amplitude is suppressed due to the finite fictitious magnetic field along z axis (Fig. 2.2(c)). The magnitude of this fictitious field is proportional to the coupling energy. In the ideal case, input states remain the same. Thus, this pulse performs the conditional gate operation, i.e., the state of the target qubit is flipped only when the control qubit is in the |0> state.

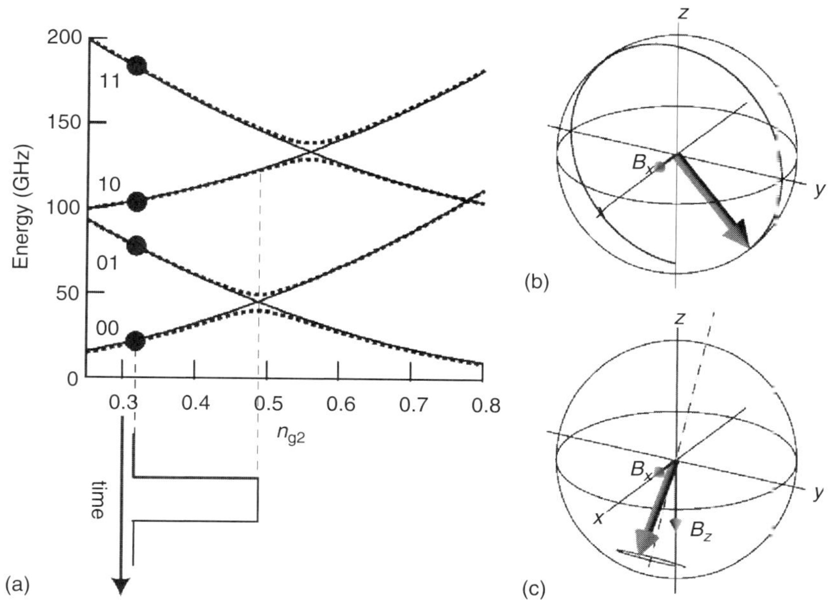

FIGURE 2.2. (a) Energy bands calculated from the hamiltonian (1). Dashed lines represent the eigenenergies and solid lines represent the charging energies of the states shown in the figure. They are plotted as functions of n_{g2}, while n_{g1} is fixed as 0.18. The rectangular-shape pulse represents the conditional gate operation. (b), (c) Single qubit Bloch sphere picture of the target qubit. When the control qubit is in the $|0>$ state (b), the operation pulse realizes the resonance for the target qubit, hence Bloch vector rotates around x axis with maximum amplitude. On the other hand, when the control qubit is in the $|1>$ state (c), the operation pulse does not realize the resonance and there is a finite fictitious magnetic field along z axis, which leads to the suppressed oscillation amplitude.

2.2.3 *Experimental Setup*

The sample chip is mounted in a shielded copper box and is cooled down to the base temperature of the dilution refrigerator (~ 40 mK). The overall wiring is almost the same as that reported in Ref. [22], except that now we have two coaxial cables for the transmission of fast voltage pulses. As the high-speed pulse generator, we use an Anritsu MP1758A pulse-pattern generator. It can create an arbitrary digital pattern with Non-Return Zero (NRZ) pulses for up to four channels. Although the digital pattern and the output level can be set independently for each channel, the base frequency, which is tunable up to 12.5 GHz, is common for all channels. Thus, once the base frequency is fixed, say, at 10 GHz, all the digital patterns are created with the unit length of 100 ps.

In the experiment reported in Ref. [20], two of these channels are used to control the two qubits. Figure 2.3(a) shows one of the pulse sequences utilized in Ref. [20]. It is the real output signal from the pulse generator measured at the top of the cryostat. The outputs of channels 1 and 2 are sent to Pulse gate 1 (control qubit) and Pulse gate 2 (target qubit), respectively. The inset shows the corresponding bit pattern to operate the generator. Here, the base frequency is set to 11.8 GHz, thus one bit corresponds to 85 ps and the repetition time is $85 \times 1505 = 127925$ ps ≈ 128 ns.

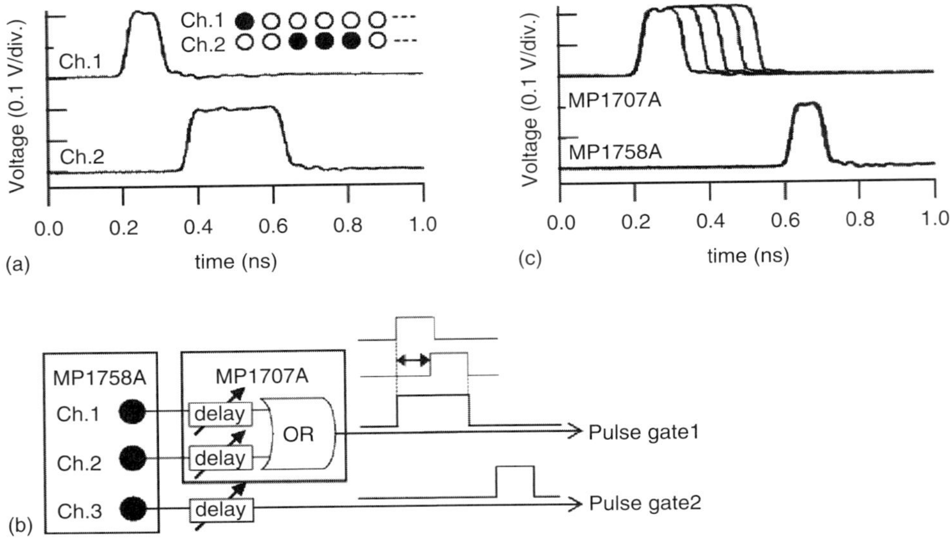

FIGURE 2.3. (a) Output signal of the Anritsu MP1758A pulse-pattern generator. The inset shows the corresponding bit pattern to generate this pulse pattern. The nominal length of one bit is set as 85 ps. This pulse sequence is used to demonstrate conditional gate operation, although the amplitudes of the pulses must be properly adjusted. (b) Schematic diagram to illustrate how to use the Anritsu MP1707A pulse controller with pulse-pattern generator. The length of the output pulse of MP1707A can be changed by the internal phase shifters, while keeping the frequency of MP1758A constant. (c) An example of the pulse pattern produced by the setup is described in (b). Here the base frequency of MP1758A is fixed as 12.5 GHz. The nominal length of the pulse from MP1758A (lower signal) is fixed as 80 ps, while the length of the pulse from MP1707A (upper singnal) is changed from 100 to 300 ps with 50 ps steps.

The first pulse applied to Pulse gate 1 creates a superposition state of the control qubit. Therefore, the state $\alpha|00> + \beta|10>$ is prepared as an input state. The second pulse is applied to Pulse gate 2 and it corresponds to the conditional gate operation explained in the previous section. Accordingly, this pulse is expected to transform the input state to $\alpha|01> + \beta|10>$. In Ref. [20], we swept a magnetic field to change the coefficients α and β. The SQUID geometry of the control qubit enables us to control E_{J1} by a magnetic field. Since E_{J1} gives the oscillation frequency during the first pulse, α and β can be changed by a magnetic field, while keeping the pulse length constant.

An alternative (and probably more natural) way to change α and β is to change the length of the first pulse. For that purpose, we use an Anritsu MP1707A pulse controller. It is a logic device with tunable delay lines. The setup of this device combined with MP1758A is shown in Fig. 2.3 (b). Two outputs of MP1758A are used as inputs of MP1707A. They are logically operated (logical OR) and one output is produced by MP1758A as shown in the figure. Since the delay for the inputs can be set with a step of increment less than 1 ps, the length of the output pulse from MP1707A can also be changed by the same step. Figure 2.3(c) shows an example of the pulse pattern produced by this setup. We can change the length of the first pulse, while keeping that of the second fixed. If we set the length of the second pulse properly, we can demonstrate the conditional gate operation by this pulse sequence.

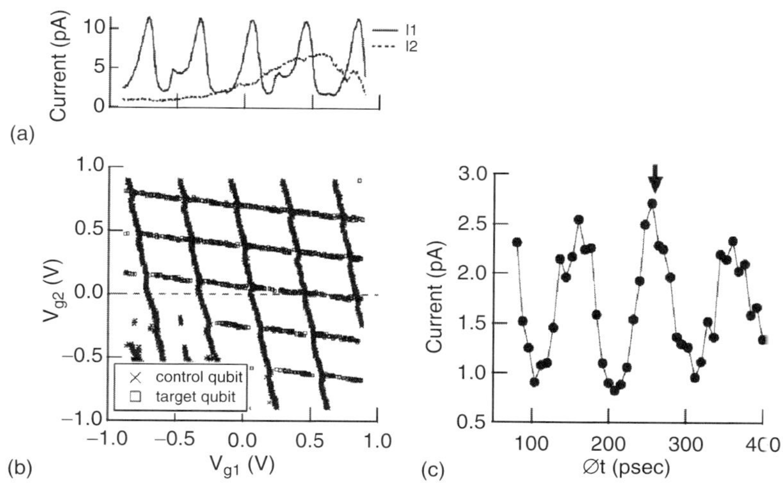

FIGURE 2.4. (a) Gate modulation of the current of the control qubit (I_1) and the target qubit (I_2). Here, the gate voltage on d.c. gate 1 is swept, while that on d.c. gate 2 is kept as zero (the dashed line in Fig. 2.4(b)). (b) Positions of JQP peaks plotted in the $V_{g1} - V_{g2}$ plane. The irregular periodicity observed around $V_{g1} = V_{g2} = -0.5$ is probably due to the change of the charge numbers in the coupling island. (c) Dependence of I_2 on the length of the pulse applied to Pulse gate 2. V_{g1} and V_{g2} are properly chosen so that this oscillation occurs at the degeneracy point for |00> and |01> states. The arrow represents the length of the pulse for the conditional gate operation.

2.3 Results and Discussion

2.3.1 *Operation Point*

To demonstrate conditional gate operation, we need first to determine the operation point by tuning two d.c. gate voltages. Figure 2.4 (a) shows the gate modulation of the current of the control qubit (I_1) and the target qubit (I_2). The gate voltage on d.c. gate 1 is swept, while d.c. gate 2 is kept unbiased. 1*e*-periodic JQP peaks and 2*e*-periodic current steps are observed as shown in the figure.[22] We measured the same gate modulation with different voltage on d.c. gate 2 and plot the position of JQP peaks in $V_{g1} - V_{g2}$ plane (Fig. 2.4(b)). Step-like features are observed in every line of peaks, from which we can estimate the strength of the capacitive coupling between the qubits. Note that the JQP peaks have 1*e* periodicity. Therefore, the honeycomb structure in Fig. 2.4(a) is smaller than that of Fig. 2.1(b) by four times in the area, apart from the difference of the dimension of two axes.

To determine the operation point, we sweep two d.c. gate voltages simultaneously along the line of $V_{g2} = V_{g1} + V_o$ (V_o: constant) with the application of fixed pulse voltage on either Pulse gate 1 (V_{p1}) or Pulse gate 2 (V_{p2}). By adjusting the value of V_o, we can find proper value of V_{g1} and V_{g2}, at which point the system is brought to the resonance for |00> and |10> by V_{p1} or the resonance for |00> and |01> by V_{p2}. Figure 2.4(c) shows the dependence of I_2 on the length of the pulse applied to Pulse gate 2 at properly chosen V_{g1} and V_{g2}. The coherent oscillation with the frequency of $E_{J2}/\hbar$ is observed. The length of the pulse for conditional gate operation should be fixed at the peak of this oscillation, 255 ps, for example.

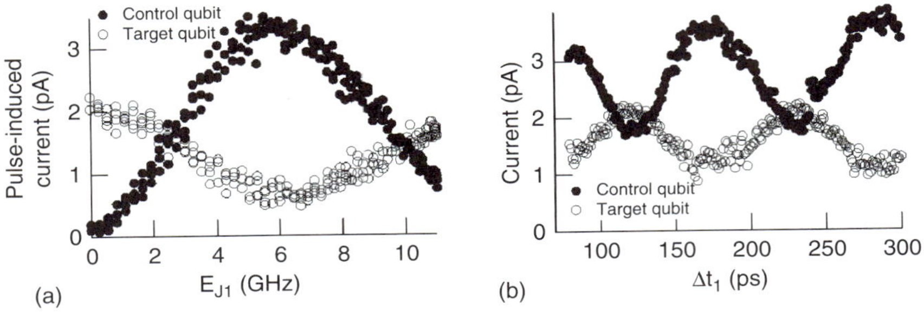

FIGURE 2.5. (a) Pulse-induced current as a function of the control qubit under the application of the pulse sequence shown in Fig. 2.3(a). (b) Output currents of the two qubits as a function of the length of the pulse for the input preparation shown in Fig. 2.3(c).

2.3.2 *Conditional Gate Operation*

Now that we determined the pulse for the conditional gate operation, we can perform this operation for various input states by applying the pulse sequences shown in Fig. 2.3. In Fig. 2.5(a), the pulse-induced current under the application of the pulse sequence shown in Fig. 2.3(a) is plotted as a function of E_{J1}. As discussed in Section 2.2.3, this pulse sequence should produce the state $\alpha|01> +\beta|10>$, which gives the output current of $I_1 \propto |\beta|^2$ and $I_2 \propto |\alpha|^2 = 1 - |\beta|^2$. The anti-correlation between the two output currents observed in Fig. 5(a) is consistent with this expectation. Quantitatively, the modulation amplitude of the current of the target qubit agrees well with the result of the numerical simulation, which takes into account the finite rise and fall time of the pulse ($\sim$ 40 ps).[20] The amplitude gets smaller from the theoretical maximum $2e/T_r = 2.5$ pA mainly due to the imperfection of both the input preparation and the gate operation. On the other hand, the modulation amplitude of the current of the control qubit exceeds 2.5 pA. We cannot yet clearly explain this, but probably this is due to the extra pulse-induced current channel (different from the JQP process), whose amplitude depends on the Josephson energy. This is the imperfection of our present read-out scheme.

The above result indicates that, at least qualitatively, the operation pulse really works as a conditional gate. However, it would be better if the coefficients α and β can be controlled by the pulse length, not by a magnetic field, for more complicated operation in the future. Figure 2.5(b) shows I_1 and I_2 under the application of the pulse pattern shown in Fig. 2.3(b). The sample used in this measurement is different from that used in Ref. (nature), although the overall structure is the same. Here, by using Anritsu MP1707A pulse controller, the width of the second pulse Δt_2 is fixed as 80 ps, while that of the first pulse Δt_1 is swept from 80 to 300 ps. The delay between the two pulses Δt_{12} is not fixed, because the final result is not sensitive to the phase evolution during Δt_{12}, as long as Δt_{12} is much shorter than the energy relaxation time T_1. The magnetic field is fixed so that E_{J1} takes its maximum value (41 μeV). In Fig. 2.5(b), the two output currents are plotted as a function of Δt_1. I_1 oscillates with the frequency of 10 GHz and I_2 shows clear anti-correlation with I_1. This is the demonstration of the conditional gate operation with a different way of input preparation.

2.4 Conclusions

The concept and experimental details concerning the conditional gate operation using capacitively-coupled superconducting charge qubits have been described. The two-qubit solid-state circuit was controlled by applying a sequence of pulses and the conditional gate operation as for the amplitude of the quantum state demonstrated. For the realization of the quantum C-NOT gate, it is also needed to examine the phase evolution during the gate operation. In addition, it is desirable that the strength of the coupling can be externally controlled. Several schemes of the tunable coupling, which are applicable to the present system, have been theoretically proposed[23–25] and the experiments are now underway.

Acknowledgments. We thank B. L. Altshuler, D. V. Averin, S. Ishizaka, J. Lantz, S. Lloyd, F. Nori, T. Tilma, C. Urbina, G. Wendin and J. Q. You for many fruitful discussions.

References

1. Y. Makhlin, G. Schön, and A. Shnirman, Quanrum-state engineering with Josephson-junction devices, *Rev. Mod. Phys.* **73**, 357 (2001).
2. Y. Nakamura, Yu. A. Pashkin, and J. S. Tsai, Coherent control of macroscopic quantum states in a single-Cooper-pair box, *Nature* **398**, 786–788 (1999).
3. D. Vion *et al.*, Manipulating the quantum state of an electrical circuit, *Science* **296**, 886–889 (2002).
4. Y. Yu, S. Han, X. Chu, S. I. Chu, and Z. Wang, Coherent temporal oscillations of macroscopic quantum states in a Josephson junction, *Science* **296**, 889–892 (2002).
5. J. M. Martinis, S. Nam, J. Aumentado, and C. Urbina, Rabi oscillations in a large Josephson-junction qubit, *Phys. Rev. Lett.* **89**, 117901 (2002).
6. I. Chiorescu, Y. Nakamura, C. J. P. M. Harmans, and J. E. Mooij, Coherent quantum dynamics of a superconducting flux qubit, *Science* **299**, 1869–1871 (2003).
7. O. Astafiev, Yu. A. Pashkin, T. Yamamoto, Y. Nakamura, and J. S. Tsai, Single-shot measurement of the Josephson charge qubit, *Phys. Rev. B* **69**, 180507(R) (2004).
8. I. Siddiqi *et al.*, An rf-driven Josephson bifurcation amplifier for quantum measurements, cond-mat/0312623.
9. A. Lupacu, C. J. M. Verwijs, R. N. Schouten, C. J. P. M. Harmans, and J. E. Mooij, Nondestructive readout for a superconducting flux qubit, cond-mat/0311510.
10. E. Collin, G. Ithier, A. Aassime, P. Joyez, D. Vion, and D. Esteve, NMR-like control of a quantum bit superconducting circuit, cond-mat/0404503.
11. Y. Nakamura, Yu. A. Pashkin, T. Yamamoto, and J. S. Tsai, Charge echo in a Cooper-pair box, *Phys. Rev. Lett.* **88**, 047901 (2002).
12. K. W. Lehnert, K. Bladh, L. F. Spietz, D. Gunnarsson, D. I. Schuster, P. Delsing, and R. J. Schoelkopf, *Phys. Rev. Lett.* **90**, 027002 (2003).
13. T. Duty, D. Gunnarsson, K. Bladh, and P. Delsing, Coherent dynamics of a Josephson charge qubit, *Phys. Rev. B* **69**, 140503(R) (2004).
14. R. W. Simmonds, K. M. Lang, D. A. Hite, D. P. Pappas, and J. M. Martinis, Decoherence in Josephson Qubits from Junction Resonances, cond-mat/0402470.
15. O. Astafiev, Yu. A. Pashkin, Y. Nakamura, T. Yamamoto, and J. S. Tsai, private communication.
16. Yu. A. Pashkin *et al.*, Quantum oscillations in two coupled charge qubits, *Nature* **421**, 823–826 (2003).
17. A. J. Berkley *et al.*, Entangled macroscopic quantum states in two superconducting qubits, *Science* **300**, 1548–1550 (2003).
18. J. B. Majer, F. G. Paauw, A. C. J. ter Haar, C. J. P. M. Harmans, and J. E. Mooij, Spectroscopy on two coupled flux qubits, cond-mat/0308192.
19. A. Izmalkov *et al.*, Experimental evidence for entangled states formation in a system of two coupled flux qubits, cond-mat/0312332.
20. T. Yamamoto, Yu. A. Pashkin, O. Astafiev, Y. Nakamura, and J. S. Tsai, Demonstration of conditional gate operation using superconducting charge qubits, *Nature* **425**, 941–944 (2003).

21. H. Pothier, P. Lafarge, C. Urbina, D. Esteve, and M. H. Devoret, Single-electron pump based on charging effects, *Europhys. Lett.* **17**, 249–254 (1992).
22. Y. Nakamura and J. S. Tsai, Quantum-state control with a single-Cooper-pair box, *J. Low Temp Phys.* **118**, 765–779 (2000).
23. J. Q. You, J. S. Tsai, and F. Nori, Controllable manipulation and entanglement of macroscopic quantum states in coupled charge qubits, *Phys. Rev. B* **68**, 024510 (2003).
24. D. V. Averin and C. Bruder, Variable electrostatic transformer: Controllable coupling of two charge qubits, *Phys. Rev. Lett.* **91**, 057003 (2003).
25. J. Lantz, M. Wallquist, V. S. Shumeiko, and G. Wendin, Josephson junction qubit network with current-controlled interaction, cond-mat/0403285.

3

Coupling and Dephasing in Josephson Charge-Phase Qubit with Radio Frequency Readout

ALEXANDER B. ZORIN

A. B. Zorin, Physikalisch-Technische Bundesanstalt, Braunschweig, Germany.

Abstract

The Cooper pair box qubit of a two-junction-SQUID configuration enables the readout of the qubit states by probing the effective Josephson inductance of the SQUID. This is realized by coupling the qubit to a high-Q tank circuit which induces a small alternating supercurrent in the SQUID loop. The effect of a small (but finite) geometrical inductance of the loop on the eigenstates of the system is figured out. The effect of qubit dephasing due to quadratic coupling to the tank circuit is evaluated. It is shown that the rate of dephasing in the vicinity of the magic points is relatively low unless the Josephson junctions forming the qubit are rather dissimilar. In the vicinity of the avoided level-crossing point such dephasing is always significant.

3.1 Introduction

The readout device is a critical component of any potential quantum computing circuit. For the Josephson qubits there is a number of sensitive cryogenic devices available (SQUIDs, switching Josephson junctions, single electron transistors and traps, etc.) enabling the readout of the qubit state. However, operation of these devices is usually associated with a significant exchange of energy between detector and qubit, so in order to avoid fast decoherence the detector must be reliably decoupled from the qubit at the time of quantum manipulation. Recently, the class of Josephson qubit detectors based on the measurement of the reactive component of electrical signals related to nonlinear behavior of the Josephson inductance has been extensively studied.[1,2,3,4] Due to specific coupling to the qubit variables and non-dissipative characteristics of the Josephson supercurrent, these circuits can have a much weaker backaction, and therefore, cause lesser decoherence. Moreover, these circuits may possibly enable quantum nondemolition measurements of a Josephson qubit.[5]

In this paper, we consider the charge-phase qubit[1] comprising a macroscopic superconducting ring including two small Josephson junctions with a small island in between (see Fig. 3.1), i.e., a Cooper pair box[6] of SQUID configuration. This setup is, in principle, similar to that of

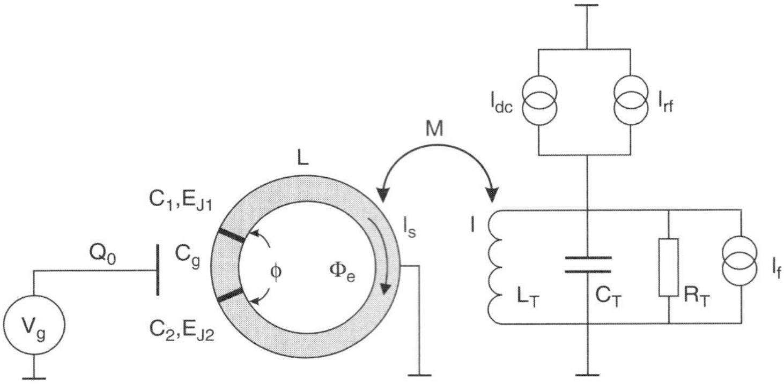

FIGURE 3.1. Electric circuit diagram of the charge-phase qubit. The core of the qubit is the superconducting ring of small inductance L including two small tunnel junctions of capacitances C_1 and C_2 and Josephson coupling energies E_{J1} and E_{J2}, respectively. The small island between the junctions is polarized by charge on small gate capacitance C_g. The ring is inductively (mutual inductance M) coupled to an rf-driven parallel tank circuit. Voltage across the tank circuit is read out by a (presumably cold) preamplifier which is not shown.

Quantronium,[7] in which, however, a larger additional Josephson junction was inserted in the superconducting loop for the purpose of readout by switching this junction into a finite-voltage state. In our case, the ring of inductance L is inductively coupled to the tank circuit that enables both phase control and readout of the qubit. The special convenience of this simple setup is that it can function even without uncoupling the tank circuit from the qubit. It has recently been shown that operation of this qubit in the so-called magic points made it possible to decouple (in linear approximation) both the diagonal and the off-diagonal components of the variables of the measuring system (environment) from the qubit.[8] Moreover, the resonance shape of the noise spectrum allowed significant reducing relaxation, and in principle, performing single shot measurements. Here we analyze the quadratic effect of this noise and show that dephasing caused by this effect in the vicinity of the magic operation points is small, provided the qubit is sufficiently symmetric.

3.2 Qubit Parameters and the Model

For the vanishingly small value of inductance L the behavior of the charge-phase qubit is primarily determined by the relationship between the energies associated with the island and the junctions. These are the Coulomb energy of the island, $E_c = e^2/2C_\Sigma$, and the Josephson coupling energies of the junctions, $E_{J1,J2} = \hbar I_{c1,c2}/2e$, where $C_\Sigma = C_1 + C_2 + C_g$ is the island's total capacitance including the capacitances of the individual junctions C_1 and C_2 and the gate capacitance $C_g \ll C_{1,2}$; $I_{c1,c2}$ are the nominal values of the Josephson critical currents. We assume that

$$(\Phi_0/2\pi)I_{c0} = E_{J0} \equiv (E_{J1} + E_{J2})/2 \simeq E_c, \tag{3.1}$$

where $\Phi_0 = h/2e$ is the flux quantum, E_{J0} and $I_{c0} = (I_{c1} + I_{c2})/2$ are the average Josephson coupling energy and the average critical current, respectively. The coupling energies of individual junctions are almost similar, $E_{J1} \approx E_{J2}$, so the values of the dimensionless parameters

$j_{1,2} \equiv E_{J1,J2}/2E_{J0} \approx 0.5$, while $j_1 + j_2 = 1$. The junction capacitances are also assumed to be almost similar, $C_1 \approx C_2$, and the dimensionless parameters $\kappa_{1,2} \equiv C_{2,1}/C_\Sigma \approx 0.5$, while $\kappa_1 + \kappa_2 = 1$. The qubit island is capacitively coupled to the gate voltage source V_g which controls the polarization charge $Q_0 = C_g V_g$. The inductively coupled coil L_T carrying current I induces the external magnetic flux $\Phi_e \equiv (\Phi_0/2\pi)\phi_e = -MI$ applied to the qubit loop and produces the sum phase bias $\phi \equiv \varphi_1 + \varphi_2 = \phi_e$. The core Hamiltonian includes the charging (kinetic) energy and Josephson (periodic potential) term,

$$H_0 = 4E_c(n - Q_0/2e)^2 - E_J(\phi)\cos\chi, \tag{3.2}$$

where the operator n of the number of excess Cooper pairs on the island is

$$n = -i\frac{\partial}{\partial\varphi} = -i\frac{\partial}{\partial\chi}, \qquad [n,\varphi] = [n,\chi] = i. \tag{3.3}$$

The operators of the island phases are $\varphi = (\varphi_1 - \varphi_2)/2$ and $\chi = \varphi + \gamma(\phi)$, where the angle deviation due to asymmetry of the junctions $\gamma = \arctan[(j_1 - j_2)\tan(\phi/2)]$. The amplitude of the effective Josephson coupling energy of two junctions is equal to

$$E_J(\phi) = \left(E_{J1}^2 + E_{J2}^2 + 2E_{J1}E_{J2}\cos\phi\right)^{1/2}. \tag{3.4}$$

The two lowest eigenvalues of energy, $E_0(Q_0, \phi)$ (ground state) and $E_1(Q_0, \phi)$ (first excited state), with corresponding eigenstates $|0\rangle$ and $|1\rangle$, form the basis suitable for qubit operation. In this basis the Hamiltonian Eq. (3.2) takes the diagonal form

$$H_0 = -(\epsilon/2)\sigma_z, \tag{3.5}$$

where energy $\epsilon(Q_0, \phi) \equiv \hbar\Omega = E_1 - E_0$ and σ_z is the Pauli matrix. The energy landscape is $2e$ periodic in Q_0 and 2π periodic in ϕ. The plots can be in found in (Refs. [7] and [8]). The reverse Josephson inductance of the qubit in the ground (excited) state $L_{0,1}^{-1}$ is determined by the local curvature of the corresponding energy surface,

$$L_{0,1}^{-1}(Q_0, \phi) = \left(\frac{2\pi}{\Phi_0}\right)^2 \frac{\partial^2 E_{0,1}(Q_0, \phi)}{\partial\phi^2}. \tag{3.6}$$

Depending on the value of the ratio E_{J0}/E_c and the relationship between the Josephson energies of the junctions, j_1/j_2, the inductance $L_{0,1}$ can take either positive or negative values in different points on the Q_0-ϕ plane. In the so-called magic points (extremum or saddle points), i.e., A ($Q_0 = 0$, $\phi = \pi$), B ($Q_0 = 0$, $\phi = 0$) and C ($Q_0 = e$, $\phi = 0$), the absolute values of $L_{0,1}^{-1}$ achieve local maxima, while in the avoided level-crossing point D ($Q_0 = e$, $\phi = \pi$) its value is the largest.[8] Specifically, in point D, where effective coupling is small, $E_J(\pi) = |E_{J1} - E_{J2}| \ll E_c$, these values for two states are equal to[9]

$$L_{0,1}^{-1}(D) = \mp\frac{2\pi}{\Phi_0}\frac{j_1 j_2}{|j_1 - j_2|}I_{c0}. \tag{3.7}$$

The drive frequency ω_{rf} (close to the resonance frequency of the tank circuit $\omega_T = (L_T C_T)^{1/2} \ll k_B T/\hbar$) is much lower than the qubit frequency Ω, so the induced classical oscillations of phase ϕ are adiabatic. Due to coupling to the ring, the effective inductance seen by the tank circuit is[10]

$$L_{\text{eff}}^{(0,1)} = L_T - \frac{M^2}{L + L_{0,1}} \approx L_T - \frac{M^2}{L_{0,1}}. \tag{3.8}$$

As a result, the resonance frequencies take the distinct values for the ground and excited states, $\omega_{0,1} = (L_{\rm eff}^{(0,1)} C_T)^{-1/2}$, and this property is used for the radio frequency readout of the qubit.[1]

3.3 Effect of Finite Inductance of the Ring

The problem of the inductance effects in the persistent current flux qubits had been addressed by Crankshaw and Orlando in Ref. [11] where the corrections to the energy level values of these qubits were found. Recently, Maassen van den Brink had analyzed the finite inductance effect in the three-Josephson-junction qubits considering the self-flux as a "fast variable".[12] He had shown that for small inductances this effect is merely reduced to renormalization of the individual Josephson coupling energies entering the system Hamiltonian.

In the case of the charge-phase qubit, the sum phase ϕ is also no longer a good variable. Instead, its combination with the normalized circulating current (which is a quantum-mechanical operator) is controlled by the external flux Φ_e,

$$\phi + \frac{2\pi}{\Phi_0} L I_s(\phi, \chi) = \phi_e. \tag{3.9}$$

The operator of the circulating current can be presented as[8]

$$I_s = I_{\|}(\phi)\cos\chi + I_{\perp}(\phi)\sin\chi. \tag{3.10}$$

In the given basis, the operator $\cos\chi$ is diagonal and $\sin\chi$ is off-diagonal, so the amplitudes of the longitudinal ($I_{\|}$) and transversal ($I_{\perp}$) components of current I_s are

$$I_{\|} = \frac{8\pi}{\Phi_0} j_1 j_2 \sin\phi \, \frac{E_{\rm J0}^2}{E_{\rm J}(\phi)}, \tag{3.11}$$

$$I_{\perp} = \frac{8\pi}{\Phi_0}\left[(j_1 - j_2)(\kappa_1 j_1 + \kappa_2 j_2) + 2 j_1 j_2 (\kappa_1 - \kappa_2)\cos^2\frac{\phi}{2}\right]\frac{E_{\rm J0}^2}{E_{\rm J}(\phi)}. \tag{3.12}$$

Note that the operator I_s is diagonal for arbitrary values of parameters Q_0 and ϕ only if the junctions are identical, i.e., $j_1 = j_2$ and $\kappa_1 = \kappa_2$, so $I_{\perp} = 0$.

The additional magnetic energy term in the system Hamiltonian is

$$H_{\rm m} = (\Phi_0/2\pi)^2(\phi - \phi_e)^2/2L = L I_s^2(\phi, \chi)/2. \tag{3.13}$$

The kinetic energy term associated with the charging of the chain of the qubit junctions connected in series, $H_{\tilde{C}} = (\Phi_0/2\pi)^2 \tilde{C}\dot{\phi}^2/2$, where $\tilde{C} = C_{\rm qubit} + C_{\rm stray}$. Although the capacitance between the ends of the SQUID loop $C_{\rm stray}$ depends on the sample layout, it clearly dominates over the first term, $C_{\rm qubit} = C_1 C_2 / C_\Sigma$ which is typically about 1 fF.[13] The resonance frequency $\omega_L = (L\tilde{C})^{1/2}$ of the oscillator formed by L and $\tilde{C}$ is therefore relatively low. Assuming $\hbar\omega_L \sim k_B T$, but still $\omega_L \gg \omega_{\rm rf}$ we recover classical behavior of ϕ. In this case it is sufficient to take into account only the magnetic term associated with the circulating current given by Eq. (3.13).

The phase variable ϕ is a single-valued function of ϕ_e for arbitrary values of Q_0, provided L is sufficiently small,

$$L|L_{0,1}^{-1}| \le L|L_{0,1}^{-1}(D)| = \frac{j_1 j_2}{|j_1 - j_2|}\beta_L < 1, \tag{3.14}$$

where the screening parameter is

$$\beta_L = \frac{2\pi}{\Phi_0} L I_{c0}. \tag{3.15}$$

One can see that for an almost symmetric qubit ($j_1 \approx j_2$) the condition of smallness of inductance L, i.e., $\beta_L < 4|j_1 - j_2|$, is more stringent than that for ordinary rf SQUIDs, i.e., $\beta_L < 1$.[14]

To derive the dependence $\phi(\phi_e)$, one can solve Eq. (3.9) by iteration. The first step gives the expression

$$\phi \approx \phi_e - (2\pi/\Phi_0) L I_s(\phi_e, \chi). \tag{3.16}$$

Then the corresponding one-dimensional Schrödinger equation,

$$\left\{ -4E_c \left(\frac{\partial}{\partial \chi} - i\frac{Q_0}{2e} \right)^2 - E_J(\phi_e)\cos\chi - \frac{L}{4}\left[I_\parallel(\phi_e)\cos\chi + I_\perp(\phi_e)\sin\chi \right]^2 \right\} \psi = E\psi, \tag{3.17}$$

is again of the type describing a quantum particle moving in a periodic potential. In the case of $L = 0$, this potential is harmonic and the equation is of the Mathieu type.[15] In the case of nonzero L, the potential energy includes a (small) second harmonic admixture. The eigenvalues and eigenfunctions of this equation are found by numerical methods.

Figure 3.2 shows the two lowest eigenvalues of energy (the qubit levels) computed from the Schrödinger equation Eq. (3.17). For the given parameters of the qubit and not very small value of β_L, the subsequent terms of iteration give substantially smaller correction. One can see that similar to the conclusion of Ref. [11], the first order effect of inductance reduces to the lowering of both energy surfaces, E_0 and E_1. The correction to the level spacing Eq. (3.5) is, however, very small: $\epsilon \to \epsilon + \Delta\epsilon_L$, $\Delta\epsilon_L \ll \epsilon$. As can be seen from Fig. 3.2, the effect of the deformation of the energy surfaces is most appreciable in the region of small effective Josephson coupling

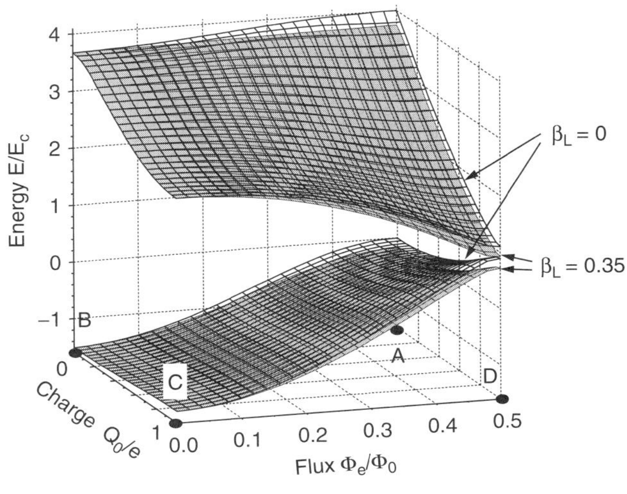

FIGURE 3.2. Energies E_0 and E_1 calculated for the zero (wire frames) and nonzero (surface plots) values of screening parameter β_L for the case of mean Josephson coupling $E_{J0} \equiv \frac{1}{2}(E_{J1} + E_{J2}) = 2E_c$ and the asymmetry parameter $a \equiv j_1 - j_2 = (E_{J1} - E_{J2})/(E_{J1} + E_{J2}) = 0.1$. Black dots mark on the Q_0-Φ_e plane the locations of the magic points A, B, and C and the avoided level-crossing point D

combined with charge degeneracy, e.g., in the vicinity of point D. Around the magic points A and especially B and C, which are most suitable for qubit operation, the change of local curvature of the surfaces is rather small.

3.4 Qubit Dephasing

The term in the system Hamiltonian which describes coupling of the qubit and the variable $\delta\Phi_e = MI_f$ associated with the external flux noise induced in the ring by the tank circuit is given by

$$H_{\text{coupl}} = -I_s\,\delta\Phi_e, \tag{3.18}$$

or, in the qubit basis Eq. (3.5),

$$H_{\text{coupl}} = -\frac{1}{2}U_z\sigma_z - \frac{1}{2}U_y\sigma_y = -\frac{1}{2}\left[(c_{11}-c_{00})I_{\parallel}(\phi_e)\sigma_z + 2s_{01}I_{\perp}(\phi_e)\sigma_y\right]\delta\Phi_e, \tag{3.19}$$

where corresponding matrix elements are denoted by

$$c_{00} = \langle 0|\cos\chi|0\rangle, \quad c_{11} = \langle 1|\cos\chi|1\rangle, \quad \text{and} \quad s_{01} = |\langle 0|\sin\chi|1\rangle|. \tag{3.20}$$

The remarkable property of the coupling described by Eq. (3.19) implies that the longitudinal term, $U_z \propto I_{\parallel}$, vanishes in the optimal points A, B, C and D, i.e., where $\sin\phi_e = 0$ (see Eq. (3.11)).[16] The transversal term, $U_y \propto I_{\perp}$, in these points is, however, nonzero, and its low-frequency ($\ll \Omega$) noise component can dephase the qubit due to renormalization of the level spacing,[17] $\epsilon \to (\epsilon^2 + U_y^2)^{1/2} \approx \epsilon + U_y^2/2\epsilon$, yielding the diagonal term

$$-\frac{1}{2}X\sigma_z \equiv -\frac{U_y^2}{2\epsilon}\sigma_z = -\frac{2}{\epsilon}s_{01}^2 I_{\perp}^2(\phi_e)(\delta\Phi_e)^2\sigma_z. \tag{3.21}$$

The problem of qubit dephasing due to quadratic longitudinal coupling to the environment was addressed by Makhlin and Shnirman in Ref. [17] where they focused on the experimentally relevant $1/f$ and the Johnson–Nyquist noise power spectra. Here this model is extended to the case of a structured bath (low-frequency resonance tank circuit).

For white noise I_f associated with losses in resistance R_T, the power spectrum density $S_\Phi(\omega)$ of variable $\delta\Phi_e$ is simply found from the network consideration. Neglecting small detuning of the tank due to coupling to the qubit ring ($\omega_0 \approx \omega_1 \approx \omega_T$) we arrive at the expression,

$$S_\Phi(\omega) = M^2\frac{k_BT}{\pi R_T}\frac{\omega_T^4}{(\omega^2-\omega_T^2)^2 + \omega^2\omega_T^2 Q^{-2}}, \qquad -\infty < \omega < \infty, \tag{3.22}$$

where $Q = \omega_T C_T R_T = R_T/\omega_T L_T$ is the quality factor. Assuming that the noise is Gaussian we find the power spectrum density of fluctuations of the variable $(\delta\Phi_e)^2$ (see, for example, Ref. [18]),

$$S_{\Phi^2}(\omega) = 2\int_{-\infty}^{+\infty} d\omega' S_\Phi(\omega')S_\Phi(\omega'-\omega). \tag{3.23}$$

Taking this integral at $\omega \to 0$ yields the value

$$S_{\Phi^2}(0) = \frac{1}{\pi}M^4Q^3\omega_T\left(\frac{k_BT}{R_T}\right)^2, \tag{3.24}$$

which determines the power spectrum density $S_X(0)$ of variable X,

$$S_X(0) = \frac{2}{\epsilon^2} s_{01}^4 I_\perp^4 S_{\Phi^2}(0), \tag{3.25}$$

and the rate of "pure" dephasing in the Bloch–Redfield approximation,[19] $\Gamma_\varphi = S_X(0)/\hbar^2$. Finally, the qubit quality factor $Q_\varphi \equiv \pi\Omega/\Gamma_\varphi$ can be presented as

$$Q_\varphi = \frac{\pi}{2}\left(\frac{I_{c0}}{s_{01} I_\perp}\right)^4 (k^2 Q\beta_L)^{-2} \frac{\epsilon^3 \hbar\omega_T}{(E_{J0} k_B T)^2} Q, \tag{3.26}$$

where $k = M/(LL_T)^{1/2}$ is the dimensionless coupling coefficient.

To evaluate Q_φ for typical parameters of the qubit (leading for the operation points A, B and C to the value $(\epsilon/E_{J0})^3 \sim 10$) let us assume that the quality factor of the tank circuit is $Q \sim 300$ while the product $k^2 Q\beta_L \sim 10$ which ensures sufficient resolution in determining the resonance frequencies $\omega_{0,1}$. Taking the value of ratio $E_{J0}\hbar\omega_T/(k_B T)^2 \sim 0.1$ (with $E_{J0} = 100\,\mu$eV, $\omega_T/2\pi = 100$ MHz and $T = 1$ K) we get an estimate: $Q_\varphi \sim 10^2 (I_{c0}/s_{01} I_\perp)^4$.

In the optimal points A and D, the amplitude $I_\perp$ (see Eq. (3.12)) is equal to

$$(I_\perp)_{A,D} = 2(\kappa_1 j_1 + \kappa_2 j_2) I_{c0} \approx I_{c0}, \tag{3.27}$$

while in points B and C

$$\begin{aligned}(I_\perp)_{B,C} &= 2\left[(j_1 - j_2)(\kappa_1 j_1 + \kappa_2 j_2) + 2 j_1 j_2 (\kappa_1 - \kappa_2)\right] I_{c0} \\ &\approx (j_1 - j_2 + \kappa_1 - \kappa_2) I_{c0}.\end{aligned} \tag{3.28}$$

The values of the matrix element s_{01} in points A and B are equal to[8]

$$(s_{01})_{A,B} \approx \frac{|j_1 \mp j_2|}{8\sqrt{2}} \frac{E_{J0}}{E_c}, \tag{3.29}$$

while in points C and D

$$(s_{01})_{C,D} \approx 0.5. \tag{3.30}$$

Summarizing, the order-of-magnitude estimates for parameter Q_φ in different operation points are

$$(Q_\varphi)_A \sim 10^4 (j_1 - j_2)^{-4}, \tag{3.31}$$

$$(Q_\varphi)_B \sim 10^4 (j_1 - j_2 + \kappa_1 - \kappa_2)^{-4}, \tag{3.32}$$

$$(Q_\varphi)_C \sim 10 (j_1 - j_2 + \kappa_1 - \kappa_2)^{-4}, \tag{3.33}$$

$$(Q_\varphi)_D \sim 10. \tag{3.34}$$

For the Josephson junctions with a reasonably high symmetry in their parameters, say $\pm 10\%$ (e.g., $j_1 - j_2 = 0.1$), the expected value of Q_φ in points A, B and C greatly exceeds the level 10^5 which should be sufficient for realization of active schemes for compensation of decoherence. In contrast to these figures, the value achieved in the avoided level-crossing point D is relatively low.

3.5 Conclusion

Qubit dephasing due to nonlinear coupling in the vicinity of magic operation points is relatively weak if asymmetry of the junction parameters is sufficiently small and the effective temperature of the tank circuit sufficiently low. Since this temperature is associated with the back-action noise of a preamplifier, its noise figure should be sufficiently low, as, for example, in the case of a cold semiconductor-based amplifier or a dc-SQUID-based amplifier.[20] Further improvement (if necessary) of the charge-phase qubit with radio-frequency readout can be achieved by using a stage ensuring variable coupling between the ring and the tank circuit. This could be, for example, a flux transformer with a variable transfer function.[21]

In conclusion, sufficiently symmetric parameters of the junctions make it possible to effectively protect the charge-phase qubit against dephasing caused by its coupling to the radio-frequency readout circuit. The dominating mechanisms which may still cause appreciable dephasing in this kind of qubit are most likely related to intrinsic characteristics of the Josephson junctions (see, e.g., Refs. [22, 23]).

Acknowledgments. I would like to thank Alec Maassen van den Brink, Apostol Vourdas and Denis Vion for helpful comments. This work was partially supported by the EU through the SQUBIT-2 project.

References

1. A. B. Zorin, Cooper-pair qubit and Cooper-pair electrometer in one device, *Physica C* **368**, 284–288 (2002).
2. E. Il'ichev, A.Yu. Smirnov, M. Grajcar, A. Izmalkov, D. Born, N. Oukhanski, Th. Wagner, W. Krech, H.-G. Meyer, and A.M. Zagoskin, Radio-frequency method for investigation of quantum properties of superconducting structures, cond-mat/0402559 and references therein.
3. A. Lupaşcu, C. J. M. Verwijs, R. N. Schouten, C. J. P. M. Harmans, and J. E. Mooij, Nondestructive readout for a superconducting flux qubit, cond-mat/0311510.
4. I. Siddiqi, R. Vijay, F. Pierre, C.M. Wilson, M. Metcalfe, C. Rigetti, L. Frunzio, and M.H. Devoret, An RF-driven Josephson bifurcation amplifier for quantum Measurements, cond-mat/0312623.
5. D. V. Averin, Quantum nondemolition measurements of a qubit, *Phys. Rev. Lett.* **88**, 207901 (2002).
6. V. Bouchiat, D. Vion, P. Joyez, D. Esteve and M. H. Devoret, Quantum coherence with a single Cooper pair, *Phys. Scr.* **T76**, 165-170 (1998).
7. D. Vion, A. Aassime, A. Cottet, P. Joyez, H. Pothier, C. Urbina, D. Esteve and M. H. Devoret, Manipulating the quantum state of an electrical current, *Science* **296**, 886–889 (2002).
8. A. B. Zorin, Josephson charge-phase qubit with radio frequency readout: coupling and decoherence, *Zh. Éksp. Teor. Fiz.* **125**(6), 1423–1435 (2004); [*JETP* **98**(6), 1250–1261 (2004)].
9. W. Krech, M. Grajcar, D. Born, I. Zhilyaev, Th. Wagner, E. Il'ichev, and Ya. Greenberg, Dynamic features of a charge qubit closed by a superconducting inductive ring, *Phys. Lett. A* **303**, 352–357 (2002).
10. R. Rifkin and B. S. Deaver, Jr., Charge-phase relation and phase-dependent conductance of superconducting point contacts from rf impedance measurements, *Phys. Rev. B* **13**(9), 3894–3901 (1976).
11. D. S. Crankshaw and T. P. Orlando, Inductance effects in the persistent current qubit, *IEEE Appl. Supercond.* **11**(1), 1006–1009 (2001).
12. A. Maassen van den Brink, Hamiltonian for coupled qubits, cond-mat/0310425.
13. In the rf-SQUID circuits comprising Josephson weak-links instead of tunnel junctions a sufficiently small value of effective capacitance $\tilde{C}$ and, hence high ω_L may cause quantum behavior of phase ϕ; see, for example, R. Whiteman, V. Schöllmann, M. Everitt, T. D. Clark, R. J. Prance, H. Prance, J. Diggins, G. Buckling, and J. F. Ralf, Adiabatic modulation of a superconducting quantum interference device (SQUID) ring by an electromagnetic field, *J. Phys.: Condens. Matter* **10**, 9951–9968 (1998).

14. P. K. Hansma, Superconducting single-junction interferometers with small critical currents, *J. Appl. Phys.* **44**(9), 4191–4194 (1973).
15. *Handbook of Mathematical Functions,* ed. by M. Abramowitz and I. A. Stegun (U.S. GPO, Washingtin, D.C., 1972), Chapter 20.
16. In particular, quantum manipulation at a remarkably low dephasing rate of the Quantronium qubit was performed in operation point C.[7]
17. Yu. Makhlin and A. Shnirman, Dephasing of solid-state qubits at optimal points, cond-mat/0308297.
18. S. A. Akhmanov, Yu. E. Dyakov and A. S. Chirkin, *Vvedenie v Statisticheskuyu Radiofiziku i Optiku (Introduction to Statistical Radiophysics and Optics),* (Nauka, Moscow, 1981), Chapter 5 - in Russian.
19. A. Shnirman, Yu. Makhlin and G. Schön, Noise and decoherence in quantum two-level systems, *Phys. Scr.* **T102**, 147–154 (2002).
20. M. Mück, J. B. Kycia and J. Clarke, Superconducting quantum interference device as a near-quantum-limited amplifier at 0.5 GHz, *Appl. Phys. Lett.* **78**(7), 967–969 (2001).
21. C. Cosmelli, M. G. Castellano, F. Chiarello, R. Leoni, D. Simeone, G. Torrioli, and P. Carelli, Controllable flux coupling for the integration of flux qubits, cond-mat/0403690.
22. E. Paladino, L. Faoro, G. Falci, and R. Fazio, Decoherence and $1/f$ noise in Josephson qubits, *Phys. Rev. Lett.* **88**, 228304 (2002).
23. D. J. Van Harlingen, T. L. Robertson, B. L. T. Plourde, P. A. Reichardt, T. A. Crane, and J. Clarke, Decoherence in Josephson-junction qubits due to critical current fluctuations, cond-mat/0404307.

4

The Josephson Bifurcation Amplifier for Quantum Measurements

I. SIDDIQI, R. VIJAY, F. PIERRE, C.M. WILSON, L. FRUNZIO, M. METCALFE, C. RIGETTI, AND M.H. DEVORET

Departments of Applied Physics and Physics
Yale University, USA
irfan.siddiqi@yale.edu

Abstract

A new type of amplifier whose primary purpose is the readout of superconducting quantum bits is constructed. It is based on the transition of an RF-driven Josephson junction between two distinct oscillation states near a dynamical bifurcation point. The main advantages of this new amplifier are speed, high-sensitivity, low back-action, and the absence of on-chip dissipation. Using pulsed microwave techniques, bifurcation amplification in nanofabricated Al junctions is demonstrated and the performance predicted by theory is verified.

4.1 Introduction

Josephson first noted that the superconducting tunnel junction can be viewed as a non-linear, non-dissipative, electrodynamic oscillator.[1] We exploit this non-linearity to produce a new type of high-sensitivity amplifier, the Josephson Bifurcation Amplifier (JBA). No shunt resistors are required in the amplification scheme here, and it is thus possible to take advantage of the elastic character of the junction and eliminate on-chip dissipation, thereby minimizing the back-action of the amplifier. The combination of high-sensitivity and minimal back-action makes the JBA well-suited for measurements on quantum systems such as superconducting qubits, and make it a strong candidate for reaching the quantum noise limit.

The operation of the JBA is represented schematically in Fig. 4.1. The central element is a Josephson junction, shunted with a lithographic capacitor, whose critical current I_0 is modulated by an input signal (input port). Coupling between the junction and the input signal can be achieved through different schemes, examples of which involve placing the JBA in a SQUID loop[2] or in parallel with a SSET.[3] The junction is driven with a pure AC signal $i_{\mathrm{RF}} \sin(\omega t)$ in the microwave frequency range fed via a transmission line through a circulator (drive port). In the underdamped regime, for certain values of ω and i_{RF}, two possible oscillation states which differ in amplitude and phase (denoted "0" and "1") can coexist. The reflected component of the drive

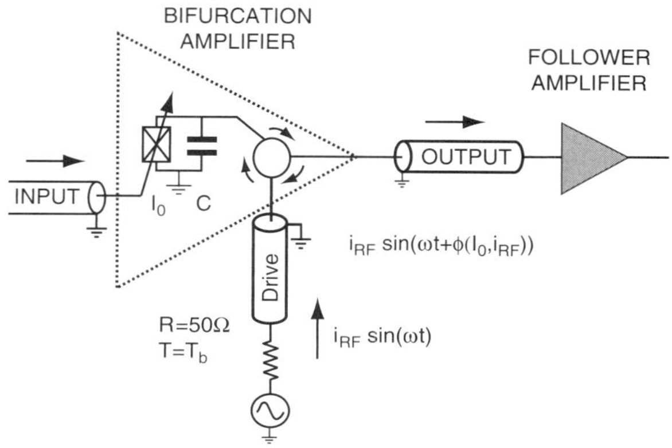

FIGURE 4.1. Schematic diagram of the Josephson bifurcation amplifier. A junction with critical current I_0, parametrically coupled to the input port, is driven by an RF signal which provides the power for amplification. In the vicinity of the dynamical bifurcation point $i_{\mathrm{RF}} = I_B$, the reflected signal phase ϕ depends critically on the input signal.

signal, measured through another transmission line connected to the circulator (output port), is a convenient signature of the junction oscillation state. At the bifurcation point where switching between oscillation states occurs, the system becomes infinitely sensitive, in the absence of thermal and quantum fluctuations, to variations in I_0. At finite temperature, the energy stored in the oscillation can always be made larger than thermal fluctuations by increasing the scale of I_0, thus preserving sensitivity. Small variations in I_0 are transformed into readily discernible changes in the escape rate Γ_{01} from state 0–1. Back-action is minimized in this arrangement since the only fluctuations felt at the input port arise from the fluctuations of the 50 Ω drive port whose dissipative elements are physically separated from the junction via a transmission line of arbitrary length and can therefore be thermalized efficiently to base temperature. Additionally, the frequency band over which the back-action contributes is narrow, and well controlled.

In Section 4.2, simplified expressions adapted from the theory of activated escape in a driven non-linear oscillator[4] are presented. Details of the devices and the measurement setup are presented in Section 4.3. Experimental results are given in Section 4.4, and concluding remarks are in Section 4.5.

4.2 Theory

The tunnelling of Cooper pairs manifests itself as a non-linear inductance that shunts the linear junction self-capacitance C_{J}, formed by the junction electrodes and the tunnel oxide layer. The constitutive relation of the non-linear inductor can be written as $I(t) = I_0 \sin \delta(t)$, where $I(t)$, $\delta(t) = \int_{-\infty}^{t} \mathrm{d}t' V(t')/\varphi_0$ and $V(t)$ are the current, gauge-invariant phase-difference and voltage corresponding to the inductor, respectively, while the parameter I_0 is the junction critical current. Here $\varphi_0 = \hbar/2e$ is the reduced flux quantum. The dynamics of the junction are given by the time evolution of δ, which exhibits the motion of a phase particle in a cosine potential $U(\delta) = -\varphi_0 I_0 \cos(\delta)$. For small oscillation amplitude about the potential minima, the frequency of oscillation is given for zero DC bias current by the plasma frequency $\omega_{P0} = 1/\sqrt{L_J C_J}$ where $L_J = \varphi_0/I_0$ is the effective junction inductance. As the oscillation amplitude increases, the potential "softens" and ω_P decreases, an effect which has been measured in both the classical and quantum regime.[5–8] A more dramatic non-linear effect manifests itself if the junction is driven with an AC current $i_{RF} \sin \omega t$ at a frequency ω slightly below ω_{P0}. If the quality factor $Q = C_J \omega_{P0}/Re[Z^{-1}(\omega_{P0})]$ is greater than $\sqrt{3}/2\alpha$, where $Z(\omega_{P0})$ is the impedance of the junction electrodynamic environment and $\alpha = 1 - \omega/\omega_{P0}$ the detuning parameter, then the junction switches from one dynamical oscillation state to another when i_{RF} is ramped above a critical value I_B.[9] For $i_{RF} < I_B$, the oscillation state is low-amplitude and phase-lagging while for $i_{RF} > I_B$, the oscillation state is high-amplitude and phase-leading. This generic non-linear phenomenon, which we refer to as "dynamical switching", is reminiscent of the usual "static switching" of the junction from the zero-voltage state to the voltage state when the DC current bias exceeds the critical current I_0.[10] However, an important distinction between dynamical and static switching is that in dynamical switching, the phase particle remains confined to only one well of the junction cosine potential, and the time-average value of δ is always zero. The junction never switches to the voltage state, and thus no DC voltage is generated. Also, for dynamical switching, the current I_B depends both on Q and on the detuning α.

In presence of the microwave drive $i_{RF} \sin(\omega t)$, the oscillations in the junction phase can be parameterized using in-phase and quadrature phase components $\delta(t) = \delta_{\parallel} \sin(\omega t) + \delta_{\perp} \cos(\omega t)$ (higher harmonics of oscillation are negligible).The two oscillation states appear as two points in the $(\delta_{\parallel}, \delta_{\perp})$ plane and are denoted by vectors labelled 0 and 1 (see Fig. 4.2). The error-current[11] that describes the generalized force felt by the system is also plotted in the $(\delta_{\parallel}, \delta_{\perp})$ plane. Its value goes to zero at the attractors corresponding to states 0 and 1 and also at a third extremum which is the dynamical saddle point. Also shown in Fig. 4.2 is the calculated escape trajectory[12] from state 0 (dashed) and the corresponding retrappping trajectory[13] into state 1 (solid line). Fig. 4.2 has been computed for $\alpha = 0.122$, $Q = 20$ and $i_{RF}/I_B = 0.87$. These values correspond to typical operating conditions in our experiment. The dynamical switching from state 0 to 1 is characterized by a phase shift given here by $\tan^{-1}[(\delta_{\parallel}^1 - \delta_{\parallel}^0)/(\delta_{\perp}^1 - \delta_{\perp}^0)] = -139°$. Using the

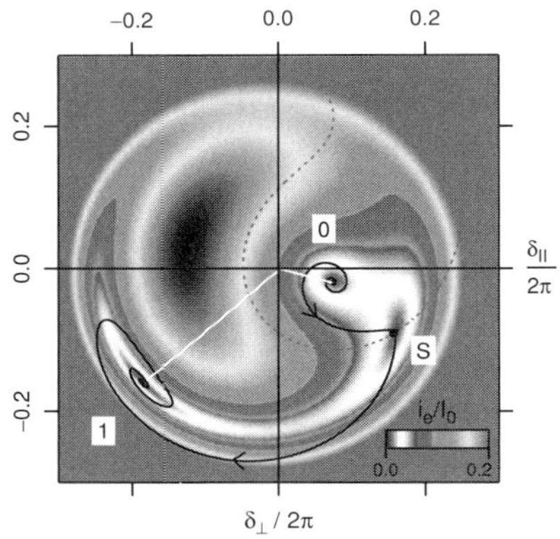

FIGURE 4.2. Poincare section of an RF-driven Josephson junction in the bistable regime ($\alpha = (1 - \omega/\omega_p) = 0.122$, $i_{RF}/I_B = 0.87$). The two stable oscillation states, labelled by 0 and 1, are indicated by white line segments. Point S which lies on the separatrix is the saddle point at which the escape trajectory from state 0 (dashed line) meets the retrapping trajectory into state 1 (solid line).

junction phase-voltage relationship and the transmission line equations, we can calculate the steady-state magnitude and phase of the reflected microwave drive signal. The change in the oscillation of δ results in a shift of the reflected signal phase $\Delta\phi_{01} = 89°$. Since there is no source of dissipation in the junction chip, there should be no change in the magnitude of the reflected signal power, even though $\sqrt{(\delta_{\parallel}^{1} - \delta_{\parallel}^{0})^2 + (\delta_{\perp}^{1} - \delta_{\perp}^{0})^2} \neq 0$.

Both static and dynamical switching can be described by an Arhennius law in which the escape rate $\Gamma_{01} = (\omega_{\text{att}}/2\pi)\exp(-\Delta U/k_b T)$ is written as the product of an attempt frequency $\omega_{\text{att}}/2\pi$ and a Boltzman factor which contains the potential barrier height ΔU and the system temperature T. For the case of a DC current bias, the cosine potential, near the switching point, is approximated as a cubic potential with height $\Delta U_{\text{st}} = (4\sqrt{2}/3)\ \varphi_0 I_0\ (1 - i_{\text{DC}}/I_0)^{3/2}$ where i_{DC} is the bias current. The attempt frequency is the plasma frequency ω_{P}. In the absence of fluctuations, the characteristic current at which switching occurs is I_0. For the AC driven junction, the dynamical switching from oscillation state 0 to 1 can be cast in a similar form using the model of a particle in a cubic metapotential.[4] In this case, the effective barrier height is, to lowest order in $1/(\alpha Q)$, $\Delta U_{\text{dyn}} = u_{\text{dyn}}(1 - (i_{\text{RF}}/I_B)^2)^{3/2}$ with $u_{\text{dyn}} = 64\hbar/(18e\sqrt{3})\ I_0\ \alpha(1-\alpha)^3$. The attempt frequency in the metapotential is given by $\omega_a = \omega_{a0}\,(1 - (i_{\text{RF}}/I_B)^2)^{1/2}$ with $\omega_{a0} = 4/(3\sqrt{3}\,RC)\,(\omega_{\text{p}} - \omega)^2$. The bifucrcation current I_B where the 0 state ceases to exist is given by $I_B = 16/(3\sqrt{3})\ \alpha^{3/2}(1-\alpha)^{3/2}\ I_0$.

4.3 Devices and Setup

Typical junction fabrication parameters limit the plasma frequency to the 20–00 GHz range where techniques for addressing junction dynamics are inconvenient. The junction is choosen to be shunt by a capacitive admittance to lower the plasma frequency by more than an order of magnitude and attain a frequency in 1–2 GHz range (microwave L-band). In this frequency range, a simple on-chip electrodynamic environment with minimum parasitic elements can be implemented, and the hardware for precise signal generation and processing is readily available. In the first step of sample fabrication, a metallic underlayer – either a normal metal (Au, Cu) or a superconductor (Nb) – was deposited on a silicon substrate to form one plate of the shunting capacitor, followed by the deposition of an insulating Si_3N_4 layer. Using e-beam lithography and double-angle shadow mask evaporation, we subsequently fabricated the top capacitor plates along with a micron sized $Al/Al_2O_3/Al$ tunnel junction. The critical current of the junction was in the range $I_0 = 1 - 2\,\mu$A. By varying both the dielectric layer thickness and the pad area, the capacitance C was varied between 16 and 40 pF.

The junction + capacitor chip is placed on a microwave circuit-board and is wire-bonded to the end of a coplanar stripline which is soldered to a coaxial launcher affixed to the side wall of the copper sample box. We anchor the RF leak-tight sample box to the cold stage of a ^{3}He refrigerator with base temperature $T = 280$ mK. The measurement setup is schematically shown in Fig. 4.3. Microwave excitation signals are coupled to the sample via the – 13 dB side port of a directional coupler. The reflected microwave signal passes through the direct port of the coupler, and is amplified first using a cryogenic 1.20 – 1.85 GHz HEMT amplifier with noise temperature $T_N =$ 4 K. A DC bias current can be applied to the junction by way of a bias tee. Cryogenic attenuators, isolators, and specially developed dissipative microstrip filters on the microwave lines in addition to copper-powder and other passive filters have been used[6] on the DC lines to shield the junction from spurious electromagnetic noise. In the first set of experiments which probe the plasma

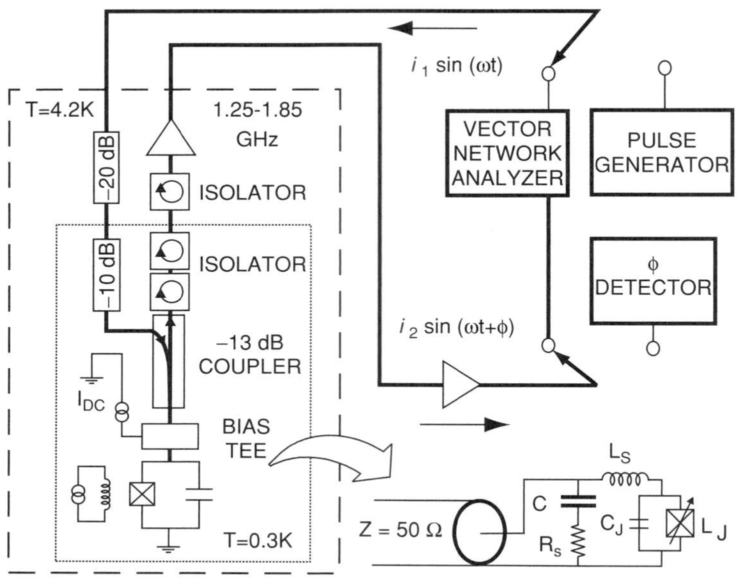

FIGURE 4.3. Schematic of the measurement setup. Thick lines correspond to 50 Ω coaxial transmission lines. The network analyzer is used for CW measurements. For probing the dynamics, the sample is switched to the pulse generator and phase detector. A lumped element model for the junction chip and measurement line is shown in the lower right. The two ideal current sources actually represent external sources.

resonance, a vector network analyzer was used to both source a CW microwave signal and to analyze the reflected power.[14] The dynamics of the transition between the two oscillation states was then probed using microwave pulses,[15] generated by the amplitude modulation of a CW source with a phase-locked arbitrary waveform generator with 1 ns resolution. For the pulsed experiments, the reflected signal was mixed down to 100 MHz and digitally demodulated using a 2 GS/s digitizer to extract the signal phase ϕ.

4.4 Results

Probed the drive current dependence of the reflected signal phase ϕ (i_{RF}) was first probed by by applying a 4 μs long symmetric triangular shaped pulse with a peak value 0.185 I_0. The demodulated reflected signal was divided into 20 ns sections, each yielding one measurement of ϕ for a corresponding value of i_{RF}. The measurement was repeated 6×10^5 times to obtain a distribution of $\phi(i_{\mathrm{RF}})$. In Fig. 4.4, the mode of the distribution is plotted as a function of i_{RF}/I_0. For $i_{\mathrm{RF}}/I_0 < 0.125$, the bifurcation amplifier is always in state 0, ϕ is constant and assigned a value of 0°. As the drive current is increased above $i_{\mathrm{RF}}/I_0 = 0.125$, thermal fluctuations are sufficiently large to cause transitions to the 1 state. In the region between the two dashed lines at $i_{\mathrm{RF}}/I_0 = 0.125$ and $i_{\mathrm{RF}}/I_0 = 0.160$, ϕ displays a bimodal distribution with peaks centered at 0° and 74° with the latter corresponding to the amplifier in the 1 state. The dotted line in Fig. 4.4

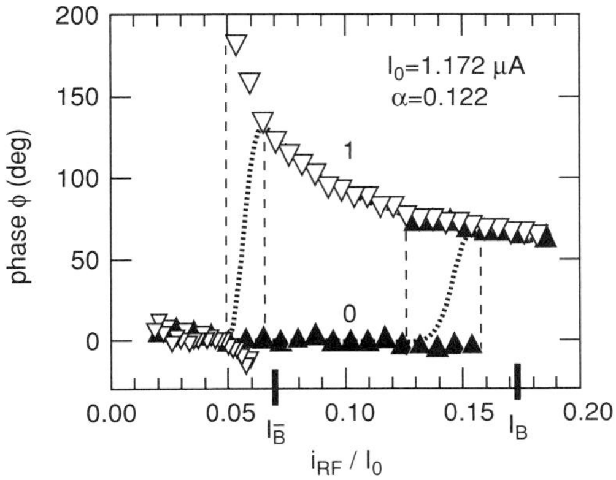

FIGURE 4.4. Hysteretic variation of the reflected signal phase ϕ with drive current i_{RF}/I_0. Symbols denote the mode of ϕ, with up and down triangles corresponding to increasing and decreasing $i_{RF} = I_B$, respectively. The dotted line is $\langle\phi\rangle$. The calculated bifurcation points, $I_{\bar{B}}$ and I_B, are marked on the horizontal axis. The 0 and 1 phase states are reminiscent of the superconducting and dissipative states of the DC current biased junction.

is the average reflected signal phase $\langle\phi\rangle$. When i_{RF}/I_0 is increased above 0.160, the system is only found in state 1. In the decreasing part of the i_{RF} ramp, the system does not start to switch back to state 0 until $i_{RF}/I_0 = 0.065$. The critical switching currents I_B for the $0 \rightarrow 1$ transition and $I_{\bar{B}}$ for the $1 \rightarrow 0$ transition, calculated from numerical simulations to treat the inductance of wire bonds, are denoted with ticks in Fig. 4.4, and are in good agreement with experiment. The hysteresis $I_{\bar{B}} < I_B$ is a consequence of the asymmetry in the escape barrier height for the two states. Thus, the $0 \rightarrow 1$ transition at $i_{RF} = I_B$ is nearly irreversible, allowing the bifurcation amplifier to latch and store its output during the integration time set by the sensitivity of the follower amplifier.

To determine the sensitivity of the bifurcation amplifier, the switching in the vicinity of the $0 \rightarrow 1$ transition has been characterized in detail. The system with two different readout pulse protocols were excited. In the first protocol, the drive current was ramped from 0 to its maximum value in 40 ns and was then held constant for 40 ns before returning to 0. Only the final 20 ns of the constant drive period were used to determine the oscillation phase with the first 20 ns allotted for settling of the phase. Histograms taken with a 10 MHz acquisition rate are shown in Fig. 4.5. In the upper panel, the two peaks corresponding to states 0 and 1 can easily be resolved with a small relative overlap of 10^{-2}. The width of each peak is consistent with the noise temperature of our HEMT amplifier. In this first method, the latching property of the system has not been exploited. In the second protocol for the readout pulse, there was again a ramp for 40 ns and a setting time of 20 ns was allowed, but then the drive current was reduced by 20% and the reflected signal was measured for 300 ns. In that latter period, whatever state was reached at the end of the initial 60 ns period is "latched" and time is spent just increasing the signal/noise ratio of the reflected phase measurement. As shown in the lower panel of Fig. 4.5, the two peaks are now fully separated, with a relative overlap of 6×10^{-5} allowing a determination of the state 1 probability with an accuracy better than 10^{-3}. This second protocol would be preferred only for

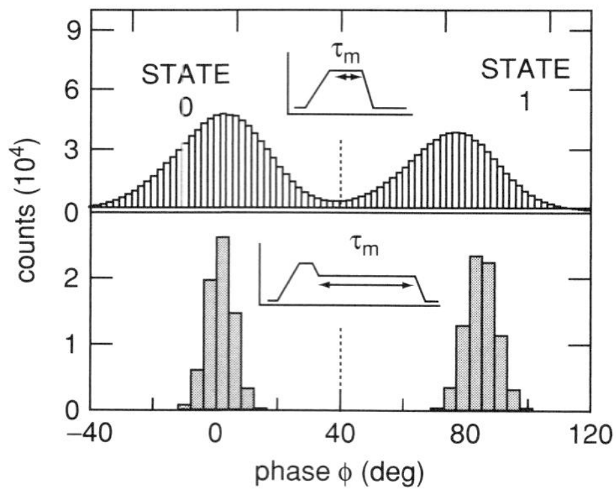

FIGURE 4.5. Histograms of the reflected signal phase ϕ at $i_{RF}/I_0 = 0.145$. The upper histogram contains 1.6×10^6 counts with a measurement time $\tau_m = 20$ ns. The lower panel, taken with the latching technique, has 1.5×10^5 counts with a measurement time $\tau_m = 300$ ns. Data here has been taken under the same operating conditions as in Fig. 4.4. The dashed line represents the discrimination threshold between the 0 and 1 state.

very precise time-resolved measurements of I_0 or for applications where a low-noise follower amplifier is impractical.

A third experiment was performed to study the state 1 switching probability P_{01} (i_{RF}) for different values of the temperature T and I_0, the latter being varied with a magnetic field applied parallel to the junction plane. Using the first readout protocol and the discrimination threshold shown in Fig. 4.5, we obtain the switching probability curves shown in Fig. 4.6. Defining the discrimination power d as the maximum difference between two switching probability curves which differ in I_0 we find that at $T = 280$ mK, $d = 57\%$ for $\Delta I_0/I_0 = 1\%$ – the typical

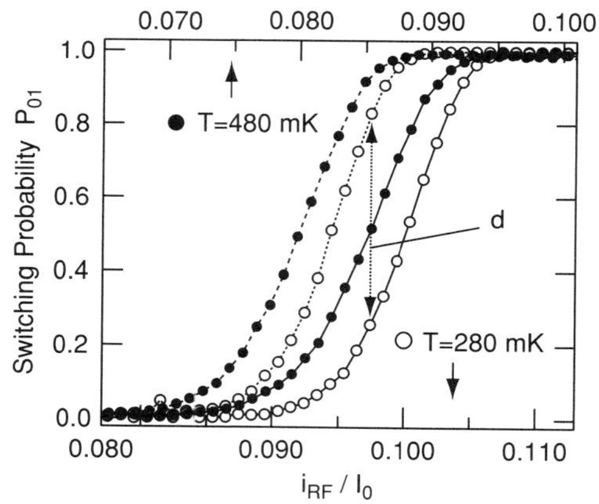

FIGURE 4.6. Switching probability curves at $T = 280$ mK and $T = 480$ mK as a function of the drive current i_{RF}. The discrimination power d is the maximum difference between two curves at the same temperature which differ by approximately 1% in I_0.

variation observed in a superconducting charge-phase qubit.[16] The switching probability curves should shift according to $(\Delta I_B/I_B)/(\Delta I_0/I_0) = 3/(4\alpha) - 1/2 + O(1/(\alpha Q)^2)$, which for our case takes the value 5.6. In Fig. 4.6, the curves are shifted by 6%, which agrees well with this prediction. For the case of the DC current biased junction, similar curves would shift only by 1% since the switching current is I_0 itself. Comparable discrimination power using DC switching has only been achieved in these devices at $T \leq 60$ mK. As the temperature is increased, the switching probability curves broaden due to increased thermal fluctuations and the discriminating power decreases: at $T = 480$ mK, $d = 49\%$.

Finally, the escape rate $\Gamma_{01}(i_{RF}, I_0, T)$ was determined as a function of i_{RF} by measuring the time dependence of the switching probability, using a method previously applied to the determination of the static switching rates to the voltage state.[17] After the initial ramp (40 ns) and settling period (20 ns), the reflected signal phase was extracted every 20 ns for a duration of 1 µs. By repeating this measurement, switching probability histograms were generated which was analyzed as $P_{01}(t) = 1 - \exp(-\Gamma_{01} \cdot t)$. To obtain the escape rate at different temperatures, two different techniques were used. In the first method, the temperature of the cryostat was varied and a magnetic field was used to keep the critical current constant at $I_0 = 1.12$ µA. In the second method, I_0 was kept fixed at 1.17 µA, and a 1 − 2 GHz white noise source irradiating the junction was used to increase the effective temperature. In Fig. 4.7 we show the drive current dependence of the escape rate as $(\ln(2\pi\omega_a/\Gamma_{01}))^{2/3}$ plotted versus i_{RF}^2 for two different sample temperatures. Data in this format should fall on a straight line with a slope $s(T)$ proportional to $(u_{dyn}/k_B T)^{2/3}$. A trace taken at $T = 500$ mK is also shown in Fig 4.7.

In parallel with these dynamical switching measurements, static switching measurements were run to obtain an escape temperature T_{st}^{esc}. Due to insufficient filtering in our RF amplifier line outside the measurement band, T_{st}^{esc} exceeded T by 60 mK. Using u_{dyn}^{calc} and $s(T)$ we can cast the results of the dynamical switching measurements into a dynamical escape temperature $T_{dyn}^{esc} = u_{dyn}^{calc}/k_B\, s(T)^{3/2}$. We plot T_{dyn}^{esc} versus T_{st}^{esc} in the inset of Fig. 4.7. The agreement is very good, and only deviations at the highest temperatures are observed. Analyzing the dynamical switching

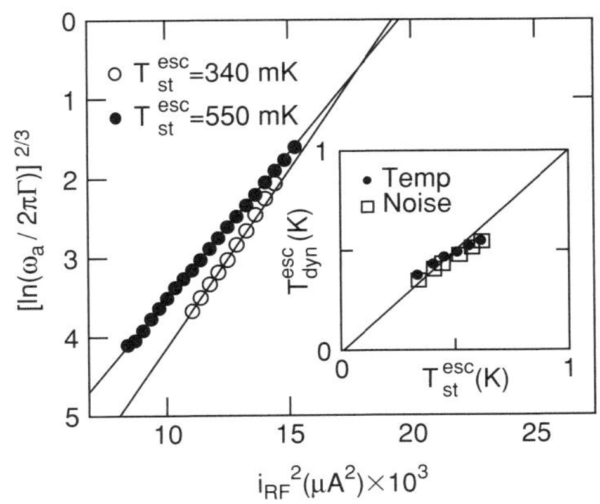

FIGURE 4.7. Escape rate as a function of drive power for two different operating temperatures with $I_0 = 1.12\,\mu$A. The inset shows the relationship between dynamic and static escape temperatures when varying either the sample temperature or the injected noise power.

data with T_{st}^{esc} in place of T, we extract a value of $u_{dyn} = 10.7$ K from the $T = 280$ mK data with $I_0 = 1.17\,\mu$A while the calculated value keeping higher order terms in $1/\alpha Q$ is $u_{dyn}^{calc} = 11.0$ K.

4.5 Conclusion

With the JBA operating at $T_{st}^{esc} = 340$ mK, it is possible in a total time ≤ 80 ns, to obtain a critical current sensitivity of $S_{I_0}^{1/2} = 3.3 \times 10^{-6}\,\text{A} \cdot \text{Hz}^{-1/2}$. This value is in agreement with the prediction from the analytical theory $S_{I_0}^{1/2} = \eta(i_{RF}/I_0, \alpha)\,(k_B T/\varphi_0) \cdot \tau_m^{1/2}$, where $\eta \approx 1.4$ near the bifurcation point and $\varphi_0 = \hbar/2e$. The advantage of the bifurcation amplifier over SQUIDs[18] resides in its extremely low back-action. Since there is no on-chip dissipation, the only source of back-action is the matched isolator load, which is efficiently thermalized at $T = 280$ mK. An important point is that in the JBA, only fluctuations from the load that occur in a narrow band centered about the plasma frequency contribute to the back-action, whereas in the SQUID noise from many high frequency bands is also significant. Finally, the bifurcation amplifier does not suffer from quasiparticle generation associated with hysteretic SQUIDS[2] and DC current-biased junctions[3] which switch into the voltage state. Long quasiparticle recombination times at low temperatures limit the acquisition rate of these devices while the recombination process itself produces excess noise for adjacent circuitry.[19]

In conclusion, the JBA is competitive with the SQUID for applications where low back-action is required. Its speed, suppression of on-chip dissipation, and latching make it ideal for the read-out of superconducting qubits. At temperatures such that $T_{dyn}^{esc} \leq 60$ mK, the discrimination power would be greater than 95%, hence permitting stringent tests of Quantum Mechanics, like the violation of Bell's inequalities.

Acknowledgments. We would like to thank D. Prober, E. Boaknin, M. Dykman, L. Grober, D. Esteve, D. Vion, S. Girvin and R. Schoelkopf for discussions and assistance. This work was supported by the ARDA (ARO Grant DAAD19-02-1-0044), the NSF (Grant ITR DMR-0325580, DMR-0072022), and the Keck foundation.

References

1. B. Josephson, *Rev. Mod. Phys.* **36**, 216 (1964).
2. I. Chiorescu, Y. Nakamura, C.J.P.M. Harmans, and J.E. Mooij, *Science* **299**, 5614 (2003).
3. A. Cottet, D. Vion, A. Aassime, P. Joyez, D. Esteve, and M.H. Devoret, *Physica C* **367**, 197 (2002).
4. M.I. Dykman and M.A. Krivoglaz, *Physica A* **104**, 480 (1980).
5. M.H. Devoret, D. Esteve, J.M. Martinis, A. Cleland, and J. Clarke, *Phys. Rev. B* **36**, 58 (1987).
6. J.M. Martinis, M.H. Devoret, and J. Clarke, *Phys. Rev. B* **35**, 4682 (1987).
7. B. Yurke, L.R. Corruccini, P.G. Kaminsky, L.W. Rupp, A.D. Smith, A.H. Silver, R.W. Simon, and E.A. Whittaker, *Phys. Rev. A* **39**, 2519 (1989).
8. T. Holst and J. Bindslev Hansen, *Physica B* 165–166, 1649 (1990).
9. L.D. Landau and E.M. Lifshitz, *Mechanics* (Reed, Oxford, 1981).
10. T.A. Fulton and L.N. Dunkleberger, *Phys. Rev. B* **9**, 4760 (1974).
11. R.L. Kautz, *Phys. Rev. A* **38**, 2066 (1988).
12. M.I. Dykman and M.A. Krivoglaz, *JETP* **50**, 30 (1980).

13. A.P. Dmitriev, M.I. D'yakonov, and A.F. Ioffe, *JETP* **63**, 838 (1986).
14. I. Siddiqi, R. Vijay, F. Pierre, C.M. Wilson, L. Frunzio, M. Metcalfe, C. Rigetti, R.J. Schoelkopf, M.H. Devoret, D. Vion, and D.E. Esteve, arXiv:cond-mat/0312553 v1 21 Dec 2003.
15. I. Siddiqi, R. Vijay, F. Pierre, C.M. Wilson, M. Metcalfe, C. Rigetti, L. Frunzio, and M.H. Devoret, arXiv:cond-mat/0312623 v1 23 Dec 2003.
16. D. Vion, A. Aassime, A. Cottet, P. Joyez, H. Pothier, C. Urbina, D. Esteve, and M. Devoret, *Science* **296**, 886 (2002).
17. C. Urbina, D. Esteve, J.M. Martinis, E. Turlot, M.H. Devoret, H. Grabert, and S. Linkwitz, *Physica B* **169**, 26 (1991).
18. M. Muck, J.B. Kycia, and J. Clarke, *Appl. Phys. Lett.* **78**, 967 (2001).
19. J. Mannik and J.E. Lukens, arXiv:cond-mat/0305190 v2 6 Nov 2003.

5

Current-Controlled coupling of superconducting charge qubits

M. WALLQUIST, J. LANTZ, V.S. SHUMEIKO, AND G. WENDIN

Department of Microtechnology and Nanoscience, Chalmers University of Technology, S-412 96 Gothenburg, Sweden

Keywords: Quantum computing, Qubits

Abstract

Theoretical investigation of a chain of loop-shaped charge qubits, interacting via joint Josephson junctions is performed. Neighbouring qubits are coupled to each other only when the bias current through the joint Josephson junction is switched on. The maximum coupling strength is estimated and how to perform a universal set of quantum gates is shown.

5.1 Introduction

In order to perform universal quantum computation, it is necessary to couple qubits in such a way that single-qubit and two-qubit operations can be performed.[1] Coupling of loop-shaped superconducting charge qubits using joint Josephson junctions at the loop intersections was proposed.[2] A bias current through the coupling junction results in circulating currents in the loops, whose interaction creates an effective controllable qubit-qubit coupling. The advantage of this method of coupling compared to other proposals is the possibility to perform gate operations at the charge degeneracy point, where the dephasing effect of the charge fluctuations is minimal. Another advantage is the scalability of the coupling, and also that current, rather than magnetic field is used as a controlling parameter, which will not disturb remote qubits.

The system of two coupled qubits will be discussed first, and how to optimize the coupling strength and perform gate operations will then be explained. Then we discuss the effect of measurement junctions inserted into the qubit loops, and generalize to a chain with an arbitrary number of qubits.

5.2 Current-Controlled Coupling of Two Qubits

Let us consider two superconducting charge qubits–single Cooper pair boxes–whose electrodes form loops.[3,4,5] The basic idea is to couple the loops via joint Josephson junctions at the loop intersections. The two-qubit system is presented in Fig. 5.1. The classical Lagrangian of the circuit is written, following the prescription of Ref. [6],

$$L = \sum_{i=1}^{2} \left(\frac{C}{2(2e)^2} \left(\dot{\phi}_{1i}^2 + \dot{\phi}_{2i}^2 \right) + \frac{C_g}{2(2e)^2} \left(2eV_{gi} - \dot{\phi}_{1i} \right)^2 + E_J(\cos\phi_{1i} + \cos\phi_{2i}) \right) + \left(\frac{C^b}{2(2e)^2} \dot{\phi}_b^2 + E_J^2 \cos\phi_b + \frac{1}{2e} I_b \phi_b \right). \tag{5.1}$$

Here the first three terms refer to the qubits and contain charging and Josephson energies of the qubit junctions, while the last term refers to the coupling junction. It is assumed that the self-inductances of the loops are small and that no external magnetic field is applied to the loops.

The qubit junctions have Josephson energy E_J and capacitance C, while the coupling junction has Josephson energy E_J^b and capacitive energy $E_C^b = (2e)^2/2C^b$. The coupling junction is much larger than the qubit junctions: $\lambda = E_J/E_J^b \ll 1$ and $C \ll C^b$. The coupling junction is chosen to be in the phase regime, $E_J^b \gg E_C^b$. In this regime, the phase ϕ_b of the coupling junction is confined in a potential well, showing small amplitude oscillations $\delta = \phi_b - \phi_{b,0}$ around the stationary value $\phi_{b,0}$ determined by the bias current, $\sin\phi_{b,0} = I_b/I_c^b$.

The Hamiltonian of the truncated system derived from the Lagrangian Eq. (5.1) has the form,

$$H = \sum_{i=1}^{2} \left(\frac{E_C}{2}(1 - 2n_{gi})\sigma_x^i - E_J \cos\frac{\phi_{b,0}}{2}\sigma_z^i \right) + E_C^b \hat{p}^2 + \frac{E_J^b}{2} \cos\phi_{b,0}\delta^2 + \lambda E_J^b B\delta. \tag{5.2}$$

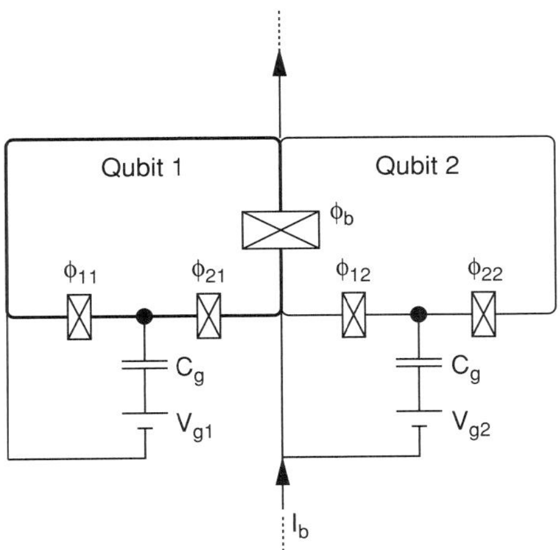

FIGURE 5.1. A system of two coupled charge qubits. V_{gi} controls the individual qubits whereas I_b controls the coupling of the two qubits.

In this equation, $E_C = (2e)^2/(2C_\Sigma)$, $C_\Sigma = 2C + C_g$, $n_{gi} = C_g V_{gi}/2e$ and $2B = \sin(\phi_{b,0}/2)$ $(\sigma_z^1 + \sigma_z^2)$. The current basis for the qubits was chosen, $|+\rangle = 1/\sqrt{2}(|1\rangle + |0\rangle)$ and $|-\rangle = 1/\sqrt{2}(|1\rangle - |0\rangle)$, rather than the charge basis $|0\rangle$ and $|1\rangle$, and expanded the coupling junction potential with respect to the small δ. The last term in Eq. (5.2) is responsible for the qubit-oscillator coupling. It should be noticed that this term is zero when $I_b = 0$.

When a bias current is applied, the oscillator becomes displaced. Since the displacement depends on the state of both the qubits (through B), the result is a switchable qubit–qubit interaction. Assuming that the bias current is changed slowly, then, if the qubits stay at the degeneracy point, $n_{gi} = 1/2$, the oscillator is not excited. Moreover, if the oscillator frequency is not in exact resonance with the qubits, then rf pulses used for the single qubit operations will not excite the oscillator either. Thus, for sufficiently low temperature, the oscillator can be assumed to always stay in the ground state, and averaging over the oscillator ground state can be performed.

After the oscillator degree of freedom has been integrated out, what is left is an effective two-qubit Hamiltonian,

$$H = \sum_{i=1}^{2} \left(\frac{E_C}{2}(1 - 2n_{gi})\sigma_x^i - E_J \cos\frac{\phi_{b,0}}{2}\sigma_z^i \right) + H_{int}, \tag{5.3}$$

$$H_{int} = -\frac{E_J^2}{4E_J^b}\frac{\sin^2\frac{\phi_{b,0}}{2}}{\cos\phi_{b,0}}\sigma_z^1\sigma_z^2. \tag{5.4}$$

Taking higher order terms with respect to the small parameter $\omega_b/E_J^b \ll 1$ into consideration $(\omega_b = \sqrt{2E_C^b E_J^b})$, we find that there is additional weak uncontrollable interaction resulting from squeezing of the oscillator by the qubits.

5.3 Maximum Coupling Strength

The stronger the qubit interaction, the shorter will be the time needed to perform operations. The maximum coupling strength depends on how large bias current can be applied and how much the oscillator frequency can be decreased (by adding shunt capacitances to the coupling junction). There are four constraints to consider: The first constraint prevents the coupling junction from Macroscopic Quantum Tunneling (MQT), which implies that the bias current must not be close to the critical value. Using the standard equation for the MQT rate[7] we find,

$$\frac{E_J^b}{\omega_b}\frac{(\cos\phi_{b,0})^{5/2}}{\sin^2\phi_{b,0}} \gg 1. \tag{5.5}$$

The second constraint originates from approximating the coupling junction by a harmonic oscillator, which is valid only if $\max[\delta] \ll 1$, or,

$$\frac{\omega_b^2}{(E_J^b)^2\cos\phi_{b,0}} \ll 1. \tag{5.6}$$

For the same reason, we require the oscillator displacement to be small. Thus, we get the third constraint,

$$\frac{E_J}{E_J^b\cos\phi_{b,0}} \ll 1. \tag{5.7}$$

During the qubit–qubit interaction, the coupling junction potential is tilted, which means that the oscillator eigenstates change. In order not to excite the oscillator during the operations, the interaction strength η must be varied adiabatically with respect to the oscillator period. Thus, we get the fourth constraint,

$$\eta \sim \frac{E_J^2}{E_J^b \cos\phi_{b,0}} \ll \omega_b\sqrt{\cos\phi_{b,0}}. \tag{5.8}$$

Comparing the constraints at large bias current, we see that MQT, Eq (5.5), puts the strongest constraint on the bias current. Eq (5.8) limits the possibility to decrease the oscillator frequency; combining it with the MQT-condition results in a constraint on ω_b at maximum bias current,

$$\omega_b > E_J\left(\frac{E_J}{E_J^b}\right)^{1/4}. \tag{5.9}$$

The maximum coupling strength is achieved with the largest possible bias current, Eq (5.5), and the smallest possible ω_b, Eq (5.9),

$$\eta_{max} \sim E_J\sqrt{E_J/E_J^b}. \tag{5.10}$$

5.4 Operating the System

During one- and two-qubit gate operations, it is important to stay as close as possible to the degeneracy point in order to minimize qubit dephasing due to fluctuations of charge and flux in the vicinity of the qubits.[5,8] When the bias current is turned off and a qubit is parked at its charge degeneracy point, due to the energy level splitting $2E_J$ a phase gate is performed. Hadamard gates are performed by inducing Rabi oscillations with microwave radiation applied to the voltage gate. Keeping the qubits at their charge degeneracy points and applying a bias current for a certain time results in a Controlled *Phase (CPHASE) gate:

$$\begin{aligned} |11\rangle &\rightarrow i|11\rangle \\ |10\rangle &\rightarrow |10\rangle \\ |01\rangle &\rightarrow |01\rangle \\ |00\rangle &\rightarrow i|00\rangle. \end{aligned} \tag{5.11}$$

The CPHASE gate can be combined with single qubit gates to perform a CNOT gate according to the scheme in Fig. 5.2. Hence the system can perform a universal set of gates.

5.5 Coupling via Measurement Junctions

In the Saclay experiment,[5] the state of the qubit has been measured by including a large Josephson junction ($E_C \ll E_J$) in the qubit loop. This can also be done in the present set up, see Fig. 5.3, and the measurement of the state of each qubit can be performed by sending dc or ac current through the corresponding measurement junction. To diminish the disturbance of the second qubit and avoid the generation of large amounts of quasiparticles, it is preferable to use the ac drive read-out with comparatively small amplitude.[9] The presence of the measurement junctions in the circuit has an additional advantage, allowing coupling of the qubits. Biasing of the measurement junction with a small dc current will generate a persistent current in the corresponding qubit loop.

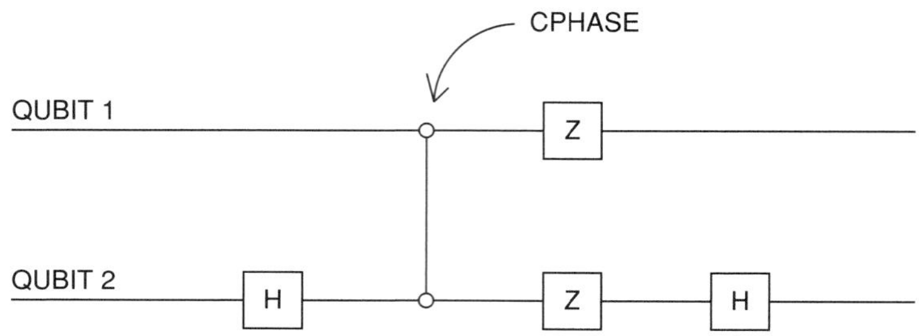

FIGURE 5.2. A CNOT operation using single qubit gates (Hadamard [H] and phase [Z] gates) and CPHASE.

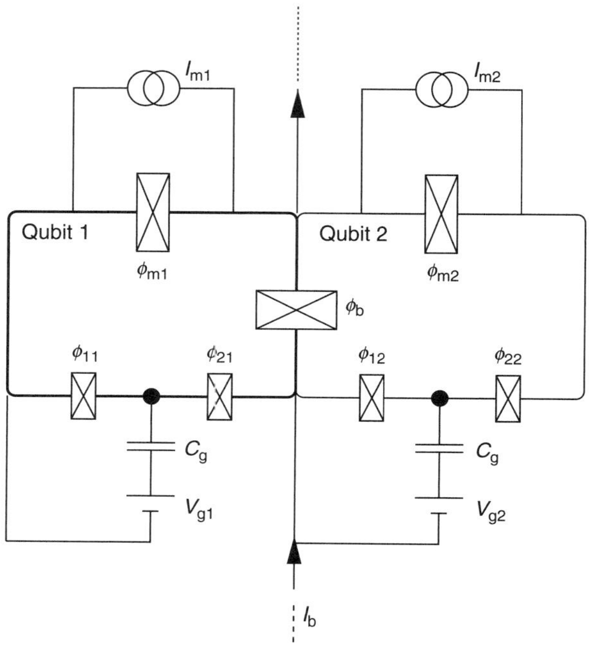

FIGURE 5.3. Bias currents I_{mi} across large Josephson junctions can be used for reading out the current state of qubit i. Applying both I_{m1} and I_{m2} results in controllable qubit–qubit interaction.

When both junctions are biased, the currents generated in both loops will be mixed in the coupling junction, providing the qubit–qubit coupling. Thus, in this case, the bias current line I_b is not needed and the effective interaction becomes

$$H_{\text{int}} = \frac{E_J^2}{4E_J^b} \sin\frac{\phi_{m1,0}}{2} \sin\frac{\phi_{m2,0}}{2} \sigma_z^1 \sigma_z^2, \tag{5.12}$$

where $\phi_{mi,0}$ is the stationary point of the ith tilted measurement junction.

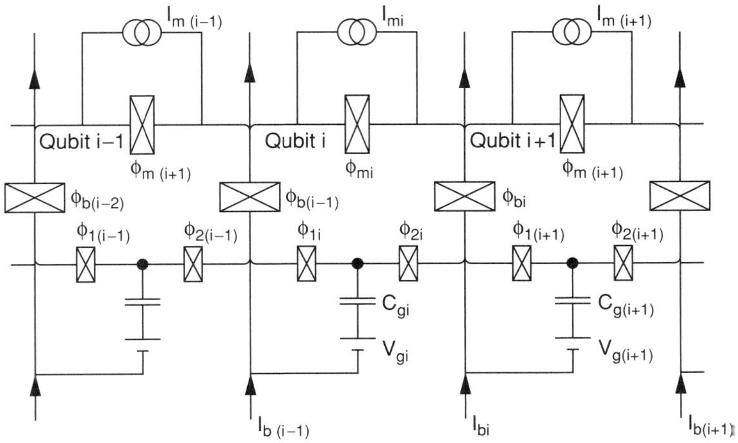

FIGURE 5.4. A system of N coupled charge qubits. V_{gi} controls the individual qubits whereas I_{bi} controls the coupling of qubits i and $i+1$.

5.6 Extension to Arbitrary Number of Qubits

The two-qubit coupling can be generalized to a chain of N coupled qubits, see Fig. 5.4. Every qubit is then coupled to its nearest neighbors via coupling junctions with individual bias current lines. The coupling junctions behave as oscillators which are weakly coupled (of order $\lambda = E_J/E_J^b \ll 1$) to one another.

When the bias currents are turned off, the qubits are independent (neglecting the small residual interaction mentioned in Section 5.2) and can be individually controlled using the gate voltages. The bias current I_{bi} controls the interaction between qubits i and $i+1$. It is possible to couple a number (depending on the total number of qubits, N) of qubit pairs at the same time; however, there must be a distance between them in order to get two-qubit coupling only. If qubits i and $i+1$ are coupled, then the nearest pairs that can be coupled simultaneously are, to the right $(i+3)$ & $(i+4)$, and, to the left $(i-3)$ & $(i-2)$. To be able to do all multi-qubit gates, it is enough to be able to do single qubit gates and CNOT gates[1] on nearest neighbor qubits [10]. Thus our network of coupled qubits can perform a universal set of quantum gates.

Acknowledgments. This work has been supported by EU IST-FET-SQUBIT and SQUBIT-2, the Swedish Research Council, and the Swedish Foundation for Strategic Research.

References

1. A. Barenco *et al.*, Elementary gates for quantum computation, *Phys. Rev. A* **52**, 3457 (1995).
2. J. Lantz *et al.*, Josephson junction qubit network with current-controlled interaction cond-mat/0403285 (2004).
3. Y. Nakamura *et al.*, Coherent control of macroscopic quantum states in a single-Cooper-pair box, *Nature* **398**, 786 (1999).

4. Y. Makhlin *et al.*, Josephson-junction qubits with controlled couplings, *Nature* **398**, 305 (1999).
5. D. Vion *et al.*, Manipulating the quantum state of an electrical circuit, *Science* **296**, 886 (2002).
6. M. Devoret, *Quantum Fluctuations in Electrical Circuits*, private communication (1995).
7. U. Weiss, *Quantum Dissipative Systems* (World Scientific, Singapore, 1999).
8. E. Collin *et al.*, NMR-like control of a quantum bit superconducting circuit, *Phys. Rev. Lett.*, cond-mat/0404503, (2004).
9. I. Siddiqi *et al.*, An RF-driven Josephson bifurcation amplifier for quantum measurements, cond-mat/0312623, (2003).
10. N. Schuch and J. Siewert, Programmable networks for quantum algorithms, *Phys. Rev. Lett.* **91**, 027902 (2003).

6

Direct Measurements of Tunable Josephson Plasma Resonance in the L-Set

MIKA A. SILLANPÄÄ, LEIF ROSCHIER, AND PERTTI J. HAKONEN

Low Temperature Laboratory, Helsinki University of Technology,
Otakaari 3 A, Espoo P.O.Box 2200 FIN-02015 HUT Finland

Abstract

Phase dynamics has been measured in a driven mesoscopic Josephson oscillator where the resonance is tuned either by magnetic flux or by gate charge modulation of the Josephson inductance. Phenomena are analyzed in terms of a "phase particle picture", and by numerical circuit simulations. An analogy to switching of a DC-biased junction into voltage state is discussed. Operation principle of the recently demonstrated Inductive Single-Electron Transistor (L-SET) based on the driven oscillator is reviewed. The obtained charge sensitivity implies that a performance comparable to the best rf-SETs has already been reached with the novel device.

6.1 Introduction

Quantum measurement in the solid state has been shown to be feasible as several impressive realizations of qubits based on mesosocopic superconducting tunnel junctions have emerged.[1–4] Sensitive measurement of physical quantities close to the limit set by the uncertainty principle is, on the other hand, an important issue in its own right.

The Single-Electron Transistor (SET) is a basic mesoscopic detector, sensitive to electric charge on a gate capacitor. In order to gain advantage of the inherently large bandwidth $(R_{\mathrm{SET}}C_{\Sigma})^{-1} \sim 10\,\mathrm{GHz}$ of the SET charge detector, basically two new technologies have been developed where the SET is read using an LC oscillator built from "macroscopic" components but coupled directly to it. The radio-frequency SET (rf-SET)[5] is based on gate modulation of the Q-value of the oscillator.

Because of limitations due to the dissipative nature of the rf-SET, a principally non-dissipative inductive SET (L-SET) technique has been developed.[6] The L-SET is based on reactive readout of the Josephson inductance of a Superconducting SET (SSET) in a resonator configuration.[7] Due to correlated Cooper pair tunneling, it does not exhibit shot noise or excessive dissipation.

In this chapter first the operating principle of the L-SET charge detector was reviewed. Then we concentrate on discussing classical dynamics of the phase φ under the microwave drive, and we present new experimental data. In particular, we discuss dynamical effects in the L-SET resonator which resemble the switching of a DC-biased Josephson junction into voltage state by drive or noise.

6.2 The L-Set Circuit

In the absence of DC bias voltage, a SSET has the Hamiltonian

$$H = \frac{(q - q_g)^2}{2C_\Sigma} - 2E_J \cos(\varphi/2)\cos(\theta), \tag{6.1}$$

where q_g is the gate charge, and φ is the phase difference across the SSET, assumed to be a classical variable here due to an environment having impedance much smaller than $R_Q \simeq 6.5$ kΩ. E_J is the single-junction Josephson energy, and the charging energy is related to sum capacitance by $E_C = e^2/(2C_\Sigma)$. Eigenvalues of this well-known Hamiltonian form bands $E_n(\varphi, q_g)$ (see Fig. 6.1).

At the lowest band $n = 0$, the energy E_0 increases approximately quadratically as a function of the phase φ when moved to either direction from the minimum at $\varphi = 0$. This type of dependence is characteristic of an inductor. The effective Josephson inductance of the SSET is then

$$L_J^* = \frac{\Phi_0^2}{2\pi}(E_J^*)^{-1}, \tag{6.2}$$

where the Josephson coupling has the effective value

$$E_J^* = \frac{\partial^2 E_0(q_g, \varphi)}{\partial \varphi^2}. \tag{6.3}$$

Here, $\Phi_0 = h/(2e)$ is the flux quantum.

Since the energy band, and consequently, the inductance, depend on the gate charge q_g, the L-SET electrometer is built so that the resonance frequency of an LC tank circuit connected to a SSET is tunable by q_g. This allows in principle a purely reactive readout.

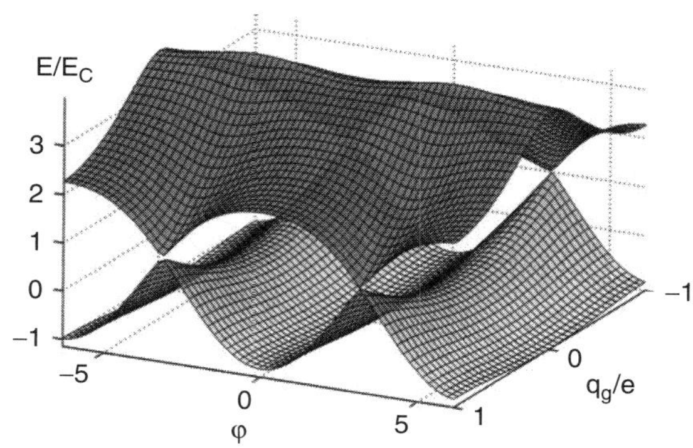

FIGURE 6.1. Two lowest bands $n = 0$ and $n = 1$ from Eq. 6.1 for $E_J/E_C = 1.7$ (sample 2) plotted as function of the gate charge q_g and the phase φ across the SSET.

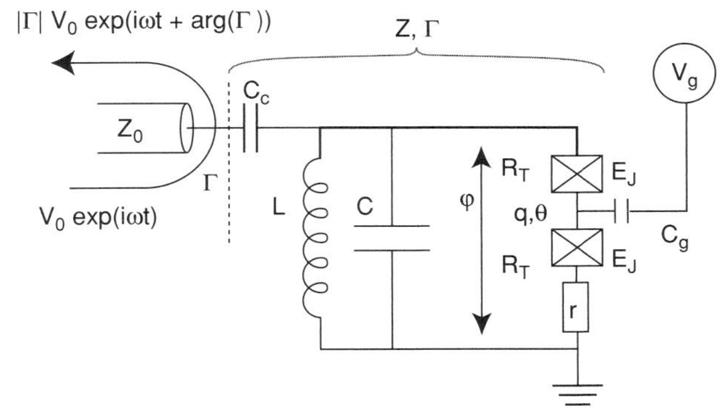

FIGURE 6.2. Schematics of the L-SET resonance circuit coupled to microwave feedline. The resonance is read by measuring amplitude or phase of the reflection coefficient $\Gamma = (Z - Z_0)/(Z + Z_0)$, where $Z_0 = 50\Omega$, and $Z(L_J^*)$ is the resonator impedance (including C_c).

The L-SET circuit we use is shown in Fig. 6.2 where the SSET is coupled in parallel to an LC oscillator resonant at the frequency $f_0 = 1/(2\pi)(LC)^{-1/2}$, roughly at 600 MHz. The total system has the gate-dependent plasma resonance at $f_p = 1/(2\pi)(L_{\text{tot}}C)^{-1/2} > f_C$, where $L_{\text{tot}} = L \parallel L_J^*$. The bandwidth $\Delta f \simeq f_p/Q_e$, where Q_e is the coupled quality factor, is typically in the range of tens of MHz.

The resistor r in series with the SSET is a model component for dissipation. As compared to the more standard way of drawing a resistor in parallel with the resonator, we got here a better agreement with the non-linear dynamics (see Section 6.6.3).

Charge detection is performed by measuring a change of amplitude or phase of the voltage reflection coefficient Γ, when the setup is irradiated by microwaves of frequency roughly f_p. In the best sample so far,[8] we have measured charge sensitivity $3 \times 10^{-5} e(\text{Hz})^{-1/2}$ over a bandwidth of about 100 MHz.

6.3 Plasma Oscillations in L-Set

Dynamics of the L-SET oscillator can be analyzed in terms of a potential $E_p = E_n(\varphi, q_g) + E_L$ due to the Josephson inductance and the shunting L, respectively, at the ground band approximately

$$E_p = -E_0(q_g)\cos(\varphi) + \Phi_0^2/(8\pi^2 L)\varphi^2. \tag{6.4}$$

At small oscillation amplitude, the phase particle experiences harmonic oscillations around $\varphi = 0$, whose frequency f_p is controlled by gate-tuning of the Josephson inductance (Fig. 6.3 (a)). This mode of operation, where the L-SET works as a charge-to-frequency converter, is the "harmonic mode".

Once the oscillation amplitude is increased close to the critical current at $\varphi = \pm\pi/2$, the $\cos(\varphi)$ term changes the local curvature of the potential, and hence, the resonance frequency changes. At very high amplitude, Fig. 6.3 (b), the cosine wiggle becomes averaged out. Thus, we expect a change of resonant frequency from f_p to f_0 roughly at an AC current of critical

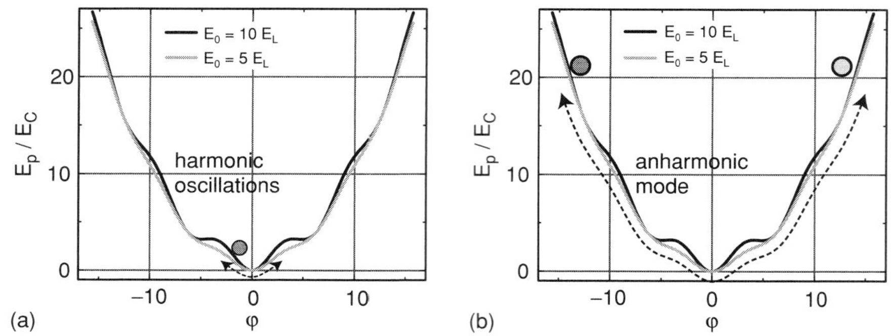

FIGURE 6.3. Illustration of the fictitious phase particle in a sinusoidally modulated parabolic potential.

current magnitude. This change of resonance frequency when the sample is probed by *critical power* P_c is reminiscent of a DC-biased Josephson junction switching into a voltage state.[9]

At large excitations above P_c, the highly nonlinear oscillator experiences complicated dynamics which does not in general allow analytical solutions. Numerical calculations over a large range of parameters, however, show consistently that the system response depends on L_J^* also in this case.[8] This mode of operation of the L-SET charge detector is called the "anharmonic" mode.

6.4 Simulation Scheme

We simulated the transition from linear to non-linear oscillations in the L-SET circuit with the Aplac circuit simulation program which contains an implementation of a Josephson junction element. The SSET was modeled as a single tunable junction. We used the method of harmonic balance where amplitudes of the first and three upper harmonics were optimized to create an approximate solution. Screenshot of the simulation schematics is shown in Fig. 6.4.

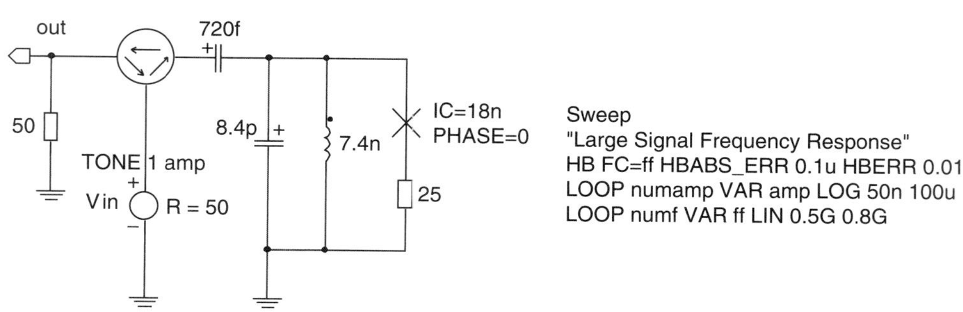

FIGURE 6.4. Screenshot from Aplac program used to simulate classical dynamics in the L-SET circuit (to aid convergence, 1 kΩ across the SSET was also used).

TABLE 6.1. Parameters of the two samples and of their tank LC oscillators as discussed in the text. $E_J = h\Delta/(8e^2\frac{1}{2}R_{SET})$ is the single-junction Josephson energy, and $E_C = e^2/(2C_\Sigma)$ is the charging energy. For sample 1, SQUID structure of the individual SET junctions allowed tuning of the E_J/E_C ratio.

Sample	R_{SET} (kΩ)	E_J (K)	E_C (K)	L_J^* (nH)	L (nH)	C (pF)	C_c (pF)	Q_e
1	4.2	3.5...0	0.17	6	3	23	0.72	13
2	9.6	1.6	0.92	16	7.4	8.4	0.72	18

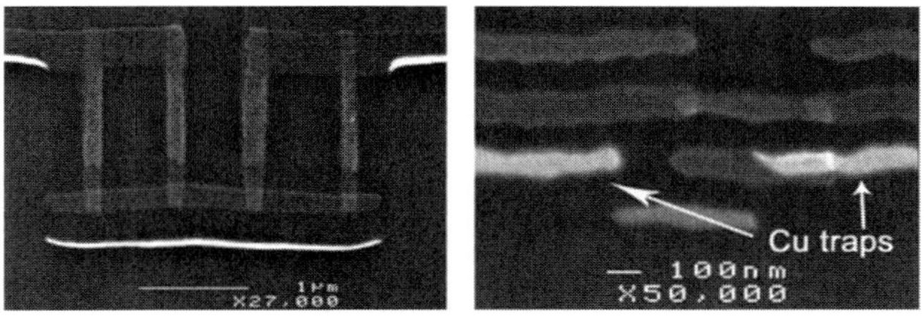

FIGURE 6.5. SEM micrographs of the samples discussed; sample 1 with SQUID junctions (left), and sample 2 (right) which had quasiparticle traps which were in contact to the electrodes 5 µm from junctions.

6.5 Experiment

The measurements were done in a powerful Leiden Cryogenics MiniKelvin dilution refrigerator. At the base temperature of 20 mK, we used a microwave circulator which had about 20 dB backward isolation to cut back-action noise from the 4 Kelvin preamplifier.[10]

The reflected probing excitation was amplified with a chain of amplifiers having a total of 5 K noise temperature, and detected with a network analyzer or a spectrum analyzer.

We studied two samples (Table 6.1. and Fig. 6.5) where sample 1 had a tunable E_J. Since it had $E_C \simeq k_B T$, where the temperature was probably set by leakage through the circulator, its response had a hardly detectable gate modulation. Accordingly, we were able to study the plasma resonance in an almost classical junction.

Sample 2 (discussed in Ref. [6], see Section 6.6.1) was a sensitive detector, with a 15 MHz gate shift of the resonance frequency.

6.6 Results and Discussion

6.6.1 *Charge detection*

For sample 2, $E_C \gg k_B T$, and hence, phase was properly localized in the potential. Accordingly, we got a rather good agreement for gate modulation of the resonance frequency (Fig. 6.6), where the resonance moves 15 MHz with respect to gate charge. This agrees well with the expectation of only 15 % modulation of L_J^* based on the rather high $E_J/E_C \simeq 1.7$ of this sample. The minimum (w.r.t. gate) of L_J^*, fitted best with 18 nH is about 20 % higher than expected. This is probably due to a small amount of phase noise, or partially due to inaccuracies in determination of the parameters.

Charge sensitivity was measured using amplitude readout. In the harmonic mode we got $s_q = 2.0 \times 10^{-3} e(\mathrm{Hz})^{-1/2}$ at the maximum power $P_c \simeq -116$ dBm, corresponding to 20 fW

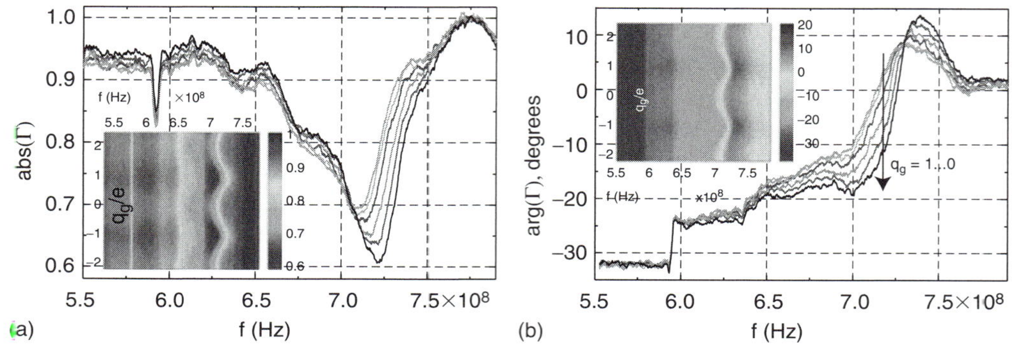

FIGURE 6.6. Sample 2; measured frequency response at $-125\,\mathrm{dBm}$ for successive gate values $q_g/e = 1 \ldots 0$, (a) amplitude, (b) phase. The insets portray the complete $2e$ periodicity.[6]

dissipation in the whole resonator circuit. Note that due to Cooper pair tunneling, the power is not dissipated in the SSET island. In the anharmonic mode, significantly better sensitivity of $\delta q = 1.4 \times 10^{-4} e(\mathrm{Hz})^{-1/2}$ was obtained.

In the best sample so far,[8] having $E_J/E_C \simeq 0.3$ we have measured $3 \times 10^{-5} e(\mathrm{Hz})^{-1/2}$ over a bandwidth of about 100 MHz. This kind of performance is comparable to the best rf-SET's,[11] though power dissipation is more than two orders of magnitude lower.

6.6.2 *Harmonic Oscillations*

In Fig. 6.7 we plot data from sample 1 which behaves almost as a classical junction with negligible charging effect. By applied magnetic flux we tune simultaneously E_J of both SQUID-shaped junctions of the SSET, and hence, L_J^*. Periodicity of the resonance frequency with respect to applied magnetic flux is evident.

A prominent feature in Fig. 6.7 is the absence of reflected signal at "critical" points where the resonance frequency switches to f_0. This implies a strong dissipation which is visible as coupling to Z_0. Pronounced dissipation at the critical points is representative of strong fluctuations at the artificial "phase transition" points.

In Fig. 6.8 we plot f_p measured at a low excitation, roughly 1/10 of critical current peak amplitude. The data are fitted to theoretical flux Φ modulation of L_J^*. A symmetric SQUID has been assumed, where $E_J \propto \cos(\pi \Phi/\Phi_0)$ which affects the E_J/E_J ratio and thus L_J^* (see Eqs. 5.2 and 6.3). It is clear that something else is happening at low E_J, where the resonance meets f_0 more rapidly than expected. Rounding of the cusp in experimental data cannot be explained by asymmetry in the SQUIDS either, since then the resonance would deviate from true f_0 which is not the case.

We argue this effect is due to "premature switching" caused by noise in the oscillator. The effect is again in analogy to physics in a DC-biased junction, namely, noise-induced switching and delocalization of phase. In contrast to running of phase in a tilted washboard, the average of

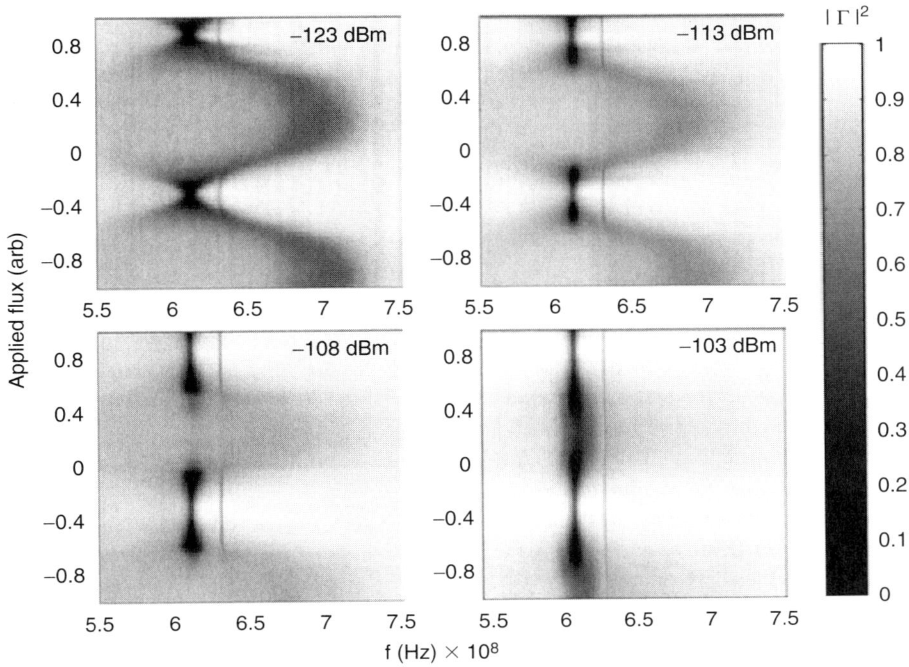

FIGURE 6.7. Flux modulation of the power reflection ($|\Gamma|^2$) for sample 1 in order of increasing probing excitation (the graph labelled −103 dBm corresponds approximately to critical power P_c at maximum E_J).

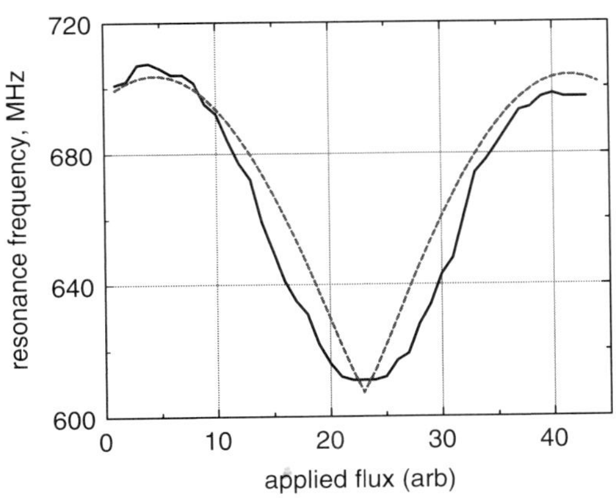

FIGURE 6.8. Measured resonance frequency for sample 1 extracted from the first graph in Fig. 6.7 (−123 dBm). The dotted line is a fit based on bare flux modulation of the E_J/E_C ratio.

phase stays at zero in our case. Noise-induced switching happens when peak phase fluctuations reach $\varphi_c \simeq \pi/2$. This condition corresponds to random motion of the phase particle at the bottom of the potential of Fig. 6.3 (a) with rms amplitude $\sim\varphi_c/2.5$.

Noise excites the phase particle in the potential E_p given by Eq. 6.4 up to an energy $E_p = k_B T_{\rm eff}$, where $T_{\rm eff}$ is the electron temperature set by noise. Thus, approximately $k_B T_{\rm eff} = \Phi_0^2(\varphi_c/2.5)^2/(8\pi^2 L_{\rm tot})$. Since in Fig. 6.8 the switching happens at the flux $\simeq 0.8 \times \pi\Phi_0/2$, we get an estimate $T_{\rm eff} \simeq 0.5$ K which is likely caused by leakage outside the band of the circulator.

6.6.3 *Switching and nonlinear oscillations*

In this section, the discussion is based on experimental data from sample 2. In Fig. 6.9, at $P_c \simeq -116$ dBm, the resonant frequency switches from the broad plasma resonance, centered at 720 MHz, into a narrower tank resonance at 613 MHz. The wavelike texture at $-105, \ldots, -90$ dBm is due to the $\cos(\varphi)$ Josephson-potential.

Changes in coupling, e.g., that Z goes through critical coupling at -105 dBm, and sharpening of the resonance above P_c, are due to an increase of the internal Q-value Q_i from 20 up to several hundreds. Since $Q_i \simeq \omega_0 L_J^*/r$, and since L_J^* acquires a large effective value at a high drive $\gg 2\pi$ due to cancelling of positive and negative contributions, the supposed dissipation of r in series with the SSET has less effect at $P \gg P_c$.

The source of the dissipation modelled by the resistor r, probably situated in the SSET itself, is presently unknown, but it may be related to quasiparticle poisoning. It also limits the quality factor up to 20 approximately.

Also plotted in Fig. 6.9 is an Aplac simulation for the circuit. It presents a qualitative agreement with experiment and predicts roughly correctly the end of the linear regime of plasma resonance. However, in experiment the switching is markedly sharp, which we attribute to the effect of higher bands of the SSET.[12]

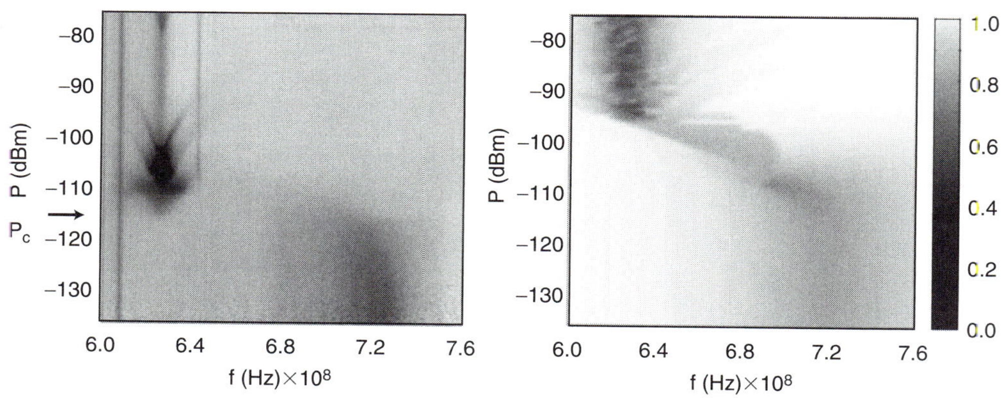

FIGURE 6.9. Measured frequency response ($|\Gamma|$, left) for sample 2 as a function of excitation power. "Critical power" P_c (see text) is marked. Gate DC level was tuned approximately to the minimum of L_J^*, i.e., the maximum of f_p. On the right is shown a corresponding Aplac simulation run using the circuit of Fig. 6.4.

Acknowledgments. It is a pleasure to thank T. Heikkilä, G. Johansson, R. Lindell, H. Seppä, and J. Viljas for collaboration. This work was supported by the Academy of Finland and by the Large Scale Installation Program ULTI-3 of the European Union (Contract HPRI-1999-CT-00050).

References

1. Y. Nakamura, Yu. A. Pashkin, and J. S. Tsai, Coherent control of macroscopic quantum states in a single-Cooper-pair box, *Nature* **398**, 786 (1999).
2. D. Vion *et al.*, Manipulating the quantum state of an electrical circuit, *Science* **296**, 886 (2002).
3. Y. Yu, S. Han, X. Chu, S. Chu, and Z. Wang, Coherent temporal oscillations of macroscopic quantum states in a Josephson junction, *Science* **296**, 889 (2002).
4. J. M. Martinis, S. Nam, J. Aumentado, and C. Urbina, Rabi oscillations in a large Josephson-junction qubit, *Phys. Rev. Lett.* **89**, 117901 (2002).
5. R. J. Schoelkopf *et al.*, The radio-frequency Single-Electron Transistor (rf-SET): A fast and ultra-sensitive electrometer, *Science* **280**, 1238 (1998).
6. M. A. Sillanpää, L. Roschier, and P. J. Hakonen, Inductive single-electron transistor, *Phys. Rev. Lett.*, to appear (2004); cond-mat/0402045.
7. A. B. Zorin, Radio-frequency bloch-transistor electrometer, *Phys. Rev. Lett.* **86**, 3388 (2001).
8. M. A. Sillanpää *et al.*, private communication.
9. A similar effect was recently observed by I. Siddiqi *et al.*, *Direct Observation of Dynamical Switching between Two Driven Oscillation States of a Josephson Junction*, cond-mat/0312553.
10. L. Roschier and P. Hakonen, Design of cryogenic 700 MHz amplifier, *Cryogenics* **44**, 783 (2004).
11. A. Aassime, G. Johansson, G, Wending, R. J. Schoelkopf, and P. Delsing, Radio-frequency single-electron transistor as readout device for qubits: Charge sensitivity and back-action, *Phys. Rev. Lett.* **86**, 3376 (2001).
12. In a more classical junction, the transition is less sharp; Mika A. Sillanpää, Leif Roschier, and Pertti J. Hakonen, Dynamics of the Inductive Single-Electron Transistor, submitted to *Proceedings of the Vth Rencontres de Moriond in Mesoscopic Physics* (2004).

7

Time Domain Analysis of Dynamical Switching in a Josephson Junction

JOACHIM SJÖSTRAND, HANS HANSSON, ANDERS KARLHEDE
JOCHEN WALTER, AND DAVID HAVILAND

Department of Physics, Stockholm University, AlbaNova University Center,
SE-10691 Stockholm, Sweden
Nanostructure Physics, Royal Institute of Technology, AlbaNova University Center,
SE-10691 Stockholm, Sweden

Abstract

A Josephson junction (JJ) when is connected to an external current source via a twisted pair with impedance Z, presents a probability distribution P of switching of the JJ, as a function of the current pulse from the source. Due to quantum effects and various noise sources, such as resistive leads at finite temperatures, this distribution will have a finite width. In a qubit system, such as the Quantronium, it is necessary that this width is small in order to clearly distinguish between different quantum states, and eventually realise the single shot measurement.

7.1 Introduction

Josephson junction physics is presently experiencing a renaissance, where many new and exciting experiments on small capacitance Josephson junction circuits are demonstrating quantum dynamics with measurements in the time domain.[1–5] In this field, an engineering approach to the design of quantum circuits is presently being applied to the long term goal of realizing a solid-state, scalable technology for implementation of a quantum bit processor. The intrest in quantum computation is fueled by a theoretical dream of massively parallel computation with quantum two-level systems, or qubits (for a nice introduction, see the book by Nielsen and Chuang[6]). A central question for these Josephson junction experiments relates to the optimal design for qubit readout, or detection of the quantum state of the circuit. Quantum state readout has been succesfully performed by switching current measurements.

In this chapter numerical simulation is used to analyze a particular type of readout–the switching of a Josephson junction subject to frequency dependent damping. The system model here is motivated by experiments, and comparisons are made with measured data. The switching process is a complex transition between two dynamical states of the circuit, and the numerical simulations allows to study the speed and resolution of the switching process, so that we can investigate the use of this switching process as a detector for quantum state readout of a Cooper Pair Transistor (CPT).

The classical description of the Josephson junction reduces to an analysis of the non-linear dynamics of the Josephson phase variable, which is conveniently discussed in terms of a fictitious "phase particle" moving in a washboard or $\cos(\phi)$ Josephson potential.[7] The time evolution of the phase variable depends entirely upon the particular circuit in which the junction is embedded. The simple and most tractable model is the so-called Resistively and Capacitively Shunted Junction model (RCSJ) of Stewart[8] and McCumber[9], where an ideal Josephson tunneling element is connected in parallel with an ideal capacitor (the tunnel junction capacitance), an ideal frequency independent impedance (a resistor), and driven by an external current source. More difficult to describe, and more relevant to experiments with small capacitance Josephson junctions, is the phase dynamics when the parallel impedance becomes frequency dependent, and when fluctuations due to the finite temperature of the dissipative elements are included. With small capacitance Josephson junctions we are often in a situation where the frequency dependent shunting impedance gives overdamped phase dynamics at high frequency, and underdamped phase dynamics at low frequency. The generic features of such a model have been studied extensively with computer simulations by e.g., Kautz and Martinis.[10] In this model the system can be in two distinct dynamical states for some given bias currents. In the "phase diffusion" state, a small finite voltage appears over the junction when the noise term causes a diffusive motion of the phase particle in a tilted washboard potential. In the "free running" state, a much larger voltage is sustained across the junction. A switching of the junction, or transition between these two dynamical states, occurs when a critical velocity of the phase particle is reached, so that it can overcome a "dissipation barrier" resulting from the frequency dependent damping.[7,10–12] In the present chapter, the probability for this dynamical switching, as a function of time and amplitude, for a particular type of bias pulse is presented.

The bias pulse, which is shown in Fig. 7.1(a), consists of a short switch pulse of amplitude I_p and duration τ_p, followed by a long hold level with amplitude I_{hold} and duration τ_{hold}. The idea behind this pulse is to quickly accelerate the phase particle above the critical velocity for dynamical switching, in order to force the switching process to happen as quickly as possible. The hold level is used to maintain the free running state with high voltage, long enough so that the voltage can be measured with an amplifier at room temperature. The hold level must be set high enough so that the phase particle will not be retrapped in the phase diffusion state, and yet low enough not to induce a late switching event. This type of bias pulse exploits the latching nature of the circuit to realize a sample-and-hold measurement strategy, in order to overcome

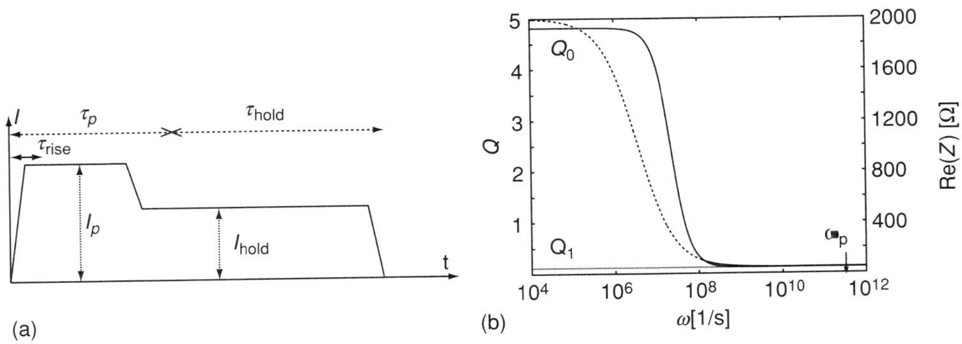

FIGURE 7.1. (a) The bias pulse and hold, and (b) the frequency dependence of the shunt impedance $Re[Z]$ (dashed line) and of the quality factor Q (solid line).

measurement bandwidth limitations resulting from filters, lossy cables, and high gain amplifiers. The latch gives us the added advantage of having a binary detection of switching, with two distinct voltage outputs for switch or no-switch, from one bias pulse.

The frequency dependent damping used in our numerical simulations is shown in Fig. 7.1(b), where we have plotted the real impedance $Re[Z]$ of the bias circuit (dashed line), and the Q factor of the junction (solid line) as function of frequency (see sections 7.2 and 7.3 for junction and model parameters). The model used for the shunting impedence is a simple RC shunt (see section 7.3) which reproduces the qualitative feature of the actual experimental situation–that of overdamped phase dynamics at high frequency ($Q \ll 1$), crossing over to underdamped phase dynamics at low frequency ($Q \gg 1$). The frequency scale associated with the switch pulse ($1/\tau_p$) can in principle be adjusted so that the junction is overdamped, and no hysteresis or latching behavior of the circuit would result for the switch pulse alone. However, the frequency scale of the hold pulse ($1/\tau_{\rm hold}$) is such that underdamped dynamics is realized, and a hysteresis or latching of the junction voltage is possible.

7.2 The Experiment

The sample used in the experiments is actually a Superconducting Quantum Interference Device (SQUID), which acts as an effective single junction and allows us to continuously tune the critical current, I_0, from its maximum value to zero by applying a magnetic field perpendicular to the loop. Here, however, we have only simulated data taken at a single value of I_0. The SQUID was fabricated by two angle evaporation of Al through a shadow mask which was defined by e-beam lithography. A micrograph of the sample is shown in the inset of Fig. 7.2(a) together with a schematic of the experimental set up. An Arbitrary Waveform Generator (AWG) biases the sample through an attenuator, a bias resistor via twisted pair cables of constantan wire of length 2.1 m and DC resistance of 135 Ω. The twisted pair acts as a lossy transmission line with a high frequency impedance of $\sim$ 50 Ω. We have measured this value at frequencies up to 500 MHz. The losses in the line are such that reflections or standing waves were only very weakly visible. At higher frequency the actual impedance seen by the junction depends critically on the geometry of the sample mount between the leads and the junction, which we have not measured

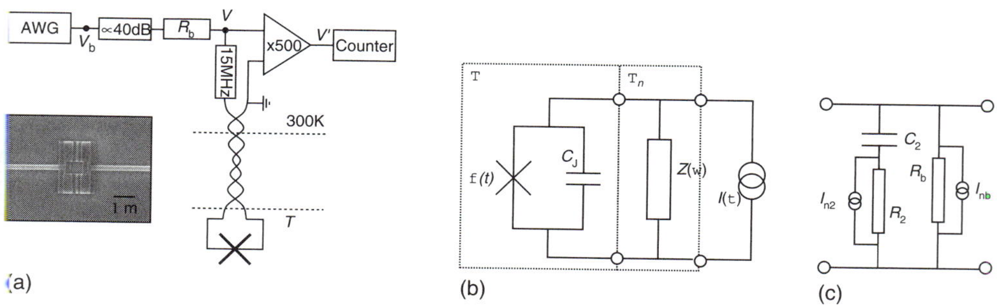

FIGURE 7.2. (a) The experimental setup and sample are shown in the inset. (b) The ideal version of the system consists of the Josephson junction (modelled by an ideal part and a capacitor C_J), connected to a current-source via some impedance $Z(\omega)$. (c) The impedance Z used here. Note that the noise currents due to the resistors have been included.

or modeled here. No additional cold filters were added at low temperature. The low frequency impedance seen by the junction is given by the bias resistor, $R_b = 2\ k\Omega$. The voltage across the sample is measured at the biasing end of the twisted pair cable through a gain 500 amplifier with bandwidth 100 kHz. The limited bandwidth of this amplifier does not allow us to observe the fast rise of the voltage when the junction switches to the free running state, the simulations however allow us to model the junction voltage in time. The bias resistor is chosen so that the voltage which builds up on the junction is kept well below the gap voltage $V_{2\Delta} = 400\ \mu eV$, in order to avoid quasiparticle tunneling in the junction and the associated dissipation. A digital counter registers a switching event when the output of the amplifier exceeds a trigger level. The switching probability is measured by counting the number of switching events and dividing by the number of applied bias pulses[13] (typically 10^4).

The capacitance of the SQUID junctions (parallel combination) is $C_J \approx 7.2$fF, as estimated from the measured junction area and the specific capacitance of 45 fF/μm^2. This gives a charging energy $E_C = 4e^2/2C_J = 45\ \mu eV$, or $E_C/k_B = 0.52$ K. The normal state resistance of the junction is 1.18 kΩ giving a critical current $I_0 = 265$ nA and Josephson energy $E_J = \hbar I_0/2e = 540\ \mu eV$, or $E_J/k_B = 6.3$ K, or $E_J/E_C \approx 12$. This large ratio, together with the low tunneling impedance of the junction, means that we are justified in treating the phase as a classical variable. Measurements were made in a dilution refrigerator with base temperature of 25 mK, well below the superconducting transition temperature of Al, $T_c \approx 1.2$ K. The temperature is varied in the range $0.025 \le T \le 0.7$ K as measured by a ruthenium oxide resistive thermometer in thermal equilibrium with the Cu sample mount. However, the effective temperature describing the noise in the theoretical model, is expected to be higher than the measured temperature as this noise is generated by dissipative elements located at higher temperatures.

The AWG is limited to a rise-time $\tau_{\text{rise}} = 25$ ns. We have measured that this rise time is effectively transmitted to the sample, although some of the switch pulse amplitude is lost for the shortest pulses due to dispersion in the twisted pairs. The switch pulse has magnitude I_p and length τ_p (including $2\tau_{\text{rise}}$) followed by a hold level of magnitude $I_{\text{hold}} < I_p$ and duration τ_{hold}. Every such bias pulse is followed by a wait time τ_w, with $I = 0$, where the phase retraps and the junction returns to thermal equilibrium.

7.3 The Model

A model circuit for the measurement set up, shown in Fig. 7.2(b), consists of an ideal Josephson junction with the current and voltage given by the Josephson relations, in parallel with the ideal capacitance of the tunnel junction C_J. This ideal tunnel junction is biased by a linear circuit with impedance $Z(\omega)$ in parallel with an ideal current source. (The model could have equivalently been cast in terms of a series combination of an impedance and voltage source). The model used for the impedance consists of a series combination of a resistor $R_2 = 50\Omega$ and capacitor $C_2 = 0.14$ nF, in parallel with the bias resistor $R_b = 2\ k\Omega$ (shown in Fig. 7.2(c)). With these values we match the measured impedance of the twisted pair fairly well. This type of model, with high impedance for low frequencies and low impedance for high frequencies (Fig. 7.1(b)), was first proposed by Ono *et al.*[14], and it is the perhaps simplest possible model of a frequency dependent environment that is known to correctly model the qualitative features of the switching process. The sample is at a low temperature ($\Gamma = k_B T/E_J$ in dimensionless form) and the resistors are at an unknown noise temperature, $\Gamma_n = k_B T_n / E_J$, which give rise to the noise currents $I_{nb,2}$.

Applying Kirchoff's rules and the Josephson relations to the circuit in Fig 7.2(b), one obtains a differential equation for $\phi(t)$. Defining the quantities $t' = t/t_s$ where $t_s = \hbar/2eI_0Z_0$ and $Z_0 = Z(0)$, $\omega' = t_s\omega$, $i = I/I_0$ (throughout this text upper case I is used for currents in amps and lower case i for (dimensionless) currents in units of I_0) and $Q_0 = Z_0\sqrt{2eI_0C_J/\hbar}$, one can write the equation in dimensionless form,

$$Q_0^2\frac{d^2\phi}{dt'^2} = -\frac{du_i}{d\phi} + i_n - Z_0\int_0^\infty \frac{d\phi}{dt'}(t'-\tau')y(\tau')d\tau', \qquad (7.1)$$

where $u_i(\phi) = -\cos\phi - i\phi$ is the tilted washboard potential, i_n is the noise current due to Z with a correlation function obeying the fluctuation-dissipation theorem[15] $\langle i_n(t_1')i_n(t_2')\rangle = \Gamma_n\frac{Z_0}{\pi}\int_{-\infty}^{+\infty} ReY(\omega')\cos\omega'(t_1'-t_2')d\omega'$, and the last term is the friction term in the form of a convolution integral between the voltage and the Fourier transform of the admittance, $y(t') = \int_{-\infty}^{+\infty} Y(\omega')e^{i\omega' t'}d\omega'$ with $Y = 1/Z$. Note that if $Z = R$, one recovers the RCSJ model.

Using our specific model for Z, we can write Eq. (7.1) as three coupled first order differential equations—a form more suitable for numerical analysis. We find

$$\begin{aligned}
\frac{d\phi}{dt'} &= v\\
\frac{dv}{dt'} &= \frac{1}{Q_0^2}\Big[i + i_{nb} + i_{n2} - \sin\phi - v + (v_C - v)(Q_0/Q_1 - 1)\Big] \qquad (7.2)\\
\frac{dv_C}{dt'} &= \frac{\rho}{Q_0^2}\Big[v - v_C - \frac{i_{nb}}{(Q_0/Q_1 - 1)}\Big],
\end{aligned}$$

where we have defined the following dimensionless parameters;

$$v = \frac{V}{R_b I_0}, \quad v_C = \frac{V_C}{R_b I_0}, \quad \rho = \frac{R_b C_J}{R_2 C_2},$$

$$Q_0 = R_b\sqrt{\frac{2eI_0C_J}{\hbar}}, \quad Q_1 = \frac{1}{1/R_b + 1/R_2}\sqrt{\frac{2eI_0C_J}{\hbar}},$$

where v is the dimensionless voltage over the junction and v_C is the dimensionless voltage over C_2. i_{nj} is the noise current due to resistor R_j ($j = b, 2$), with correlation function $\langle i_{nj}(t_1')i_{nj}(t_2')\rangle = 2\Gamma_n R_b/R_j\delta(t_2'-t_1')$ where we for simplicity have assumed that R_b and R_2 are at the same temperature. The generic features of the solutions to Eq. (7.2) are known and have been investigated thoroughly by e.g., Kautz and Martinis.[10]

The frequency dependent quality factor, plotted in Fig. 7.1(b), is

$$Q(\omega) = Q_0\frac{1 + Q_0^2\rho^{-2}(\omega/\omega_p)^2}{1 + Q_0^3 Q_1^{-1}\rho^{-2}(\omega/\omega_p)^2} \qquad (7.3)$$

where $\omega_p = \sqrt{2eI_0/\hbar C_J}$ is the plasma frequency (indicated in Fig. 7.1(b)). This expression is derived using the following definition; $Q(\omega) = \sqrt{2eI_0C_J/\hbar G^2(\omega)}$ where $G = Re[1/Z]$.

We solve Eq. (7.2) using a fourth order Runge–Kutta algorithm. The noise curents are $i_{nb,2} = N\sigma_{b,2}$, where $N \in [-1, 1]$ is a Gaussian distributed random number, $\sigma_{b,2} = \sqrt{2\Gamma_n R_b/\Delta t' R_{b,2}}$

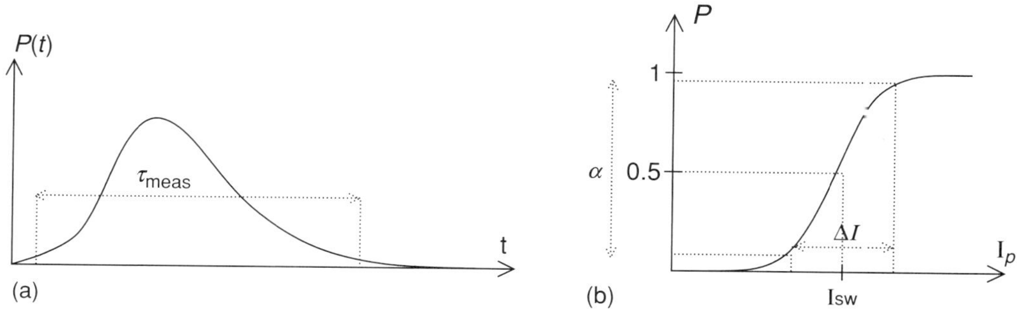

FIGURE 7.3. (a) A probability distribution of switching as a function of time, with the measurement time defined as the width of 98% of this distribution. (b) A graphical representation of the discriminating power α and the current resolution ΔI.

is the standard deviation and $\Delta t'$ is the dimensionless timestep in the numerical routine. This last factor is present since it is the mean value of $i_{\mathrm{nb},2}$ during a time $\Delta t'$ that enters Eq. (7.2).

We have simulated the switching process using $\tau_p = 0.1, 1.0$ and 10 μs. If the junction switches, there will be a finite voltage $\langle V \rangle > V_{\mathrm{trig}}$ over the junction (where $\langle V \rangle$ denotes the mean value over time and V_{trig} is the trigger voltage) and the counter will click. The scale of $\langle V \rangle$ is determined by R_{b} and I_{hold}. The finite voltage is latched for the duration of the hold pulse due to the hysteretic behavior of the non-linear circuit. In the no-switch case, the phase particle either fluctuates thermally ($\langle V \rangle = 0$) or diffuses down the washboard ($\langle V \rangle < V_{\mathrm{trig}}$)–in both cases the counter will not click, assuming V_{trig} is chosen high enough.

7.4 Results and Discussion

Two important quantities for the use of this switching process as a detector of the quantum state of a circuit, are the measurement time and the resolution. The measurement time can be determined by recording when the switching events occur, i.e., at what time the simulated voltage reaches the prescribed trigger level. Due to bandwidth limitations in the actual experiment, we can not measure this time. However, we can simulate many switching events to determine the probability, $P(t)\mathrm{d}t$, that the switching event occurs between the time t and $t - dt$. A typical distribution $P(t)$ is shown in Fig. 7.3(a). We define the measurement time to be the width of 98% of this distribution, neglecting the first and last percent. The resolution of the detector is determined from the probability, $P(I_p)\mathrm{d}I_p$, that a switching event will occur with a switch pulse of amplitude I_p. This probability distribution can be both simulated and measured, and a typical curve is shown in Fig. 7.3(b). The resolution of our detector ΔI, is the ability of the detector to discriminate between two different values of the switching current (which will differ depending on the quantum state of a circuit). We have (arbitrarily) chosen a discriminating power $\alpha = 0.8$ (meaning that 20% of the events will be miscounted) to determine the resolution ΔI.

Fig. 7.4 shows a simulated $P(t)$ for three different values of the hold level, i_{hold}. The magnitude of the noise was set to $\Gamma_{\mathrm{n}} = k_B T_{\mathrm{n}}/E_{\mathrm{J}} = 0.47$ (corresponding to $T_{\mathrm{n}} = 3.0$ K with E_{J} from sec. 2) and the voltage trigger to $V_{\mathrm{trig}} = 35$ μV, which is safely above the phase diffusion voltage.

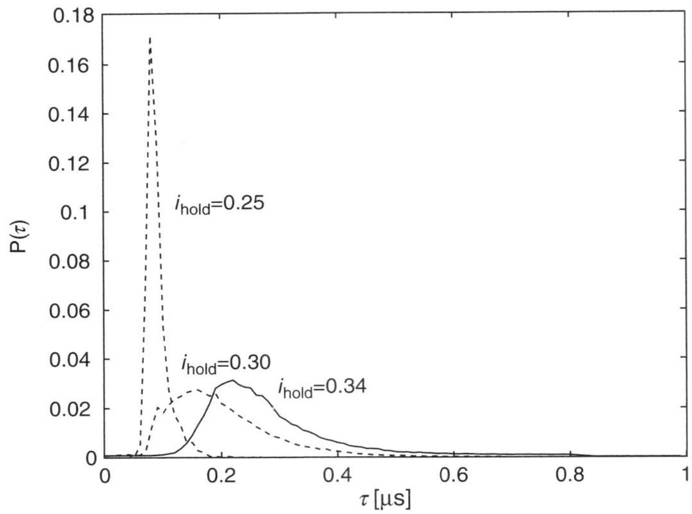

FIGURE 7.4. This shows $P(t)$ during one pulse, for $dt = 0.01$ μs, $P = 0.50$ and $V_{trig} = 35$ μV, with $\Gamma_n = 0.47$ and $i_{hold} = 0.34$ (solid line), $i_{hold} = 0.30$ (dashed line) and $i_{hold} = 0.25$ (dash-dotted line).

The switch pulse had duration $\tau_p = 0.1$ μs and the hold pulse had duration $\tau_{hold} = 0.9$ μs. The amplitude of the switch pulse i_p was adjusted for each curve so that $P(I_p) = 0.5$. In this simulation it was first determined that a pulse consisting of the hold level itself (i.e., when $i_p = i_{hold}$) did not induce any switching ($P < 0.001$). We can see from the solid curve of Fig. 7.4 that the hold level $i_{hold} = 0.34$ induces late switching events, with the actual switching taking place during the hold time, resulting in a measurement time that exceeds τ_p. Lowering the hold level to $i_{hold} = 0.3$ (dashed curve of Fig. 7.4) causes the switches to occur earlier in time, with a slight reduction of τ_{meas}. If the hold level is lowered to $i_{hold} = 0.25$ (dash-dotted curve of Fig. 7.4) we find that the distribution $P(t)$ becomes sharply peaked, with $\tau_{meas} \approx \tau_p$.

The simulations clearly show that τ_{meas} increases with i_{hold}. We find $\tau_{meas} = 0.10, 0.42, 0.44$ μs for $i_{hold} = 0.25, 0.30, 0.34$ respectively. These values of τ_{meas} are approximately the same for different switching probabilities in the region $i_p = i_{sw} \pm \Delta i/2$. For $i_{hold} > 0.34$, τ_{meas} will increase, since the hold pulse alone will begin to induce switching events. For $i_{hold} < 0.25$, τ_{meas} will continue to be close to τ_p, although for very low i_{hold} it will become hard to distinguish between switch and no-switching events. We have also run this simulation with lower strength of the noise term, $\Gamma_n = 0.11$, and we observe $\tau_{meas} \approx 0.1 - 0.15$ μs for $0.40 \leq i_{hold} \leq 0.63$. As expected, a weaker noise term will give a more rapid measurement time over a wider range of hold levels.

From this simulation we can conclude the following: (1) Even in the presence of significant noise ($\Gamma_n = 0.47$) it is possible to find a switch pulse amplitude and hold level which give a rapid dynamical switching that occurs during the switch pulse without a significant number of late switches. (2) Simply adjusting the hold level for zero switching when $i_p = i_{hold}$ is not good enough to achieve optimal measurement time. Because one can not observe the escape process in the experiment, simulations of this kind are necessary to verify that the hold level is sufficiently low.

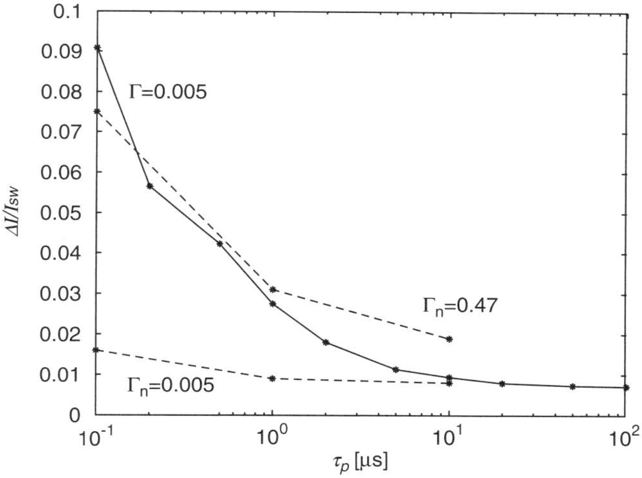

FIGURE 7.5. This shows $\Delta I/I_{sw}$ as a function of τ_p. The dashed curves are simulated data with $\Gamma_n = 0.005$ and 0.47, and the solid is measured data with $\Gamma = 0.005$ ($T = 0.030$ K).

Several switching curves $P(I)$ of the type shown schematically in Fig. 7.3(b) were simulated for various values of the temperature and switching pulse parameters, as they were adjusted in the actual experiment. In Fig. 7.5 we show a comparison between the simulated and measured resolution $\Delta I/I_{sw}$, as a function of the switch pulse duration τ_p. Here, I_{sw} is the value of the switch pulse amplitude which gives 50% switching probability, $P(I_p) = 0.5$. It is difficult to compare the simulated and experimental value of I_{sw}, because in the experiment we measure the voltage pulse applied to the bias resistor. Dispersion in the twisted pairs, and parasitic capacitance of the sample mount, make it difficult to know the actual amplitude of the current pulse applied to the junction.

The simulated values shown in Fig. 7.5 are calculated for two different noise tempertures. The value $\Gamma_\mathrm{n} = 0.005$ (or $T_\mathrm{n} = 30$ mK for the particular junction parameters considered), corresponds to the measured temperature of the sample mount during the experiment. For this value of the noise temperature we see that the experimental and simulated values agree for longer τ_p, while for shorter pulses the simulation gives much higher resolution than the experiment. The noise temperature has been adjusted, and reasonable agreement is found over the range of τ_p studied for $\Gamma_\mathrm{n} = 0.47$ ($T = 3$ K). This temperature seems excessively high, but note that in the simulation we have put R_b and R_2 at the same temperature. It would be more realistic to take R_b at a higher, and R_2 at a lower temperature. In turn, this would increase i_nb and decrease i_n2 in Eq. (7.2), causing a partial cancellation effect, giving an intermediate effective temperature. The exact consequences remains though to be investigated.

Both the experimental and theoretical curves of Fig. 7.5 show decreasing resolution (increasing ΔI) for shorter measurement time (smaller τ_p). Partly this is due to that i_{sw} increases with decreasing τ_p for given Γ_n, but also the statistical fluctuation of the mean noise amplitude should be bigger for a short pulse than for a long pulse. Thus, as expected, there is a trade off between measurement speed and measurement resolution.

7.5 Summary

In summary, computer simulations of the dynamical switching process of a Josephson junction subject to noise and to frequency dependent damping is performed. The model for damping is appropriate for measurements of small capacitance Josephson junctions, where the phase dynamics is overdamped at high frequencies, and underdamped at low frequencies. A special bias current pulse, consisting of a short switch pulse followed by a longer hold level is used, appropriate for binary detection of the switching process. The switching process is characterized in terms of speed, τ_{meas}, and resolution, $\Delta I/I_{sw}$. The simulations show that it is possible to achieve a minimum switching time given by the duration of the fast switching pulse, even in the presence of strong noise. To achieve this fast measurement, the hold level must be set appropriately low. As expected, there is a trade off between measurement speed and resolution.

Acknowledgments. We wish to thank D. Vion for interesting discussions. This work was partially supported by the EU project SQUBIT, in the 5th framework QIPC program, by the Swedish Research Council and by the Swedish SSF.

References

1. Y. Nakamura, Yu. A. Pashkin, and J. S. Tsai, *Nature* **398**, 786 (1999).
2. D. Vion *et al.*, *Science* **296**, 886 (2002).
3. I. Chiorescu, Y. Nakamura, C. J. P. M. Hormans, and J. E. Mooij, *Science* **299**, 1869 (2003).
4. J. M. Martinis, S. Nam, J. Aumentado, and C. Urbina, *Phys. Rev. Lett.* **89**, 117901 (2002).
5. T. Duty, D. Gunnarsson, K. Bladh, and P. Delsing, *Phys. Rev. B* **69**, 140503 (2004).
6. M. Nielsen and I. Chuang, *Quantum Computation and Quantum Information*, (Cambridge University Press, Cambridge, 2000).
7. M. Tinkham, *Introduction to Superconductivity*, (McGraw-Hill, 1996).
8. W. C. Stewart, *Appl. Phys. Lett.* **12**, 277 (1968).
9. D. E. McCumber, *J. Appl. Phys.* **39**, 3113 (1968).
10. R. L. Kautz and J. M. Martinis, *Phys. Rev. B* **42**, 9903 (1990).
11. D. Vion, M. Götz, P. Joyez, D. Esteve, and M. H. Devoret, *Phys. Rev. Lett.* **77**, 3435 (1996).
12. P. Joyez, D. Vion, M. Götz, M. H. Devoret, and D. Esteve, *J. Superconductivity* **12**, 6 (1999).
13. Further experimental details can be found in J. Walter, S. Corlevi, D. B. Haviland, cond-mat/0403467.
14. R. H. Ono, M. W. Cromar, R. L. Kautz, R. J. Soulen, J. H. Colwell, and W. E. Fogle, *IEEE Trans. Magn. MAG* **23**, 1670 (1987).
15. See e.g. G.-L. Ingold, in *Quantum Processes and Dissipation* (Wiley-VCH, 1998).

8

Cooper Pair Transistor in a Tunable Environment

S. CORLEVI, W. GUICHARD, AND D. B. HAVILAND

Nanostructure Physics, Royal Institute of Technology, AlbaNova, 106 91 Stockholm, Sweden

Abstract

We report an experimental study on the effect of high impedance environment on a Cooper pair transistor (CPT). The CPT consists of two small capacitance Josephson tunnel junctions in series with a gate electrode coupled through a capacitance Cg to a central island. In small-capacitance CPT the charging energy EC $= e^2/2C$, where C $\sim 2C_J + C_g$ is the total capacitance of the island, becomes relevant at low temperature, and charging effects influence the transport properties.

8.1 Introduction

Devices based on small-capacitance Josephson junctions have been widely studied in the last two decades as model systems displaying macroscopic quantum phenomena. Presently, the possible application of such devices for a solid-state quantum bit drives the research field. One of the most studied superconducting single-charge devices is the Cooper Pair Transistor (CPT), also known as superconducting-single electron transistor (S-SET). The CPT consists of two small-capacitance Josephson tunnel junctions, of capacitance C_J, connected in series and a gate electrode coupled to the central island through a capacitance Cg. In a small-capacitance CPT, the charging energy $E_{C\Sigma} = e^2/2C_\Sigma$, where $C_\Sigma \cong 2C_J + C_g$ is the total capacitance of the island, becomes relevant at low temperature, and charging effects influence the transport properties. In typical measurements of the CPT, which are performed in a low impedance environment, the charging effects are observed as gate voltage modulation of the critical current. However, in a high impedance environment, a Coulomb blockade and its modulation with the gate voltage can also be observed, as is demonstrated in this chapter.

The difficulty to observe Coulomb blockade in a single junction and also in a CPT, is related to the problem of creating a very high impedance electromagnetic environment (Re$[Z(\omega)] \gg R_Q$, where $R_Q = h/4e^2 \approx 6.45\text{k}\Omega$ is the quantum resistance)[1] at very high frequencies of the order of Josephson junction plasma frequency ($\omega_p = (2\pi I_C/\phi_0 C)^{1/2} \approx 10^{11}\text{s}^{-1}$, where C is the junction capacitance, I_C the critical current and $\phi_0 = h/2e = 2 \times 10^{-15}$ Wb the flux quantum). If no special care is taken, the impedance of the environment is of the order of the free space impedance $Z_0/2\pi = 60\Omega$, and the classical Josephson effect is realized instead of the Coulomb blockade, even when the charging energy is of the same order of Josephson energy $E_J = \phi_0 I_C/2\pi$. When the impedance of the environment is increased, large quantum fluctuations of the phase result, and the current voltage (I–V) characteristic of the CPT does not show a supercurrent feature, but a Coulomb blockade feature and a region of negative differential resistance, which is the signature of coherent tunneling of single Cooper pairs.[2]

Several experiments have been focused on investigating the properties of CPT in a low impedance environment.[3,4,5] In these experiments it was shown how the 2e-periodic modulation of the supercurrent in the gate-induced charge could be suppressed by quasiparticles. The problem of nonequilibrium quasiparticles tunneling on to the island, or "quasiparticle poisoning", has been studied, and some methods have been developed to reduce it. Quasiparticle poisoning can be suppressed when normal metal leads, acting as quasiparticles traps, are connected close to the CPT.[4] In a more recent study[5] it was shown how the profile of the superconducting gap of the island and the electrodes of a CPT may dramatically influence the effect of nonequilibrium quasiparticles. When the superconducting gap of the island of the CPT is higher than the gap of the electrodes, a potential barrier is obtained, which reduces the unpaired quasiparticle lifetime on the island. A higher superconducting gap was obtained when the aluminium used to create superconducting tunnel junctions is doped with oxygen during the deposition. To make the superconducting gap of the island higher than that of the leads, the island is formed during the first step of the two-angle Al deposition process, when the "dirty" Al is deposited, while the electrodes are formed in the second deposition, when pure Al is evaporated.

In this paper we present our experimental results on the transport properties of a CPT in a high impedance environment created by nanometer-scale SQUID arrays. In other studies, the high impedance of the environment has been obtained by tunable two dimensional electron gas beneath the CPT structure[6,7] or small chromium resistors located on chip close to the CPT.[8,9] In our experiments, as in those of Watanabe[10,11], the high impedance environment is obtained by biasing the CPT with four SQUID arrays. The main advantage of this method is that the effective impedance of the leads can be tuned *in situ* by applying a magnetic field perpendicular to the SQUIDs loop. This tunability allows us to clearly separate out the effect of the environment on tunneling from spurious effects, because in our design the junctions "under test" defining the CPT do not have a SQUID geometry, and thus are unaffected by the relatively small magnetic fields (the flux $\Phi = \phi_0/2$ at 5mT) needed to tune the SQUIDs.

8.2 Sample Fabrication and Measurement Techniques

The tunnel junctions ($Al/Al_2O_3/Al$) are fabricated on a SiO_2/Si substrate using e-beam lithography and the standard shadow evaporation technique. In Fig. 8.1 a SEM micrograph of the sample is presented. The junctions of the CPT are designed to have an area of 0.01 μm^2 for sample 9 and 0.02 μm^2 for sample 6 and 7. For all the samples each array has 70 SQUIDs, with each junction having an area of 0.03 μm^2 and a loop size of $\approx 0.18\,\mu m^2$. Table 8.1 presents characteristic parameters for the measured samples. The normal state resistance of the CPT and of the SQUID arrays is calculated from the slope of the $I - V$ curve in the normal state, measured at high magnetic field, $B = 125$mT. In Table 8.1, the value R_N represents the resistance of one SQUID of the biasing arrays. Assuming a symmetric CPT with identical junctions each having a resistance r_N, the Josephson energy for the CPT could be calculated according to the expression $E_J = (R_Q/r_N)(\Delta/2)$, where $\Delta \approx 200\,\mu eV$ is the superconducting gap of aluminium. The

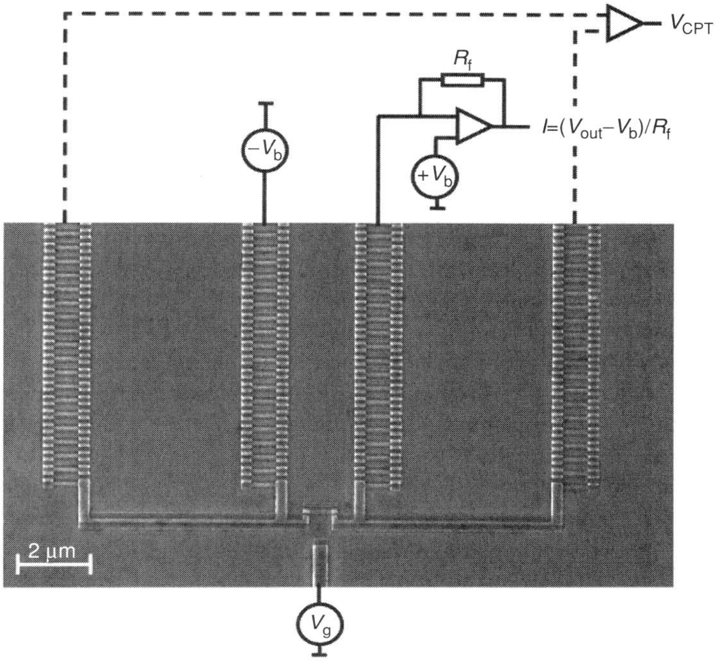

FIGURE 8.1. SEM micrograph of the sample and scheme of the biasing circuit. The CPT (in the center) is biased by four SQUID arrays that enable a four point configuration measurement.

TABLE 8.1. List of a few characteristic parameters of the measured samples.

	C_Σ (fF)	$E_{C\Sigma}$ (µeV)	r_N (kΩ)	E_J (µeV)	$E_J/E_{C\Sigma}$	R_N (kΩ)
Sample 6	1.53	52.3	6.4	100	1.9	1.8
Sample 7	1.2	66	22.5	28.7	0.43	8.5
Sample 9a	1.1	74	11.7	55.4	0.75	3.1
Sample 9b	0.65	123	9	71.5	0.58	3

capacitance of the CPT junctions C_J has been estimated from the junction area measured from SEM micrographs, using the specific capacitance value 45 fF/µm^2. The gate capacitance C_g is measured from the period of the gate voltage modulation in the normal state, and for all the samples has the value 3.3 aF. The charging energy of the CPT is calculated from the sum capacitance $C_\Sigma = 2C_J + C_g$ and is also given in Table 8.1.

The samples are mounted in a RF-tight copper box and measured in a dilution refrigerator with a base temperature of 15mK. The leads to the sample are twisted pairs of constantan wire, and no low temperature filtering is implemented in the cryostat. However, when the arrays are in the high impedance state, they act as good filters for electromagnetic fluctuations guided to the sample via the leads. The I–V characteristics of the CPT are measured in a four-point configuration as shown in Fig. 8.1. One pair of SQUID arrays is used to apply a symmetric bias and measure the current, while the other pair of arrays is used to measure the voltage across the CPT. The current is measured with a current amplifier (SR 570 or OPA 111) and the voltage with a high input impedance differential voltage amplifier (INA110).

8.3 Squid Arrays Characterization

A one-dimensional array of nanometer-scale SQUIDs can be considered as an array of single Josephson junctions with tunable Josephson coupling. In these small SQUIDs, the inductance of the loop and the junction's critical current are well in the limit $L_{\text{loop}} \ll \phi_0/2\pi I_{\text{C}}$, where an external magnetic field applied to the SQUID loop will simply modulate the critical current, or the Josephson energy of the SQUID as

$$E_{\text{J}} = \frac{\Delta\phi_0}{4eR_N}\left|\cos\left(\frac{\pi BA_{\text{loop}}}{\phi_0}\right)\right|.$$

The one-dimensional SQUID array can be modeled as a 1D network of Josephson junctions with tunable E_J, with each island coupled to ground through a capacitance C_0, as shown in Fig. 8.2a. The two Josephson energies E_{J1}, E_{J2}account for the possible difference in loop size of two adjacent SQUIDs. Neglecting the quasiparticle resistance, a Josephson junction can be viewed as a lumped element constituted by a parallel combination of a Josephson element and the capacitance of the SQUID. In the limit $I \ll I_{\text{C}}$, when the Josephson current–phase relation can be linearized, a Josephson junction is equivalent to an effective inductance[14] $L_{\text{J}} = \phi_0/2\pi I_{\text{C}} = \phi_0^2/4\pi^2 E_{\text{J}}$, as presented in Fig. 8.2b.

This picture can be further simplified considering that for frequencies much smaller than the plasma frequency, $\omega_{\text{p}} = 1/(L_{\text{J}}C)^{1/2}$, the contribution to the impedance $Z(\omega)$ due to the capacitance can be neglected. Thus, in this limit, the one-dimensional SQUID array can be described as a series of tunable Josephson inductances coupled to ground through a capacitance C_0, as shown in Fig. 8.2c. The impedance of such a 1D network, which is infinitely long, is independent of frequency (transmission line impedance) in the limit $\omega \ll \omega_{\text{p}}$, and is given by

$$Z_A = \sqrt{\frac{L_{\text{J1}} + L_{\text{J2}}}{2C_0}},$$

where $L_{\text{J1}}(B)$ and $L_{\text{J2}}(B)$ have different period of flux modulation. The impedance $Z_A(B)$ is calculated and plotted together with the measured zero bias resistance R_0 of the SQUID arrays in Fig. 8.3.

As the field is increased, the arrays undergo a superconductor-insulator quantum phase transition[12] that results in a periodic modulation of the zero bias resistance R_0. The double-peak features in the modulation of R_0 can be explained by our impedance model, which takes

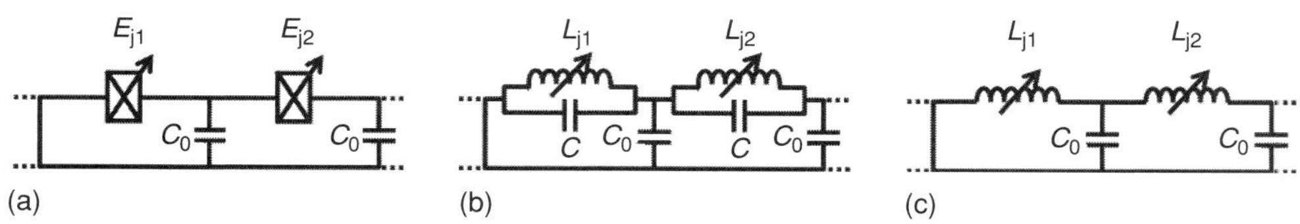

FIGURE 8.2. (a) Schematic of two adjacent SQUID in a 1D SQUID array. (b) Equivalent lumped elements for small phase difference across the SQUID junctions. (c) Equivalent lumped elements in the approximation $\omega \ll \omega_p$.

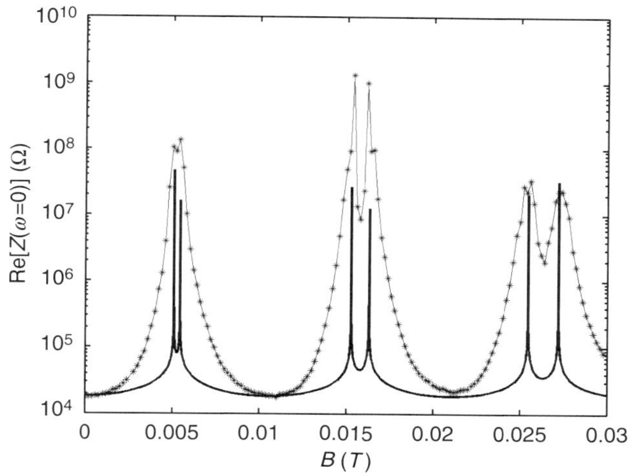

FIGURE 8.3. Line∗symbol: Measured zero bias resistance for the two biasing SQUID arrays of sample 9a as function of magnetic field. Solid line: Calculated $Z_A(\omega = 0)$ for an infinite array modeled as a series of Josephson inductances with the parameters of sample 9a. In the fit the capacitance to ground is C_0 = 0.01 fF. The values of the loop area of the SQUIDs, A_1 and A_2 are chosen so that the peaks in magnetic field coincide.

into account the possible difference in loop size between two adjacent SQUIDs. For this sample the difference in loop size was found to be $\approx 7\%$. Moreover, the value chosen for the fitting parameter C_0= 0.01 fF well agrees with the estimated value for the stray capacitance of similar SQUID arrays.[12]

It is interesting to note that the values of Z_A and R_0 roughly coincide, but the modulation of $Z_A$is much more abrupt around the minimum of the SQUID critical current than the measured behavior of R_0. However, it is difficult to draw any conclusions from this comparison, as the zero bias resistance strictly represents the DC component of the impedance of the environment (R_0 = Re[$Z(\omega = 0)$]), which would coincide with Z_A only in the infinite array limit. Nevertheless, we can use the measured R_0 to characterize the environment. The tunability of R_0 over several orders of magnitude ($10^4\Omega < R_0 < 10^8\Omega$) provides the possibility of studying the same CPT in a variety of different environments. Note that for each sample, the supercurrent of the SQUID arrays at zero field presents a large finite resistance ($R_0 \approx$ 10–20 kΩ, so that all our measurements are performed in the regime Re[$Z(\omega = 0)$] $> R_Q$.

8.4 Measurement of the CPT

In Fig. 8.4 the different stages of the I–V characteristic of the CPT (Fig. 8.4a) are presented for different magnetic fields and therefore for different environments (Fig. 8.4b). As the biasing leads become highly resistive ($R_0 >$ 1 MΩ), the fast discharge of the CPT junctions through the external circuit is prevented, and the I–V curve of the CPT starts to show a Coulomb blockade feature. When $R_0 \approx$ 20 MΩ, the Coulomb blockade region is fully developed, and the I–V curve

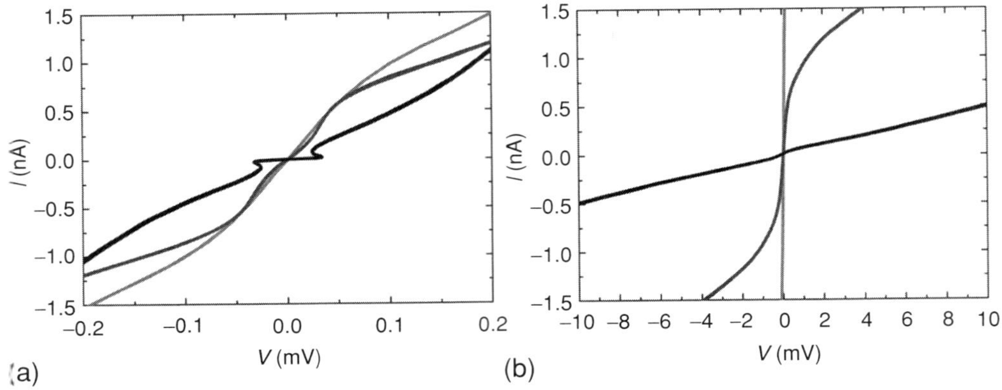

FIGURE 8.4. I–V characteristics of sample 9a: (a) I–V curves of the CPT for different values of R_0. (b) Corresponding I–V curves of two SQUID arrays in series. The values for R_0 are: 55 kΩ, (gray curves), 450 kΩ (dark gray curves), 20 MΩ (black curves).

shows a back-bending feature at low voltages. According to theory[2,13], this region of negative differential resistance is related to the coherent single Cooper pair tunneling.

The blockade voltage for the onset of current through the CPT is sensitive to the average charge on the gate capacitance ($Q_g = V_g C_g$).[15] In order to study this effect, we fixed the bias voltage across the SQUID arrays and measured the current and voltage across the transistor as function of V_g. By tuning the magnetic field, we study the gate modulations in different impedance environments. The presence of current or voltage oscillations, as the gate is swept at low frequency (DC), is an indication of the effective type of DC bias applied to the transistor. When $R_0 < 50$ kΩ, there is no evidence of oscillations in the voltage, which indicates that the sample is voltage-biased at low frequency. When $R_0 > 100$ MΩ, the sample is effectively current-biased at low frequency and the current oscillations are very small. In the region with 50 kΩ $< R_0 <$100 MΩ, current and voltage oscillations coexist as is shown in Fig. 8.5. It is interesting to note how the period of the oscillations in gate charge changes as the impedance of the environment is increased. As

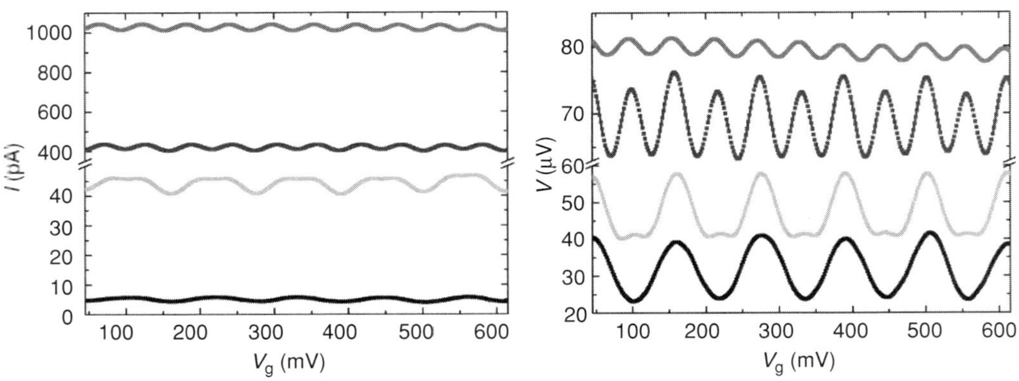

FIGURE 8.5. Current and voltage modulations as function of V_g for different R_0 values. From top to bottom: 55 kΩ, 450 kΩ, 10 MΩ, 20 MΩ. The bias voltage across the SQUID arrays is fixed at 50 μV.

the magnetic field increases the Josephson inductance of the SQUID arrays, the Q_g dependence changes from e-periodic to $2e$-periodic. The change in periodicity is observed in both current and voltage oscillations and the $2e$-periodicity is fully developed when the I–V curve of the CPT presents a sharp Coulomb blockade feature. For these samples no special precautions were taken during the sample fabrication and measurements to avoid quasiparticle poisoning, in fact, no normal metal leads are used and no low-temperature filters are implemented in our cryostat.

We have observed the same e to $2e$ transition for two CPTs fabricated simultaneously with a mirror symmetry in a similar way to that of Aumentado *et al.*, where the gap profiling was used to enhance $2e$-periodicity of the CPT. The fact that we can see a clean e to $2e$ transition, even for samples where the island of the CPT is not formed in a "dirty" Al evaporation, is a good indication that the nonequilibrium quasiparticles present in the system and responsible for the e-periodicity can be effectively reduced by high impedance leads.

In conclusion, we presented our experimental results on the effect of high impedance environment on the transport properties in a small-capacitance CPT. The CPT is biased by four SQUID arrays which allow the tuning *in situ* of the effective impedance of the electromagnetic environment. As the impedance of the leads is increased, the I–V characteristic of the CPT presents a sharp Coulomb blockade feature, with a region of negative differential resistance, which is the signature of coherent tunneling of single Cooper pairs. Moreover, we showed that when the SQUID arrays are in the insulating state, the gate-induced modulations of the current and voltage change from e-periodic to $2e$-periodic, suggesting that high impedance leads effectively reduce nonequilibrium quasiparticle poisoning.

References

1. G.-L. Ingold, Y. V. Nazarov, in *Single Charge Tunneling*, ed. by H. Grabert and M. H. Devoret (Plenum, New York, 1992), Chap. 2.
2. K. K. Likharev and A. B. Zorin, *J. Low Temp. Phys.* **59**, 347 (1985).
3. M. T. Tuominen, J. M. Hergenrother, T. S. Tighe, and M. Tinkham, *Phys. Rev. Lett.* **69**, 1997 (1992).
4. P. Joyez, P. Lafarge, A. Philippe, D. Esteve, and M. H. Devoret, *Phys. Rev. Lett.* **72**, 2458 (1994).
5. J. Aumentado, M. W. Keller, J. M. Martinis, and M. H. Devoret, *Phys. Rev. Lett.* **92**, 66802 (2004).
6. J. B. Kycia, J. Chen, R. Therrien, Ç. Kurdak, K. L. Campman, A. C. Gossard, and J. Clarke, *Phys. Rev. Lett.* **87**, 017002 (2002).
7. F. K. Wilhelm, G. Schön, and G. T. Zimányi, *Phys. Rev. Lett.* **87**, 136802 (2001).
8. D. B. Haviland, Yu. Pashkin, and L. S. Kuzmin, *Physica B* **203**, 347 (1994).
9. S. V. Lotkhov, S. A. Bogoslovsky, A. B. Zorin, and J. Niemeyer, *Phys. Rev. Lett.* **91**, 197002 (2003).
10. M. Watanabe, and D. B. Haviland, *Phys. Rev. Lett.* **83**, 3721 (1999).
11. M. Watanabe, *Phys. Rev. B.* **69**, 094509 (2004).
12. D. B. Haviland, K. Andersson, and P. Ågren, *J. Low Temp. Phys.* **118**, 733 (2000).
13. G. Schön, A.D, Zaikin, *Phys. Rep.* **198**, 237 (1990).
14. K. K. Likharev, *Dynamics of Josephson junctions and circuits*, (Gordon and Breach Science Publisher, 1986), Chap. 1.
15. A. B. Zorin, S. V. Lotkhov, Yu. A. Pashkin, H. Zangerle, V. A. Krupenin, T. Weimann, H. Scherer, and J. Niemeyer, *J. Supercond.* **12**, 747 (1999).

9

Phase Slip Phenomena in Ultra-Thin Superconducting Wires

M. Zgirski, K.-P.Riikonen, T. Holmqvist, M.Savolainen, V. Touboltsev, and K. Arutyunov

University of Jyväskylä, Nanoscience Centre, Department of Physics,
PB 35, 40014, Jyväskylä, Finland

Abstract

We present results on phase-slip phenomena in a superconducting wire which can be considered as quasi-one dimensional (1D) if its characteristic transverse dimension $\sqrt{\sigma}$ (σ being the cross section) is smaller than the coherence length Λ(T). The shape of the bottom part of the resistive transition R(T) of a 1D superconducting strip is described by the model of phase slips activation. If the wire is infinitely long, then there is always a finite probability that a small part of the sample is instantly driven normal.

9.1 Introduction

As continuous miniaturization of electronic circuits is progressing we should better understand physics of very narrow wires which obviously are basic constituents of such circuits. This work concentrates on experimental study of the shape of the bottom part of resistive transition $R(T)$ of very narrow superconducting wires. Efforts to observe quantum phase slips which prevent wires from going to superconducting state and according to theory can produce finite resistance even at temperatures close to absolute zero is presented here.

9.2 Theoretical

When a wire made of a superconducting material is cooled down and approaches critical temperature (T_c) superconducting phase starts to nucleate reducing its resistance. It can be observed as a rounded top of superconducting transition $R(T)$. On the other hand when the same wire being in superconducting state is heated up to critical temperature some of its parts can be driven normal increasing its resistance. It can be observed as a rounded bottom of superconducting transition. Nucleation of superconducting and normal phases above and below critical temperature respectively is stimulated by thermal fluctuations. To observe the contribution of normal phase nucleation wire must be one-dimensional (1D): a single normal domain blocks a supercurrent. This condition is satisfied if the wire's characteristic transverse dimension $\sqrt{\sigma}$ (σ being the cross section) is smaller than the coherence length $\xi(T)$. Coherence length is the smallest length over which superconductivity can appear, or the smallest length over which superconductivity can be destroyed. Condition of one-dimensionality guaranties that if a normal section appears somewhere along superconducting wire it must block supercurrent and produce finite resistance.

The shape of the bottom part of the resistive transition $R(T)$ of a 1D superconducting channel is described by the model of thermally activated phase slips first introduced by Langer and Ambegaokar[1]. When superconducting wire is sufficiently close to critical temperature there is always a finite probability that its small part is instantly driven normal due to thermal fluctuation. The smallest volume of a wire that can be driven normal is $\Omega = \sigma \cdot \xi$. Within the Ginzburg–Landau formalism macroscopically coherent state of a superconductor is described by order parameter $\Delta = |\Delta|e^{i\varphi}$. The squared modulus of the order parameter is proportional to density of Cooper pairs, while the gradient of its phase is proportional to supercurrent. When thermal fluctuation destroys superconductivity in volume Ω modulus of the order parameter goes to zero and the phase 'slips' by 2π. Process of evolution of the complex order parameter during phase slippage is schematically presented in Fig. 9.1. Utilizing Josephson relation $V \sim \mathrm{d}\phi/\mathrm{d}t$, each phase slip event generates a voltage pulse on the wire which gives rise to measurable resistance and can be expressed as $\Delta V = I \cdot R_{\mathrm{N}} \cdot (\xi/L)$, where R_{N} is the normal state resistance, I is the measuring current and L is the length of the wire. A single phase slip lasts too short to be observed (characteristic lifetime is $\tau_o \sim h/\Delta$, Δ being the energy gap). All we can measure is a voltage averaged over macroscopic time. The higher the rate of the phase slips Γ is, the higher is the time-averaged voltage $< V >= \Delta V \cdot \tau_o \cdot \Gamma$. Effective resistance $R_{\mathrm{eff}}(T)$ is the ratio between this time averaged voltage and the measuring current.

The free energy of a current-carrying 1D superconductor is described by so called 'tilted washboard' potential (Fig. 9.2). The black full circle denotes a thermodynamically metastable state of a macroscopically coherent superconductor corresponding to a local minimum of the

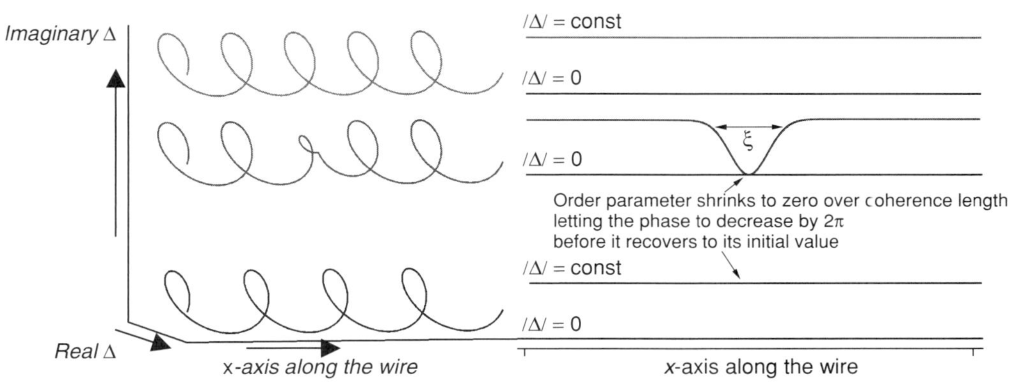

FIGURE 9.1. The phase slip process can be understood in terms of helix loosing its one loop. The topmost helix presents uniform coherent distribution of order parameter. This coherent distribution is responsible for dissipationless transport. When phase slip event occurs magnitude of the order parameter locally goes to zero and coherence is broken. It produces resistance. After a short time $\tilde{1}/Delta$ order parameter rebuilds, but helix losses one loop decreasing the difference in phase between two ends of wire by 2Π (bottom helix)

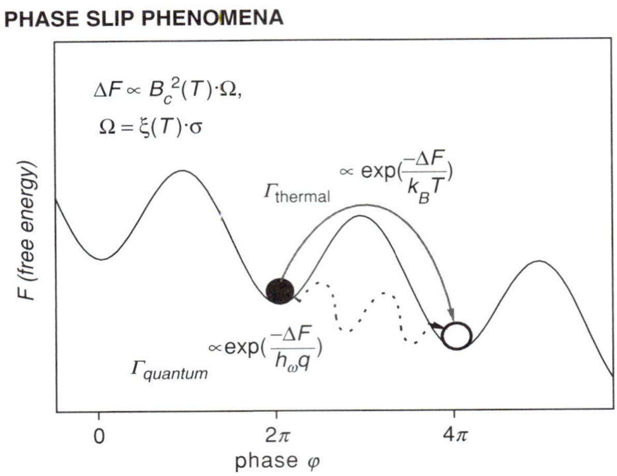

FIGURE 9.2. Dependence of the free energy F vs. phase φ of the order parameter $\Delta = |\Delta|e^{i\varphi}$ of a 1D current-carrying superconductor. The system (represented by a circle) can change its quantum state in two ways: (1) by thermally activated phase slips (solid arrow); (2) by quantum tunneling (dotted arrow). Γ stands for the rates of the corresponding processes.

free energy. The system can relax to a state of lower energy reducing its phase by 2π. So far we have described only thermally activated relaxation. Actually, there are two possible ways for a superconductor to decrease its energy. First, due to thermal excitation it can 'jump' over the energy barrier ΔF to a new potential minimum[1]; solid arrow in Fig. 9.2 (thermally activated phase slips). Second, the system can tunnel through the energy barrier[2]; dotted arrow in Fig. 9.2 (so called quantum phase slip). In the limit of infinitely small supercurrent the height of the barrier ΔF is the energy difference between normal and superconducting states of a system of the smallest possible size $\Omega = \xi\sigma \cdot \Delta F \sim B_c^2 \cdot \Omega$, B_c being the critical magnetic field. It can be shown[1], that in case of thermally activated phase slips the effective resistance $R_{\text{eff}}(T) \sim \exp(-\Delta F/k_B T)$.

Compared to thermal phase slips, which appear only sufficiently close to critical temperature, the rate of Quantum Phase Slips (QPS) is almost temperature independent and they should produce finite resistance even at temperatures close to absolute zero. The rate of QPS is governed by ratio $r = R_Q/(R_N \cdot \xi/L)$. Expression in denominator is the normal state resistance of the smallest possible length of a wire over which superconductivity can be destroyed and $R_Q = h/4e^2 = 6.47k\Omega$ is superconducting quantum resistance. The longer the wire is the bigger is the number of independent superconucting domains L/ξ in which quantum phase slip can appear. In case of a short wire, where only one phase slip can happen at a time, it can be shown[2] that the rate of quantum phase slips is $\Gamma_{\text{QPS}} = A \cdot r/\tau_0 \cdot (L/\xi) \cdot \exp(-A \cdot r)$, where $\tau_0 \sim h/\Delta$ is the duration of a phase slippage and $A \sim 1$. The effective resistance can be expressed as: $R_{\text{eff}} = A \cdot R_Q \cdot (L/\xi) \cdot \exp(-A \cdot r)$, so it only depends on number of independent superconducting domains and the ratio of the normal state resistance of such a domain to the superconducting quantum resistance. The presented simplified model holds for a not too long wire. If the wire length L is much larger then ξ multiple phase slip event can happen at a time. As far as coherence length is the function of temperature $\xi(T)$, probability of quantum phase slippage is temperature dependent close to T_c.

Several experimental attempts have been made to detect the QPS mechanism in narrow superconducting channels: ultra-thin-In and Pb-In wires[3] and Mo-Ge film evaporated on top of a carbon nanotube[4]. The results are very intriguing, but the matter is far from being settled. The motivation of this work was to study the manifestation of the QPS mechanism in the *same* superconducting Al wire as a function of its cross section.

9.3 Experimental

A method of progressive reduction of transverse dimensions of e-beam lift-off fabricated nanostructures by ion-beam sputtering has been developed.[5] The method enables galvanomagnetic measurements of the same sample in between the sessions of sputtering. Sputtering can be considered as an erosion of the surface due to bombardment of primary ions. The method is very promising, for it lets us directly follow changes in superconductive transition in a 1D superconductor along with successive reduction of its cross section. The effect of sputtering of Al wires is presented in Fig. 9.3. In this research worse low energetic (1keV) Ar plasma is used,

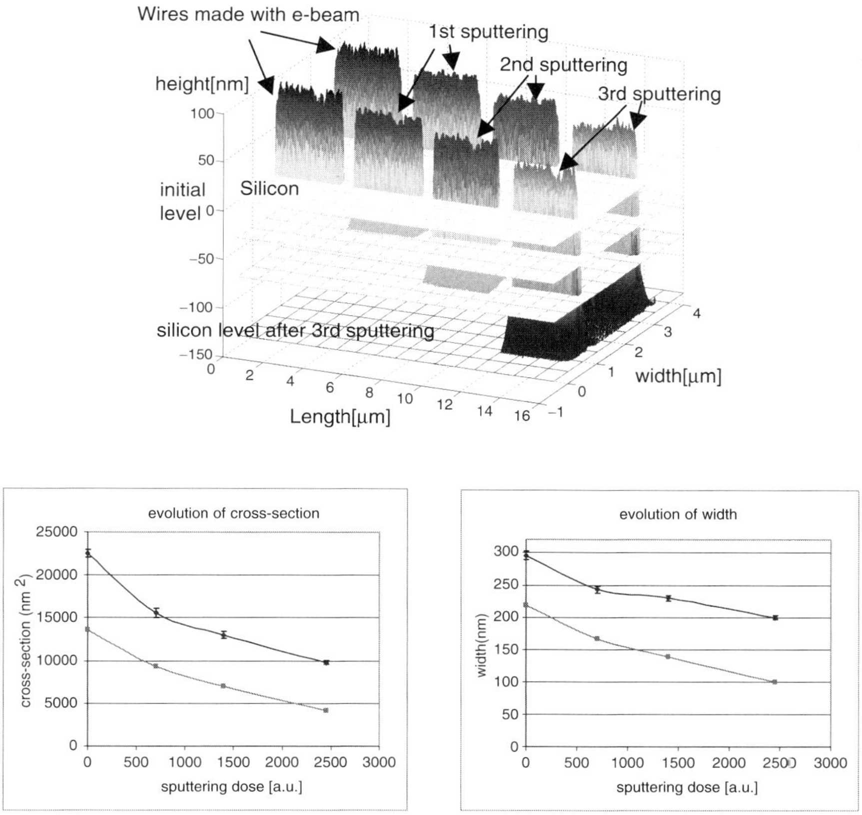

FIGURE 9.3. Effect of sputtering of Al wires. While sputtering the wire cross-section shrinks. Planes in 3D picture indicate silicon substrate levels after successive sessions of sputtering. As silicon is sputtered faster then aluminum finally wire is situated at the top of silicon pedestal. The topmost plane (height = 0) separates Si/SiO from Al. Pictures obtained with SPM in tapping mode. Inset shows evolution of the wire width and the wire cross-section.

which can penetrate original material only approximately 1 nm deep letting inner part of material untouched. For comparison, in Al the thickness of natural oxide at the surface is 2–3 nm.

After a wire is fabricated with e-beam lithography its homogeneity is verified with SPM. Only uniform wires are accepted and their superconducting transitions are measured. Four-probe measurement method with typical AC current density $j \sim 100\text{A/cm}^2$ is used. Afterwards wires are sputtered, analyzed with SPM and measured again. This scheme is repeated until wires are broken.

9.4 Experimental Evidence of Quantum Phase Slips?

In Fig. 9.4, we present data obtained for 5 µm long Al wire produced with e-beam lithography. The shape of $R(T)$ transition of the as-fabricated wire can be fitted by the thermal activation model.[1] After few sputtering sessions when wire reached the effective diameter $\sqrt{\sigma} < 35$ nm a low temperature 'foot' develops, which can not be explained in terms of thermally activated phase slips but can be associated with the QPS phenomena (Fig. 9.4.). Fitting to theory presented in introduction is possible but with parameter much smaller then predicted[2]: $A = 0.011$ instead of $A \sim 1$. However not all our experiments confirm QPS existence, although we have produced much thinner wires (Fig. 9.5.). According to theory of QPS[2] for thinner (more resistive wires) 'quantum foot' should be more pronounced. We found that wire with resistance of 250Ω per µm of length showed no sign of QPS (Fig. 9.5, wire after 5th sputtering). Its transition can be fully explained in terms of thermally activated phase slips. It contradicts the result presented in Fig. 9.4, where the wire with much smaller resistance of $15\Omega/\mu$m exhibits QPS-like behavior.

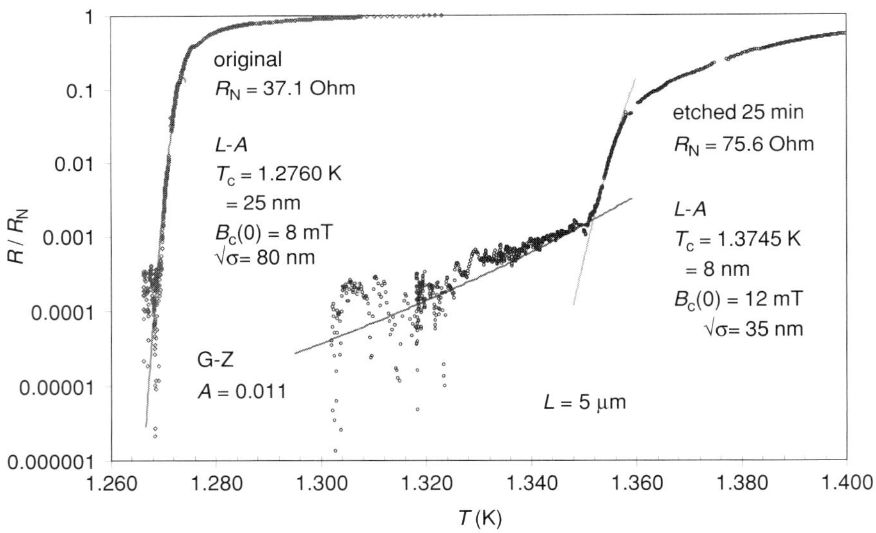

FIGURE 9.4. Resistive transition R(T) of the same 5 μm long Al wire before sputtering (left curve) and after (right). High-temperature part of both transitions can be described by the thermal phase slip activation mechanism (solid lines), while for the thinner wire the bottom 'foot' can be fitted by the QPS model. B_c and T_c were obtained from experiment. σ (wire's cross–section), l (mean-free path) are fitting parameters to the Langer–Ambegaokar model[1].

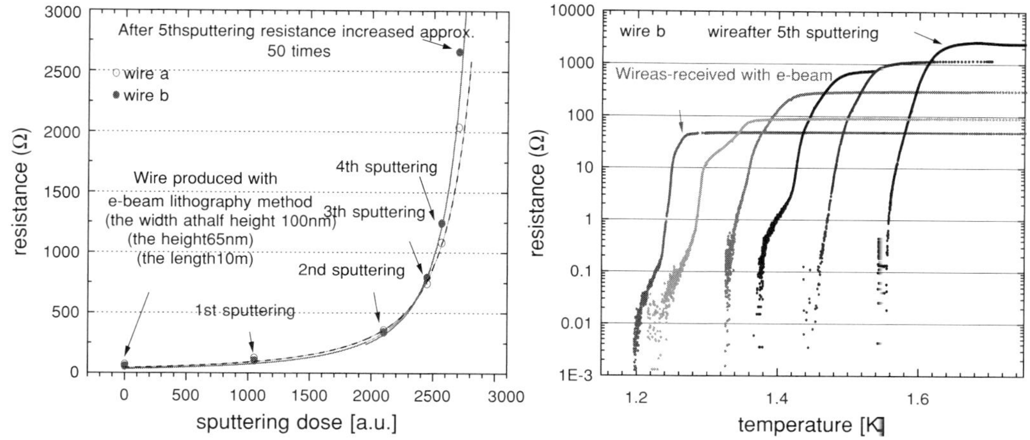

FIGURE 9.5. The increase of wire's resistance (left picture) and evolution of superconducting transition (right) after successive sessions of sputtering. Please note 'smoothing effect' of sputtering on the shape of the $R(T)$ transition, indicating the improvement of the wire homogeneity. No sign of quantum phase slip can be stated. In left picture the evolution of resistance for two wires (wire a and wire b) is given to show reproducibility of obtained results. In the right picture the evolution of superconducting transition for wire b is presented.

Further experiments are required to establish existence of QPS phenomenon and conditions for its observation in thin superconducting wires.

Acknowledgments. This work has been supported by the EU Commission "ULTRA-1D" project NMP4-CT-2003-505457 "Experimental and theoretical investigation of electron transport in ultra-narrow 1-dimensional nanostructures" and the Academy of Finland under the Finnish Centre of Excellence Program 2000-2005 (Project No. 44875, Nuclear and Condensed Matter Program at JYFL).

References

1. J. S. Langer, V. Ambegaokar, *Phys. Rev.* **164**, 498 (1967); D.E. McCumber, B.I. Halperin, *Phys. Rev. B* **1**, 1054 (1970)
2. A. Zaikin, D. Golubev, A. van Otterlo, and G. T. Zimanyi, *PRL* 78, 1552 (1997); A. Zaikin, D. Golubev, A. van Otterlo, and G. T. Zimanyi, *Uspexi Fiz. Nauk* 168, 244 (1998); D. Golubev and A. Zaikin, *Phys. Rev. B* 64, 014504 (2001)
3. N. Giordano and E. R. Schuler, Phys. Rev. Lett. 63, 2417 (1989); N. Giordano, *Phys. Rev. B* 41, 6350 (1990); *Phys. Rev. B* **43**, 160 (1991); *Physica B* 203, 460 (1994)
4. A. Bezryadin, C. N. Lau and M. Tinkham, *Nature* **404**, 971 (2000); C. N. Lau, *et. al.*, *Phys. Rev. Lett.* **87**, 217003 (2001)
5. M. Savolainen, V.Touboltsev, P.Koppinen, K.-P.Riikonen and K.Yu.Arutyunov, *Appl. Phys. A* **79**, 1769 (2004), cond-mat/0311383

10

Dynamics of a Qubit Coupled to a Harmonic Oscillator

P. BERTET[1], I. CHIORESCU[1,3], K. SEMBA[2], C. J. P. M. HARMANS[1], AND J. E. MOOIJ[1]

[1] Kavli Institute for Nanoscience, Delft, The Netherlands
[2] NTT Basic Research Laboratories, Atsugi, Japan
[3] present address: Michigan State University, East Lansing, USA

Abstract

We present experimental results showing the coherent dynamics of a flux-qubit coupled to the plasma mode of its measuring DC-SQUID. Resonances corresponding to transitions where both the qubit and the oscillator energy state are changed were observed in the excitation spectrum. We also measured Rabi oscillations on these transitions. We used this coupling to measure the effective temperature of the oscillator and to probe the qubit state after a Rabi oscillation. Our measurements indicate that the intrinsic efficiency of the Rabi pulses is 90%, much larger than the signal visibility (35%). We conclude that this limited visibility is likely due to relaxation during the measurement.

10.1 Introduction

Many recent experiments have shown that it is possible to manipulate the quantum state of mesoscopic superconducting circuits containing Josephson junctions [1, 2]. One of the driving forces in the field is trying to realize and controllably couple artificial two-level systems (qubits) in order to build a quantum computer [3]. Single-qubit operations and realization of a quantum gate between two charge qubits have been reported [4]. The strong coupling between a charge qubit and a one-dimensional microwave resonator has also been very recently observed [5]. Here we experimentally investigate the coupling between a flux-qubit and a harmonic oscillator embodied in the plasma mode of the measuring DC-SQUID [6]. The coupling between a two-level system and a harmonic oscillator has been shown to give rise to many interesting effects in ion trap and Cavity Quantum Electrodynamics experiments [7]. The two-level system can be used to generate and probe non-classical states of the oscillator. Reciprocally, the oscillator has been used as a "catalyst" to produce entanglement between multiple qubits. The results presented here constitute a first step towards the realization of similarly complex quantum-state engineering experiments in solid-state physics.

Following this introduction, the qubit-oscillator system is described.. In Section 3, the sample fabrication, the setup and the characterization of the parameters of both individual subsystems is discussed. In Section 4, the experimental evidence for the coupling, starting with spectroscopic measurements is presented followed by the data showing Rabi oscillations between coupled |qubit, oscillator> states. In the conclusion, how this coupling is used to measure intrinsic properties of the oscillator probed by the qubit is presented. Inversely, the qubit state after a Rabi oscillation employing the coupled qubit-oscillator dynamics has also been probed.

10.2 The System

First the principle of the flux-qubit will be briefly described; a more complete analysis can be found in Ref. [8]. The flux-qubit is a superconducting loop intersected with three Josephson junctions (an Atomic Force Micrograph of the sample is shown in Fig. 10.1a. The qubit loop is shown surrounded by a black square). The loop is biased by a magnetic flux Φ_x imposing a phase drop γ_Q across the junctions. When γ_Q is close to π (that is when Φ_x is close to $\Phi_0/2$), the loop has two low-energy eigenstates which we will call $|0>$ (ground state) and $|1>$ (excited state), well separated from the higher-energy states, which constitute the basis states of our qubit. These are linear combinations of two states $|\uparrow\rangle$ and $|\downarrow\rangle$which carry a persistent current respectively $+I_p$ and $-I_p$ and are coupled by tunneling. In the $|\uparrow\rangle,|\downarrow\rangle$ basis, the effective Hamiltonian reads $H_q = -h(\varepsilon\sigma_z/2\Delta\sigma_x/2)$, where $h\Delta/2$ is the tunneling matrix element, $\varepsilon = 2I_p\delta\Phi/h$ and $\delta\Phi = \Phi_x - \Phi_0/2$. The Larmor frequency of the qubit is $\nu_L = \sqrt{\Delta^2 + \varepsilon^2}$ [8].

The qubit is inductively coupled to a DC SQUID measurement system (larger loop in Fig. 10.1a containing two large junctions). How the SQUID is used to detect the qubit state will be explained later; here the focus is on its specific dynamics. The SQUID has two degrees of freedom, corresponding to the two phases γ_1 and γ_2 across each Josephson junction. It is convenient to introduce the internal and external phases defined respectively as $\gamma_{\text{int}} = (\gamma_1 - \gamma_2)/2$ and $\gamma_{\text{ext}} = (\gamma_1 + \gamma_2)/2$. The internal phase has a very fast dynamics and will be considered here as frozen. We will also neglect the effect of the SQUID self-inductance and consider it as a single Josephson junction whose critical current $I_C(\Phi'_x)$ depends on the flux through the SQUID loop Φ'_x. The external phase dynamics is determined by the external impedance to which the SQUID is connected

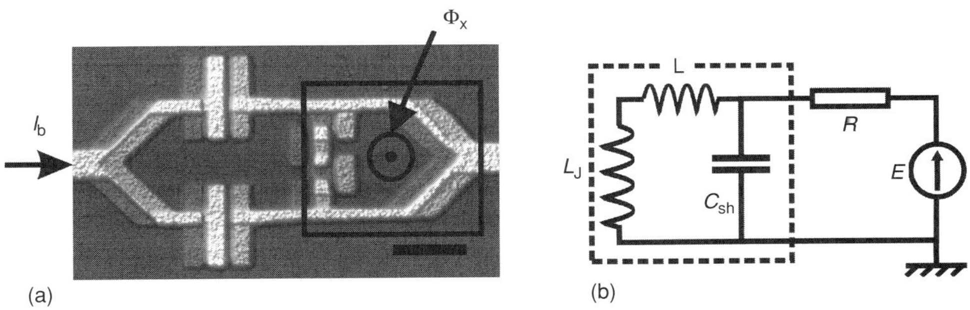

FIGURE 10.1. a. Atomic-Force Micrograph of the sample. The square box encloses the qubit loop, which is threaded by a magnetic flux Φ_x generated by a coil. b. Electrical model of the sample. The SQUID is modeled by an inductor $L_J$and is shunted by a capacitor C_{sh} and a resistor R.

(see Fig. 10.1b). In our experiments the SQUID was connected to a shunt capacitor C_{sh}, via superconducting lines which we model as an inductor L (the self-inductance of the SQUID loop is also included in L). The capacitor is connected to the measuring apparatus which present an impedance $Z(\omega)$ to the device. With such environment, the external phase of the SQUID has the dynamics of an oscillator of resonant frequency $\nu_p = 1/2\pi\sqrt{(L+L_J)C_{sh}}$. For bias conditions (Φ'_x, I_b) and small amplitude oscillations, the Josephson inductance L_J has a value $L_J = \Phi_0/2\pi I_C(\Phi'_x)\sqrt{1-(I_b/I_C(\Phi'_x))^2}$ and the oscillator is harmonic, its quality factor being $Q_p = 2\pi\nu_p \,\mathrm{Re}(Z(y\pi\nu_p))C$. We introduce the plasmon creation and annihilation operators $a$and a^+defined as a function of the operators representing the total phase difference across the SQUID and the superconducting lines γ and the charge stored on the capacitor Q by the relation $a = (\Phi_0/2\pi)\sqrt{1/2(L+L_J)h\nu_p}\gamma + i/\sqrt{2Ch\nu_p}Q$($a^+$ is the hermitian conjugate operator for a). The Hamiltonian describing the external phase dynamics reads then $H_{OSC} = h\nu_p(a^+a + 1/2)$.

The coupling between the qubit phases and the external phase of the SQUID has been investigated within the circuit theory presented in [9]. It can be shown [10] that the interaction Hamiltonian has the form $H_i = \lambda\sigma_z(a + a^+)$where λ is a coupling constant which depends on the exact inductance matrix of the circuit and on the bias conditions. We note however that this expression of the coupling does not account for the plasma frequency dependence on the qubit state [11, 12] since the Josephson inductance of the SQUID is considered constant in that analysis. The dynamics of the whole coupled system was studied [13] for the case where the qubit is driven by an oscillating magnetic flux. The main finding of [13] was that in addition to the "bare" qubit resonance at $\nu = \nu_L$, additional resonances appear in the spectrum at frequencies $\nu = (\nu_L - m\nu_p)/n$, where n and m are integers. This corresponds to transitions between $|0, 0>$ and $|1, 1>$.

10.3 Sample Fabrication, Setup and Characterization

The sample fabrication involved several steps of electron-beam lithography and subsequent metal evaporation. Starting with the gold pads and on-chip bias resistors (R = 400Ω) followed by the evaporation of the aluminum layer (which is superconducting at the temperature of our measurements) forming the bottom plate of our shunt capacitor, and finally the aluminum layer forming the top plates of the capacitors together with the Josephson junction circuit. The junctions were realized by the shadow evaporation technique. The sample was enclosed in a RF-tight copper box thermally anchored to the mixing chamber of our dilution fridge (base temperature 25 mK). It was connected to the measurement electronics through copper-powder filters installed in the current injection and voltage measurement coaxial lines, also thermalized at 25 mK. Microwave pulses could be applied via an antenna close to the sample. Experimental sequences presented in this paper invariably proceed in the following way: microwave pulses of varying frequency, length and amplitude are applied to the sample (two successive pulses of different frequencies can be generated), followed by a bias current pulse up to a value I_{bmax}. Because of the coupling between the SQUID and the qubit, the SQUID critical current depends on the qubit state, I_C^0 for the state $|0>$ and I_C^1 for the state $|1>$. If I_{bmax} is properly chosen, the switching probability of the SQUID is then proportional to the qubit eigenstate occupancy. The experimental sequence described here allows then to probe the qubit state occupancy after the coherent manipulations by repeating it typically 10,000 times and deriving the SQUID switching probability [2].

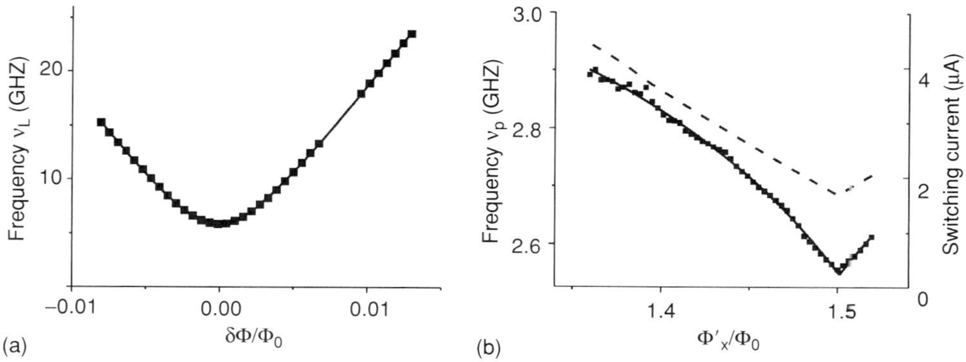

FIGURE 10.2. a. Qubit Larmor frequency as a function of the magnetic flux Φ_x; the squares are experimental data and the line is a fit yielding the parameters presented in the text. b. Plasma mode frequency as a function of the magnetic flux through the SQUID loop Φ'_x. The solid line is a fit to the data with the values of $C_{sh}$and L presented in the text. The dashed line shows the SQUID switching current (experimental points).

The parameters of the qubit were determined by spectroscopic measurements. A 500 ns long microwave pulse of variable frequency νg gs applied. When the microwave frequency matches the qubit frequency ($\nu = \nu_L$), a peak or dip in the SQUID switching probability is observed. The Larmor frequency is plotted as a function of $\delta\Phi$ in Fig. 10.2a. From these data the tunneling matrix amplitude Δg 5.85 GHz and the persistent-current $I_q = 270$ nA is deduced. The oscillator was also characterized spectroscopically: a high-power microwave pulse in presence of a bias current I_b and for a variable external flux Φ'_x was applied. When the microwave frequency matches the plasma mode frequency ($\nu = \nu_p(\Phi'_x)$), the switching probability is enhanced (resonant activation). The plasma mode resonance is unambiguously identified by the dependence of the plasma mode resonance on the external flux Φ'_x. The experimental curve $\nu_p(\Phi'_x)$ is shown in Fig. 10.2b (squares). Knowing the critical current dependence on the external flux $I_c(\Phi'_x)$ from switching current measurements (dashed curve in Fig. 10.2b), these data were used to fit the values of C_{sh} and L (the fit is the solid curve in Fig. 10.2b). $C_{sh} = 12 \pm 2$pF and $L = 170 \pm 20$pH were found, close to the values expected from the circuit design. The width of the resonances (between 20 and 50 MHz depending on the bias point) indicates a quality factor between 50 and 150.

10.4 Coupled Dynamics

10.4.1 Spectroscopy

The excitation spectrum of the coupled system was first studied. Large power microwave pulses were applied and the subsequent switching probability of the SQUID was measured. In addition to the qubit bare resonant frequency ν_L we observed two resonances at respectively $\nu_R = \nu_L - \nu_P$ and $\nu_B = \nu_L + \nu_P$ (Fig. 10.3a. Note that the plasma frequency is expected to change only very little on this scale). These resonances correspond to transitions where both the qubit and the

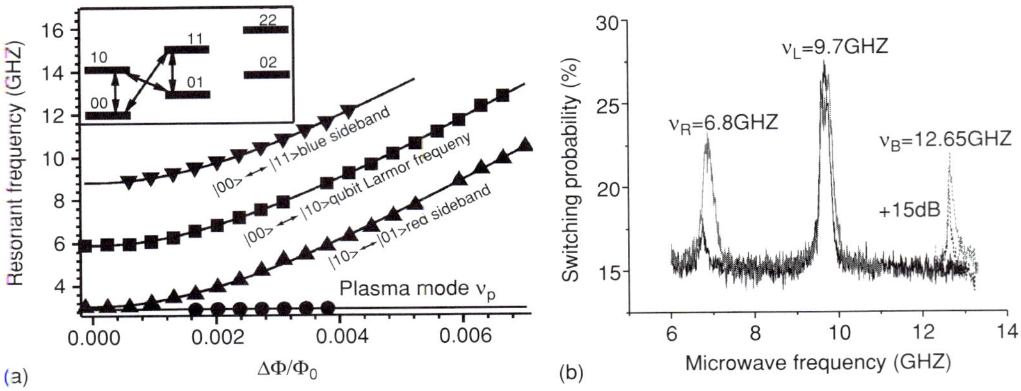

FIGURE 10.3. a. Resonance frequencies as a function of the magnetic flux $\Delta\Phi = \Phi_x - \Phi_0/2$. b. Typical spectrum showing the qubit resonance at ν_L, and the side band transitions at ν_r and ν_b (black line). The grey line shows the excitation spectrum immediately after exciting the plasma mode with a microwave pulse at ν_p.

oscillator states are changed [6]. They will be called in the following red- and blue-sideband resonances by analogy to ion-trap experiments.

A typical spectrum is shown in Fig. 10.3b (black lines). A peak in the switching probability is observed when the microwave frequency matches ν_R, ν_L, and ν_B (it is noted that in order to excite the blue sideband transition, 15dB more microwave power has to be applied, probably due to a lower transmission in this frequency range. The blue sideband peak is thus shown in dashed lines). The red sideband peak has smaller amplitude than the two other resonances. We can qualitatively account for this by the following argument: microwave irradiation on the red sideband induces transitions between the states $|0, n>$ and $|1, n-1>$. Thus, the state $|0, 0>$ can not be excited since the oscillator starts from the ground state and can't lose an energy quantum any more. On the contrary, microwave irradiation at ν_B could excite the system from $|0, 0>$ to $|1, 1>$. The lower amplitude of the red sideband peak compared to the blue sideband is thus an indication that the oscillator is in its ground state with a large probability, or in other words that its effective temperature T_{osc} is lower than $h\nu_p/k$. In the following, that the red sideband to estimate T_{osc} can be actually used is shown. Note also that the red sideband transitions could be used to actively cool the oscillator before the measurements, similarly to the "sideband-cooling" scheme used in ion-trap experiments to cool the ion's motion to the ground state [7]. A similar scheme has also been proposed to cool down a nanomechanical oscillator coupled to a superconducting qubit [14]. In the present experiments however, cooling is not required since the frequency ν_p is high enough so that even at thermal equilibrium, the oscillator is in its ground state with a large probability (see below).

To check the validity of this argument, we applied a first microwave pulse at the plasma frequency to excite the oscillator, and measured the system's spectrum with a second pulse (grey curve in Fig. 10.3b). The microwave pulse excites the oscillator in a coherent state, thus lowering the occupation of the state $|0, 0>$ and making the red sideband transition possible. A clear enhancement of the red sideband peak was observed (together with a shift towards higher frequency which could possibly be explained by radiative effects although we don't have quantitative data to support this interpretation), whereas the blue sideband and bare qubit transitions are little affected. This supports our interpretation of the sideband transitions in terms of processes where both the qubit and the oscillator state are modified.

Let us stress an important point for the understanding of the curves presented here. Away from the resonances, the switching probability is the same with (black curve) or without (grey curve) exciting the oscillator. This might be surprising, given the fact that exciting the SQUID plasma mode should in principle enhance its switching probability. Experimentally, the damping time of the oscillator turns out to be shorter than the time needed for the measurement, so that at the end of the measurement pulse the oscillator is in thermal equilibrium again (for pulses of larger microwave power however, an increase of the switching probability due to the first pulse was observed.). As a result, the switching probability of the SQUID in the conditions of the experiments presented here is only sensitive to the qubit state, whose relaxation time is longer than the oscillator, and not to the oscillator state.

10.4.2 Dynamics

When we applied short microwave pulses on the sideband transitions, we observed oscillations in the SQUID switching probability, showing that the qubit energy state occupancy is oscillating (Fig. 10.4) [6]. The oscillation frequency increases with the microwave amplitude. We claim that these are Rabi oscillations on the $|0,0\rangle - |1,1\rangle$ transition. They are damped in a characteristic time of 2 ns, and are superimposed on a general increase of the switching probability. This increase results from the relaxation of the oscillator towards its ground state: if we apply a long microwave pulse on the blue sideband transition, the system will end up with certainty in the state $|1,0\rangle$, a process somewhat analogous to optical pumping in atomic physics.

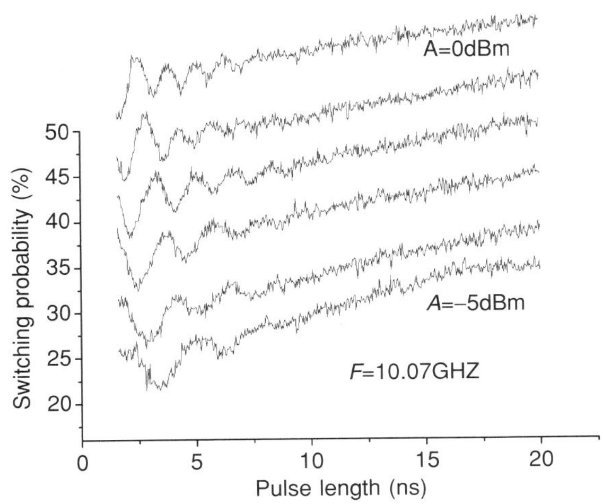

FIGURE 10.4. Rabi oscillations on the blue sideband for different microwave amplitudes (F = 10.07 GHz).

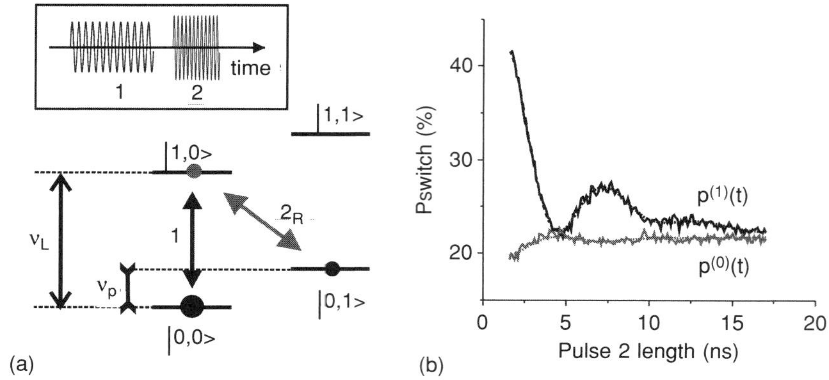

FIGURE 10.5. Measuring the oscillator temperature. a. Principle of the method: a first microwave pulse at the qubit Larmor frequency prepares the qubit in state $|1>$ or $|0>$. A second one induces transitions on the red sideband. b. Red sideband oscillations for the qubit prepared in state $|1>$ (black curve $p^{(1)}(t)$) and in $|0>$ (grey curve, $p^{(0)}(t)$). The black and grey dashed lines are numerical adjustments performed as indicated in the text.

10.5 Using the Coupled Dynamics to Probe the System

10.5.1 *Measuring the Oscillator Temperature*

The coupled dynamics to measure an intrinsic property of the oscillator was employed: its temperature T_{osc}. The principle of our method is as follows. The coherent oscillations obtained by applying microwave pulses on the red sideband (frequency ν_R) with two different initial conditions were compared: first, qubit in the ground state (curve $p^{(0)}(t)$), and then qubit in the excited state (curve $p^{(1)}(t)$). A simple analysis shows that $p^{(1)}(t)$ should be deduced from $p^{(0)}(t)$ by the simple expression: $p^{(1)}(t) = 1 - e^{h\nu_p/kT_{osc}} p^{(0)}(t)$, which thus allows us to determine T_{osc}. We measured $p^{(1)}(t)$ by applying first a π pulse at the qubit Larmor frequency, and then a pulse on the red sideband of variable duration (as schematized in Fig. 10.5a). The curve $p^{(0)}(t)$ was simply measured in exactly the same conditions by suppressing the first microwave pulse.

Both curves are shown in Fig. 10.5b. First $p^{(1)}(t)$ was fitted (solid black curve in Fig. 10.5a, the fit is a dashed line) with a damped sine plus an exponentially decaying curve, the latter describing phenomenologicaly the pumping into the state |0,0> as mentioned above. Then $p^{(0)}(t)$ was fitted (solid gray line in Fig. 10.5b) with two adjustable parameters A and B given the formula: $p^{(0)}(t) = Ap^{(1)}(t) + B$. The fit (dashed gray line) gives a very good agreement with the data for $A = -0.10 \pm 0.01$. We thus interpret $-A$ as being the factor $e^{-h\nu_p/kT_{osc}}$. This yields a temperature $T_{osc} = 60 \pm 5$ mK.

10.5.2 *Probing the Qubit State*

A major problem in these experiments is the proper detection of the qubit state. The detection scheme based on the switching of the SQUID conditioned to the qubit state should yield a detection efficiency of at least 85% according to the design. Still, Rabi oscillations yield a limited contrast of about 35% (however, note that we reached contrasts of 60% with a previous sample [2]).

It is important to know whether this limited amplitude is due to an imperfect quantum state preparation of the qubit during the Rabi oscillations, or to an imperfect detection scheme. The results presented below strongly suggest that the limited contrast arises mainly because of excited state relaxation during the measurement process.

The method used here in order to probe the qubit state after a Rabi oscillation is explained in Fig. 10.6a. It consists of probing the qubit ground state population after the Rabi pulse by applying a second microwave pulse on the blue sideband. Assuming that the oscillator is initially in its ground state, the system is in state $|\Psi >= \cos(\theta_1/2)\ |0,0> + \sin(\theta_1/2)\ |1,0>$ after the first Rabi pulse (θ_1 is the Rabi angle), and thus in the state $|0,0>$ with a probability $p(0,0) = \cos^2(\theta_1/2)$. If the system is in $|0,0>$, the second pulse induces Rabi oscillations with the state $|1,1>$ resulting in an oscillating switching probability ; but if the system is in $|1,0>$, no oscillation should be seen since the oscillator can not lose an additional energy quantum in its ground state. So, the amplitude of the blue sideband Rabi oscillations is then directly proportional to the probability for finding the qubit in its ground state after the qubit Rabi pulse $p(0) = \cos^2(\theta_1/2)$.

The blue sideband oscillations are shown in Fig. 10.6b for different Rabi angles θ_1. The most surprising feature of these curves is that virtually no oscillation is seen after a π pulse. This means that the ground state is very efficiently depopulated, i.e., that the π pulse has a very high efficiency. Note also that this suppression of the blue-sideband oscillations is remarkably well reversed after a 2π pulse as expected. On the other hand, after a 17π pulse, oscillations with significant amplitude are observed, indicating that the ground state is now partially populated due to decoherence.

To make this quantitative, the frequency, amplitude, phase and damping time of the oscillations was fitted for $\theta_1 = 0$. Then all the other curves with only one adjustable parameter was fitted: a scaling factor $V(\theta_1)$. Fig. 10.5c shows the evolution of V as a function of the length of the first microwave pulse, together with the SQUID switching probability. The curve $V(\theta_1)$ shows the same behavior as the Rabi oscillations measured by the SQUID (same phase and damping time), which strongly supports the interpretation that $V(\theta_1)$ actually measures the probability to

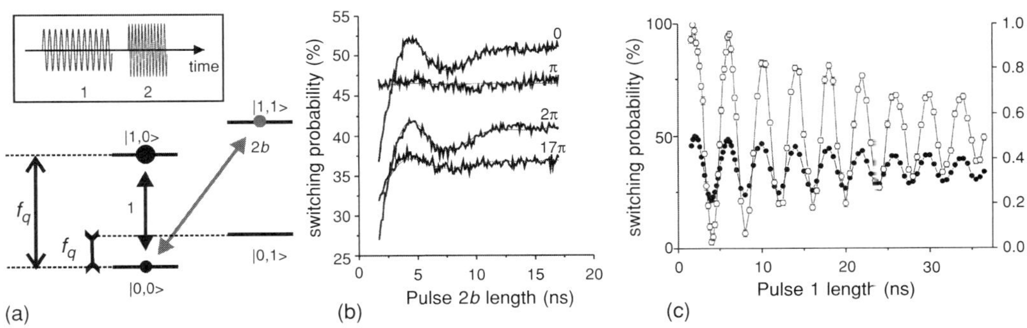

FIGURE 10.6. Probing the qubit state after a Rabi oscillation. **a**. Principle of the method: a first microwave pulse at the qubit Larmor frequency induces a Rabi rotation with an angle θ_1. A second one induces blue sideband oscillations. **b**. Blue sideband oscillations for different Rabi angles $\theta_1 = 0, \pi, 2\pi, 17\pi$. The solid lines are fits sharing all the same parameters except the amplitude of the oscillation, characterized by the parameter $V(\theta_1)$. **c**. Parameter $V(\theta_1)$ as a function of the length of pulse 1 (open circles). The curve in closed circles shows the Rabi oscillations measured in exactly the same conditions by the usual switching probability measurement.

find the qubit in its ground state. Still, there is a large discrepancy in the amplitude of these two curves: whereas the switching probability of the SQUID oscillates with amplitude of 30%, $V(\theta_1)$ displays a 100% amplitude modulation.

This analysis needs to be refined by taking into account the finite temperature of the oscillator, and thus the finite occupation probability of the state $|0, 1>$ at the beginning of the experiment. The measured value of $T_{osc} = 60$ mK to find an overall efficiency of 90% for the π pulse was used. From these measurements, it seems thus that the qubit strongly relaxes during the measurement process, so that the measured amplitude of the Rabi oscillations is only 30%. (These results were confirmed by an alternative measurement technique which was demonstrated on the same sample to yield contrasts of 68% for the Rabi oscillation [12])

That Rabi oscillations have not been measured with a contrast of 90% can be stressed; simply a third level has been used to probe the population of the ground state and to deconvolute the effects of imperfect quantum state preparation and imperfect detection, indicating a good (around 90%) state preparation and a limited detection efficiency. To reach this conclusion, a physically reasonable assumption had to be made. It was assumed that the processes which limit the visibility of the bare qubit Rabi oscillations and the blue sideband oscillations are temporally uncorrelated. As an example, it was assumed that a bistable fluctuator leads to loss of visibility. If the fluctuator has a characteristic time longer than the duration of both microwave pulses (a few tens of nanoseconds), it would sometimes shift the qubit away from resonance and thus would induce a correlated loss of visibility of both oscillations. One would still observe a suppression of the blue sideband oscillations after a π pulse, although the excited state population would be in that case effectively limited to much less than 90%.

10.6 Conclusion

In designing and discussing quantum circuits, one has to take into account the quantum dynamics of all the variables of the system. In the present experiments, the variable used for the detection of a flux qubit (the external phase of a DC SQUID) has the dynamics of a harmonic oscillator of frequency ν_p. Its coupling to the qubit variables leads to additional resonances at frequencies $\nu_L + \nu_p$ and $\nu_L - \nu_p$. These resonances correspond to processes where both the qubit and the oscillator state are modified. The observation of these resonances is reported here. A fast-decaying quantum dynamics on these sideband transitions was observed. By applying two successive microwave pulses of different frequencies, it was shown that these sidebands lead to conditional dynamics of the qubit-oscillator system. This dynamics to measure intrinsic properties of the oscillator such as the temperature is used here. It was also used to probe the qubit state after a Rabi oscillation, and it was shown that the limited visibility of these oscillations was very likely limited by relaxation during the measurement process. These results are a first step towards complex manipulations of entangled states, like the one performed in ion-trap or cavity QED experiments.

References

1. Y. Nakamura, Yu. A. Pashkin, and J. S. Tsai, *Nature* (London) **398**, 786 (1999); D. Vion et al., *Science* **296**, 886 (2002); J. M. Martinis, S. Nam, J. Aumentado, and C. Urbina, *Phys. Rev. Lett* **89**, 117901 (2002); T. Duty, D. Gunnarsson, K. Bladh and P. Delsing; *Phys. Rev. B* **69**, 140503 (2004); J. Claudon, F. Balestro, F.W.J. Hekking and O. Buisson, arXiv:cond-mat/0405430 (2004).

2. I. Chiorescu, Y. Nakamura, C. J. P. M. Harmans, and J. E. Mooij, *Science* **299**, 1869 (2003); published online 10.1126/science.1081045 (2003).
3. Y. Makhlin, G. Schoen, and A. Shnirman, *Rev. Mod. Phys.* **73**, 357 (2001).
4. Y. Pashkin et al., *Nature* **421**, 823 (2003); T. Yamamoto et al., *Nature* **425**, 941 (2003).
5. A. Wallraff et al., cond-mat/0407325 (2004).
6. I. Chiorescu et al., private communication.
7. J.-M. Raimond et al., *Rev. Mod. Phys.* **73**, 565 (2001); D. Leibfried et al., *Rev. Mod. Phys.* **75**, 281 (2003).
8. J. E. Mooij et al., *Science* **285**, 1036 (1999); T.P. Orlando et al., *Phys. Rev. B* **60**, 15398–15413 (1999).
9. G. Burkard, R. Koch and D. P. DiVicenzo, *Phys. Rev. B* **69**, 064503 (2004).
10. G. Burkard et al., private communication.
11. A. Lupascu et al., arXiv:cond-mat/0311510 (2003).
12. P. Bertet et al., arXiv:cond-mat/0405024 (2004).
13. M. C. Goorden, M. Thorwart and M. Grifoni, cond-mat/0405220 (2004).
14. I. Martin, A. Shnirman, L. Tian and P. Zoller, arXiv:cond-mat/0310229 (2004).

11

Josephson junction Materials Research Using Phase Qubits

R. W. Simmonds[1], D. A. Hite[1,2], R. McDermott[1,3], M. Steffen[1,3], K. B. Cooper[1,3], K. M. Lang[1,4], John M. Martinis[1,3], and D. P. Pappas[1]

[1] National Institute of Standards and Technology*, 325 Broadway St. Boulder, CO 80305
[2] Physics Department, Western State College of Colorado, Gunnison, Colorado 81231
[3] Physics Department, University of California, Santa Barbara, CA 93106
[4] Physics Department, Colorado College, Colorado Springs, CO 80903

Keywords: Josephson junctions, Quantum computing, Qubits

Abstract

At present, the performance of superconducting qubits is limited by decoherence. Strong decoherence of phase qubits is associated with spurious microwave resonators residing within the Josephson junction tunnel barrier.[1] In this Chapter three different fabrication techniques for producing tunnel junctions that vary the properties of the superconductor-insulator interface are investigated. Through experimental measurements, the junction and corresponding qubit quality is characterized. A strong correlation between the morphology of oxidized base electrodes and the lowering of subgap currents in the junction I–V characteristics is found, while there is no noticeable improvement in the performance of fabricated phase qubits. Thus, "traditional" indicators of junction performance may not be enough to determine qubit performance. However, truly crystalline insulating barriers may be the key to improving Josephson junction based qubits.

11.1 Introduction

The most significant obstacle to the realization of a practical superconducting quantum computer is decoherence. It is found that the "quality" of Josephson junctions strongly affects the coherent performance of Josephson phase qubits.[1] Specifically, are discovered spurious resonant states that couple to phase qubits, causing decoherence. In fact, there is evidence that all Josephson junction-based qubits might be suffering from these material defects.[2] It is reported by measurements that by changing the processing of fabricated tunnel junctions, in particular, through epitaxially grown base electrodes, one can improve the "traditional" indicators of Josephson tunnel junction performance, such as low subgap conductance. Although we find considerable improvement is found in the subgap currents, there is no noticeable improvement in the performance of fabricated phase qubits. This shows that "traditional" indicators of junction performance may not be enough to determine qubit performance. Furthermore, changing the properties of the superconductor–insulator interface, while still using amorphous barriers, does not significantly improve qubit performance. These results motivate the study of tunnel junctions formed by crystalline insulating barriers, which may be the key component in producing superconducting qubits of high quality.

11.2 Josephson Phase Qubits

A high-impedance current bias and measurement scheme for controlling a Josphson phase qubit has been developed, while providing sufficient isolation from the external environment. A Josephson junction is included in a superconducting loop of inductance L, as shown in Fig. 11.1(a), which is described in more detail in Ref.[1] Microwave current lines are capacitively coupled to the junction, while a dc bias coil is placed some distance from the "qubit loop". For an applied flux $\phi = \Phi/\Phi_o$, where $\Phi_o = h/2e$ is the flux quantum, the potential energy $U(\delta)$ stored in the Josephson junction as a function of the superconducting phase difference δ is

$$U(\delta) = -\frac{\Phi_o}{2\pi} I_o \cos(\delta) + \frac{\Phi_o^2}{2L}(\phi - \frac{\delta}{2\pi})^2 \tag{11.1}$$

As shown in Figure 11.1(b), ϕ is chosen so that the qubit states, $|0\rangle$ and $|1\rangle$, are formed in the left (~cubic) well and the $|1\rangle$ state is measured by an induced tunneling event to states in the (~quadratic) right well, changing the flux in the qubit loop by roughly a flux quantum. This large flux difference allows for easy readout using a pulsed dc SQUID.

The $0 \rightarrow 1$ qubit transition frequency ω_{10} is measured spectroscopically[3] by applying a microwave drive current $I_{\mu w}$ at frequency ω and subsequently measuring the occupation probability of state $|1\rangle$ using a "fast" qubit state measurement technique.[4] This is done for a range of static flux biases ϕ applied to the qubit loop. It is observed in Figure 11.2 that the expected decrease in the transition frequency as the current through the junction approaches the critical current. Spurious resonators that are characteristic of energy level repulsion predicted for quantum mechanically coupled systems is also observed (see the inset of Fig. 11.2). These extra resonators have a distribution in splitting size and frequency. They are indicative of nanoscopic two-level systems within the junction's insulating barrier. Away from any spurious resonators,

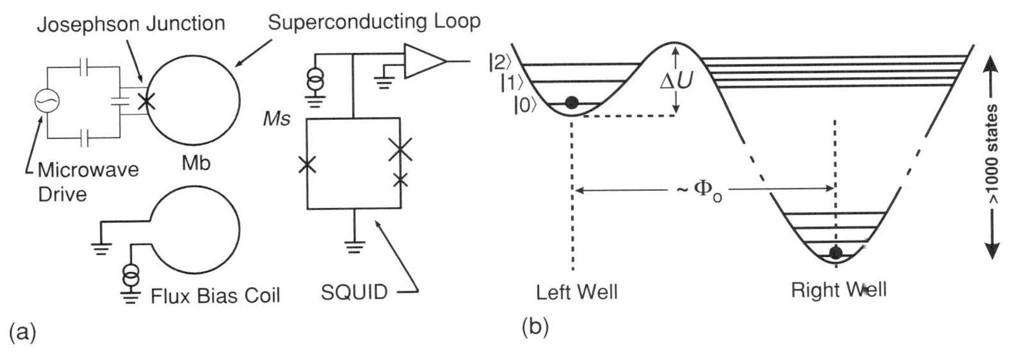

FIGURE 11.1. (a) The phase qubit schematic circuit. (b) The potential $U(\delta)$ at a particular ϕ.

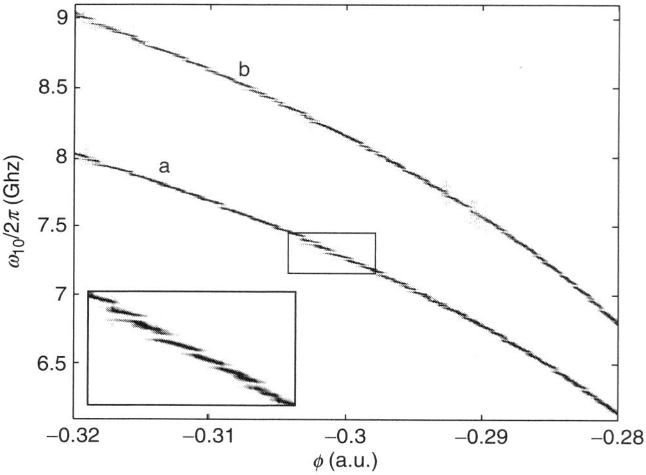

FIGURE 11.2. Qubit spectroscopy for of (a) a "ion mill process" qubit and (b) a "standard trilayer process" qubit.

is observed coherent Rabi oscillations between the $|0\rangle$ and $|1\rangle$ state,[1] while near a spurious resonator, undesirable coupled interactions between the qubit and the resonator[4] is found. These resonators can be connected with measurements of the tunnel junction current–voltage (I–V) characteristic as discussed in Section 11.4.

11.3 Junction Fabrication Processes

State-of-the-art Josephson junctions employing superconducting (Al or Nb) electrodes and native-oxide tunnel barriers are typically fabricated using sputter deposition onto thermally oxidized Si wafers. For most applications, such as SQUID technologies, there has never been a necessity to improve the crystalline quality of the junction to correspondingly improve the device performance.[5] However, the discovery of spurious high-frequency resonant states within Josephson tunnel barriers, which are undesirable defects for quantum computing applications, suggests that more effort should be employed in order to improve junction fabrication. Here, detail recent efforts to improve Josephson junction "quality" through different fabrication techniques is presented. Ultimately, it could be tested whether the crystalline quality in the microstructure of the junctions correlates to improved low-frequency transport measurements, lower levels of 1/f critical current noise, and finally, qubit performance.

In what follows, the "ion mill" and "sputtered trilayer" junctions are produced on 3-inch thermally oxidized Si(100) wafers in a clean room sputter chamber with a base pressure of 7×10^{-8} Torr, while the "evaporated trilayers" are deposited in a UHV chamber with base pressure 8×10^{-11} Torr on Si(111) chips, 2.5×2.5 cm^2. The Al films are grown by dc magnetron sputter deposition using 5 mTorr Ar with the substrate at room temperature. The oxidation parameters are typically 10 Torr of oxygen pressure for a 10 m exposure at room temperature. This procedure leads to the preferred oxide thickness, giving junction critical-current densities of ~15–20 A/cm^2, appropriate for the phase qubits. The base electrodes for all devices discussed here have a thickness of 200 nm, while the top Al films are from 60 to 200 nm thick.

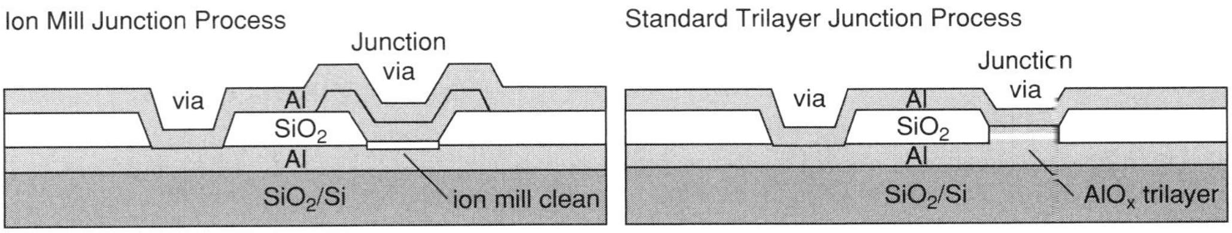

(a) (b)

FIGURE 11.3. The (a) ion mill process (IMP) and (b) standard trilayer process (STP) for producing Josephson tunnel junctions.

11.3.1 The "Ion Mill Junction Process"

The "Ion Mill Process" (IMP) for Josephson junction fabrication,[6] as shown in Figure 11.3(a), is a simple and convenient method for producing tunnel junctions. It is started by sputtering a base Al wiring layer, which is then defined using a wet acid etch. This surface is then covered with $\sim$ 350 nm of SiO_2 for insulation. The area of the tunnel barrier is defined by etching a via through the SiO_2. After breaking vacuum, the native oxide on this exposed surface is then removed from the base electrode by ion milling at 800 V with 0.1 mA/cm^2 for 1 m (creating an atomically *rough* surface), and the amorphous tunnel barrier is grown by thermal oxidation at room temperature. Following oxide growth, a fresh layer of Al is deposited to form the tunnel junction. Most of this layer is then wet etched away except for an area defining the junction. Subsequent SiO_2 etching, ion milling, and deposition steps define vias for contacting to the base wiring layer and the tunnel junction.

11.3.2 The "Standard Trilayer Junction Process"

We have developed a "Standard Trilayer Process" (STP) that produces tunnel junctions with superconductor–insulator interfaces that are smoother than the IMP junctions. As shown in Figure 11.3(b), the oxide tunnel barrier is grown *in situ* on a freshly sputtered Al base layer without breaking vacuum, then the tunnel barrier is immediately capped with another sputtered Al layer. The Josephson junction area is defined by ion milling the top two layers of the trilayer to create a mesa. The base wiring layer is then defined using a wet etch. This surface is then covered with $\sim$ 350 nm of SiO_2 for insulation. Subsequent processing steps define vias for electrical contacts between the top wiring layer, the trilayer junction, and the bottom wiring layer.

11.3.3 The "Evaporated Trilayer Junction Process"

The "Evaporated Trilayer Process" (ETP) grows Josephson junctions in a unique way using advanced growth techniques in order to produce atomically smooth superconductor–insulator interfaces with fewer structural defects. Typically, metal films grown on semiconductor surfaces follow a three-dimensional growth mode. It is often found that interdiffusion and reactive epitaxy dominate in the early stages of growth for many metal-on-semiconductor systems. This results in further clustering and hinders the production of smooth films and abrupt interfaces. Sputter deposition used for the STP junctions can produce Al films with lower rms roughness; however, due to high deposition rates and kinetic limitations, these films are dominated by crystal defects and are polycrystalline at best.

Here, clean Si(111) substrates are used, flash annealed at 1500 K, and then standard cleaning procedures[7] are used to obtain a (7×7) reconstructed surface. This is used surface because it is clean, highly ordered, thermodynamically stable, and allows the epitaxial growth of Al.[8] Furthermore, it has been shown that low-temperature deposition of Al on Si(111)-(7×7), followed by annealing, produces films with extremely sharp and very well defined surfaces.[9] This is in contrast to films grown under room-temperature conditions,[10] but consistent with the novel growth mechanisms responsible for atomically flat metal films on semiconductor substrates.[11]

It is started by evaporating an epitaxial Al seed layer ($\sim$ 5 nm thick) on the Si(111)-(7×7) surface, held at 120 K during deposition and subsequently annealed at 475 K. Low-Energy Electron Diffraction (LEED) is used to determine the crystalline quality of the substrate, seed layer, and subsequent films. Once the Al seed layer is prepared, we then complete the base electrode through the homoepitaxy of Al on the seed layer at room temperature. Although this base electrode is 200 nm thick, Fig. 11.4(c) shows that it still exhibits a LEED pattern of fairly high quality. The evaporated trilayer is then completed by thermal oxidation at room temperature, followed by evaporation of the final top electrode. Each step of the growth was monitored with Auger Electron Spectroscopy (AES), and LEED. After the oxidation there was no LEED pattern to report, as expected, due to the amorphous nature of the native oxide. AES data indicated clean metal films with no detectable amounts of contaminants. In addition, *ex situ* Atomic Force Microscopy (AFM) measurements have indicated smoother oxides when compared to surfaces produced after sputter deposition, and significantly smoother than those evaporated onto

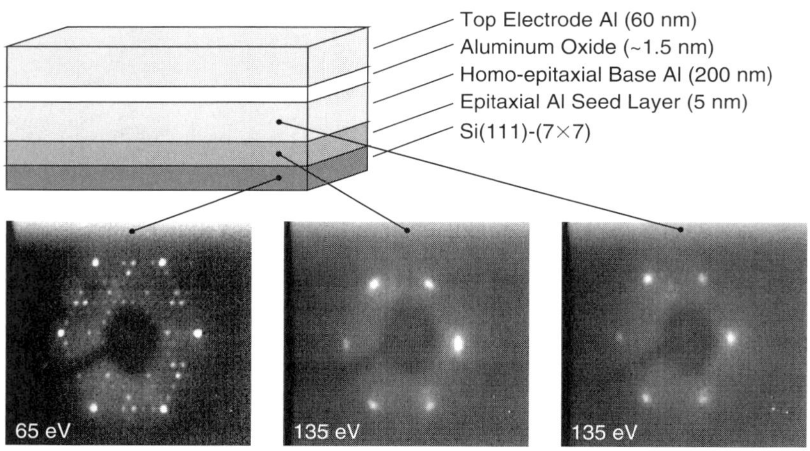

FIGURE 11.4. Schematic diagram of different layers in the $Al/AlO_x/Al/Si(111)$ evaporated trilayer growth sequence. The LEED images show the quality of the epitaxial seed layer and the base electrode. The LEED images were obtained with electron energies of 65 eV, 135 eV, and 135 eV respectively. The oxide film and top electrode are amorphous.

non-epitaxial substrates. The subsequent steps for producing devices from the ETF are the same as those described earlier for the STP.

11.4 Josephson Junction and Qubit Characterization

I–V curves for tunnel junctions are interpreted in the context of the Landauer tunneling model[12, 13] for junction conductance. The conductance G of tunnel junctions at voltages above the superconducting gap voltage ($2\Delta/e$) is expressed as a sum over the individual conductance channels, $G = G_o \sum_{i=1}^{N} \tau_i$, where $G_o = 2e^2/h$ and τ_i is the transmission probability of the ith channel. At low voltages, nth order multiple Andreev reflections form steps in the quasiparticle branch of the I–V characteristic of magnitude $2/\tau_i$. Let us assume that the current is carried by N channels, each with an average transmission τ, then the largest step appearing in the I–V characteristic is given by $\sim 2/\tau$. Let us now imagine that these individual channels fluctuate "on" and "off" as switches, so that the average size of a critical-current fluctuation is $\delta I_o \approx 2\Delta G_o \tau/e$. If the two level resonators in the tunnel barrier couple to the qubit through the critical current, then the size of each resonant level splitting is directly proportional to δI_o. By correlating measurements of the subgap conductance from I–V characteristics (Section 11.4.1 below) and qubit spectroscopy (Section 11.2) the relationship between what has traditionally been considered "high quality" Josephson junctions[14] and the corresponding quality of phase qubits is found.

11.4.1 Measurements of Junction I–V Characteristics

For direct comparison, all samples are processed into hysteretic tunnel junctions (of the type used in phase qubits with $\beta_c \gg 1$) through standard photolithographic techniques. The I–V characteristics of the junctions are measured* at temperatures below 50 mK. I–V characteristics from IMP, STP, and ETP junctions are shown in Fig. 11.5. Each curve has been normalized by its critical current I_o in order to facilitate a relative comparison of subgap currents. The I–V from the STP junction shows a current step size at $2\Delta/e \approx 380\mu$V which is roughly larger by a factor of 6 than for the IMP junction. Even more pronounced is the ETP junction, whose subgap currents are lower still by almost another order of magnitude. The reduced subgap current of the STP and ETP junctions indicates a relatively large number of conduction channels, each with relatively low transmissivity. However, the distribution of τ_i's, which can fluctuate at GHz frequencies, may or may not be contributing strongly to the measured, average τ If their influence *did* strongly determine τ, then one would expect that the fluctuation of the junction critical current due to these resonant channels will be reduced, so that the maximum level splitting size found in spectroscopy data for trilayer qubits should be reduced. This would then allow us to predict the quality of superconducting qubits through a "traditional" quality check of just the Josephson tunnel junctions themselves (and there is a vast amount of this information in the current literature for different Josephson junction technologies).

*We use a nonlinear current bias made from diodes and resistors in order first to exceed the critical current, then to sample more densely quasiparticle currents, smaller by several orders of magnitude, in the subgap region. To ensure that the features of the I–V characteristic are not distorted by high-frequency interference, these measurements were performed using RC low-pass filters cooled at 4 K.

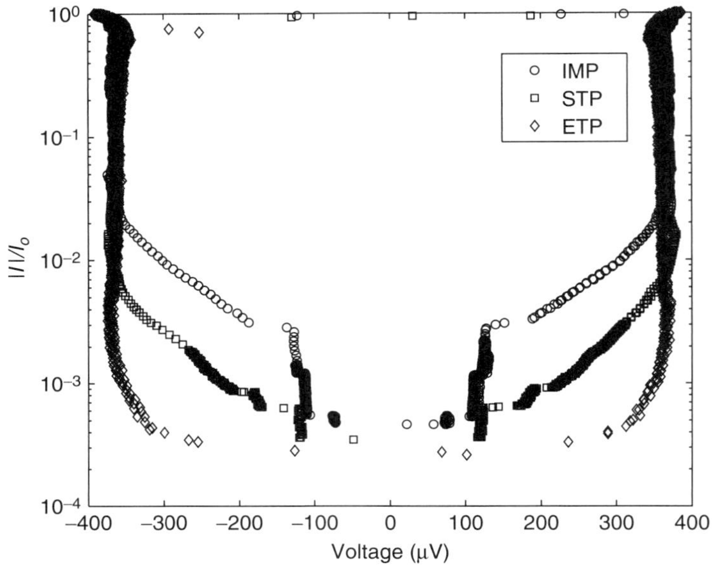

FIGURE 11.5. I–V characteristics, normalized current $|I|/I_o$ vs. voltage, for tunnel junctions with area $\sim 70\ \mu m^2$ comparing the IMP ($I_o \sim 20\ \mu A$), STP ($I_o \sim 10\ \mu A$), and ETP ($I_o \sim 10\ \mu A$).

11.4.2 Measurements of Qubit Spectroscopy

In Fig. 11.2, we show two samples of spectroscopic data from two qubits with the exact same geometrical design but with IMP and STP junctions used in each respective device.† In Fig. 11.5, the I–V characteristics show a difference of almost an order of magnitude in the subgap conductance for these two junction fabrication methods. From a "traditional" point of view, the trilayer junctions are considered "higher quality" junctions. However, spectroscopic characterization of the phase qubits, as seen in Fig. 11.2, shows *no* significant difference in the maximum level splitting amplitude or in the number of resonators per frequency range. Furthermore, measurements of the $|1\rangle$ state energy relaxation time (~ 400 ns), the Rabi oscillation visibility ($\sim 50\%$), and the Rabi oscillation decay time (~ 80 ns) were typical for both types of qubits. This suggests, if the microscopic model is still correct, the number of high-frequency resonators that one sees in the spectroscopy, which are very few compared to the total number of resonators including those at very low frequency, do not contribute significantly to the properties of the measured I–V characteristic. Thus, our "traditional" notions of junction quality obtained through low-frequency measurements are not sufficient for characterizing the quality of Josephson phase qubits operated at GHz frequencies. These spectroscopic measurements are crucial for identifying defects not seen using standard low-frequency techniques.

†Unfortunately, thus far, we have not been able to compare qubit spectroscopy results for the ETP junctions because of low junction yield over a single chip. However, we have tested numerous STP qubits and three similar IMP qubits.

11.5 Concluding remarks

Superconducting tunnel junctions are grown using three different processing techniques: IMP, STP, and ETP. The last uses an atomically flat metallic seed layer of Al on Si(111)-(7 × 7) substrates in order to produce an epitaxial base electrode with an atomically smooth interface. Test junctions were fabricated using all three techniques, while qubits were successfully produced using the first two processing techniques (IMP and STP). Low-frequency transport measurements (I–V characteristics) as well as high-frequency measurements (qubit spectroscopy) were made below 50 mK. I–V characteristics for all three samples have been compared and it is concluded that there is a strong correlation between the morphology of atomically smooth thermally oxidized base electrodes and the lowering of subgap currents. However, for the first two processing techniques (IMP and STP), qubit measurements showed that these "traditional" indicators did not assure the fabrication of high-quality Josephson phase qubits. Future tests will focus on the crystalline quality of the insulating barrier itself. This may play the pivotal role in eliminating spurious resonators in tunnel junctions and producing high-quality superconducting quantum bits.

Acknowledgments. We thank John Clarke, B. Plourde, M. Devoret, V. Shumeiko S. Nam, and K. Osborn. This work is supported by ARDA under Contract No. MOD717304. *Contribution of the U. S. Government not subject to copyright.

References

1. R. W. Simmonds, K. M. Lang, D. A. Hite, D. P. Pappas, John M. Martinis, Decoherence in josephson phase qubits from junction resonators, *Phys. Rev. Lett.* **93**, 077003 (2004).
2. John Clarke, Hans Mooji, Michel Devoret (private communication).
3. J. M. Martinis, M. H. Devoret, J. Clarke, Energy-level quantization in the zero-voltage state of a current-biased Josephson junction, *Phys. Rev. Lett.* **55**, 1543 (1985).
4. K. B. Cooper, Matthias Steffen, R. McDermott, R. W. Simmonds, Seongshik Oh, D. A. Hite, D. P. Pappas, John M. Martinis, Observation of quantum oscillations between a Josephson phase qubit and a microscopic resonator using fast readout, cond-mat/0405710 *accepted by Phys. Rev. Lett.* (2004).
5. A. I. Braginski, J. Talvacchio, M. A. Janocko, and J. R. Gavaler, Crystalline oxide tunnel barriers formed by thermal oxidation of aluminum overlayers on superconductor surfaces *J. Appl. Phys.* **60**, 2058 (1986).
6. K. M. Lang, S. Nam, J. Aumentando, C. Urbina, and John M. Martinis, Banishing quasiparticles from Josephson- junction qubits: Why and how to do it, *IEEE Trans. Appl. Supercond.* **13**, 989 (2003).
7. B. S. Swartzentruber, Y.-W. Mo, M. B. Webb, and M. G. Lagally, Scanning tunneling microscopy studies of structural disorder and steps on Si surfaces, *J. Vac. Sci. Technol. A* **7**, 2901 (1989).
8. H. J. Wen, M. Dhne-Prietsch, A. Bauer, M. T. Cuberes, I. Manke, and G. Kaindl, Thermal annealing of the epitaxial Al/Si(111)7 × 7 interface: Al clustering, interfacial reaction, and Al-induced p^+ doping, *J. Vac. Sci. Technol. A* **13**, 2399 (1995).
9. L. Aballe, C. Rogero, P. Kratzer, S. Gokhale, and K. Horn, Probing interface electronic structure with overlayer quantum-well resonances: Al/Si(111), *Phys. Rev. Lett.* **87**, 156801 (2001).
10. L. Aballe, C. Rogero, S. Gokhale, S. Kulkarni, and K. Horn, Quantum-well states in ultrathin aluminium films on Si(111), *Surf. Sci.* **482–485**, 488 (2001).

11. L. Floreano, D. Cvetko, F. Bruno, G. Bavdek, A. Cossaro, R. Gotter, A. Verdini, and A. Morgante, Quantum size effects in the low temperature layer-by-layer growth of Pb on Ge(001), *Prog. Surf. Sci.* **72**, 135 (2003).
12. E. N. Bratus, V. S. Shumeiko, G. Wendin, Theory of subharmonic gap structure in superconducting mesoscopic tunnel contacts, *Phys. Rev. Lett.* **74**, 2110 (1995)
13. D. Averin, A. Bardas, AC Josephson effect in a single quantum channel, *Phys. Rev. Lett.* **75**, 1831 (1995)
14. Theodore Van Duzer and Charles W. Turner *Principles of Superconductive Devices and Circuits* (Prentice Hall, New Jersey, 1999) pp. 194–195.

12

Energy level spectroscopy of a bound vortex-antivortex pair

A. KEMP[1], M.V. FISTUL[1], A. WALLRAFF[1,2], Y. KOVAL[1], A. LUKASHENKO[1], B.A. MALOMED[3], AND A.V. USTINOV[1]

[1] Physikalisches Institut III, University of Erlangen-Nuremberg, D-91058, Erlangen, Germany
[2] Department of Applied Physics, Yale University, New Haven, CT 06520, USA
[3] Department of Interdisciplinary Science, Faculty of Engineering, Tel Aviv University, Tel-Aviv 69978, Israel

Keywords: Macroscopic quantum tunneling, Quantum coherence

Abstract

We report microwave spectroscopy of energy levels of a bound Josephson Vortex–Antivortex (VAV) state in an annular Josephson junction. The bound VAV pair is formed in the narrow long junction by applying an in-plane magnetic field. The dissociation of the bound state, which is associated with a switching of the junction from the superconducting to the resistive state, is induced by ramping up a bias current applied to the junction. This process can be mapped to the escape of a particle from a potential well. Above the crossover temperature, the pair dissociation is thermally activated, below it is induced by quantum tunnelling, which is inferred from switching current measurements. In the quantum regime, the dependence of the energy level spectrum on field and bias current is probed using microwave spectroscopy.

Macroscopic quantum phenomena in Josephson coupled systems have attracted considerable attention both theoretically and experimentally.[1–9] Most of the studied systems, such as dc-biased Josephson Junctions (JJ), Superconducting Quantum Interference Devices (SQUIDs) and small Josephson-junction arrays, are based on a small number of point-like Josephson junctions. A few degrees of freedom, namely the Josephson phases, suffice to describe the dynamics of these systems. At low temperatures, quantum-mechanical effects such as macroscopic quantum tunnelling, energy level quantization,[2–5] and coherent oscillations[6–9] of the Josephson phase have been observed.

In the case of spatially extended Josephson systems, such as large parallel arrays or Josephson-junction ladders, the analysis and observation of quantum dynamics is more complex. A particularly interesting system in that respect is the *quasi-one-dimensional long Josephson junction* in which the phase continuously varies in space. In the classical limit the phase dynamics in the long JJ is described by the sine-Gordon equation. This system supports well-known nonlinear excitations,[10] for example Josephson vortices (magnetic fluxons) and vortex-antivortex pairs, that can interact with inhomogeneities or linear (Josephson-plasma) modes. The classical dynamics of such excitations are well established.[10] In particular, the escape induced by thermal fluctuations of the Josephson phase from a metastable state has been investigated both experimentally and theoretically.[11,12] Recently, the ac-induced escape of a vortex in the presence of thermal fluctuations has been studied in detail by numerical simulations.[13] A number of macroscopic quantum-mechanical effects have been predicted for Josephson vortices,[14–19] however, only few have been observed experimentally until now. In particular, macroscopic quantum tunnelling of many vortices[20,21] and more recently, also tunnelling and level quantization of an individual vortex[22] have been observed.

A next natural step in the study of *macroscopic quantum effects* in long JJs is to investigate the dissociation of a *single vortex-antivortex pair*. The crossover of the dissociation process from the thermal to the quantum regime has already been theoretically analyzed and experimentally observed.[23] A characteristic of the dynamics of the bound VAV pair is its small oscillation frequency in the effective binding potential, a property which has been theoretically analyzed in Ref.[23] In the quantum limit, the small oscillation frequency determines the energy level separation. As shown for the single-vortex case,[22] resonant microwave absorption can be used to map the level structure of the system by observing the enhancement of the escape rate. In this study microwave spectroscopic measurements of the energy levels of a VAV pair is reported.

A *single* VAV pair appears naturally in a long *annular* JJ subject to an external magnetic field H[24,25] applied in the plane of the junction, see Fig. 12.1(a)–(c). For experiments, we have fabricated annular junctions of diameter $d = 100\,\mu\text{m}$ and widths $w = 0.5\,\mu\text{m}$ and $w = 1\,\mu\text{m}$ by etching a sputtered Nb/AlO$_x$/Nb trilayer patterned using electron-beam lithography.[26] The critical current density of the junctions is 220 A/cm^2 at 20 mK. The Josephson length is $\lambda_J \approx 30\,\mu\text{m}$, resulting in a normalized junction length of $L \equiv \pi d/\lambda_J \approx 10.5$. The 0.5 μm-wide junction was used to observe the thermal to quantum crossover,[23] and the 1 μm-wide junction was used for spectroscopic measurements. A measurement of the magnetic field dependence of the switching current for the 0.5 μm-wide annular junction is shown in Fig. 1d. In the field range $|H| < 1.5$ Oe (main central lobe), the switching of the JJ from the superconducting to the resistive state occurs through the *dissociation* of a single field-induced VAV pair confined in the potential well formed by external magnetic field and dc bias current. This process is confirmed by direct numerical simulations of the perturbed sine-Gordon model for an annular long JJ[24,25] subject to an in-plane dc magnetic field H. The numerically found magnetic-field dependence of the switching current is in excellent agreement with measurements, see the solid line in Fig. 12.1(d). Simulations of the magnetic field distribution in the junction clearly show the nucleation and subsequent dissociation of the VAV pair, see Fig. 12.1(c). Fluctuations, thermal and quantum, induce oscillations of the confined pair. At high temperatures, the dissociation takes place by thermal activation over the confinement barrier. At low temperatures, *macroscopic quantum tunnelling* through the barrier occurs. At fields $|H| > 1.5$ Oe the system becomes bistable as a well-separated VAV pair is spontaneously created in the junction. This state is perfectly reproduced by our numerical calculations, see the first side lobes in Fig. 12.1(c).

According to the theoretical derivation in Ref.,[23] the phase distribution in the junction can be written as $\varphi(x,t) = \pi/2 + \xi(x,t)$, where $\xi(x,t)$ is a perturbation describing a small-amplitude VAV pair,

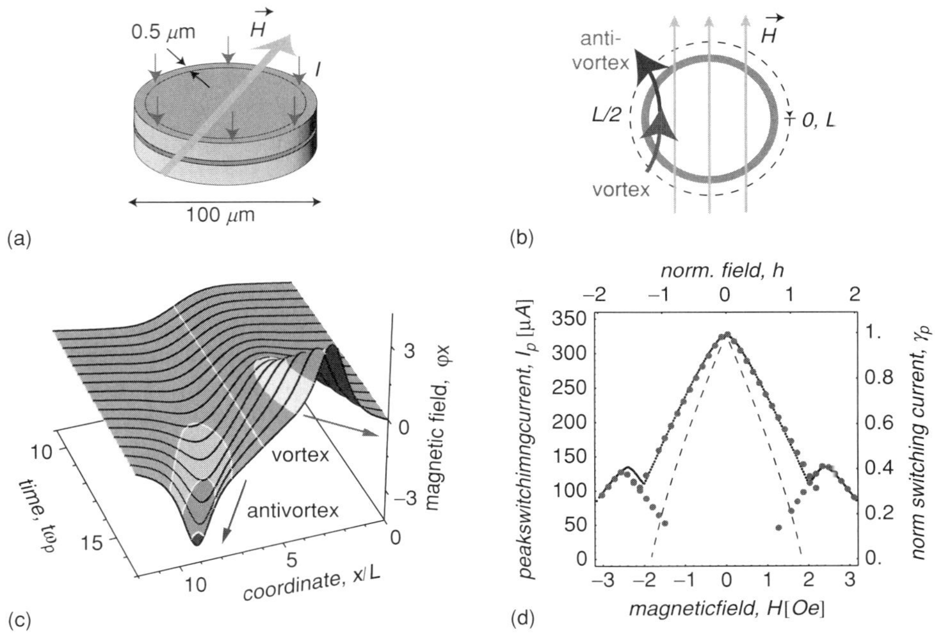

FIGURE 12.1. (a) Schematic view of a long annular Josephson junction of diameter 100 μm *without trapped vortices* in an in-plane external magnetic field H with uniform bias current I. (b) Generation of a confined vortex-antivortex pair with center coordinate at $x = L/2$. (c) Numerically simulated evolution of the magnetic field distribution, as the Josephson junction switches to the resistive state. The emerging vortex and antivortex move in opposite directions. (d) Magnetic field dependence of switching currents: experimental data at $T = 100$ mK (circles), numerical calculation for $L = 10.5$ (solid line), and theoretical prediction[23] for $L \gg 2\pi$ (dashed line).

$$\xi(|x - x_1|, A) = \sqrt{2\delta}\left[\frac{3}{\cosh^2\left(\frac{|x-x_1|+A}{2}(2\delta)^{1/4}\right)} - 1\right]. \tag{12.1}$$

Here, $\delta \equiv (1 - I/I_{c0}) \ll 1$ and $x_1(t)$ is the center-of-mass coordinate of the bound pair. The distance between the vortex and the antivortex $A(t)$ is allowed to vary in time. The magnetic field and dc bias create the pinning potential for such a state. Since we consider dc magnetic fields only, we neglect the dynamics of the center-of-mass coordinate $x_1(t)$. Hence the effective energy of the JJ,

$$E(A) = \frac{m_{\mathrm{eff}}(A)}{2}\dot{A}^2 + U_{\mathrm{pot}}(A)\,, \tag{12.2}$$

is a function of A and $\dot{A} \equiv \mathrm{d}A/\mathrm{d}t$. Expressions for $U_{\mathrm{pot}}(A)$ and $m_{\mathrm{eff}}(A)$ are derived in.[23]

In the absence of fluctuations, the critical current, $\delta_c(h) = 2h/3$, is found by minimization of the energy $E(A)$ in A. The corresponding critical amplitude A_0 is determined by the condition $\sinh\left((2\delta)^{1/4}A_0/2\right) = 1$. It is noted that the critical current decreases linearly with magnetic field, in contrast to the case of *linear* long JJs, where similar considerations yield $\delta_c^{\mathrm{lin}}(h) \propto h^{4/3}$. Both results are valid in the case of uniform bias-current distribution and for $L/2\pi \gg 1$.

In the presence of thermal or quantum fluctuations, the dissociation of the pinned VAV pair occurs at a random value of bias δ exceeding $\delta_c(h)$. Assuming weak fluctuations, $\delta - \delta_c(h) \ll \delta_c(h)$, we expand the energy about $A = A_0$ ($\delta A \equiv A - A_0$) as

$$E(A) = \frac{\chi h^{5/4}(\delta \dot{A})^2}{2} + \frac{3^{3/2}\sqrt{h}}{2}(\delta - \delta_c(h))(\delta A) - \frac{h^2}{6}(\delta A)^3 , \tag{12.3}$$

where χ is a numerical coefficient of order unity. Thus, the problem of the fluctuation-induced dissociation of a bound VAV pair is mapped to the well-known problem of particle escape from a cubic potential. The dissociation rate depends on the height of the effective potential barrier,

$$U_{\text{eff}}(\delta) = 2 \cdot 3^{5/4} h^{-1/4} (\delta - \delta_c(h))^{3/2} . \tag{12.4}$$

At high temperatures, the dissociation is driven by thermal activation over this barrier. Using well known theory describing the particle escape from such a potential well,[1–4] we find the switching rate of a long Josephson junction from the superconducting to the resistive state to be

$$\Gamma_T(I) \propto \exp[-U_{\text{eff}}/k_B T]. \tag{12.5}$$

Thus, at high temperatures the standard deviation σ of the critical current distribution should increase with temperature and weakly depend on magnetic field as $\sigma_T \propto T^{2/3} h^{1/6}$. Notice that σ_T *increases* with H, in contrast to small Josephson junctions[1,2] where $\sigma_T \propto (I_c(H))^{1/3}$ decreases with H.

We experimentally investigated the fluctuation-induced dissociation of the VAV pair by measuring the temperature and magnetic field dependence of the statistical distribution P of the switching currents $I < I_{c0}$, using techniques described in Refs.[11,16,27] In Fig. 12.2(a), the temperature dependence of the switching current distribution measured at $H = 0$ is shown. At high temperatures the $P(I)$ distribution is temperature-dependent. At low temperatures a saturation of the distribution width is observed. In Fig. 12.2(b), the standard deviation σ of $P(I)$ is plotted versus bath temperature T for two values of magnetic field. The standard deviation σ is to a good approximation proportional to $T^{2/3}$, and is larger for the higher field, as predicted above. As clearly seen in Fig. 12.2(b), σ decreases with temperature and saturates below a crossover temperature of $T^\star \approx 100\,\text{mK}$. At $T < T^\star$ the dissociation of the VAV pair occurs through

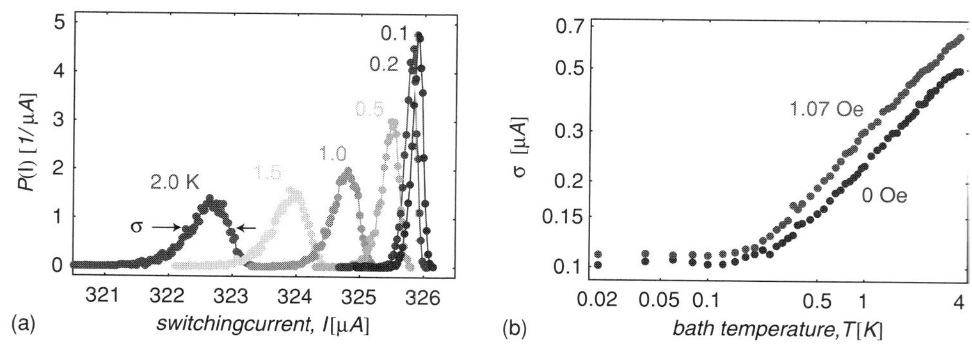

FIGURE 12.2. (a) Switching-current distributions $P(I)$ at $H = 0$ for the indicated bath temperatures T. (b) Distribution of the standard deviation σ of $P(I)$ versus the bath temperature for the two indicated values of magnetic field.

macroscopic quantum tunnelling. The crossover temperature $T^\star \simeq \hbar\omega(\delta)/(2\pi k_B)$ is determined by the small oscillation frequency $\omega(\delta)$ of the VAV pair. In the quantum regime, this frequency $\omega(\delta) = 3^{3/8}/\sqrt{\chi}\,(\delta - \delta_c(h))^{1/4}$ determines the oscillatory energy levels $E_n \simeq \hbar\omega(\delta)(n + 1/2)$ of the pinned VAV state. A further discussion of the thermal and quantum dissociation rates and their specific scaling with H is given in Ref.[23] Using the techniques demonstrated in Ref.[22] for single vortices, we have performed microwave spectroscopic measurements of the VAV pair energy levels. The measurements were performed at temperatures between 15 mK and 20 mK by exciting the VAV pair into the first excited state by means of microwave radiation. A Lorentzian peak in the decay rate Γ centered around the resonant current for a given spectroscopy frequency is observed. Rewriting the small oscillation-frequency dependence as

$$\nu(I, H) = \nu_0(H)\left(1 - \frac{I}{I_c(H)}\right)^{1/4} \tag{12.6}$$

highlights the small-junction-like scaling of $\nu(I, H)$ with $I/I_c(H)$ at H = const and $I/I_c(H)$ close to unity. The factor $\nu_0(H)$ is the magnetic-field-dependent frequency of internal oscillations of the VAV pair at zero bias current. For a number of different values of magnetic field, we have measured the dependence of the resonance current on microwave frequency. Experimental data for zero magnetic field and $H = 0.2$ Oe, which compare well to Eq. (12.6), are displayed in Fig. 12.3. Least-squares fits to the data for different magnetic fields yield both $I_c(H)$ and $\nu_0(H)$.

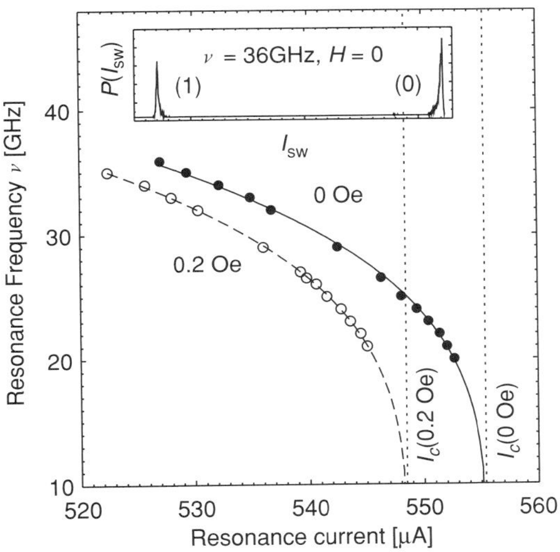

FIGURE 12.3. Resonance current versus resonance frequency. Experimental data for zero magnetic field (filled circles) and 0.2 Oe (open circles) are compared to the model (solid line and dashed line, respectively). The respective critical currents are indicated by the dotted vertical lines. In the inset, the original switching-current probability distribution $P(I_{sw})$ is indicated for one of the data points using the same current scale as the main figure. The peak (1) corresponds to the escape from the resonant level, the peak (0) to the escape from the ground state.

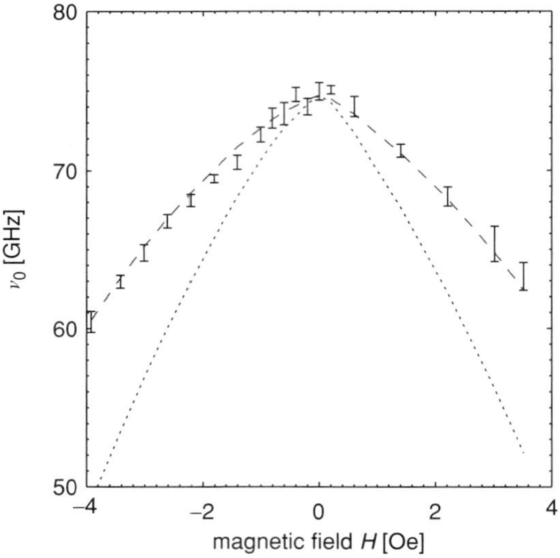

FIGURE 12.4. Field dependence of zero bias oscillation frequency ν_0 extracted from measured data (solid error bars), $\nu_0(H)$ calculated from $I_c(h)$ using the long-junction model (dashed line) and the small-junction model (dotted line).

The theoretical dependence of $\nu_0(H)$ on $I_c(H)$ for VAV dissociation is given by

$$\nu_0(H) = \frac{\omega_0}{2\pi}\frac{3^{3/8}}{\sqrt{\chi}}\left(\frac{I_c(H)}{I_{c0}}\right)^{1/4}, \tag{12.7}$$

where $\omega_0/2\pi$ is the plasma frequency of the junction at $H = 0$. This expression differs from the one expected for the phase escape in a small Josephson junction $\nu_0(H) \propto \sqrt{I_c(H)/I_{c0}}$. In Fig. 12.4 the experimentally determined magnetic field dependence of the zero bias oscillation frequency $\nu_0(H)$ is fitted into Eq. (12.7). A characteristic frequency $3^{3/8}\omega_0/(2\pi\sqrt{\chi})$ of 74.7 GHz is determined, from the least-squares fits. We note that the field dependence of the zero bias oscillation frequency extracted from experimental data is in good agreement with the predictions based on our model for VAV dissociation. For comparison, the expected oscillation frequency for homogeneous escape of the phase is shown by the dotted line in the same plot, see Fig. 12.4. The prediction for homogeneous escape of the phase clearly disagrees with the experimental data.

Summary

It is shown that the switching of an annular Josephson junction from the superconducting to the resistive state in the presence of magnetic field occurs through the nucleation and subsequent dissociation of a single vortex–antivortex pair. At low temperatures we observe the pair dissociation by quantum tunnelling. Using microwave spectroscopy, we have shown in this work that the resonantly stimulated dissociation process is in excellent agreement with the model derived in

Ref.[23] The resonant enhancement of the dissociation rate is a manifestation of the energy-level quantization.

Acknowledgments. We thank A. Price for critical reading of the manuscript, A. Abdumalikov for useful discussions and J. Lisenfeld and C. Coqui for technical help. We acknowledge the partial financial support of this project by the Deutsche Forschungsgemeinschaft (DFG).

References

1. M. Tinkham, *Introduction to Superconductivity*, (McGraw-Hill International Editions, 1996).
2. M. H. Devoret, J. M. Martinis, and J. Clarke, Measurements of macroscopic quantum tunneling out of the zero-voltage state of a current biased Josephson junction, *Phys. Rev. Lett.*, **55**(18), 1908 (1985).
3. J. M. Martinis, M. H. Devoret, and J. Clarke. Experimental tests for the quantum behavior of a macrospcopic degree of freedom: The phase difference across a Josephson junction, *Phys. Rev. B*, **35**(10), 4682 (1987).
4. M. H. Devoret, D. Esteve, C. Urbina, J. Martinis, A. Cleland, and J. Clarke, in *Quantum Tunnelling in Condensed Media*, (North-Holland, Amsterdam, 1992).
5. A. Wallraff, T. Duty, A. Lukashenko, and A. V. Ustinov, Multiphoton transitions between energy levels in a current-biased Josephson tunnel junction, *Phys. Rev. Lett.*, **90**, 037003 (2003).
6. Y. Yu, S. Han, X. Chu, S. Chu, and Z. Wang, Coherent temporal oscillations of macroscopic quantum states in a Josephson junction, *Science*, **296**, 889 (2002).
7. J. M. Martinis, S. Nam, J. Aumentado, and C. Urbina. Rabi oscillations in a large Josephson-junction qubit, *Phys. Rev. Lett.*, **89**, 117901 (2002).
8. C. H. van der Wal, A. C. J. Haar, F. K. Wilhelm, R. N. Schouten, C. J. P. M. Harmans, T. P. Orlando, S. Lloyd, and J. E. Mooij. Quantum superposition of macroscopic persistent-current states. *Science*, **290**, 773 (2000).
9. J. R. Friedman, Vijay Patel, W. Chen, S. K. Tolpygo, and J. E. Lukens. Quantum superposition of distinct macroscopic states, *Nature*, **406**, 43 (2000).
10. A. V. Ustinov, Solitons in Josephson junctions, *Physica D*, **123**, 315 (1998).
11. M.G. Castellano, G. Torrioli, C. Cosmelli, A. Costantini, F. Chiarello, P. Carelli, G. Rotoli, M. Cirillo, and R.L. Kautz. Thermally activated escape from the zero-voltage state in long Josephson junctions, *Phys. Rev B*, **54**(21), 15417 (1996).
12. M. V. Fistul, M. G. Castellano, M. Cirillo, G. Torrioli, A. Wallraff, and A. V. Ustinov. Escape of a Josephson vortex trapped in an annular Josephson junction, *Physica B*, 284–288, 585–586 (2000).
13. N. Gronbech-Jensen and M. Cirillo. Ac-induced thermal vortex escape in magnetic-field-embedded long annular Josephson junctions, (2004), cond-mat/0404721.
14. T. Kato and M. Imada, Macroscopic quantum tunneling of a fluxon in a long Josephson junction, *J. Phys. Soc. Jpn.*, **65**(9), 2963 (1996).
15. A. Shnirman, E. Ben-Jacob, and B.A. Malomed. Tunneling and resonant tunneling of fluxons in a long Josephson junction, *Phys. Rev. B*, **56**(22), 14677 (1997).
16. A. Wallraff, Yu. Koval, M. Levitchev, M. V. Fistul, and A. V. Ustinov, Annular long Josephson junctions in a magnetic field: Engineering and probing the fluxon potential, *J. Low Temp. Phys.*, **118**(5/6), 543 (2000).
17. K. Maki, Creation of soliton pairs by electric fields in charge-density-wave condensates, *Phys. Rev. Lett.*, **39**, 46 (1977).
18. I. V. Krive, B. A. Malomed, and A. S. Rozhavsky, Tunnel creation of kink–antikink pairs from an inhomogeneous vacuum by an external field, *Phys. Rev. B*, **42**, 273 (1990).
19. T. Kato, Dimensional crossover by a local inhomogeneity in soliton-pair nucleation, *J Phys. Soc. Jpn.*, **69** 2735 (2000).

20. H. S. J. van der Zant, F. C. Fritschy, T. P. Orlando, and J. E. Mooij, Dynamics of vortices in underdamped Josephson-junction arrays, *Phys. Rev. Lett*, **66**, 2531 (1991).
21. A. Oudenaarden, S.J.K. Vardy, and J.E. Mooij, One-dimensional localization of quantum vortices in disordered Josephson junction arrays, *Phys. Rev. Lett.*, **77**(20), 4257 (1996).
22. A. Wallraff, A. Lukashenko, J. Lisenfeld, A. Kemp, Y. Koval, M. V. Fistul, and A. V. Ustinov, Quantum dynamics of a single vortex, *Nature*, **425**, 155 (2003).
23. M. V. Fistul, A. Wallraff, Y. Koval, A. Lukashenko, B. A. Malomed, and A. V. Ustinov, Quantum dissociation of a vortex-antivortex pair in a long Josephson junction, *Phys. Rev. Lett*, **91**, 257004 (2003).
24. N. Gronbech-Jensen, P.S. Lomdahl, and M.R. Samuelsen. Bifurcation and chaos in dc-driven long Josephson junctions. *Phys. Rev. B*, **43**(16), 12799 (1991).
25. N. Martucciello and R. Monaco, Static properties of annular Josephson tunnel junctions, *Phys. Rev. B*, **54**(13), 9050 (1996).
26. Yu. Koval, A. Wallraff, M. Fistul, N. Thyssen, H. Kohlstedt, and A. V. Ustinov, Narrow long Josephson junctions, *IEEE Trans. Appl. Supercond.*, **9** 3957 (1999).
27. A. Wallraff, A. Lukashenko, C. Coqui, A. Kemp, T. Duty, and A. V. Ustinov, Switching current measurements of large area Josephson tunnel junctions, *Rev. Sci. Inst.*, **74**, 3740 (2003).

13

Adiabatic Quantum Computation with Flux Qbits

VALENTINA CORATO, CARMINE GRANATA, BERARDO RUGGIERO, AND PAOLO SILVESTRINI*

*Valentina Corato and Paolo Silvestrini, Dipartimento di Ingegneria dell'Informazione, Seconda Università di Napoli, I-81031, Aversa, Italy and Istituto di Cibernetica "E. Caianiello" del CNR, I-80078, Pozzuoli, Italy. Carmine Granata and Berardo Ruggiero, Istituto di Cibernetica "E. Caianiello" del CNR, I-80078, Pozzuoli, Italy. Email: v.corato@cib.na.cnr.it

Abstract

We propose a superconducting device for adiabatic quantum computation with flux qbits. This system includes rf SQUIDs coupled via a superconducting flux transformer especially designed for obtaining an anti-ferromagnetic coupling. Moreover we have characterized a niobium-based vertical Josephson interferometer to be included in a superconducting loop for applications to quantum computation using flux qbits. The most interesting feature of this device is that the Josephson current is precisely modulated by a small transversal magnetic field parallel to the superconducting loop plane, from a maximum to zero with fine control and precision. This vertical interferometer can be used to independently control the off-diagonal Hamiltonian terms of flux qbits and/or to control the flux transfer function of a superconducting transformer for inter-qbits coupling.

13.1 Introduction

Quantum entanglement and controlled manipulation of arrays of quantum bits are the main phenomena under investigation in any physical implementation of quantum information processing devices. Quantum Computation (QC)[1] requires data encoding into nontrivial multiparticle superpositions of some selected basis states, and the controlled evolution of the initial quantum states into final states that encode the solution to the problem. The coupling to the environment produces decoherence[2] that tends to destroy the quantum computation, considering that QC relies on complicated interference effects.

In fact, reliable transmission of quantum states, controlled quantum evolution, implementation of read-out systems as well as quantum coherence preservation at the level required for Quantum Algorithms (QA) to work, are the main challenges of current research activities all over the world in the quantum information area. Manipulation of quantum states is often obtained by tailored sequences of resonant rf-pulses of precise area[3]. In this case the main topic under study is that deviation from precise pulse area and resonance mismatch may lead to relevant errors. Since quantum computers are very susceptible to making errors, fault tolerant protocols are necessary for operation of large-scale quantum computers[4]. Therefore it would be highly advantageous to weave fault tolerant into the design of quantum hardware. In this direction, adiabatic quantum evolution is a powerful tool to manipulate quantum devices, either when operating on single qbit (as NOT operation)[5] or two-qbits gate (like C-NOT)[6,7], as well as to control entangled states of many coupled qbits to implement complicated quantum algorithms.

Adiabatic Quantum Computation (AQC)[8] is in fact a recently proposed general approach to solve NP-hard combinatorial minimization problems. Whereas a conventional quantum algorithm is implemented as a sequence of discrete unitary operation involving many energy levels of the computer and complicated superposition of them, the adiabatic algorithm works by keeping the quantum computer close to the instantaneous ground state of a Hamiltonian which varies adiabatically in time. As a consequence it has an inherent robustness against errors and decoherence, illustrating the principle that it may be possible to design suitable quantum hardware, which effectively resists to noise. Any problem which can be cast as the minimization of an energy function can be potentially solved by adiabatic evolution, provided we can design and construct a complex quantum device whose Hamiltonian corresponds to the function to minimize. Then the algorithm can be implemented by adiabatically changing the external overall magnetic field, interpolating between a starting Hamiltonian of decoupled qbits, whose ground state is easily reachable, to the final Hamiltonian whose ground state encodes the desired solution to the problem[9].

Obviously, any new hardware for QC must be constructed via incremental improvements and integration of existing technologies. In this chapter we propose a superconducting device, consisting of a set of persistent current qbits whose inductive coupling is externally controlled with reasonable accuracy to realize the required time-dependent Hamiltonian. The experimental characterization of the single critical components of the whole device is presented, with emphasis on the flux transformer realizing the critical antiferromagnetic coupling. The special strength of superconducting qbits is the reliance of lithographic fabrication, reasonably scalable in a close future to more than 1000 qbits, as well as the possibility to integrate various traditional and mesoscopic systems to give many types of control and better read-out, while the effect of the rather strong coupling to the environment, causing dephasing and relaxation, is minimized by the use of the adiabatic technique.

13.2 Adiabatic Quantum Computation

The device presented in this study is inspired by the fact that it is NP-hard to calculate the ground state of a 2D antiferromagnetically coupled Ising model in a magnetic field in which each Ising spin is coupled to at least 3 others[10]. In other words, if G is a planar graph with vertices V and edges Esuch that each vertex is connected by edges to $\geq$3 other vertices, then it is NP-hard to find the ground state of

$$H = \sum_{j \in V} \sigma_j + \sum_{(j,k) \in E} \sigma_j \sigma_k \tag{13.1}$$

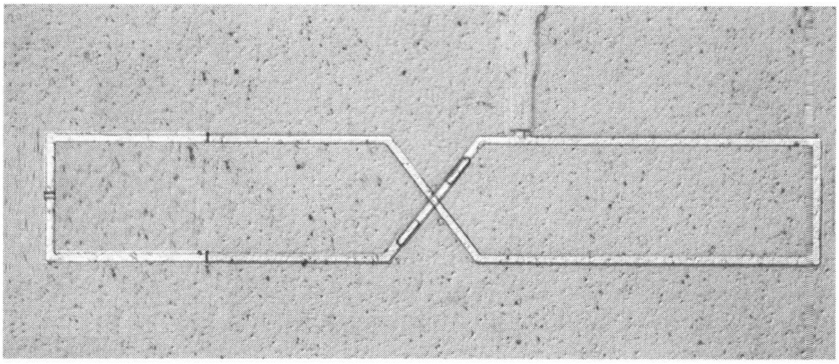

FIGURE 13.1. Picture of the superconductive flux transformer providing an anti-ferromagnetic coupling between two SQUIDs (device made at CNR-IC, Naples).

where σ_j is the Pauli spin operators along the z-axis (set by the direction of an uniform magnetic field): $\sigma_j \in \{-1, 1\}$ for all $j \in V$ Eq.1 represents an Ising model from the family

$$H = \sum_{j \in V} \varepsilon_j \sigma_j + \sum_{(j,k) \in E} \Lambda_{kj} \sigma_j \sigma_k \tag{13.2}$$

with equal antiferromagnetic coupling strength and the energy difference between the two low-lying states equal to the coupling energy; $\varepsilon_j = \Lambda_{kj}$. SQUIDs qbits can be inductively coupled, either simply designing them one close to the others or via superconducting flux transformers. The natural inductive coupling, as well as all the ones proposed so far in the litterature[11], is ferromagnetic, but the flux transformer (see Fig. 13.1) we have designed, fabricated and experimentally characterized provides actually an inductive antiferromagnetic coupling whose strength can be controlled in the design within a wide range of useful values. By this kind of flux transformer, a Josephson device consisting of inductively coupled persistent current QBITs offers a straightforward way to realize a hardware for AQC interpolating between a weakly coupled starting Hamiltonian, whose ground state is easy to obtain, and the strongly coupled Hamiltonian Eq.(13.1), whose ground state is hard to obtain by classical computation.

13.3 AQC with an Array of Flux Qbits

The set of equations for an array of underdamped flux linked rf-SQUIDs is

$$\Phi - \Phi_e = LI \tag{13.3}$$

where L is a matrix representing the self- and mutual- inductances, while Φ, I, and Φ_e are meant as vectors representing all the fluxes Φ_j and currents I_j in the j-th superconducting rings, as well as the external fluxes Φ_j^e. The currents I_j flowing in the j-th ring is expressed in terms of the Josephson critical supercurrent I_j^c, the capacitance C_j and the effective resistance R_j of the Josephson junction interrupting the superconducting ring. Taking advantage that L is symmetric, by the reciprocity of mutual inductance, we can invert L obtaining another symmetric matrix L^{-1}. In extremely under damped systems, neglecting the dissipation, Eq.(13.3) can be now cast in the form[9]:

$$(\Phi_0/2\pi)^2\, C_j \phi_j = -\frac{\partial U}{\partial \phi_j} \tag{13.4}$$

$$U = \frac{1}{2}(\Phi_0/2\pi)^2 \sum_{j,k} L^{-1}_{jk}\left(\phi_k - \phi^e_k\right)\left(\phi_j - \phi^e_j\right) + \Phi_0/2\pi \sum_j I^c_j \cos\phi_j \tag{13.5}$$

where ϕ_j and $\phi^e_j$are normalized flux in units of the flux quantum $\Phi_o/2\pi$ measured with respect to $\Phi_o/2$. With this choice, it must be kept in mind that $\phi^e_j = 0$ corresponds to a nonzero applied field on the j-th SQUID ring. The external fluxes ϕ^e can be adjusted or varied to find the favorable operating points for the device as well as to perform the necessary manipulation of the Hamiltonian. For non-interacting systems (namely if the mutual inductance are zero, $L_{jk} = 0$ for $j = k$), each SQUID is described by a double well potential, and the external magnetic flux controls the energy difference between the minima, the symmetric situation being for $\phi_e = 0$. Each logical state is represented by a wave function localized in a distinct potential well, and corresponds to distinguishable flux states trapped in the ring with current flowing in opposite directions, say $|0\rangle$ clockwise and $|1\rangle$ anticlockwise. When the energy difference ε of the minima of the two different wells is small with respect to the oscillation frequency ω around the minima, $\varepsilon \ll \omega$, these states become coupled and the wave functions spread over both the wells, the coupling being maximum in resonance conditions ($\varepsilon = 0$), while the energy eigenstates tend to be localized in one of the well away from resonance as ε is increasing. The coupling of the states can be described by the tunneling amplitude $\Delta(\varepsilon)$, and the Hamiltonian of any SQUID qbit reduces to the regular two-state form in the basis of these logical states:

$$H = \frac{1}{2}\left[\varepsilon\left(|0\rangle\langle 0| - |1\rangle\langle 1|\right) - \Delta\left(|0\rangle\langle 1| + |1\rangle\langle 0|\right)\right] = \frac{1}{2}\left(\varepsilon\sigma_z + \Delta\sigma_x\right) \tag{13.6}$$

In other words, if all the mutual inductances as well as ϕ_e are equal to zero, all the logical states of the qbits are symmetric. Now consider when the inductive coupling among qbits is turned on in the potential Eq.(13.5) and imagine varying $\phi_j$at fixed ϕ_k. A coupling term $\varepsilon^{\text{coupl}}_{j,k} = L^{-1}_{jk}(\phi_k - \phi^e_k)(\phi_j - \phi^e_j)$ is added such that there is a bias for ϕ_j even though $\phi^e_j = 0$ and as a consequence the logical states for the j-th qbit may be asymmetric. The direction of this bias depends if the coupling is ferromagnetic or anti-ferromagnetic. In the usual approximation for which an rf-SQUID can be considered as a two level system, namely $\varepsilon_j$and$\varepsilon^{\text{coupl}}_{j,k} \ll \omega_j$, the Eqs.(13.5)–(13.6) for the system of flux linked qbits can be cast in the form of an Hamiltonian problem of the kind:

$$H = \sum_j \varepsilon_j \sigma^z_j + \sum_i \Delta_j \sigma^x_j + \sum_{j,k} \Lambda_{kj}\sigma^z_j \sigma^z_j \tag{13.7}$$

In order to control such Hamiltonian, we should be able to modulate the tunneling amplitude of each SQUID as well as to switch on and off the magnetic coupling between neighbors SQUIDs.

13.4 Vertical Two Josephson Junctions Interferometer

Let's consider a rf SQUID, described by a double well potential:

$$U = U_{\text{J}}\left[\frac{1}{2\beta}(\phi - \phi_e)^2 + \cos\phi\right] \tag{13.8}$$

where $U_J = \left(\frac{\Phi_0}{2\pi}\right) I_c$ is the Josephson energy and $\beta = \frac{2\pi L I_c}{\Phi_0}$ is the reduced inductance.

The diagonal elements of H in Eq.(13.6) can be easily controlled by an external magnetic field B_z in the z-direction generating an external flux Φ_e (see Fig. 13.2), while the off-diagonal elements are related to the tunneling amplitude which strongly depends on the Josephson energy U_J and this parameter is difficult to be independently controlled with enough precision. A way proposed is to replace the Josephson junction with vertical two Josephson junctions interferometer whose loop area is much smaller than the one of FQ.[12] In this case an external magnetic field B_x in the x-direction generating an external flux Φ_e^{dc}, independently applied to the small loop, modulates the Josephson energy U_J, and as a consequence, the tunneling rate $\Delta(\varepsilon, U_J)$. To be effective in controlling the FQ Hamiltonian we need a wide modulation of U_J (possibly to zero) as well as a magnetic field generating Φ_e^{dc} orthogonal to the one generating Φ_e in order to reduce the interference between the two control signals. Here we propose to use a two Josephson junctions interferometer to realize such a device (see Fig. 13.2).

As a further application the modulation to zero of I_c by a transversal magnetic field B_x, can be used to switch off a flux transformer, interdicting the read-out of the phase qbit or controlling the inter-qbit coupling. In fact the flux transfer function of a superconducting transformer closed by a Josephson junction is a function of the loop inductance L and the critical current I_c, which appear in the definition of β. As well known, the total current flowing into the superconducting transformer through the Josephson junction, is a function of the external flux Φ_e and of the flux $\Phi = \Phi_e + LI$ generated into the loop and it can be written as: $I = \frac{\beta\Phi_0}{2\pi}\sin(2\pi\frac{\Phi}{\Phi_0})$ This current is coupled to the final device via a mutual inductance M, providing a flux $\Phi_{\text{out}} = MI$ The transfer function θ s then

$$\theta = \frac{\partial\Phi_{\text{out}}}{\partial\Phi_e} = \frac{M\beta}{L}\cos\left(\frac{2\pi\,\Phi}{\Phi_0}\right)\frac{\partial\Phi}{\partial\Phi_e} = \frac{M}{L}\frac{\beta\cos(2\pi\,\Phi/\Phi_0)}{1+\beta\cos(2\pi\,\Phi/\Phi_0)}$$

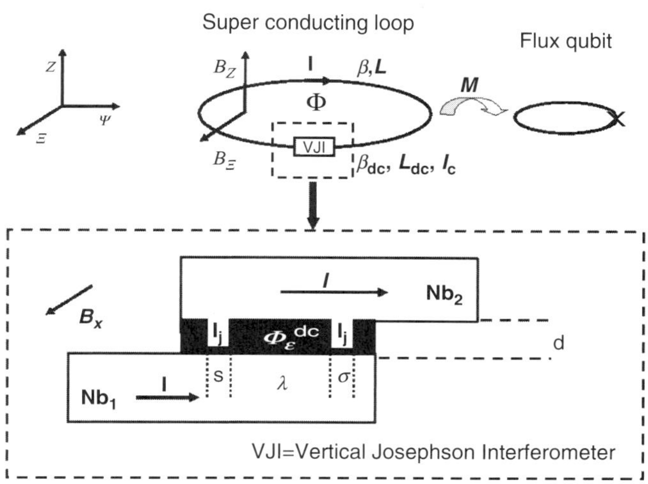

FIGURE 13.2. Sketch of the proposed device including a vertical two Josephson junctions interferometer to get a fine control of the inter-qbit or many-qbit couplings.

and in the limit $\Phi/\Phi_0 \ll 1$,θ reduces to

$$\theta = \frac{\partial \Phi_{out}}{\partial \Phi_e} \approx \frac{M}{L} \frac{\beta}{1+\beta}$$

Inserting a vertical Josephson junction interferometer instead of a simple Josephson junction in a superconducting loop, we can control θghrough a transversal magnetic field B_x by varying it adiabatically from zero to a maximum value, which for $\beta \gg 1$ tends to M/L, the full superconducting limit.

The structure here characterized is made up by a niobium-based Josephson interferometer employing two window tunnel junctions. It differs from any previous proposed ones[11,12,13] as the controlling flux is generated by a magnetic field B_x parallel to the plane of the superconducting films, and the Josephson current modulates to zero as well, provided the normalized self flux ($L_{dc} I_c/\Phi_0$) of the structure is much less than one.

The devices were fabricated with all refractory niobium technology.[14] The Nb/AlO$_x$/Nb trilayer was deposited by dc-magnetron sputtering on a 3" Si crystalline wafer in a UHV system, without breaking the vacuum. The base and top Nb electrodes were 200 nm and 35 nm thick respectively. They were deposited at a rate of 1.5 nm/sec. The Al film, about 7 nm thick, was deposited at a rate of 0.09 nm s^{-1} obtained by rotating the wafer carrier. The artificial (AlO$_x$) tunnel barrier was realised by thermal oxidation of the aluminium layer by dry oxygen at a pressure of 250 mbar. The trilayer geometric patterning was performed by a lift-off process. The geometry of the two Josephson junctions of the interferometer is obtained by a standard photolithography and a Selective Niobium Anodization Process (SNAP). In order to improve the Junction insulation and reduce the parasitic capacitance, a SiO film (350 nm thick) was deposited by thermal evaporation and patterned by a lift-off process. The wiring is obtained, after a rf-sputtering cleaning process, by depositing a Nb film (500nm thick) by dc magnetron sputtering at a rate of 3.0 nm/s, and then defined by a lift-off procedure.

We have fabricated interferometers with symmetrical window-type junctions located at a distance l = 100–200 μm and with area A_j ranging between (4 × 4) μm^2 and (40 × 40) μm^2, and junction Josephson current I_j ranging between 10 μA and 15 mA. The interferometer loop area perpendicular to the x-axis is $A_l = l \times d$, where d is the insulation layer thickness t plus twice of the London penetration depth of the niobium λ_L = 0.06 μm ($d = t + 2\lambda_L = 0.18 \div 0.30$ μm). A relevant parameter for the interferometer is $\beta_{dc} = \frac{\pi L_{dc} I_{C0}}{\Phi_0}$, where L_{dc} is the superconducting loop inductance ($L_{dc} = 2L_0$, being L_0 is the right- or left- branch of the vertical superconducting loop) and $I_{C0} = I_C(0)$=2 $I_j$is the total critical current at zero magnetic field. The critical current in the presence of a magnetic field is modulated by the external flux, as shown below u We have fabricated different interferometers with β_{dc} ranging between $0.001 < \beta_{dc} < 3$.

The device characterization is performed in a pumped liquid ^{4}He cryostat with three μ-metal shields and two copper layers to reduce electromagnetic noise. All the electrical connections to room temperature went through home-made miniature twin axial cables. The measurements were performed by using a battery-powered ramp generator as well as the pre-amplification stage.[15] The Josephson interferometer was biased through a limiting resistor with a triangular-shaped waveform at frequencies ranging from dc up to 30 Hz. The current amplitude has been measured by a differential comparator in the y axis of an oscilloscope, and the external magnetic field is provided by a solenoid assembled around the sample.

In the Fig. 13.3, we report data on the critical current I_C normalised to I_{C0}, $I_C(\phi_e^{dc})/I_C(0)$, as a function of the applied flux Φ_e^{dc}, trapped into the vertical loop area A_l, normalised to the flux quantum Φ_0, $\phi_e^{dc} = \frac{l d B_x}{\Phi_0}$, for an interferometer with a small value of $\beta_{dc} = 2.7 \times 10^{-2}$.

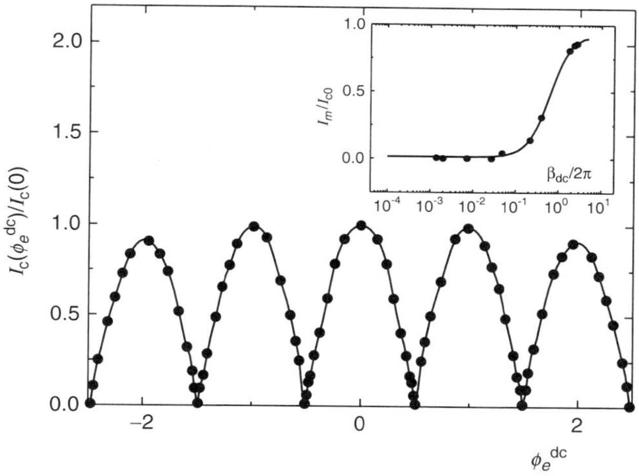

FIGURE 13.3. Data and theoretical prediction on the dependence of the Josephson critical current I_c as a function of the external magnetic flux ϕ_e^{dc},$I_c(\phi_e^{dc})$. The theoretical curves[16] are calculated for $\beta_{dc} = 2.7 \times 10^{-2}$, $I_{c0} = 2I_j = 22.0 \pm 0.68$ μA, $L_{dc} = 2L_0 = 0.8 \pm 0.2$ pH, s = 8 μm, and l = 195 μm. In the inset are also shown data on the critical current modulation depth I_m normalised to I_{c0}, I_m/I_{c0}, as function of the $\beta_{dc}/2\pi$ value for different interferometers.

This device is characterized by a total critical current $I_{c0} = 2I_j = (22.0 \pm 0.7)\mu A$, a loop inductance $L_{dc} = (0.8 \pm 0.2)$ pH and a loop area $A_l = l \times d = (0.22 \times 195)\mu m^2 = 43\ \mu m^2$. This sample presents symmetrical inductance branches ($L_0 = L_{dc}/2 = 0.4 \pm 0.1$ pH) and symmetrical junction areas ($A_{ja} = A_{jb} \cong A_{Je} = 4 \times 4\,\mu m^2$). In the Fig. 13.3 we also show the theoretical prediction[16] for $I_c(\phi_e^{dc}) = I_c(0)\left|\cos(\pi\phi_e^{dc})\right|$. As well known,[16] for interferometers with larger β_{dc} values, the critical current minima do not go to zero and we plot in the inset the minimum values I_m of the interferometrical pattern$I_c(\phi_e^{dc})$, (I_m normalised to I_{c0} is referred as the critical current modulation depth I_m/I_{c0}), as a function of $\beta_{dc}/2\pi$. The data fit very well by the theoretical prediction, and we note that in a wide range of small $\beta_{dc}/2\pi$ we can obtaing fine control down to zero for the critical current modulation depth.

13.5 Conclusions

In conclusion, in the framework of adiabatic quantum computation, we have presented a niobium-based vertical two Josephson junctions interferometer to be included in a superconducting loop for applications using flux qbits. We have reported data on magnetic measurements of the critical current of the interferometer, showing that the Josephson current is precisely controlled by a small transversal magnetic field parallel to superconducting loop plane and modulates from a maximun to zero with fine control and precision. The proposed device can be used in flux qbits to independently control the off-diagonal Hamiltonian terms (in the basis of the logical states) as well as to control the flux transfer function of a superconducting transformer closed by such a vertical interferometer in application for inter-qbits coupling.

Acknowledgments. This work is partially supported by MIUR-FIRB under the project "Nanocircuiti Superconduttivi".

References

1. D. P. DiVincenzo, Science **270**, 255 (1995); C. H. Bennet and D. P. DiVincenzo, *Nature* **404**, 247 (2000); M. A. Nielsen, I. L. Chuang, *Quantum Compuatation and Quantum Information*, Cambridge Univ. Press, Cambridge (2000).
2. A. D. Caldeira and A. J. Leggett, *Phys. Rev. Lett.* **46,** 211 (1981).
3. Y. Nakamura Y. A. Pashkin, J. S. Tsai, *Nature* (London) **398**, 786 (1999); I. Chiorescu, Y. Nakamura, C. J. P. M. Harmans, and J. E. Mooij, *Science* **299**, 1869 (2003); Yu. A. Pashkin, T. Yamamoto, O. Astafiev, Y. Nakamura, D. V. Averin, and J. Tsai, *Nature* **421**, 823 (2003).
4. A. R. Calderbank, and P. W. Shor, *Phys. Rev. A* **54**,1098 (1996); E. Knill, and R. Laflamme, *Phys. Rev. A* **55**, 900 (1997); A. M. Steane, *Nature* **399**, 124 (1999).
5. P. Silvestrini and L. Stodolsky, *Phys. Lett. A* **280**, 17 (2001).
6. V. Corato, P. Silvestrini, L. Stodolsky, J. Wosiek, *Phys. Lett. A* **309,** 206 (2003), *cond-mat/0205514.*
7. D. V Averin, *Solid State Commun.* **105**, 659 (1998); D. V. Averin, *Nature* **398**, 748 (1999).
8. E. Farhi, J. Goldstone, S. Gutmann, and M. Sipser, *Quant-Ph/0001106.*
9. V. Corato, P. Silvestrini, L. Stodolsky, J. Wosiek, *Phys. Rev. B* **68**, 2245081 (2003).
10. W. M. Kaminsky, and S. Lloyd in "*Quantum Computing and Quantum Bits in Mesoscopic Systems*", A. J. Leggett, B. Ruggiero, and P. Silvestrini, eds. Kluwer Academic/ Plenum Publishers (2004).
11. J. R. Friedman, V. Patel, W. Chen, S. K. Tolpygo, and J. E. Lukens, *Nature* **406**, 43 (2000); C. H. van der Wal, A. C. J. Ter Haar, F. K. Wilhem, R. N. Schouten, J. P. M. Harmaas, T. P. Orlando, S. Lloyd, and J. E. Mooij, *Science* **285**, 1036 (2000).
12. R. Rouse, S. Han, and J. E. Lukens, Phys. Rev. Lett. **75**, 1614 (1995); and references of the same authors therein reported.
13. J. Clarke, T. Hime, S. Linzen, B. L. T. Plourde, P. A. Reichardt, T. L. Robertson, K. B. Whaley, F. K. Wilhelm, C. E. Wu, and J. Zhang, *Macroscopic Quantum Coherence and Computing Workshop*, Naples Italy June 2004, Booklet Abstract p.1(www.mqc2.it).
14. C. Granata, C. Di Russo, A. Monaco, and M. Russo, IEEE Trans. On Appl. Supercond. **11**, 95 (2001).
15. B. Ruggiero, V. Corato, C. Granata, L. Longobardi, R. Rombetto, and P. Silvestrini, *Phys. Rev. B* **67**, 132504 (2003).
16. W. T. Tsang and T. Van Duzer, J. *Appl. Phys.* **46**, 4573 (1975); C.D. Tesche and J. Clarke, *J. of Low Temp. Phys.* **27**, 301 (1977).

14

Anomalous Thermal Escape in Josephson Systems Perturbed by Microwaves

N. Grønbech-Jensen[1], M. G. Castellano[2], F. Chiarello[2], M. Cirillo[3], C. Cosmelli[4], V. Merlo[3], R. Russo[3,5], and G. Torrioli[2]

[1] Department of Applied Science, University of California, Davis, California 95616, U.S.A.
[2] IFN-CNR and INFN, via Cineto Romano 42, I-00156 Rome, Italy.
[3] Department of Physics and INFM, University of Rome "Tor Vergata", I-00133 Rome, Italy.
[4] Department of Physics and INFN, University of Rome, "La Sapienza", I-00185 Rome, Italy.
[5] Istituto di Cibernetica "E. Caianiello", CNR, I-80078 Pozzuoli, Italy.

Abstract

By experiments and numerical simulations, thermal activation processes of Josephson tunnel junctions are investigated in the presence of microwave radiation. When the applied signal resonates with the Josephson plasma frequency oscillations, the switching current may become multi-valued temperature ranging both below and above the classical to quantum crossover temperature. Switching current distributions are obtained both experimentally and numerically at temperatures both near and far above the quantum crossover temperature. Plots of the switching currents traced as a function of the applied signal frequency show very good agreement with a simple anharmonic theory for Josephson resonance frequency as a function of bias current. Throughout, experimental results and direct numerical simulations of the corresponding thermally driven classical Josephson junction model show very good agreement.

14.1 Introduction

The Josephson tunnel junction is a physical system very well studied due to its simplicity and nonlinearity.[1] Statistical properties of Josephson junctions have been another subject of intense investigation through, e.g., measurements of the escape statistics from the zero-voltage state, successfully confirming consistency with the classic Kramers model for thermal activation from a potential well.[2,3] Escape measurements represent a powerful tool for probing the nature of the underlying potential well, and applying an ac field to a low-temperature system has been reported to produce anomalous switching distributions with two, or more, distinct dc bias currents for which switching is likely. These measurements have been interpreted as a signature of the ac field aiding the population of multiple quantum levels in a junction, thereby leading to enhancement of the switching probability for bias currents for which the corresponding quantum levels match the energy of the microwave photons. Work performed in this direction appeared first in the literature two decades ago.[4,5] These results have significantly attracted interest toward Josephson junction systems as possible basic elements in the field of quantum coherence and quantum computing,[6–10] and more recently other investigations have further indicated that the application of microwaves may not be the only condition under which level quantization can be observed in Josephson junctions.[11]

Within the framework of this research topic we recently reported experimental measurements conducted on a Josephson junction, operated well above the so-called quantum transition temperature T_*, and direct numerical simulations of the classical pendulum model, parameterized to mimic the experimental device.[12] It was found that multi-peaked switching distributions are not unique to the quantum regime (below $T_* = \hbar\omega_0/2\pi k_B$), and in fact, are manifested with the same features and under the same conditions in the classical regime as has been previously reported for low temperature measurements below T_*. With the present study an anharmonic theory is contributed that accurately captures the bias current values of the observed resonant peaks in the switching distributions as a function of the applied frequency of the microwave field. Agreement between the theory, direct numerical simulations, and experimental measurements for direct resonances as well as harmonic and sub-harmonic resonances at temperatures both well above and near T_* is demonstrated.

Figure 14.1 illustrates the process under investigation: In the classical one-degree-of-freedom single-particle washboard potential of the Josephson junction,[4] thermal excitations (shaded in the sketch) of energy $k_B T$ and the energy E_{ac} of forced oscillations due to microwave radiation, can cause the particle to escape from the potential well. This process can be traced by sweeping the current–voltage characteristics of the Josephson junction periodically. Escape from the potential well corresponds to an abrupt transition from the top of the Josephson-current zero-voltage state to a non-zero voltage state. The statistics of the switching events, in the absence of time-varying perturbations, have been shown to be consistent with Kramers' model[2] for thermal escape from a one-dimensional potential. Since the thermal equilibrium Kramers model does not include the effect of non-equilibrium force terms, the results of the switching events generated by the

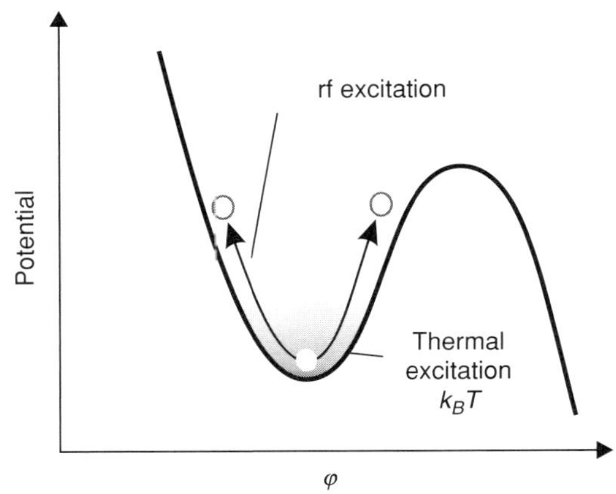

FIGURE 14.1. Sketch of the physical phenomenon under investigation: A driven oscillation energy E_{ac} superimposed onto thermal excitations, may cause a particle to escape a washboard potential.

presence of a microwave radiation on a Josephson junction can be investigated. in a thermal regime, only by a direct numerical simulation of the governing equations (RSCJ model).[1]

14.2 Theory

The RSCJ model reads,

$$\frac{\hbar C}{2e}\frac{\mathrm{d}^2\varphi}{\mathrm{d}t^2} + \frac{\hbar}{2eR}\frac{\mathrm{d}\varphi}{\mathrm{d}t} + I_c \sin\varphi = I_{dc} + I_{ac}\sin\omega_d t + N(t). \tag{14.1}$$

Here, φ is the phase difference of the quantum mechanical wave functions of the superconductors defining the Josephson junction, C is the magnitude of junction capacitance, R is the model shunting resistance, and I_c is the critical current, while I_{dc} and $I_{ac}\sin\omega_d t$ represent, respectively, the continuous and alternating bias current flowing through the junction. The term $N(t)$ represents the thermal noise-current due to the resistor R given by the thermodynamic dissipation-fluctuation relationship[13]

$$\langle N(t)\rangle = 0 \tag{14.2}$$

$$\langle N(t)N(t')\rangle = 2\frac{k_B T}{R}\delta(t-t'), \tag{14.3}$$

with T being the temperature. The symbol, $\delta(t-t')$, is the Dirac delta function. Current and time are usually normalized respectively to the Josephson critical current I_c and to ω_0^{-1}, where $\omega_0 = \sqrt{2eI_c/\hbar C}$ is the Josephson plasma frequency. With this normalization, the coefficient of the first-order phase derivative becomes the normalized dissipation $\alpha = \hbar\omega_0/2eRI_c$. It is also convenient to scale the energies to the Josephson energy $E_J = I_c\hbar/2e = I_c\Phi_0/2\pi$, where $\Phi_0 = h/2e = 2.07 \cdot 10^{-15}$ Wb is the flux-quantum. Thus, the set of Eqs (14.1)–(14.3) can be expressed in normalized form as

$$\ddot{\varphi} + \alpha\dot{\varphi} + \sin\varphi = \eta + \eta_d \sin\Omega_d\tau + n(\tau) \tag{14.4}$$

$$\langle n(\tau)\rangle = 0 \tag{14.5}$$

$$\langle n(\tau)n(\tau')\rangle = 2\alpha\theta\delta(\tau-\tau'), \tag{14.6}$$

where $\theta = \frac{k_B T}{E_J}$ is the normalized temperature. The normalized dc and ac currents are $\eta = \frac{I_{dc}}{I_c}$ and $\eta_d = \frac{I_{ac}}{I_c}$, respectively.

For small-amplitude oscillations around a stable (zero-voltage) energetic minimum we obtain the standard relationship between resonance frequency and bias current,

$$\Omega_p = \sqrt[4]{1-\eta^2}, \tag{14.7}$$

where the dissipative contribution to the resonance frequency is omitted. However, this linear resonance is not directly relevant for the dynamics leading to anomalous resonant switching. Looking at Fig. 14.1 it is obvious that a switching event will arise from probing the anharmonic region of the potential near the local energetic maximum, and we therefore must anticipate a depression of the resonance frequency at these large amplitudes. In order to quantify this notion, the following ansatz is adopted,

$$\varphi = \varphi_0 + \psi, \tag{14.8}$$

where φ_0 is a constant and ψ represents oscillatory motion. Inserting this ansatz into Eq. (14.4) (for $\theta = 0$) yields,

$$\ddot{\psi} + \sin\varphi_0 \cos\psi + \cos\varphi_0 \sin\psi = \eta + \eta_d \sin\Omega_d t - \alpha\dot{\psi}. \tag{14.9}$$

Making the single-mode assumption, $\psi = a\sin(\Omega_d t + \kappa)$, κ being some constant phase, we obtain the following

$$\sin\varphi_0 = \frac{\eta}{J_0(a)} \tag{14.10}$$

$$\Omega_{\rm res} = \sqrt{\frac{2J_1(a)}{a}\sqrt{1 - \left(\frac{\eta}{J_0(a)}\right)^2}}, \tag{14.11}$$

the functions, J_n, being the zero's order Bessel function of the first kind. Notice that $\Omega_{\rm res} \to \Omega_p$ for $a \to 0$. Since multi-peaked switching distributions must require some switching events to happen near the resonance and others to happen for larger bias currents, we can estimate that the amplitude a must be approximately given by

$$a \approx \pi - 2\sin^{-1}\frac{\eta}{J_0(a)} \tag{14.12}$$

which represents the phase distance from the energetic minimum to the saddle point. Approximating $J_0(a)$ by its Taylor expansion, we can arrive at the simple expression between the oscillation amplitude and the applied bias current η,

$$a \approx \sqrt{\frac{4}{3}[1 - \eta]}. \tag{14.13}$$

Inserting this approximate expression into Eq. (14.11) gives an explicit relationship between the anharmonic resonance and the bias current, relevant for the bias current location of the anomalous secondary peak in the switching distribution.

14.3 Experiments and Simulations

Experiments were performed on Josephson tunnel junctions fabricated according to classical Nb–NbAlOx–Nb procedures.[14] The samples had very good current–voltage characteristics and magnetic field diffraction patterns. The junctions were cooled in a ^{3} He refrigerator (Oxford Instruments Heliox system), providing temperatures down to 360mK. Microwave radiation, brought to the chip-holders by a coax cable, was coupled capacitively to the junctions, and the junction had a maximum critical current of $I_c = 143$ µA and a total capacitance of 6pF from which we estimate a plasma frequency of $\omega_0/2\pi = 42.5$ GHz. From this value of the plasma frequency the classical to quantum crossover temperature[15] $T_* = (\hbar\omega_0/2\pi k_B) = 325$ mK between classical thermal and quantum mechanical behavior can be estimated. The sweep rate of the continuous current $I_{\rm dc}$ was $i_{\rm dc} = 800$ mA/s, and we verified that the experiment was being conducted in adiabatic conditions.[11] The junction has a Josephson energy $E_J \approx 46.4 \cdot 10^{-21}$ J in the temperature range from 370 mK to 1.6 K, and effective resistance $R = 74\,\Omega$. Evaluation of the dissipation parameter was based on the hysteresis of the current–voltage characteristics of the junctions.[1] We show data for two temperatures, $T \approx 388$ mK and $T = 1.6$ K.

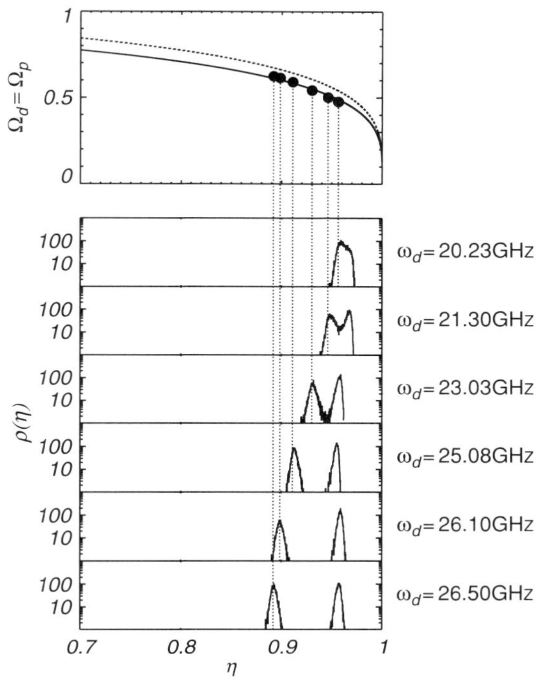

FIGURE 14.2. Experimentally obtained switching distributions, $\rho(\eta)$, for the microwave-driven junction obtained for increasing values of the drive frequency. The frequency data points in the uppermost plot are relative to the position of the secondary peak in the plots. Temperature is $T = 1.6$ K, and bias sweep rate is $\dot{I} = 800$ mA/s. Dashed curve in uppermost graph represents the linear plasma resonance of Eq. (14.7), while the solid curve represents the anharmonic resonance of Eqs. (14.11) and (14.13).

Figure 14.2 shows experimentally obtained results at $T = 1.6$ K.[12] The lower frames of the figure displays the switching distributions in bias current at different microwave frequencies. The top frame shows the relationship between the normalized current, for which the switching distributions have their resonant peak, and the applied frequency (normalized to the junction plasma frequency). Each black marker represents one of the switching distributions. Also shown in Fig. 14.2 is the linear resonance of Eq. (14.7), shown as a dashed curve, and the anharmonic resonance of Eqs. (14.11) and (14.13), shown as a solid curve. The agreement between the experimental measurements and the anharmonic theory of the classical model is near perfect for the available data points, and it is emphasize that the theoretical model of Eqs (14.11) and (14.13) has no fitted parameters to adjust in the comparison. Thus, the consistent depression of the experimental data relative to a linear resonance consideration, observed in Ref. [12], should be expected and not give rise to re-fitting the critical current or the plasma resonance frequency.

Figure 14.3 shows experimentally obtained results at $T = 388$ mK, presented in the same manner as the data in Fig. 14.2. The agreement between the experimental measurements and the anharmonic theory of the classical model is again near perfect for the available data points. The resonance curves shown in Figs 14.2 and 14.3 are identical, since we have not included the resonance dependence on the dissipation (since dissipation is very small) and since the

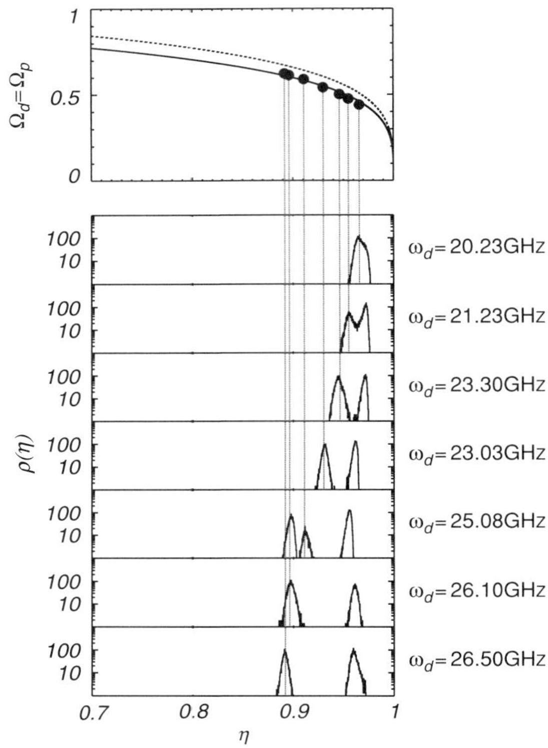

FIGURE 14.3. Same experimental situation as described in Fig. 14.2, but data is here acquired at $T = 388$ mK.

measurements indicated that plasma resonance frequency and critical current were unchanged in the investigated temperature range. By comparing figures 2 and 3, we notice that there seem to be no qualitative (and hardly any quantitative) differences between the data obtained at the two very different temperatures, even though the data of Fig. 14.2 is acquired at $T \approx 5T_*$ and the data in Fig. 14.3 represent $T \approx 1.2T_*$.

Numerical simulations of escape in a system described by Eqs (14.4)–(14.6) corresponding to the experiments with $\alpha = 0.00845, \theta = 115.4 \cdot 10^{-6}, 4.76 \cdot 10^{-4}$, and continuous bias sweep rate $\frac{d\eta}{dr} = 2.1 \cdot 10^{-8}$ have also been conducted in order to investigate the purely classical dynamics in comparison with the experimental measurements. The parameters have been chosen in agreement with the experiments discussed above. Switching distributions (each corresponding to 1,000–10,000 events), obtained for different values of the normalized drive frequency and temperature, were obtained as a function of the continuous bias, and secondary resonant peaks in the distribution were easily obtained in the classical model by adjusting the simulated microwave amplitude for a given frequency.

Figure 14.4 shows both experimental measurements and direct numerical simulations of the resonant peak location as a function of applied microwave frequency at $T = 1.6K \approx 5T_*$. Experimental data are shown as box markers and numerically obtained data are shown as circles. As in Figs. 14.2 and 14.3, dashed curves represent the linear resonance Eq. (14.7) while

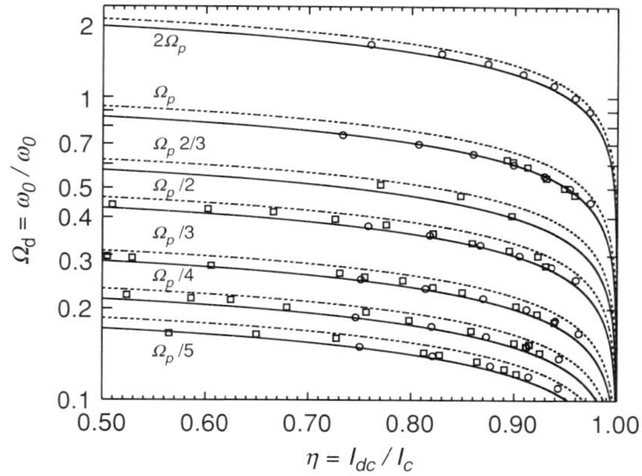

FIGURE 14.4. The functional dependencies of the driving frequency upon the location of the secondary peak in $\rho(\eta)$ obtained for subharmonic and harmonic pumping. Circles represent numerical results and squares experimental data. Parameters are as given in Fig. 14.2. Dashed curves represent the linear plasma resonance of Eq. (14.7), while the solid curves represent the anharmonic resonance of Eqs (14.11) and (14.13).

the solid curves are generated from Eqs (14.11)–(14.13). Experimental data for the fundamental resonance (labeled Ω_p) in the figure are the ones from Fig. 14.2. Close agreement between theory, experiment, and simulation is observed. Data for subharmonic resonances is also presented, and here too very close agreement between simulation and experiment is found. The theoretical harmonic and subharmonic resonance curves are the ones of Eqs (14.7), (14.11) and (14.13) multiplied with the indicated fraction in the figure. This simple theory seems to also predict the sub-harmonic microwave induced resonant peak location very well.

We finally show, in Fig. 14.5, the data similar to the ones in Fig. 14.4 taken at $T = 388$ mK $\approx 1.2T_*$. Also at this temperature we observe very close agreement between experiment, simulation, and theory, amplifying the notion that microwave induced switching and anomalous switching distributions can be understood within a classical, thermal framework.

14.4 Conclusions

In conclusion, the theory, and experiments presented here on ac-driven, thermal escape of a *classical* particle from a one-dimensional potential well have shown that resonant coupling (harmonic or subharmonic) between the applied microwaves and the plasma resonance frequency provides an enhanced opportunity for escape, and we have directly observed the signatures of such microwave-induced escape distributions in the form of anomalous multi-peaked escape statistics at two temperatures, $T = 388$ mK $\approx 1.2\ T_*$ and $T = 1.6$ K $\approx 5\ T_*$. The straightforward agreement between the classical hypothesis of anomalous distributions being directly produced by ac-induced anharmonic resonances, the results of numerical simulations of the classical pendulum model of a Josephson junction, and actual Josephson junction experiments indicate a consistent interpretation of ac-induced anomalous multi-peaked switching distributions in the classical regime of Josephson junctions.

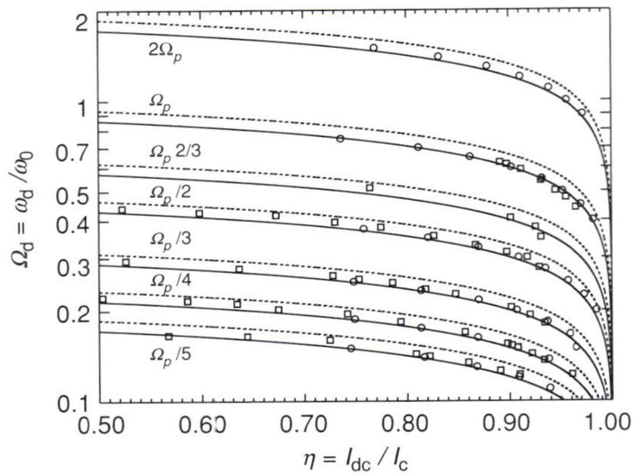

FIGURE 14.5. The functional dependencies of the driving frequency upon the location of the secondary peak as described in Fig. 14.4. The temperature is here $T = 388$ mK.

It is noted that previous experimental work on ac-induced escape distributions obtained at temperatures below T_* is consistent with the observations presented here. Those experiments have produced ac-induced peaks in the observed switching distributions, and the relevant peaks are located alongside the expected classical plasma resonance curve, which is also found here. An important observation is that the microwave-radiation frequency necessary for populating an excited quantum level ($\hbar\omega_d$) in a quantum oscillator coincides with the classical resonance frequency of the corresponding classical oscillator. Thus, the switching distributions obtained from classical and quantum mechanical oscillators may exhibit the same microwave induced multi-peak signatures, which in the classical interpretation is merely due to resonant nonlinear effects. It is evident then that multi-peaked switching distributions are not a unique signature of quantum behavior in the ac-driven Josephson junction. Similar anomalous (resonant) switching has been observed both experimentally[16] and theoretically[17] for single-fluxon behavior in long annular Josephson junctions in an external magnetic field.

Acknowledgments. This work was supported in part by the Computational Nanoscience Group, Motorola, Inc, and in part by INFN under the project SQC (Superconducting Quantum Computing). NGJ acknowledges generous hospitality during several visits to Department of Physics, University of Rome "Tor Vergata".

References

1. A. Barone and G. Paternó, *Physics and Applications of the Josephson Effect* (Wiley, New York, 1982); T. Van Duzer and C. W. Turner, *Principles of Superconductive Devices and Circuits*, 2nd ed. (Prentice-Hall, New York, 1998).
2. H. A. Kramers, *Physica* **7**, 284 (1940).
3. T. A. Fulton and L. N. Dunkelberger, *Phys. Rev. B* **9**, 4760 (1974).
4. J. M. Martinis, M. H. Devoret, and J. Clarke, *Phys. Rev. Lett.* **55**, 1543 (1985).
5. M. H. Devoret, J. M. Martinis, and J. Clarke, *Phys. Rev. Lett.* **55**, 1908 (1985).

6. B. Ruggiero, M. G. Castellano, G. Torrioli, C. Cosmelli, F. Chiarello, V. G. Palmieri, C. Granata, and P. Silvestrini, *Phys. Rev.* **B59**, **177** (1999).
7. J. M. Martinis, S. Nam, and J. Aumentado, *Phys. Rev. Lett.* **89**, 117901 (2002).
8. Y. Lu, S. Han, Xi Chu, S. Chu, Z. Wang, *Science* **296**, 889 (2002).
9. A. J. Berkley, H. Xu, M. A. Gubrud, R. C. Ramos, J. R. Anderson, C. Lobb, and F. C. Wellstood, *Physical Review* **B68**, 060502 (2003).
10. A. Wallraff, T. Duty, A. Lukashenko, and A. V. Ustinov, *Phys. Rev. Lett.* **90**, 037003 (2003).
11. P. Silvestrini, V. G. Palmieri, B. Ruggiero, and M. Russo, *Phys. Rev. Lett.* **79**, 3046 (1997).
12. N. Gronbech-Jensen, M. G. Castellano, F. Chiarello, M. Cirillo, C. Cosmelli, L. V. Filippenko, R. Russo, and G. Torrioli, *Phys. Rev. Lett.* **93**, 107002 (2004).
13. See, e.g., G. Parisi, *Statistical Field Theory* (Addison–Wesley, 1988).
14. S. Morohashi and S. Hasuo, *J. Appl. Phys.* **61**, 4835 (1987).
15. I. Affleck, Phys. *Rev. Lett.* **46**, 388 (1981).
16. A. Wallraff, A. Lukashenko, J. Lisenfeld, A. Kemp, M. V. Fistul, Y. Koval, and A. V. Ustinov, *Nature* **425**, 155 (2003).
17. N. Gronbech-Jensen and M. Cirillo, *Physical Review B* (in press, 2004) – cond-mat/0404721.

15

Realization and Characterization of a Squid Flux Qubit with a Direct Readout Scheme

F. CHIARELLO[1], P. CARELLI[2], M. G. CASTELLANO[1], C. COSMELLI[3], R. LEONI[1], F. MATTIOLI[1,4], S. POLETTO[1,4], D. SIMEONE[1,4], AND G. TORRIOLI[1]

[1] Istituto di Fotonica e Nanotecnologie – CNR, via Cineto Romano 42, 00156 Roma, and INFN, Sez. Roma;
[2] Università dell'Aquila, Monteluco di Roio, 67040 L'Aquila, and INFN, Sez. L'Aquila;
[3] Università di Roma "La Sapienza", P.le A. Moro 2, 00185 Roma, and INFN, Sez. Roma.
[4] Università di Roma 3, via della Vasca Navale 84, 00146, Roma.

15.1 Introduction

Superconducting devices based on the Josephson effect are promising candidates for the implementation of quantum computing. After the experimental demonstration of the coherent manipulation and coupling of Josephson qubits[1–8], the actual efforts concern the design and realization of new optimized elements with small decoherence, efficient readout, simple and complete coherent manipulation, integration and coupling of more qubits. In this direction we have realized a double SQUID qubit with a fully gradiometric configuration and a readout scheme based on the direct injection of current in a large junction in the SQUID loop. A preliminary experimental characterization at 4.2 K showing the capability to perform a highly efficient discrimination of the qubit flux states is presented here.

15.2 The RF Squid Qubit

An rf SQUID (radio frequency Superconducting Quantum Interference Device) consists of a superconducting loop of inductance L interrupted by a Josephson junction with critical current I_0 and capacitance C. The device can be biased by an applied magnetic flux Φ_x, and responds with a shielding current I in the loop and with a total magnetic flux$\Phi = \Phi_x - LI$. The system is described by the Hamiltonian

$$H = \frac{Q^2}{2C} + \frac{(\Phi - \Phi_x)^2}{2L} - I_0\varphi_0 \cos\left(\frac{\Phi}{\varphi_0}\right), \qquad (15.1)$$

with $\varphi_0 = \hbar/(2e)$. Charge Q on the junction capacitance is the variable conjugate to the flux Φ, so that the first term in the left side of Eq. 1 is the kinetic energy, with the capacitance C playing the role of the system mass, and the other two terms are the potential energy. When the adimensional inductance $\beta = LI_0/\varphi_0$ is larger than 1 the potential presents one or more wells. In particular, for $\Phi_x = \Phi_0/2$ the potential is symmetric with two identical wells, with plasma frequency ω, separated by a barrier of height ΔU.

The solution of the time independent Schrödinger equation for the Hamiltonian Eq (15.1) shows that for a sufficiently high barrier the first two energy levels are very close to each other with respect to the upper ones, i.e., $E_1 - E_0 \ll E_2 - E_1$. In fact the level spacing is approximately given by

$$\hbar\Delta \equiv E_1 - E_0 \cong \hbar\omega\sqrt{10.2\frac{\Delta U}{\hbar\omega}}\exp\left(-5.1\frac{\Delta U}{\hbar\omega}\right). \tag{15.2}$$

while $E_2 - E_1 \cong \hbar\omega$. For this reason, provided that no effects (temperature, noise or any non adiabatic changes) induce transitions to the upper levels, the system presents a reduced dynamics, and can be described by a base consisting of the first two eigenstates $|0\rangle$ and $|1\rangle$. Alternatively, one can consider the base formed by the distinct flux states $|L\rangle = (|0\rangle + |1\rangle)/\sqrt{2}$ and $|R\rangle = (|0\rangle - |1\rangle)/\sqrt{2}$, approximately given by gaussian curves centered on the left and right minima respectively.

This two state system can be used as a qubit, with the distinct flux states used as computational states. The qubit can be manipulated by modifying the symmetry of the potential, changing the applied flux Φ_x. For example, the system can be prepared in one of the two flux states by moving the bias flux far from the symmetry point $\Phi_0/2$, in order to have just one potential well, and then returning to the symmetric position (Fig. 15.1b). Moreover, the relative phase between the two states can be changed by applying a flux close to $\Phi_0/2$ but slightly different (Fig. 15.1c). The energy difference between the two minima is given by

$$\hbar\varepsilon = \delta\Phi\left(\Phi_x - \varphi_0/2\right)/L \tag{15.3}$$

where $\delta\Phi$ is the distance between the two minima. Then the Hamiltonian can be written in the reduced base $|L\rangle$, $|R\rangle$ by using the Pauli matrices σ_x, σ_y and σ_z

$$H = -\frac{\hbar\Delta}{2}\sigma_x - \frac{\hbar\varepsilon}{2}\sigma_z \tag{15.4}$$

With a simple rf SQUID the manipulation of the σ_z part only is possible. Let us now discuss how to improve the system in order to control the σ_x part also and have the full control on the Hamiltonian too.

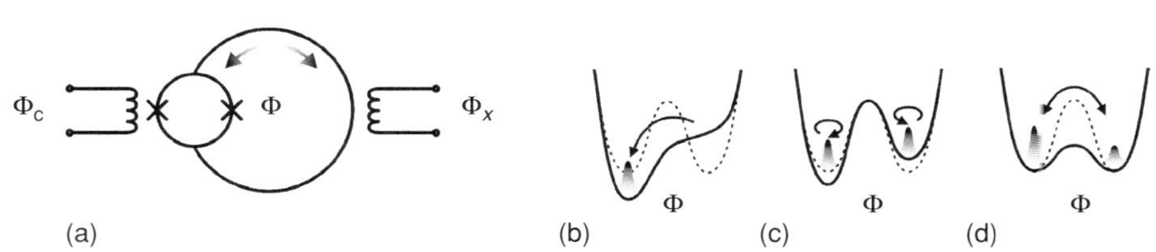

FIGURE 15.1. a. Scheme of the double SQUID qubit, with the two control fluxes Φ_x and Φ_c. b. Qubit potential $U(\Phi)$ *vs.* the flux Φ in the loop, and example of state preparation obtained by a strong asymmetry. c. Phase rotation, obtained by a slight potential asymmetry. d. Rotation between states, obtained by reducing the barrier height.

15.3 The Double Squid Qubit

The single Josephson junction in the rf SQUID can be replaced by a small dc SQUID[9], a superconducting loop of inductance L_{dc} interrupted by two identical junctions of currents i_0 and capacitances c_0, biased by an applied magnetic flux Φ_c. For $i_0 L_{dc}/\varphi_0 \ll 1$ this device behaves essentially like a single junction with capacitance $C = 2c_0$ and critical current modulated by the applied flux according to

$$I_0\left(\Phi_c\right) \simeq 2i_0\left|\cos\left(\frac{\Phi_c}{2\varphi_0}\right)\right| \tag{15.5}$$

The obtained SQUID, here indicated as double SQUID, behaves approximately like a simple rf SQUID but now with controllable $\beta\left(\Phi_c\right)$. In this way it is possible to control the level spacing $\Delta\left(\Phi_c\right)$ in order to have the complete control on the Hamiltonian in Eq (15.4) and the possibility to arbitrarily manipulate the qubit state[10]. For example, the potential can be maintained to be symmetric and the barrier very high in order to freeze the qubit state; by reducing the barrier for a desired time one can cause the coherent rotation between the two states (Fig. 15.1d). This kind of manipulations requires the capability to change the flux biases nonadiabatically with respect to Δ and ε, but adiabatically with respect to ω in order to remain in the reduced dynamics approximation. This task can be extremely difficult with standard techniques, but in principle can be simply achieved by using RSFQ logic in order to control the qubit[11,12]. In alternative, one can use microwave pulses in order to manipulate the qubit state, using static control fluxes just to obtain the correct working point.

15.4 Gradiometric Configuration

The SQUID can be designed using a gradiometric configuration in order to reduce the effective area of the loop, and consequently decreasing the pick-up of external magnetic flux. This reduction works for radiation (and static fields) that are uniform within the SQUID dimensions and that present wavelength longer than this dimensions, and it is limited by the capability to realize really symmetric loops. In our case we expect a reduction by a factor one hundred of the effective area for frequencies below the THz. This limit, calculated by considering devices with dimensions smaller than 100 μm, is higher than the typical SQUID frequencies.

For the larger loops of our SQUID we have chosen the gradiometric configuration sketched in Fig. 15.2a, consisting two identical loops of inductance L_0 in parallel to the junction, with a total effective inductance $L = L_0/2$. This design presents a second advantage: the device is surrounded by an uninterrupted superconducting loop, and if a single flux quantum (or an odd

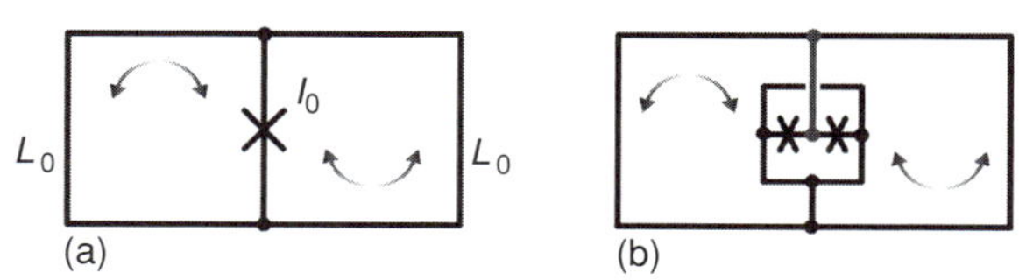

FIGURE 15.2. a. Gradiometric configuration of the rf SQUID. b. Double gradiometric configuration of the double SQUID.

number of flux quanta) is trapped within this loop during the cooling, the SQUID is automatically and permanently biased at $\Phi_0/2$, with a strong simplification of the problems related to the bias in the correct working point.

The small dc SQUID used as a tunable junction is also designed with a gradiometric configuration and the two gradiometer are along orthogonal directions, in order to reduce spurious couplings (Fig. 15.2b).

15.5 Direct Readout Scheme

The qubit flux state can be detected by an inductively coupled detector such as a SQUID magnetometer, or better, a hysteretic dc SQUID used as a discriminator that ensures a small noise contribution because of the absence of shunting resistors. In this paper we propose a different one-shot readout scheme, based on the injection of current pulses in a junction larger than the SQUID junctions (indicated with LJ, with critical current I_{0L}) and directly inserted in the flux qubit loop as in Fig. 15.3(a). This scheme, successfully used a different type of qubit (the so called "Quantronium")[4], presents the advantages to have a very simple design, to make simpler the readout operations, and to allow an extremely high efficiency in discriminating the qubit states. On the opposite, this scheme is extremely invasive, in the sense that the qubit potential is strongly affected by the readout activation, and an accurate qubit design is required in order to preserve the information to be read. We indicate this scheme with the generic expression "direct readout" in contrast with the "external, indirect readout" adopted elsewhere.

Let us describe the working principles of this system: a current I_p is externally applied across the large junction LJ, and it is added to the current circulating in the qubit loop I_q that depends on the qubit state (for simplicity we consider the non gradiometric case), so that the current effectively crossing the large junction is $I = I_p - I_q$. If this current is well below the large junction critical current I_{0L} the system remains in the superconducting state, otherwise there is a transition to the voltage state and consequently a voltage signal across the terminals. Therefore, to perform a single shot readout it is sufficient to send a current pulse of amplitude $I_p \sim I_{0L}$, and since the current I_q can be positive or negative according to the qubit right/left states, there will

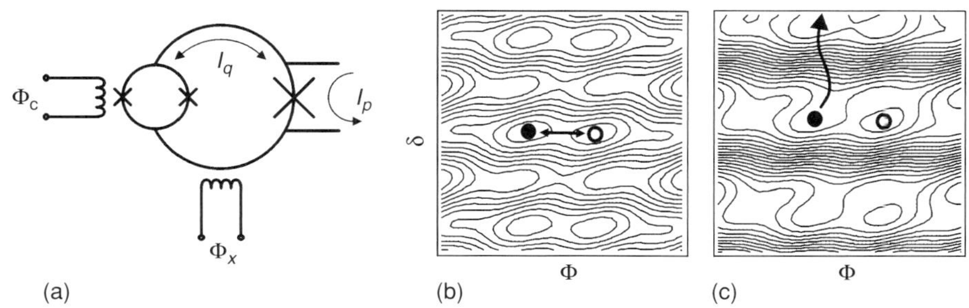

FIGURE 15.3. a. Direct readout of the flux qubit state by injecting a current I_p in the large junction LJ inserted in the loop. b. Contour plot of the qubit potential for $I_p = 0$ and $\Phi_x = \Phi_0/2$, with the indicated two wells and the corresponding flux states. c. Contour plot of the qubit potential for I_p equal to the large junction critical current and $\Phi_x = \Phi_0/2$. The right state remains in the minimum, while the left one jump to the voltage state (the potential parameters are chosen in order to highlight the effect).

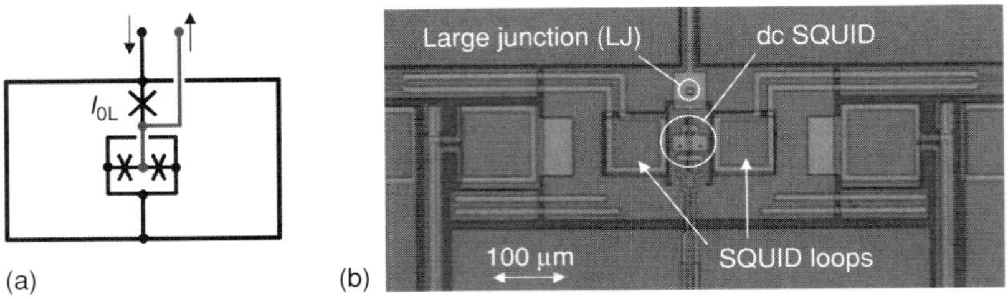

FIGURE 15.4. a. Scheme of the Direct Readout Flux Qubit (DRFQ). b. Photo of the device.

be the discrimination by observing the presence (right state) or absence (left state) of an output voltage pulse.

The complete description of this system requires the study of the full Hamiltonian, which has the following potential

$$U = \frac{(\Phi - \Phi_x)^2}{2L} - I_0\varphi_0 \cos\left(\frac{\Phi}{\varphi_0} + \delta\right) - I_p\varphi_0\delta - I_{0L}\varphi_0 \cos(\delta). \tag{15.6}$$

δ is the phase difference across the large junction LJ (Figs. 15.3b and 15.3c). Remembering that we are in the limit $I_{0L} \gg I_0$ (LJ very large), for I_p well below I_{0L} the phase δ is essentially frozen at the value $\delta = \arcsin(I_p/I_{0L})$, and the SQUID behaves like a simple rf SQUID with an additional flux bias $\Phi_{\mathrm{eff}} = \varphi_0 \arcsin\left(I_p/I_{0L}\right)$, so that the SQUID appears biased by an effective input flux $\Phi_x + \Phi_{eff}$ and respond with an effective output flux $\Phi + \Phi_{eff}$. For I_p just below I_{0L} there could either be a jump or not to the voltage state, depending on the qubit flux state, according to the qualitative description given before. The system is moved away from the correct bias point by the additional flux Φ_{eff}; to avoid the destruction of the information to be read due to this spurious bias, one can maintain the barrier high enough to prevent jumps between the states during the readout as in Fig. 15.3(c). Alternatively, one can constantly apply a dc current to the large junction, small enough to avoid the voltage transition, and add just a small readout current pulse, in order to minimize the deviation from the correct flux bias.

This direct readout scheme can be applied very easily to the fully gradiometric double SQUID described in Fig. 15.2, obtaining the device described in Fig. 15.4 and indicated as DRFQ (Direct Readout Flux Qubit).

15.6 Experimental Characterization

We have designed and realized a set of chips, microfabricated using Nb/AlOx/Nb trilayer technology, each containing a DRFQ (Fig. 15.4b) inductively coupled to a standard SQUID magnetometer and to a dc SQUID hysteretic discriminator[13], used for characterization purposes. Typical parameters of the DRFQ are $L = 85$ pH, $c_0 = 0.5$ pF, $i_0 = 2\ \mu$A, $L_{dc} = 10$ pH, $I_{0L} = 60\ \mu$A. A first characterization at $T = 4.2$ K has been performed in order to test and study the main expected features.

The current–voltage characteristic of the DRFQ, observed from the large junction terminals, is very similar to the characteristic of a simple Josephson junction, but now the switching current corresponding to the jump to the voltage state is modulated by the fluxes Φ_x and Φ_c. To measure

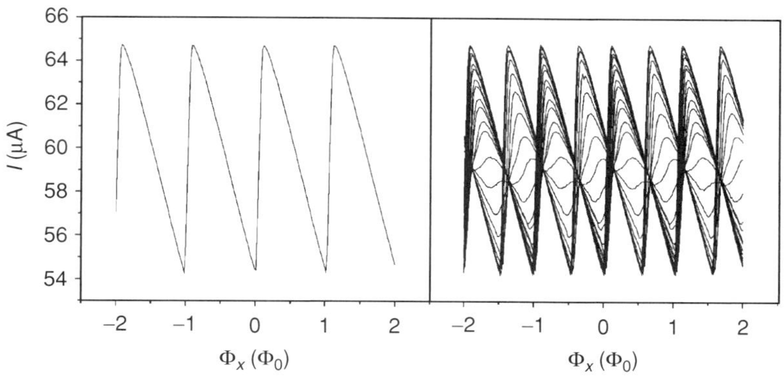

FIGURE 15.5. a. Critical current of the DRFQ measured for different fluxes Φ_x at a fixed flux Φ_c. b. Collection of different experimental curves obtained for different Φ_c.

this critical value, a current ramp is applied to LJ and the value for which the voltage transition occurs is recorded. For each fixed couple of bias fluxes Φ_c and Φ_x this is repeated for typically 1000 times, in order to estimate the mean value I of the critical current. Because of thermal activation, this average value will be slightly below the real critical current I_c; where necessary it is possible to extrapolate the correct value I_c from the distribution of the switching currents. In Fig. 15.5a is plotted the average current I experimentally measured for different values of Φ_x and for a fixed flux Φ_c. In Fig. 15.5b are plotted different curves of this kind obtained for different Φ_c, corresponding to different

$$\beta = LI_0\left(\Phi_C\right)/\varphi_0 \tag{15.7}$$

smaller β corresponds to smaller and more symmetric curves (potential with one single well), while large values are related to larger and more asymmetric ones (two or more wells in the potential, with barrier width and distance between minima increasing with β). The steepest side of the largest curves correspond to a jump between two distinct potential wells.

To test the one-shot readout capabilities of our system we send a sequence of current pulses (with typical pulse duration $\Delta t = 2\ \mu$s, repetition rate $f = 20$ kHz and amplitude $I_p = 60\ \mu$A, close to the large junction critical current) and record the voltage response to each pulse, that can be 0 or 2.7 mV depending on the rf SQUID flux state. In this way we are using the large junction to discriminate the SQUID flux state. For each couple of fluxes Φ_x and Φ_c it is acquired a sequence of $N = 100/1000$ voltage responses in order to estimate the switching probability, corresponding to the probability that the flux value exceeds the threshold (that is fixed by the I_p value). In Fig. 15.6 is plotted the experimental switching probability curve for different applied flux Φ_x and a fixed Φ_c that corresponds to the largest β.

The flat regions a and c of Fig. 15.6 correspond to a flux state clearly on the left or on the right of the detector threshold, and so to a flux state clearly discriminated. The less steep region b corresponds to a state moving in the "gray zone" of our discriminator. The steepest region d is related to a jump between two distinct flux states.

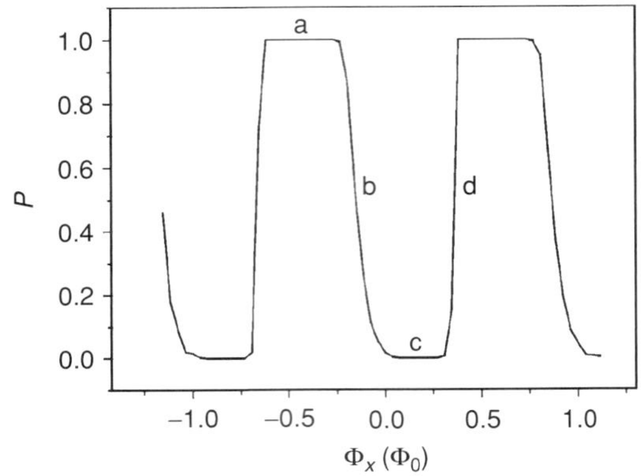

FIGURE 15.6. Switching probability obtained by applying a sequence of current pulses to the DRFQ.

15.7 Conclusions

We have proposed and realized a possible flux qubit based on a Niobium fully gradiometric double SQUID, with a direct readout scheme based on the insertion in the loop of a large Josephson junction. The first characterization performed at 4.2 K shows the capability of our device to discriminate the flux states in a very efficient way using a very simple scheme. Further work is required to study and optimize the effects of the readout on the flux state, together with a low temperature characterization, and finally, a study of the coherent behavior of this system.

This work is partially supported by INFN under the project SQC (Superconducting Quantum Computing).

References

1. J. R. Friedman, V. Patel, W. Chen, S. K. Tolpygo, and J. E. Lukens, *Nature* **406**, 43 (2000).
2. J. M. Martinis, S. Nam, and J. Aumentado, *Phys. rev. Lett.* **89**, 518 (2002).
3. Y. Yu, S. Han, X. Chu, S. Chu, and Z. Wang, *Science* **296**, 889 (2002).
4. D. Vion, A. Aassime, A. Cottet, P. Joyez, H. Pothier, C. Urbina, D. Esteve, and M. H. Devoret, *Science* **296**, 886 (2002).
5. I. Chiorescu, Y. Nakamura, C. J. P. Ma. Harmans, and J. E. Mooij, *Science* **299**, 1869 (2003).
6. Y. A. Pashkin, T. Yamamoto, O. Astafiev, Y. Nakamura, D. V. Averin, and J. S. Tsai, *Nature* **423**, 823 (2003).
7. T. Yamamoto, Y. A. Pashkin, O. Astafiev, Y. Nakamura, and J. S. Tsai, *Nature* **425**, 941 (2003).
8. C. Cosmelli, P. Carelli, M. G. Castellano, F. Chiarello, G. Diambrini Palazzi, R. Leoni, G. Torrioli, *Superconductor Science and Technology* **16**, 1337 (2003).
9. S. Han, J. Lapointe, and J. E. Lukens, *Phys. Rev. Lett.* **63**, 1712 (1989).
10. M. G. Castellano, *Fortschr. Phys.* **51**, 288 (2003).
11. R. C. Rey-de-Castro, M. F. Bocko, A. M. Herr, C. A. Mancini, M. J. Feldman, *IEEE Trans. Appl. Sup.* **11**, 1014 (2001).
12. V. K. Semenov and D. V. Averin, *IEEE Trans. Appl. Sup.* **13**, 960 (2003).
13. C. Cosmelli, F. Sciamanna, M. G. Castellano, R. Leoni , G. Torrioli, P. Carelli, F. Chiarello, *Appl. Phys. Lett.* **80**,3150 (2002)

16

A critique of the Two Level Approximation

T. Hakioğlu[1], Kerim Savran[1] and E. Meşe[2]

[1] Department of Physics, Bilkent University, Bilkent, 06800 Ankara, Turkey
[2] Department of Physics, Dicle University, Diyarbakır, Turkey

Abstract

The conditions in favor and necessity of a realistic multileveled description of a decohering quantum system is examined. Under these conditions approximate techniques to simplify a multileveled system by its first two levels is unreliable and a realistic multilevel description in the formulation of decoherence is unavoidable. In this regard, the first crucial observation is that, the validity of the two level approximation of a multileveled system is not controlled purely by *sufficiently low temperatures*. That the type of system-environment coupling and the environmental spectrum have a dominant role over the temperature is demonstrated. Particularly, zero temperature quantum fluctuations induced by the Caldeira–Leggett type linear coordinate coupling can be influential in a wide energy range in the systems allowed transitions. The second crucial observation is that the decoherence times being among the system's short time scales are found to be dominated not by the resonant but *non-resonant* processes.

16.1 Introduction

The interaction between the quantum system and the environment (which is known to be the greatest source of decoherence) is analyzed under such approximated two level systems, and models like spin-boson[1,2] and central-spin[3] models have been used widely. However, an exact justification of the widely used 2 Level Approximation (2LA) has not been made yet. There are also studies which focus on such concepts like leakage rates[4] or zero temperature fluctuations[5] that shows the doubts on the validity of the 2LA.

The validity of the 2LA is examined under the conditions which are believed to justify the approximation. In order to do this, a master equation approach is used to a generic quantum mechanical system, which is in interaction with an environment, usually named as the noise. The coupling we use is a linear coordinate type coupling which is first introduced by Caldeira and Leggett[6] and used widely in most of the realistic interaction approaches. Using the time evolution of the reduced master equation, the decoherence time scales of the system under certain conditions is observed. Later, the photon number corrections and photon number fluctuations on the environment due to the interaction with the system were calculated , using the Green's function techniques. For various system and environment parameters, it was observed that the shape of the spectral function defining the environment does not have a significant effect on decoherence, as believed. Also the dominating mechanism on decoherence is not the resonant transitions induced by environment, but rather the non-resonant (virtual) transitions.

In Section 16.2 the generic quantum mechanical system, the different spectral models and the interaction used are described, and the master equation formalism is developed. In Section 16.3 thedensity matrix solutions are examined, and in Section 16.4 the Green's function formalism used in order to calculate the photon number corrections in the environment due to interaction is developed.

16.2 System and Environment

In order to check for the validity of the 2LA, a generic Hamiltonian is used which defines a 3 Level System (3LS), an environmental bosonic bath and the interaction between them is represented by

$$\begin{aligned}\mathcal{H} = & \sum_{n=1}^{N} E_n \, |n\rangle\langle n| + \int_{\omega} d\omega \, \omega \, (a_{\omega}^{\dagger} a_{\omega} + a_{\omega} a_{\omega}^{\dagger}) \\ & + \sum_{n,r=1}^{N} \varphi_{n\,r} |n\rangle\langle r| \int d\omega \, \eta_{\omega} (a_{\omega} + a_{\omega}^{\dagger})\end{aligned} \tag{16.1}$$

where the energies E_n may be defined as desired. Since the 2LA assumes a qubit subspace well separated from the higher levels, during our calculations, we chose $E_1 = E_2 = 0.5$ and $E_3 = 2.5$ which are scaled with an absolute energy scale. The states $|n\rangle$ are the eigenvectors of the system, the matrix $\varphi_{n,r}$ defines the couplings between the levels n and r, and η_{ω} defines the spectral distribution of the environmental noise bath. During most of the calculations, the coupling matrix is taken as real, symmetric and parity conserving such as $\varphi_{1,2} = \varphi_{2,3} = 0.1$ and $\varphi_{1,3} = 0$. However, in order to perform pure two Level System (2LS) calculations we also set $\varphi_{2,3} = 0$. During these calculations, all of the energy and time scales are scaled with the same energy (time) scale that the eigenenergies are scaled.

In order to examine the effect of spectral distributions on the decoherence mechanism, we used three different environmental spectral functions which are mainly described as $I(\omega) = \eta_{\omega}^2(2n_{\omega} + 1)$. Since we want to check the validity of the 2LA, we also set the environmental temperature as $T = 0$, since 2LA is also based on the assumption $\Delta E \gg T$ where ΔE is the energy spacing between the qubit subspace and higher levels. As we set $T = 0$, the bare distribution $n_{\omega} = 0$, so $I(\omega) = \eta_{\omega}^2$. The three different spectral functions used are rectangular spectrum

$$I_{\text{rec}}(\omega) = \frac{A}{2\epsilon} \Theta(\epsilon^2 - (\omega - \omega_0)^2) \tag{16.2}$$

the Lorentzian spectrum

$$I_{\text{Lor}}(\omega) = \frac{A}{\pi} \frac{\epsilon}{(\omega - \omega_0)^2 + \epsilon^2} \tag{16.3}$$

and the power-Gaussian spectrum which is very similar to the spectrum used originally by Caldeira–Leggett[6]

$$I(\omega) = \mathcal{A}\left(\frac{\omega}{\Lambda}\right)^{1+\nu} e^{-\omega^2/4\Lambda^2}. \tag{16.4}$$

In the first two spectra, there are two spectral parameters, ϵ which defines the spectral width, and ω_0 which defines the central position of the spectrum. Both of the spectra are scaled by the total area under the spectral function $A = \int_{-\infty}^{\infty} d\omega I(\omega)$. However, in the last spectrum, the spectral width is defined with Λ, and the parameter ν in the power contribution to spectrum, defines the characteristic of the spectrum. The values of $\nu < 0$, $\nu = 0$, $\nu > 0$ define the sub-Ohmic, Ohmic and super-Ohmic spectral characteristics respectively. The constant $\mathcal{A}$ defines both the coupling to the system and the normalization of the spectral area.

The three relevant decoherence time scales, Relaxation, Dephasing and Leakage (RDL) rates are defined as $\rho_{22}^{(S)}(t)$, $|\rho_{12}^{(S)}(t)|$ and $1 - [\rho_{11}^{(S)}(t) + \rho_{22}^{(S)}(t)]$ respectively. The master equation describes the time evolution of the reduced density matrix as

$$\frac{d}{dt}\tilde{\rho}_{nm}^{(S)}(t) = -\int_0^t dt' \sum_{r,s} K_{rs}^{nm}(t,t')\,\tilde{\rho}_{rs}^{(S)}(t') \tag{16.5}$$

where

$$\begin{aligned} K_{rs}^{nm}(t,t') &= \{\mathcal{F}(t-t')\big[(\tilde{\varphi}_t\tilde{\varphi}_{t'})_{n\,r}\delta_{s,m} - (\tilde{\varphi}_{t'})_{n\,r}(\tilde{\varphi}_t)_{s\,m}\big] \\ &\quad + \mathcal{F}^*(t-t')\big[(\tilde{\varphi}_{t'}\tilde{\varphi}_t)_{m\,s}\delta_{r,n} - (\tilde{\varphi}_t)_{n\,r}(\tilde{\varphi}_{t'})_{s\,m}\big]\} \end{aligned} \tag{16.6}$$

is the non-Markovian system-noise kernel. Here $(\tilde{\varphi}_t)_{n\,r} = (\tilde{\varphi}_0)_{n\,r}\, e^{-i(E_n - E_r)t}$. The $\tilde{}$ here is used to denote the interaction picture. During these calculations, Born–Oppenheimer approximation[7] is used, and the relevant time rates are calculated at very short times. The initial state preparation throughout the calculations is made as $a|0\rangle + b|1\rangle$ and the constants are set as $a = \sqrt{0.1}$, $b = e^{i\pi/2}\sqrt{0.9}$.

16.3 Density Matrix Solutions

16.3.1 The effect of spectral distribution on Decoherence

First of all, the rectangular and Lorentzian spectra were compared with the same spectral area A and varying spectral center ω_0. For a 3LS we calculated the RDL rates with the short time behavior of the reduced density matrix. It was observed that the shape of the spectrum doesn't affect the time rates, so that both spectral functions give identical results for the same spectral area A. Also as it can be seen from Fig. 16.1 that the location of the spectrum also has a negligible effect on the rates. The results are invariant under the spectral translations. The Fig. 16.1 is plotted for rectangular spectrum.

Later, the effect of spectral area on the RDL rates was checked. As the results were again identical for the spectral functions Eqs. (16.2) and (16.3) we just show the results for the rectangular spectrum in Fig. 16.2. The effect of inclusion of higher levels to the calculations in Fig. 16.2 is examined as the rates for 2LS, 3LS and 10LS are calculated. During the calculations for the 10LS, the coupling matrix is defined as $\varphi_{n,r} = 0.1$ for $n + r = odd$ and 0 otherwise. As it can be seen from Fig. 16.2 the rates have a very strong correlations with the total spectral area, unlike

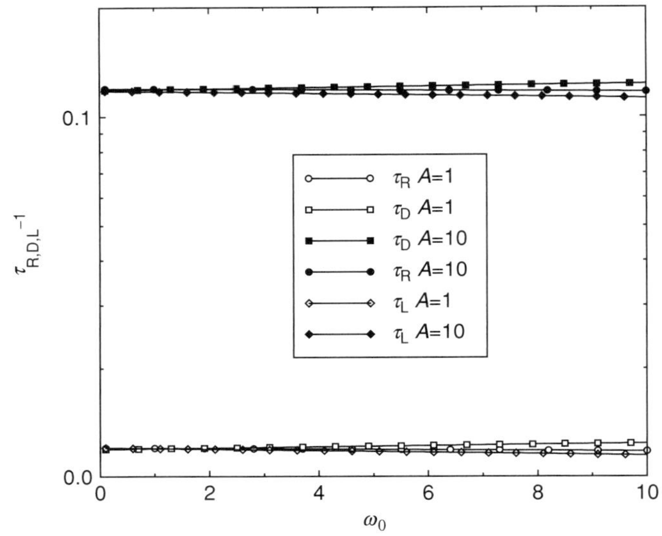

FIGURE 16.1. Dependence of the RDL rates on the spectral shifts for various spectral areas. It should be noted that the vertical axis is logarithmic.

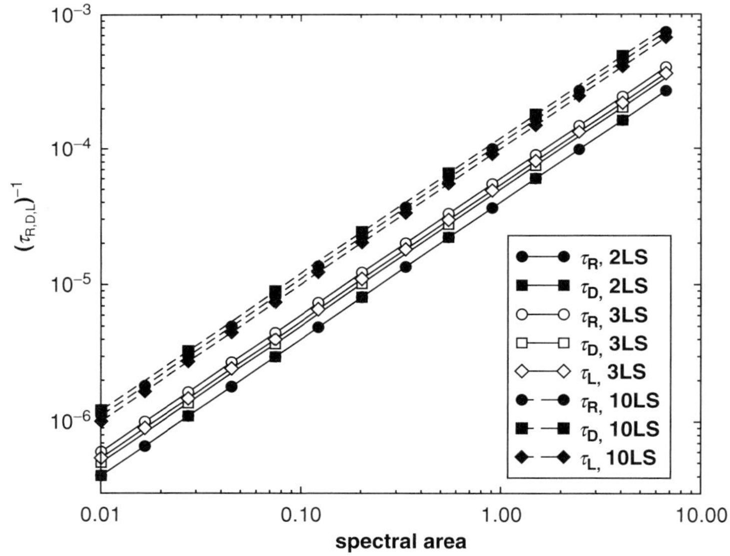

FIGURE 16.2. Dependence of the RDL rates on the spectral area for the spectral function in Eq. (16.2). Note that the both axes are logarithmic. In generating the ten level results we assumed that the system's energies above the first two levels are harmonic and given by $E_n = (n - 1/2)$, $3 \leq n$ and all the couplings are $\varphi_{n\,r} = 0.1$ for $n + r = odd$ and zero otherwise.

the spectral shape or spectral center. The slope on the log-log scale is calculated as 1 which shows a linear correlation with the area.

Lastly, in order to make an analysis about the effect of the spectral characteristic on the RDL rates, the spectral function in Eq. (16.4) is used, and for different ν parameters, the rates for varying spectral area are calculated. As it can be seen from Fig. 16.3 the behavior is again the same with varying spectral area. The dephasing and leakage rates show very simiiliar behavior as in Fig. 16.2 so only relaxation rates are shown. Interestingly, the rates turned out to be identical for different ν paramaters but same spectral areas. Opposed to the earlier belief that the spectral characteristic has influence on decoherence rates, we observed that again, the area was the dominant parameter effecting the RDL rates. However for same Λ parameter, the spectral areas were different for different ν parameters, which may be the reason for the belief that ν was effective. Another interesting point is that for same spectral area ranges for the spectral functions in Eqs. (16.2),(16.3) and (16.4) the rates were again turned out to be identical.

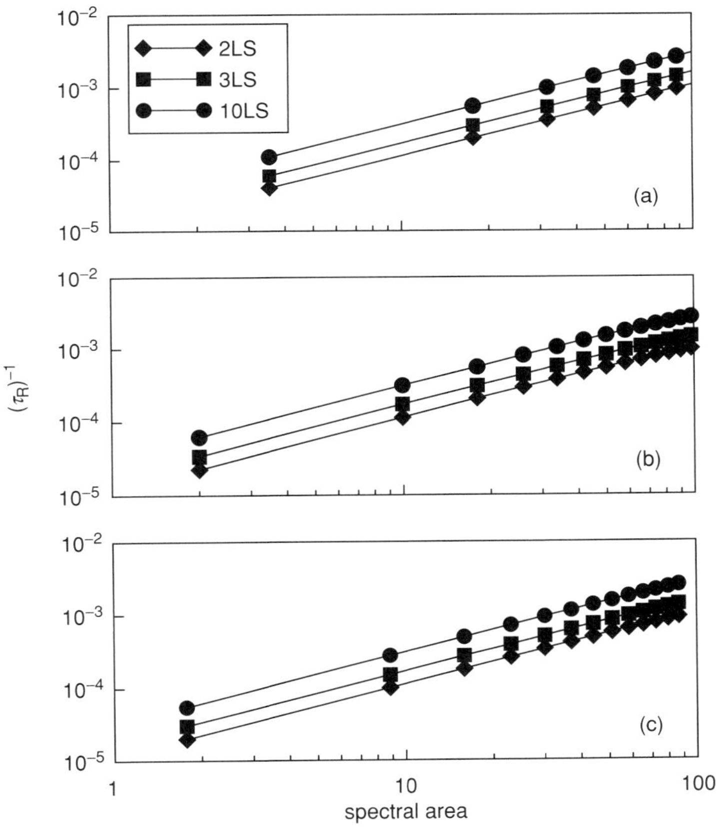

FIGURE 16.3. Dependence of the relaxation rates on the spectral area for the spectral function in Eq. (16.4) corresponding to (a) $\nu = 1$ super-Ohmic, (b) $\nu = 0$ Ohmic, (c) $\nu = -1$ sub-Ohmic spectra and for two, three and ten level systems. The trend for the dephasing and the leakage rates are very similar to Fig. 16.2 and they are not shown here. Note that the both axes are logarithmic. The energies E_n and couplings $\varphi_{n\,r}$ of the multilevel system are as indicated in Fig. 16.2. The horizontal range in (c) is common to all graphs.

The strong dependance on the spectral area is an evidence for the effect of non-resonant transitions in decoherence mechanism. As the spectral area increases, the available noise modes increase, whether resonant or not (indeed non-resonant dominantly) and rates increase accordingly. As the number of high energy modes available in the environment increase, the non-resonant transitions occur in rather shorter time scales, which is crucial considering that the decoherence is a short time effect. Another important point is that, the leakage rates to higher levels seemed to be in the same order of magnitude as RD rates, which makes the leakage a non-negligible short time effect. Even at $T = 0$ and well seperated higher levels, leakage to higher levels occur inevitably, and in short time scales too.

16.3.2 The effect of number of levels on Decoherence

The influence of the multilevels is examined by using a model coupling matrix φ in Eq. (16.1) as

$$\varphi_{nr} = 0.1\, e^{-|n-r|/\Delta} \tag{16.7}$$

for $n + r = odd$ and zero otherwise. By varying the multilevel coupling range Δ, the effect of the actively coupled levels on the RDL rates can be examined. In Fig. 16.4 the data is generated for different values of the multilevel range $\Delta = 8, 40$. Since the linear dependance on the area is previously shown, change in the spectral parameters is unnecessary in this subsection. The important observation here is that the multilevel coupling range is observed to give rise to a saturation in the RDL rates at the onset $M \simeq \Delta$. Fig. 16.4 therefore demonstrates further evidence of the observable effects in the decoherence rates arising from the finite coupling to multilevels. So as long as there is finite couplings between qubit subspace and higher levels, it is not possible to neglect the existence of higher levels.

16.4 The Photon Number Calculations

A characteristic feature of the coordinate coupling is to produce weak vacuum fluctuations as well as finite number of environmental modes at zero temperature. These corrections, can be calculated perturbatively and they give a clue about the effectiveness of the virtual (non-resonant) processes.

The corrections to environmental number of modes can be calculated using the retarded Greens function

$$G_\omega(t-t') = -i\Theta(t-t')\,\langle \mathcal{T}\tilde{a}^\dagger_\omega(t)\tilde{a}_\omega(t')\tilde{\mathcal{S}}(\infty,-\infty)\rangle \tag{16.8}$$

where $\mathcal{T}$ denotes the time ordering, the $\tilde{}$ denotes the interaction picture and the $\mathcal{S}$ matrix includes the interaction Hamiltonian in Eq. (16.1) in the interaction picture. The environmental *photon number* is than found by the standard method $n_\omega = i \lim_{t'\to t} G_\omega(t-t')$. The average is taken over the noninteracting initial state including system and the environment in the product form

$$|\rangle = (a|1\rangle + b|2\rangle) \otimes |\text{env}\rangle. \tag{16.9}$$

where $|\text{env}\rangle$ is characterized by the spectral function $I(\omega)$ at zero temperature. Evaluating the time ordered integrals arising from the expansion of the $\mathcal{S}$ matrix, and taking the limit $t' \to t$ we obtain a second order correction as

$$n^{(2)}_\omega = \sum_{s=1}^{3} \frac{2I(\omega)}{\lambda_s^2}\left(|a|^2\varphi_{0s} + |b|^2\varphi_{1s} + (a^*b + b^*a)\varphi_{1s}\varphi_{0s}\right) \tag{16.10}$$

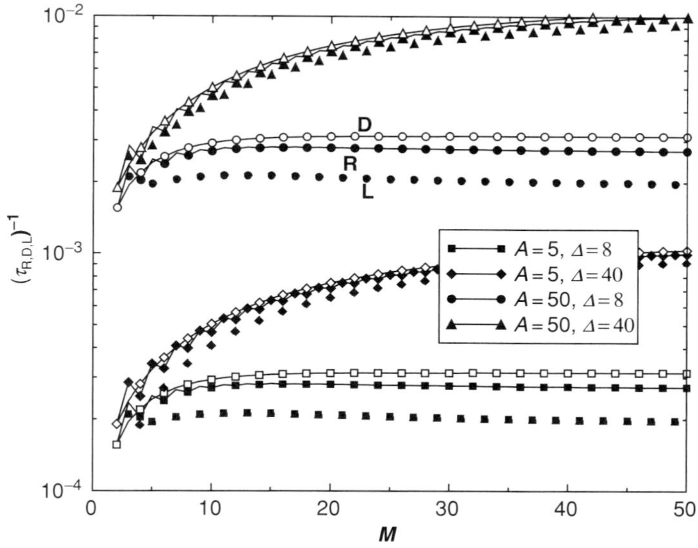

FIGURE 16.4. RDL rates at $T = 0$ against the number of levels for different multilevel coupling ranges Δ. Here the coupling matrix elements between the levels n and r are modeled by $\varphi_{n\,r} = 0.1\, e^{-|n-r|/\Delta}$ for $n + r = odd$ and zero otherwise. Note the logarithmic vertical axis. The open and solid symbols superimposed on the solid line refer to dephasing and relaxation rates respectively. The pure symbols without connecting lines refer to leakage rates. Also Δ values are 8 for squares, 40 for circles. Here, the Lorentzian spectrum is used.

where $\lambda_s = \omega + E_1 - E_s$ and and $E_1 = E_2$ as announced in Eq. (16.1). Equation (16.10) yields a divergent result for the total number of photons even for arbitrarily small system-environment couplings. The divergence is obtained independently from the type of the spectrum in Eqs. (16.2) or (16.3) and arises from the second order singularity in Eq. (16.10). To avoid this unphysical result in the second order, the second order correction is converted into an RPA sum following conventional practices in perturbative approaches.[8,9] As we use RPA, the second order pole in Eq. (16.10) is split into two first order poles which yields a finite result at the RPA level. The relevant RPA graphs are depicted in Fig. 16.5 below.

Next, we calculate the fluctuations in the number of photons Δn_ω by using a four point retarded Green's function for the environment as

$$D_\omega(t - t') = -i\Theta(t - t') \langle T\tilde{a}_k^\dagger(t)\tilde{a}_\omega(t)\tilde{a}_\omega^\dagger(t')\tilde{a}_k(t')\tilde{S}(\infty, -\infty)\rangle \tag{16.11}$$

The connected part of Eq. (16.11) directly yields the fluctuations in the photon number as $\Delta n_\omega^2 = i \lim_{t' \to t} D_\omega(t - t')$. The diagrams corresponding to the RPA scheme using Eq. (16.11) are shown in Fig. 16.5. The fluctuation in the total photon number is found by $(\Delta N)^2 = \int \mathrm{d}\omega\, \Delta n_\omega^2$ and is given by

$$(\Delta N)^2 = \sum_s \int_{-\infty}^{\infty} \frac{\mathrm{d}\omega}{(\omega + \lambda_s)} \frac{1}{1 - 4(\text{system})_s I(\omega)/(\omega + \lambda_s)} \tag{16.12}$$

which is plotted in Fig. 16.6 with respect to ϵ and parameterized by ω_0 for the two, three and ten level systems. The bare spectral area is fixed at $A = 5\pi$. The observed independence from ϵ

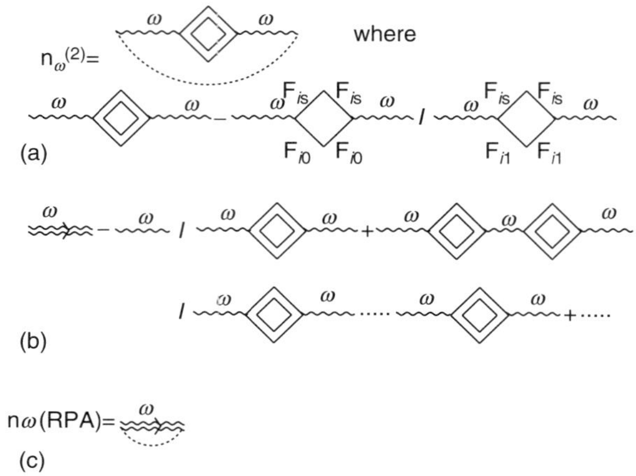

FIGURE 16.5. Contributing diagrams in the calculation of Eq. (16.11). (a) The definition of the basic second order root diagrams that are used to implement RPA. The dashed arc lines indicate the retarded limit $t' \to t + 0^-$ in the calculation of $n_\omega^{(2)}$, (b) the RPA hierarchy using the root diagram in (a), (c) the resulting RPA Green's function used in the calculation of Eq. (16.12). Here the dashed arc lines indicate again the retarded limit $t' \to t + 0^-$ in the calculation of $n_\omega^{(\mathrm{RPA})}$.

at fixed spectral area implies independence from the spectral shape. Additionally independence from ω_0 is also shown.

In Fig. 16.7 below, the ratio $\Delta N/N$ is plotted as a function of the spectral area. The figure, together with the inlet for ΔN versus spectral area, is well fitted to

$$\frac{\Delta N}{N} \simeq \frac{\text{Const.}}{\sqrt{A}} \tag{16.13}$$

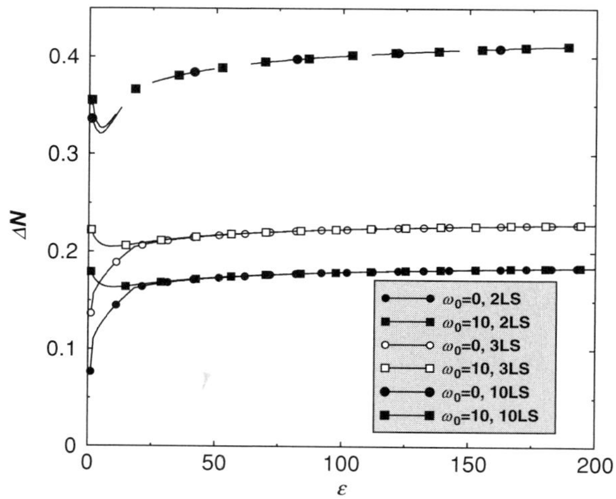

FIGURE 16.6. Fluctuations in the total photon number versus the spectral width for various ω_0 values computed for two three and ten leveled MLS. Here the bare spectral area $A = 5\pi$

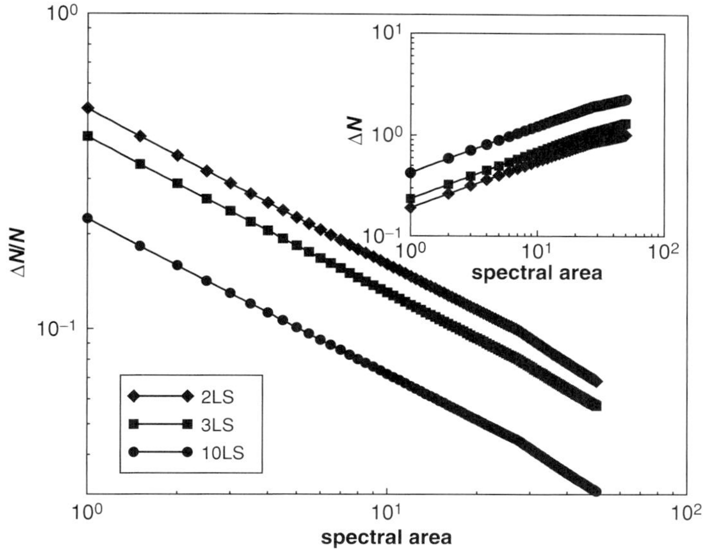

FIGURE 16.7. The ratio $\Delta N/N$ versus the spectral area (inlet: ΔN versus the spectral area).

We observe that $\Delta N \propto \sqrt{A}$ and $N \propto A$ in a large range of spectral areas. Here the positive constant Const. depends on the number of multi levels.

The results of this section are in a sense also a check for the validity of the Born–Oppenheimer approximation. The correction obtained to the spectral area as a result of the system-environment coupling is bounded by a few ten percent. However, as the number of levels increase the total fluctuations decrease relative to the increase in the total number of environmental modes. For a MLS with strong couplings between a large number of levels the validity of the Born–Oppenheimer approximation may still survive although we did not check this result explicitly.

16.5 Conclusions and Discussions

It is shown that for the most conventionally used system-environment couplings of linear coordinate type the non-resonant processes overwhelm the resonant ones in their contribution to the decoherence rates at zero temperature. In this regard, and within the studied spectral range, it is observed that the decoherence rates do not depend on the specific low/high energy properties of the spectrum, whereas, a strong (and dominantly) dependence is observed on the overall spectral area. These results are confirmed for three independent spectral profiles.

The effect of the system-environment coupling was also examined. It was observed that the number of levels which are coupled by the system-environment coupling plays a non-negligible role in decoherence time rates. It was seen that as the number of coupled levels in the system increase, both the decoherence time rates and photon number corrections increase further from the rates and corrections in a pure two level system. The observed effects for the MLS cannot be explained by an equivalent renormalization of the two leveled subsystem. In the light of these observations, we conclude that the conventional postulates of the two-level approximation do not necessarily lead to a dynamical behavior of a multileveled system largely confined to its lowest two levels. The presented results are obtained from the numerical solution of the master equation with the Born–Oppenheimer approximation, and hence, they are exact in the short time limit where the decoherence properties are studied.

References

1. A. J. Leggett, S. Chakravarty, A. T. Dorsey, Matthew P. A. Fisher, A. Garg and W. Zwerger, *Rev. Mod. Phys.* **59**, 1 (1987); Daniel Loss and David P DiVincenzo, *Exact Born Approximation for the spin-boson model*, cond-mat/0304118; T. Vorrath and T. B. B. Kramer, *Dynamics of a large spin-boson system in the strong coupling regime*, cond-mat/0111220; M. Grifoni, E. Paladino and U. Weiss, *Eur. Phys. J* **B 10**, 719 (1999).
2. C. Anastopoulos and B. L. Hu, *Phys. Rev.* **A62**, 033821 (2000).
3. M. Dube and P. C. E. Stamp, *Mechanism of Decoherence at low Temperatures, Chem. Phys. Quantum Physics of Open Systems*, [cond-mat/0102156]; P. C. E. Stamp, I. S. Tupitsyn, *Chem. Phys.* **296** 281 (2004); L. Fedichkin, A. Fedorov and V. Privman, *Proc. SPIE* **5105**, 243 (2003).
4. G. Burkard, R. H. Koch and D. P. Di Vincenzo, *Phys. Rev.* **B 69**, 064503 (2004).
5. P. Cedraschi, V. V. Ponomarenko and M. Büttiker, *Phys. Rev. Lett.* **84**, 346 (2000); M. Büttiker, *EDP Sciences*, 231 (2001).
6. A. O. Caldeira and A. J. Leggett, *Phys. Rev. Lett.* **46**, 211 (1981); A. O. Caldeira and A. J. Leggett, *Annals of Physics* **149**, 374 (1983).
7. M. Born and J. R. Oppenheimer, *Ann. Phys.* **84**, 457 (1927); A. Bohm and M. Loewe, *Quantum Mechanics* (Springer Verlag 2001).
8. T. Hakioğlu, K. Savran, *Decoherence and dephasing in multilevel systems interacting with thermal environment*, cond-mat/0311136.
9. K. Savran, T. Hakioğlu, E. Meșe, *When is the two-level approximation untenable in issues of Decoherence?*, quant-ph/0406160.

17

Josephson Junction Qubits with Symmetrized Couplings to a Resonant LC Bus

STANFORD P. YUKON

Air Force Research Laboratory/SNHE, Electromagnetics Technology Division, 80 Scott Road, Hanscom AFB, MA. 01731 U.S.A.
Email: yukon@hanscom.af.mil
Tel: 781-377-2968
Fax: 781-235-2717

Abstract

We investigate the dynamics of a scalable quantum computer employing Josephson junction (JJ) prism qubits coupled to resonant LC buses. The basic qubit states are the symmetric ground |0 and antisymmetric excited |1 states of a two triangular cell JJ prism qubit. The physical qubits are paired into logical qubits that operate in a decoherence free subspace. The circulating currents, corresponding to the |0 and |1 physical qubit states, follow the perimeter of the qubit, and are symmetric with respect to the flux threading the two cells of the qubit.

17.1 Introduction

The system for Quantum Computation (QC) that we are proposing is modeled on the ion trap system for QC[1] where the ion qubits are coupled among themselves through their collective oscillations in the ion trap, which plays the role of a resonant bus. The bus for the proposed Josephson Junction (JJ) prism qubit QC is a single mode LC resonant loop.[2,3,4] Magnetic flux from the loop threading a qubit will produce a loop-qubit interaction via the Bohm Aharonov effect. At their preferred operating point, the proposed qubits have negligible circulating persistent currents and are kept by symmetry from interacting with their coupled measurement SQUIDs. By pairing the physical qubits into logical qubits, it is possible to maintain the logical qubits in a decoherence free subspace(DFS)[5,6,7] that nulls the decoherence effects of uniform external flux perturbations, and the perturbations associated with computational Mølmer–Sørensen(MS) bichromatic gates.[8,9,10] The only time a logical qubit should interact with the environment is during the unitary gate rotation that initialize the qubits, and during the final Hadamard rotation, which allows the qubit to interact with the measurement SQUID gradiometer.

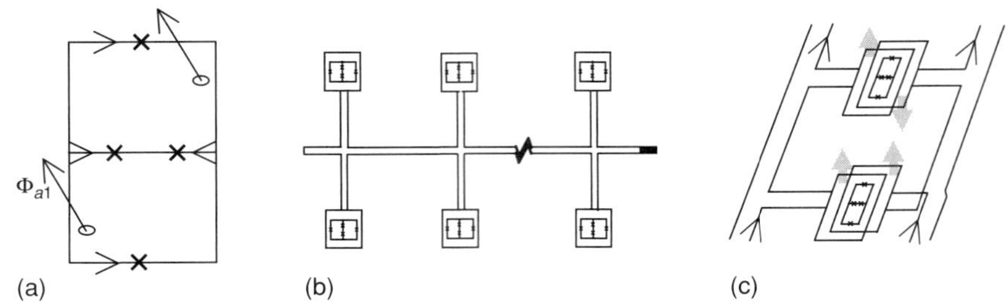

FIGURE 17.1. a. JJ prism qubit with junction critical currents J_{Ci}, gauge invariant phases ζ_i and external magnetic fluxes Φ_i indicated. b. Resonant LC bus symmetrically coupled to qubit pairs. c. Antisymmetric coupling of adjacent buses to a logical transfer qubit.

17.2 Josephson Junction Triangular Prism Qubits

The physical qubits that we shall consider are based on a simplified version of qubits considered in a previous paper.[4] We assume that the junction critical currents and the circuit dimensions are small enough so that the dimensionless self and mutual inductances of the qubit circuit in units of $L_c = \Phi_0/(2\pi I_c)$ are small and can be neglected.[11,12] The Hamiltonian for the qubit shown in Fig. 17.1 may be written in terms of the junction gauge invariant phases ζ_i, critical currents J_{Ci}, and capacitances C_i as $H = \sum_{i=1}^{4} [\frac{1}{2}\dot{\zeta}_i^2(\Phi_0/2\pi)^2/C_i - E_J(J_{Ci}/J_C)\cos(\zeta_i)]$ where the flux quantum and junction characteristic energy are written as $\Phi_0 = h/2e$ and $E_J = J_c\Phi_0/2\pi$.

The phase constraint equations for the two cells are given by $\zeta_1 - \zeta_2 - \zeta_3 = \phi_{a1}$ and $\zeta_2 - \zeta_1 + \zeta_4 = \phi_{a2}$. Defining the symmetric and antisymmetric combinations of the external magnetic fluxes and phases as $\theta = (\zeta_1 + \zeta_2)/2$, $\psi = (\zeta_1 - \zeta_2)/2$ and $\phi_s = \frac{1}{2}(\phi_{a1} + \phi_{a2})$, $\phi_a = \frac{1}{2}(\phi_{a1} - \phi_{a2})$, with $\phi_i = 2\pi\,(\Phi_i/\Phi_0)$ yields $\zeta_3 = 2\psi - \phi_a - \phi_s$ and $\zeta_4 = 2\psi - \phi_a + \phi_s$.. We allow for the possibility of differing junction areas by writing $J_{C\,1,2} = J \pm \Delta J, J_{C\,3,4} = j \pm \Delta j$, $C_{1,2} = C \pm \Delta C$, and $C_{3,4} = \bar{C} \pm \Delta\bar{C}$.

In order to obtain the unperturbed Hamiltonian and its wave function, we assume equal area junctions and circuit loops, zero external antisymmetric flux, and a constant symmetric flux given by $\Phi_s^0 = (\phi_s^0/2\pi)\Phi_0$. The Hamiltonian may be written using θ and ψ variables as

$$H_0(\theta, \psi) = \tfrac{1}{2}M_\theta\dot{\theta}^2 + \tfrac{1}{2}M_\psi\dot{\psi}^2 - E_J[2\cos(\theta)\cos(\psi) + 2\,(j/J)\cos(\phi_s^0)\cos(2\psi - \phi_a)] \quad (17.1)$$

where we have defined $M_\psi = 2C\,(1 + 4\rho)\,(\Phi_0/2\pi)^2 = (1 + 4\rho)\,\hbar^2/4E_c$, $\rho = \bar{C}/C$, $M_\theta = 2C\,(\Phi_0/2\pi)^2 = \hbar^2/4E_c$, with the capacitive energy E_C given by $E_C = e^2/2C$. A contour plot of the potential energy $V(\theta, \psi) = -2\cos(\theta)\cos(\psi) + r\cos(2\psi - \phi_a)$ is shown in Fig. 17.2a for $r = 0.675$ where $r = -2(j/J)\cos(\phi_s^0)$. Approximate eigensolutions are obtained by making a Hartree approximation, where the full two dimensional wave function is approximated by a separable wave function $\Psi(\theta, \psi) \approx \Theta(\theta)\Psi(\psi)$. Assuming that the motion in the θ direction is in its ground state, the $\cos(\theta)$ term in $V(\theta, \psi)$ may be replaced by its expectation value$\eta_H = \langle 0_\theta|\cos(\theta)\,|0_\theta\rangle$, leading to an effective qubit potential in ψ given by $V_0(\psi) = -2\eta_H\cos(\psi) + r\cos(2\psi)$. Writing $\dot{\theta}$ and $\dot{\psi}$ in terms of the momentum operators $M_\theta\dot{\theta} = p_\theta = -i\hbar\partial/\partial\theta$, $M_\psi\dot{\psi} = p_\psi = -i\hbar\partial/\partial\psi$ with $m_\psi = M_\psi E_J/\hbar^2$, the unperturbed effective 1D Hamiltonian leads to the time independent Schrödinger equation

$$-\partial^2\Psi(\psi)/\partial\psi^2 - 2m_\psi[2\eta_H\cos(\psi) - r\cos(2\psi)]\Psi(\psi) = 2m_\psi(E_i/E_J)\Psi(\psi) \quad (17.2)$$

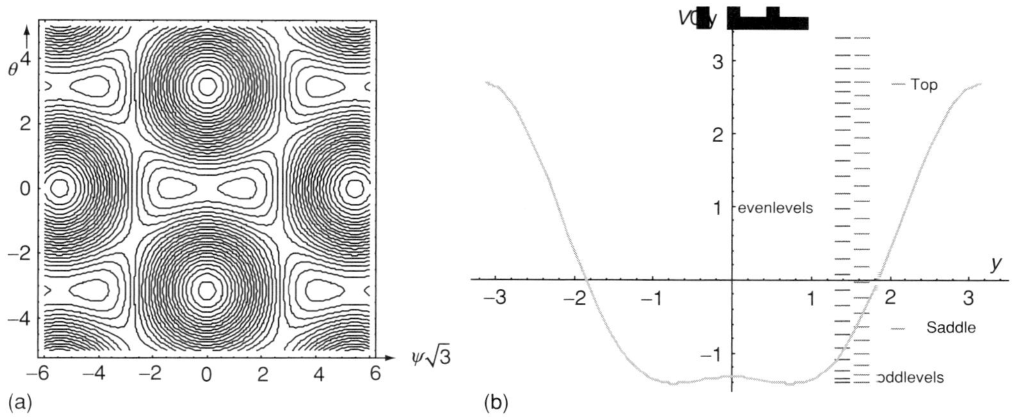

FIGURE 17.2. a. Contour plot of $V(\theta, \psi)$ for $r = 0.675$. b. Energy levels for $V(0, \psi)$ in units of E_J.

which is in the form of the Whittaker Hill equation. Solutions for the eigen functions and energies for the Whittaker Hill equation have been found by Unwin and Arscott terms of continued fractions.[13]

17.3 Coupling the Resonant LC Bus to the Qubits

The qubits are coupled among themselves via their (symmetric) mutual inductance coupling with a resonant LC bus as shown in Fig. 17.1b. The Hamiltonian for a bus oscillator with inductance L and capacitance C may be written in terms of the phase operators φ_s and p_s as $H^{osc}/(\Phi_0/2\pi)^2 = p_s^2/2M_\varphi + M_\varphi \omega_{LC}^2 \varphi_s^2/2$ with $M_\phi = C(\Phi_0/2\pi)^2$, $\omega_{LC} = 1/\sqrt{LC}$, and $M_\varphi \dot{\varphi}_s = p_s = -i\partial/\partial\varphi_s$. In terms of annihilation and creation operators $\varphi_s = \sqrt{\hbar/2M_\varphi\omega_{LC}}(a + a^\dagger)$ and $p_s = i\sqrt{M_\varphi\hbar\omega_{LC}/2}(a^\dagger - a)$. In order to carry out unitary qubit gates on the ith qubit, we include an external microwave pulse $\delta\phi_a^{(i)}(t)$ to be added to $\phi_a^{(i)0}$, as well as an external microwave or DC pulse $\delta\phi_{s,\,ext}^{(i)}(t)$ to be added to $\phi_s^{(i)0}$ along with $k^{(i)}\varphi_s$, the portion of the bus flux that interacts with the ith qubit, where $k^{(i)} = (L^{(i)}/L)$ with $L^{(i)}$ the inductance of the ith coupling loop. For the case with no defects, we may write the Hamiltonian for the ith perturbed qubit, with the definitions $c_{00} = \langle 0|\cos 2\psi|0\rangle$, $c_{11} = \langle 1|\cos 2\psi|1\rangle$, $s_{01} = \langle 0|\sin 2\psi|1\rangle$, $\varepsilon_j = \langle j|H_0|j\rangle/E_J$ as

$$
\begin{aligned}
&\hat{1}^{bus} \otimes \hat{1}^{(i)}\left(H_0^{(i)} + H_1^{(i)} + H_{osc}\right)/E_J\,\hat{1}^{(i)} \otimes \hat{1}^{bus} \\
&\quad \simeq \frac{1}{2}(\varepsilon_0+\varepsilon_1)\hat{1}^{(i)} \otimes \hat{1}^{bus} + \frac{1}{2}(\varepsilon_0-\varepsilon_1)\hat{\sigma}_z^{(i)} \otimes \hat{1}^{bus} + \hat{1}^{bus}\hbar\omega_{LC}/E_J\left(a_j^\dagger a_j + \frac{1}{2}\right)\hat{1}^{bus} \otimes \hat{1}^{(i)} \\
&\qquad + \frac{1}{2}\rho M_\theta\,\dot{\phi}_a^{(i)2}\,\hat{1}^{(i)} \otimes \hat{1}^{bus} + \frac{1}{2}\rho M_\theta\,\hat{1}^{bus}\delta\dot{\phi}_s^{(i)2}\,\hat{1}^{bus} \otimes \hat{1}^{(i)} \\
&\qquad + \hat{1}^{bus}\left\{\frac{1}{2}r(c_{00}+c_{11})\left[\left(J_0\left(\delta_a^{(i)}\right)-1\right) + \left(J_0\left(\delta_s^{(i)}\right)-1\right) - \tan\phi_s^0\,J_0\left(\delta_a^{(i)}\right)\sin\delta\phi_s^{(i)}\right]\right\}\hat{1}^{bus} \otimes \hat{1}^{(i)} \\
&\qquad + \hat{1}^{bus}\left\{\frac{1}{2}r(c_{00}-c_{11})\left[\left(J_0\left(\delta_a^{(i)}\right)-1\right) + \left(J_0\left(\delta_s^{(i)}\right)-1\right) - \tan\phi_s^0\,J_0\left(\delta_a^{(i)}\right)\sin\delta\phi_s^{(i)}\right]\right\}\hat{1}^{bus} \otimes \hat{\sigma}_z^{(i)}
\end{aligned}
$$

$$+ \left\{\rho\dot{\phi}_{\mathrm{a}}^{(i)}/(1+4\rho)2E_J\right\}\langle\Psi_0(\psi_i)|\,\partial/\partial\psi_i\,|\Psi_1(\psi_i)\rangle\,\hat{\sigma}_x^{(i)}\otimes\hat{1}^{\mathrm{bus}}$$

$$+\,\hat{1}^{\mathrm{bus}}\left\{rs_{01}(\cos\delta\phi_{\mathrm{s}}^{(i)}\sin\delta\phi_{\mathrm{a}}^{(i)}-\tan\phi_{\mathrm{s}}^{0}\sin\delta\phi_{\mathrm{a}}^{(i)}\sin\delta\phi_{\mathrm{s}}^{(i)})\right\}\hat{1}^{\mathrm{bus}}\otimes\hat{\sigma}_x^{(i)} \tag{17.3}$$

17.3.1 Logical Qubits for a Decoherence Free Subspace

The Rabi frequency $\Omega^{(i)}$ for coupling a bus excitation to the ith qubit via the interaction term $\sin\delta\phi_{\mathrm{a}}^{(i)}\sin\delta\phi_{\mathrm{s}}^{(i)}\sigma_x^{(i)} \rightarrow \delta\phi_{\mathrm{a}}^{(i)}\varphi_{\mathrm{s}}\sigma_x^{(i)}$is given by $\Omega^{(i)} = [r\tan(\phi_{\mathrm{s}}^{0})\delta_{\mathrm{a}}^{(i)}s_{10}\sqrt{\hbar\omega_{LC}L/2}\,\mu(L^{(i)}/2L)]$. Since $\Omega^{(i)} \propto 1/\sqrt{N}$, to increase $\Omega^{(i)}$ for a particular value of N it is necessary to increase either $\delta\phi_{\mathrm{a}}^{(i)}$ or $L^{(i)}$ or both.

This type of phase error (as well as any symmetric external flux fluctuations) can be nulled by using encoded logical product qubits with $\{\,|0_L\rangle = |01\rangle \equiv |0\rangle\otimes|1\rangle,\ |1_L\rangle = |10\rangle \equiv |1\rangle\otimes|0\rangle\,\}$ or by employing the two entangled Bell states $\{\,|\tilde{0}_L\rangle = (|01\rangle+|10\rangle)/\sqrt{2},\ |\tilde{1}_L\rangle = (|01\rangle-|10\rangle)/\sqrt{2}\,\}$ as logical qubits. For either type, applying $\delta\phi_{\mathrm{a}}^{(i)}$ and $\delta\phi_{\mathrm{a}}^{(i+1)}$ to the first (i) and second ($i+1$) physical qubits of a logical qubit, will result in a positive phase shift in one physical qubit canceling a negative phase shift in the other if their two Rabi frequencies $\delta\Omega^{(i)}$ and $\delta\Omega^{(i+1)}$ are equal. It can be shown that this type of cancellation will occur during MS gates as well.

The Bell state qubits have the property that $t_r = \langle\tilde{0}_L|H^{(1)}+H^{(2)}|\tilde{1}_L\rangle = (\varepsilon_1-\varepsilon_0)^{(1)} - (\varepsilon_1-\varepsilon_0)^{(2)}$ and $\langle\tilde{0}_L|H^{(1)}+H^{(2)}|\tilde{0}_L\rangle = \langle\tilde{1}_L|H^{(1)}+H^{(2)}|\tilde{1}_L\rangle$, i.e., their eigen energies are exactly equal (if they are properly formed). However if the junction or flux properties of the two physical qubits differ, the transition matrix element t_r will be non zero and there will be tunneling between the two states. Transforming to a basis in which $t_r = 0$, yields the product states with zero tunneling but $\langle 0_L|H^{(1)}+H^{(2)}|0_L\rangle - \langle 1_L|H^{(1)}+H^{(2)}|1_L\rangle = (\varepsilon_1-\varepsilon_0)^{(2)} - (\varepsilon_1-\varepsilon_0)^{(1)}$ so there may be nonzero time evolution.

17.3.2 Mølmer–Sørensen Gate

Sørensen and Mølmer have given a detailed description of a bichromatic excitation scheme that enables multiple qubit gate operations to be carried out for ion trap quantum computers[8,9,10]. Their scheme is desirable since it allows for relatively rapid gates to be carried out via an oscillator bus, while returning the bus to its pre-gate initial configuration irrespective of the initial state of the bus. For the case of logical JJ prism qubits coupled by a resonant LC bus, the MS scheme is desirable since it allows logical qubit gates to be carried out without exciting figure 8 currents. As the gate operators for the ion-trap case have been set forth[7,3] we display the needed gate operators for the slightly different case of the interaction Hamiltoniansin $\delta\phi_{\mathrm{a}}^{(i)}\sin\delta\phi_{\mathrm{s}}^{(i)}\sigma_x^{(i)}$while relaxing the condition that all qubit interactions have the same Rabi frequency. Expanding the $\sin\delta\phi_{\mathrm{a}}^{(i)}$ and $\sin\delta\phi_{\mathrm{s}}^{(i)}$terms to lowest order in $\delta\phi_{\mathrm{a}}^{(i)}$ and $\delta\phi_{\mathrm{s}}^{(i)} = \frac{1}{2}\mu\varphi_{\mathrm{s}}(L^{(i)}/L)$ and making the rotating wave approximation leads to an interaction term

$$V(t) \cong -\sqrt{2}\Omega\, J_x(\vec{\phi})[\varphi_{\mathrm{s}}(\cos(\omega_{LC}-\delta)t + p_{\mathrm{s}}(\sin(\omega_{LC}-\delta)t] = J_x(\vec{\phi})[f(t)\varphi_{\mathrm{s}} + g(t)p_{\mathrm{s}}]$$

where $J_x(\vec{\phi}) = \frac{1}{2}\sum_i[e^{i\phi_i}\hat{\sigma}_+^i + e^{-i\phi_i}\hat{\sigma}_-^i](\Omega_i/\Omega)$ with $\Omega_i = [r\tan(\phi_{\mathrm{s}}^{0})\delta_{\mathrm{a}}^{(i)}s_{10}\sqrt{\hbar\omega_{LC}L/2}\,\mu\,(L_{\mathrm{site}}/2L)]$ and Ω being the individual and average Rabi frequencies of the interacting qubits.

For the MS bichromatic excitation, $\delta\phi_a^{(i)}$ is made up of signals from two microwave sources with source phases $\phi^{(i)}$ and frequencies $\omega_\pm = \omega_{10} \pm \delta$, where $\omega_{10} = (\varepsilon_1 - \varepsilon_0)/\hbar$ and $\delta \sim \omega_{LC} \pm 2\Omega$. In Ref. 10, the exact evolution operator for $V(t)$ is shown to be given exactly by

$$U(t) = \exp(-iA(t)J_x^2)\exp(-iF(t)J_x x)\exp(-iG(t)J_x p) \tag{17.4}$$

with $F(t) = \int_0^t f(\tau)d\tau = -\sqrt{2}\Omega\sin[(\omega_{LC} - \delta)t]/(\omega_{LC} - \delta)$, $G(t) = \int_0^t g(\tau)d\tau$ and $A(t) = -\int_0^t F(\tau)g(\tau)d\tau$. For a pulse length given by $(\omega_{LC} - \delta)t_K = 2\pi K$ where K is an integer, $F(t_K) = G(t_K) = 0$ and the bus can be returned to its initial vibrational state. Choosing values for K, δ, and Ω, yields $A(t_K) = -\Omega^2 2\pi K/(\omega_{LC} - \delta)^2$ and conversely $t_K = \sqrt{-A\,2\pi K}/\Omega$. The single (logical) qubit gates can be composed as

$$\begin{aligned}\bar{X} &= 2J_x^2(\phi_{1,2} = 0) = \tfrac{1}{2}(\lambda_1^2 + \lambda_2^2)\hat{1} + \lambda_1\lambda_2\hat{\sigma}_x^{(1)}\hat{\sigma}_x^{(2)},\\ \bar{Y} &= 2J_x^2(\phi_1 = 0,\ \phi_2 = \tfrac{1}{2}\pi) = \tfrac{1}{2}(\lambda_1^2 + \lambda_2^2)\,\hat{1} - \lambda_1\lambda_2\hat{\sigma}_x^{(1)}\hat{\sigma}_y^{(2)}\end{aligned} \tag{17.5}$$

where $\lambda_i = \Omega_i/\Omega$. Operators for product and Bell logical qubits are given respectively by $\{\sigma_{Lx} = \sigma_x^{(1)}\sigma_x^{(2)},\ \sigma_{Ly} = \sigma_y^{(1)}\sigma_x^{(2)},\ \sigma_{Lz} = -i\,\sigma_{Lx}\sigma_{Ly}\}$ and $\{\tilde{\sigma}_{Lx} = -i\,\tilde{\sigma}_{Lx}\tilde{\sigma}_{Ly},\ \tilde{\sigma}_{Ly} = -\sigma_y^{(1)}\sigma_x^{(2)},\ \tilde{\sigma}_{Lz} = \sigma_x^{(1)}\sigma_x^{(2)}\}$. For the product logical qubits, the CZ gate may be written as $U_{CZ}^{(I,II)} = \sqrt{i}\exp(i\sigma_{Lz}^{(I)}\sigma_{Lz}^{(II)}\pi/4)\exp(-i\sigma_{Lz}^{(I)}\pi/4)\exp(-i\sigma_{Lz}^{(II)}\pi/4)$ where the term $\exp(i\sigma_{Lz}^{(I)}\sigma_{Lz}^{(II)}\pi/4)$ can be expressed in terms of a four (physical) qubit Mølmer–Sørensen gate $\overline{XX} = 2J_x^2(\phi_{1,2,3,4} = 0)$

$$U_{CZ}^{(I,II)} = \exp(i\pi/4)\exp(i\overline{XX}\pi/4)\exp(i\sigma_{Lz}^{(I)}\pi/4)\exp(i\sigma_{Lz}^{(II)}\pi/4) \tag{17.6}$$

When the Rabi frequencies for the four physical qubits are not identical, mixed terms like $\sigma_x^1\sigma_x^4$ will arise with non zero coefficients proportional to $\gamma_i = \lambda_i - 1$. If the Rabi frequencies for each physical qubit of a logical qubit pair can be made equal, then Ωcan be defined as the average Rabi frequency with $\gamma = \gamma_1 = \gamma_2 = -\gamma_3 = -\gamma_4$. Terms that are proportional to $\gamma_1 + \gamma_2 + \gamma_3 + \gamma_4$ will be nulled, and the remaining terms will be proportional to either $\sigma_x^{(1)}\sigma_x^{(2)}$, $\sigma_x^{(3)}\sigma_x^{(3)}$, or $\sigma_x^{(1)}\sigma_x^{(2)}\sigma_x^{(3)}\sigma_x^{(3)}$ that can be accommodated by small changes in pulse lengths. If the two Rabi frequencies$\delta\Omega^{(i)}$ and $\delta\Omega^{(i+1)}$for each logical qubit pair are equal, as is required to cancel out$\delta\Omega^{(i)}\hat{\sigma}_z^{(i)}$ and $\delta\Omega^{(i+1)}\hat{\sigma}_z^{(i+1}$phase errors generated during a gate, then $j_c^{(1)}\cos(\phi_{s0}^{(1)})\,(c_{00}^{(1)} - c_{11}^{(1)})\,[\delta_a^{(1)}]^2 = j_c^{(2)}\cos(\phi_{s0}^{(2)})\,(c_{00}^{(2)} - c_{11}^{(2)})\,[\delta_a^{(2)}]^2$ is satisfied. If $j_c^{(1)}\sin(\phi_{s0}^{(1)})\,s_{01}^{(1)}\,[\delta_a^{(1)}] = j_c^{(2)}\sin(\phi_{s0}^{(2)})\,s_{01}^{(2)}\,[\delta_a^{(2)}]$ is satisfied, then the Rabi frequencies for qubits 1 and 2 will be equal. For both constraints to hold concurrently, the equality of Eq. (17.7) must be satisfied, which is possible to achieve by varying $\delta_a^{(1)}$ (or $\delta_a^{(2)}$) if $\phi_{s0}^{(1)}$and

$$\begin{aligned}(\delta_a^{(2)}/\delta_a^{(1)}) &= \tan(\phi_{s0}^{(2)})\,s_{01}^{(2)}(c_{00}^{(1)} - c_{11}^{(1)})/\tan(\phi_{s0}^{(1)})\,s_{01}^{(1)}(c_{00}^{(2)} - c_{11}^{(2)})\\ &= f(r^{(1)}, r^{(2)})\end{aligned} \tag{17.7}$$

$\phi_{s0}^{(2)}$are fixed, as shown in Fig. 17.3.

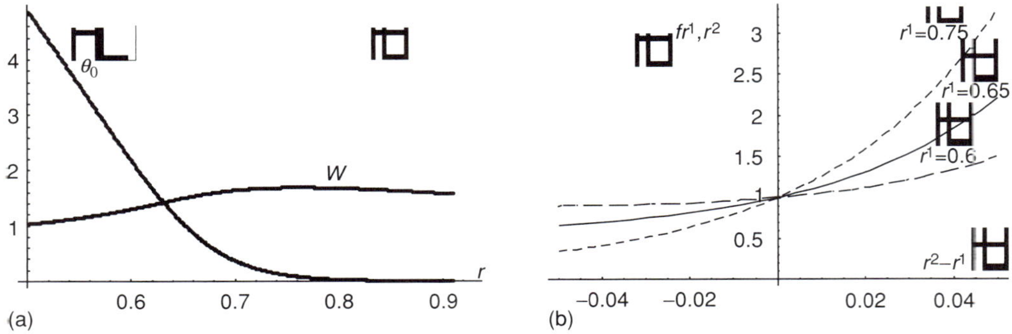

FIGURE 17.3. a. $\frac{1}{2}$Transition frequency (for $E_J/\hbar = 200\text{GHz}$) and const. × Rabi frequency Ω vs. r. b. $f(r^{(1)}, r^{(2)})$ as a function of $r^{(2)} - r^{(1)}$ for $r^{(1)} = 0.60, 0.65$, and 0.75.

17.4 Circulating Current Patterns for the Qubit States

The circulating current patterns associated with the $|0\rangle$ and $|1\rangle$ states can be derived by using $\zeta_{3,4} = 2\psi - \phi_a \mp \phi_s$ and by assuming ϕ_a to be a small perturbation, which then yields

$$J_{3,4}^{|0\rangle} = \langle 0|\, J_{C3,4} \sin\zeta_{3,4}\, |0\rangle = \mp j\kappa_0(1 \pm \cos\phi_s^0 \sin\phi_a / \sin\phi_s^0 \cos\phi_a)$$

$$J_{3,4}^{|1\rangle} = \langle 1|\, J_{C3,4} \sin\zeta_{3,4}\, |1\rangle = \mp j\kappa_1(1 \pm \cos\phi_s^0 \sin\phi_a / \sin\phi_s^0 \cos\phi_a) \tag{17.8}$$

with $\kappa_0 = \sin\phi_s^0 \cos\phi_a^0 \langle 0|\cos 2\psi|0\rangle$ and $\kappa_1 = \sin\phi_s^0 \cos\phi_a^0 \langle 1|\cos 2\psi|1\rangle$. In the Hartree approximation if $\Delta J = \Delta j = 0$ the expectation value of $J_1 = J_{C1}\sin\zeta_1$ in the $|1\rangle$ or$|0\rangle$ state of ψ is equal to zero. The circulating current patterns associated with the unperturbed $|0\rangle$ and $|1\rangle$ states are thus clockwise currents of magnitude $J_3^{|0\rangle} = -J_4^{|0\rangle} = -j\kappa_0$ and $J_3^{|1\rangle} = -J_4^{|1\rangle} = -j\kappa_1$ around the perimeter of the qubit, and are plotted versus. r in Fig. 17.3.

Since the probability density for the two superposition states $|\pm\rangle = \frac{1}{\sqrt{2}}(|0\rangle \pm |1\rangle)$ is centered on the potential well minima at $\psi = \mp\psi^*$, circulating currents for these states are close to the classical equilibrium currents. Writing $\kappa_{01} = \langle 0|\sin 2\psi|1\rangle \cos(\phi_s^0 - \phi_a^0)$, circulating currents for the $|\pm\rangle$states may be reexpressed as

$$J_3^{|+\rangle} = \langle +|\, J_{C3}\sin\zeta_3\, |+\rangle = -\tfrac{1}{2} j(\kappa_0 + \kappa_1) + j\,\kappa_{01}$$

$$J_3^{|-\rangle} = \langle -|\, J_{C3}\sin\zeta_3\, |-\rangle = -\tfrac{1}{2} j(\kappa_0 + \kappa_1) - j\,\kappa_{01}$$

$$J_4^{|+\rangle} = \langle +|\, J_{C4}\sin\zeta_4\, |+\rangle = +\tfrac{1}{2} j(\kappa_0 + \kappa_1) + j\,\kappa_{01}$$

$$J_4^{|-\rangle} = \langle -|\, J_{C4}\sin\zeta_4\, |-\rangle = +\tfrac{1}{2} j(\kappa_0 + \kappa_1) - j\,\kappa_{01}. \tag{17.9}$$

The circulating current is thus comprised of a perimeter current $\frac{1}{2} j(\kappa_0 + \kappa_1)$, which has the same sign for the $|\pm\rangle$states, and a figure eight current $j\kappa_{01}$ that changes sign for $|\pm\rangle$states.

17.4.1 Initializing the Qubits and the Effect of Critical Current and Geometric Defects

All of the qubits are initially allowed to relax to their respective ground states. For a system in which all of the cell and junction parameters are identical, every even numbered qubit is given an x rotation of 180° In order to avoid phase errors during these initializing single qubit gates, the gates should be carried out with long, small amplitude microwave pulses. Each $\{i_{odd}, i_{odd+1}\}$ pair of qubits is then designated as a product logical qubit {I, II, III...}. To initialize the Bell state logical qubits a further logical (MS) Hadamard gate is needed for each logical qubit. A test measurement should show that there is no antisymmetric flux coupled to the SQUID gradiometer detectors.

For a qubit in which there is some combination of junction critical current defects and cell geometry defects, a term in the Hamiltonian will be generated that is antisymmetric in ψ and θ

$$\Delta V = -\{2j \sin\phi_{\text{a, geom}} \cos\phi_s^0 + 2\Delta j \cos\phi_{\text{a, geom}} \sin\phi_s^0\} \sin 2\psi - 2\Delta J \sin\theta \sin\psi$$

$$= -\kappa_{\text{defect}} \sin 2\psi - 2\Delta J \sin\theta \sin\psi = \Delta V^a + \Delta V^b. \tag{17.10}$$

The effect of ΔV^a on the potential is to tilt it in the ψ direction as shown in Fig. 17.4b, while the effect of ΔV^b is to skew the potential as shown in Fig. 17.4c. For ΔV^b an accurate numerical solution for the 2D potential has been undertaken but not yet completed. We will thus approximate the perturbed wavefunctions using first order perturbation theory.

In a Hartree approximation there are four types of states when labeled by parity with respect to θ and ψ; $|m, n\rangle = \{|\text{even, even}\rangle, |\text{even, odd}\rangle, |\text{odd, even}\rangle$ and $|\text{odd, odd}\rangle\}$, where m and n are the number of nodes in θ and ψ. The only type of (separable) states contributing to the summation that will survive in a first order perturbation expansion $|n\rangle = |n^0\rangle + \sum_{k\neq n} k^0|\Delta V^b|n^0\rangle/(E_n^0 - E_k^0)$ for $|0\rangle$ and $|1\rangle$, are $|\text{odd, odd}\rangle$ and $|\text{odd, even}\rangle$ type states respectively, where $|n^0\rangle$ are unperturbed states. Keeping only the lowest energy terms $|2^0\rangle = |1, 1\rangle$ and $3^0\rangle = |1, 0\rangle$, an approximate solution may be written as $|0\rangle = |0^0\rangle + \alpha|2^0\rangle$ and $|1\rangle = |1^0\rangle + \beta|3^0\rangle$ where $\alpha = 2\Delta J \langle 2^0|\sin\theta \sin\psi|0^0\rangle/(E_2^{(0)} - E_0^{(0)})$ and $\beta = 2\Delta J \langle 3^0|\sin\theta \sin\psi|1^0\rangle/(E_3^{(0)} - E_0^{(0)})$. The matrix elements of ΔV^b are then given by $\langle 0|\Delta V^b|0\rangle = 2\alpha\langle 0^0|\Delta V^b|2^0\rangle$, $\langle 1|\Delta V^b|1\rangle = 2\beta\langle 1^0|\Delta V^b|3^0\rangle$, and $\langle 0|\Delta V^b|1\rangle = 0$. Critical current defects in junctions 1 and 2 will thus lead to a second order (in $\Delta J/J$) correction in the energies of the two qubit states, but it will not lead to transitions between them.

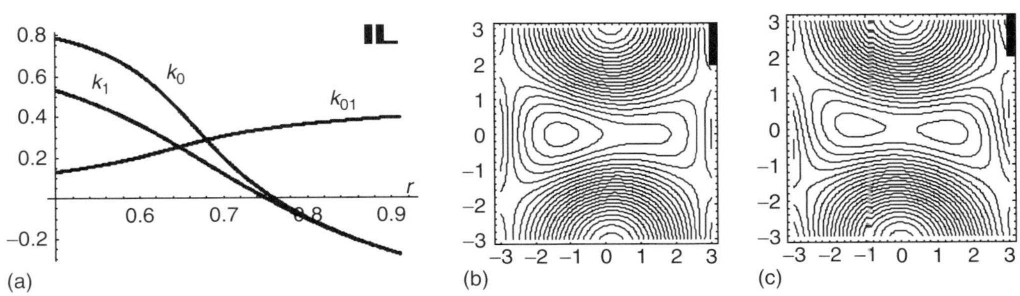

FIGURE 17.4. a. Symmetric$\{\kappa_0, \kappa_1\}$ and antisymmetric $\{\kappa_{01}\}$ current amplitudes vs. r. b. Contour plot of $V(\theta, \psi)$ for $\Delta V_a \neq 0$. **c** Contour plot of $V(\theta, \psi)$ for $\Delta V_b \neq 0$.

The most direct way of nulling the ΔV^a term is to add a small external antisymmetric 'trim' flux to the existing flux so that the resultant antisymmetric flux $\phi_a^0 = \phi_{a,\,\mathrm{geom}} + \phi_{a,\,\mathrm{trim}}$ will satisfy $\tan\phi_a^0 = -(\Delta j/j)\tan\phi_s^0$. The net result of having $\phi_a^{(0)}$determined by this equation, is that small figure eight currents that were present initially in the $|0\rangle$ and $|1\rangle$ states and could interact with the SQUID detector are now nulled, while perimeter currents in the $|0\rangle$ and $|1\rangle$ states will be slightly decreased by the last factor in Eq. (17.8) which simplifies to $(1 \pm \Delta j/j)(1 \mp \cos\phi_s^0 \sin\phi_a/\sin\phi_s^0\cos\phi_a) = [1-(\Delta j/j)^2]$. If $\Delta J/J$ is not zero, then there will be a multiplicative factor of $(1+\alpha^2 c_{11}/c_{00})$for the $|0\rangle$ state and $(1+\beta^2 c_{00}/c_{11})$ for the $|1\rangle$ state.

If the trim current is not applied, then there will be a non zero figure eight current of magnitude $\kappa_0(\Delta j + j\ \tan\phi_{a,\,\mathrm{geom}}/\tan\phi_s^0)$ for the $|0\rangle$ state and$\kappa_1(\Delta j + j\ \tan\phi_{a,\,\mathrm{geom}}/\tan\phi_s^0)$ for the $|1\rangle$ state. From Fig. 17.4a it can be seen that if the qubit is operated around r^*, then $\kappa_0$and κ_1 are small ($\sim$ equal and opposite) and Ω_{Rabi} is near its maximum at $r \approx 0.75$.

The Hamiltonian $H_0 + \Delta V$ can be diagonalized by a unitary rotation to a new basis. Since the unperturbed wave functions for the two basic states $\Psi_0(\theta,\psi) = \langle\theta,\psi|\,0\rangle$ and $\Psi_1(\theta,\psi) = \langle\theta,\psi|\,1\rangle$are both symmetric in θ, the matrix element of the last term in ΔV will be zero. Only junction defects Δj, associated with outer junctions 3 and 4, and a small antisymmetric flux ϕ_a due to different cell sizes, will then affect the operation of the qubit.

17.4.2 Connecting Buses into a Network

It is possible to couple groups of buses in an open branching network using transfer qubits as first described in Eq. (17.4). A diagram of one possible coupling scheme is shown in Fig. 17.1c where bus 1 is coupled symmetrically to the transfer qubit and bus 2 is coupled antisymmetrically in order to prevent direct mutual inductance coupling from bus 1 to bus 2. Using $\tilde{U}_{CZ}^{(I,II)}$ from Eq. (17.6), an expression for $\tilde{U}_{CN}^{(j,k)}$ may be written using the Hadamard operator $\tilde{H}_k$as $\tilde{U}_{CN}^{(j,k)} = \tilde{H}_k\,\tilde{U}_{CZ}^{(j,k)}\,\tilde{H}_k$. Writing $\tilde{H}_k = -ie^{-i(\pi/4)\tilde{\sigma}_{Ly}^{(k)}}e^{i(\pi/2)\tilde{\sigma}_{Lz}^{(k)}}$, the $\tilde{C}\tilde{N}$ operator may be expressed as $\tilde{U}_{CN}^{(j,k)} = e^{-i(\pi/4)\tilde{\sigma}_{Ly}^{(k)}}\,\tilde{U}_{CZ}^{(j,k)}e^{i(\pi/4)\tilde{\sigma}_{Ly}^{(k)}}$ which requires 4 (bichromatic) microwave pulses. Since $\{x_{(1)},\ p_{(1)}\}$ commute with $\{x_{(2)},\ p_{(2)}\}$, the expression for $A(t)$ in Eq. 4. is replaced by $A(t) \rightarrow A_{(1)}(t) + A_{(2)}(t)$. For both bus oscillators to return to their pre gate values both of the constraints $(\omega_{(1)LC} - \delta)t_K = K_{(1)}2\pi$ and $(\omega_{(2)LC} - \delta)t_K = K_{(2)}2\pi$ with integer $K_{(i)}$ must be satisfied simultaneously. For $K_{(2)} = -K_{(1)}$ a solution is $\delta = (\omega_{(1)LC} + \omega_{(2)LC})/2$, $t_K = 4\pi K_{(1)}/(\omega_{(1)LC} - \omega_{(2)LC})$, which yields $A_{(2)}(t) = -A_{(1)}(t)(-1)^{\mathrm{P}}$ where P is the parity of the bus 1-bus 2 coupling. Thus $U(t_K) = \exp(-2iA_{(1)}(t_K)\sigma_x^{(1)}\sigma_x^{(2)})$.

For the 2 logical qubit CZ gate, if one of the logical qubits is a transfer qubit, its coupling to a second bus will engender a single logical qubit rotation that is in the proper direction for the CZ gate. Thus if logical qubit 3–4 is a transfer qubit connected to a second bus, the expression $\exp(i\pi\overline{XX}/4) = \exp(-i\pi/4)\exp(-i\pi\hat{\sigma}_x^{(1)}\hat{\sigma}_x^{(2)}\hat{\sigma}_x^{(3)}\hat{\sigma}_x^{(3)}/4)$ picks up a factor $\exp[i\pi\hat{\sigma}_x^{(3)}\hat{\sigma}_x^{(3)}/4]$ obviating the necessity of a separate gate. If there are two transfer qubits involved, no single logical qubit gates are needed at the transfer qubit sites. This reduces the number of separate microwave pulses for a CN gate with one bus transfer from 5 to 3 and also reduces the time needed to carry out the gate by $2t_K$.

To carry out a CN gate over a number of buses, it is necessary to first carry out a sequence of $\tilde{U}_{CN}^{(j,t)}$'s starting from the control qubit, through all of the intervening transfer qubits to the target qubit, followed by a final sequence of $\tilde{U}_{CN}^{(i,i')}$ from the original qubit to the last transfer qubit in order to reset the transfer qubits to their original ground states. The sequence involving

2 transfer qubits would thus be $\tilde{U}_{CN}^{(i,i')}$ between $(\boldsymbol{j}, \boldsymbol{t}_1)$, $(\boldsymbol{t}_1, \boldsymbol{t}_2)$, $(\boldsymbol{t}_2, \boldsymbol{k})$ followed by resetting transfer qubits with $\tilde{U}_{CN}^{(i,i')}$ applied to $(\boldsymbol{j}, \boldsymbol{t}_1)$ and $(\boldsymbol{t}_1, \boldsymbol{t}_2)$. These maneuvers may be chained across as many coupled buses as required. The wave equation for the entire system may be written as $|\Psi\rangle = \sum_n c_n |\Psi_N^n\rangle$ with $|\Psi_N^n\rangle = \prod_{i=1}^N |\tilde{\chi}_{\alpha_i}^{(i)}\rangle \otimes \prod_{j=1}^{N_{tr}} |\tilde{\chi}_{\beta_j}^{(j)}\rangle \otimes \prod_{k=1}^{N_{bus}} |\chi_{(k),\gamma_k}\rangle$, where the qubits and bus are restricted to the two lowest energy states $\alpha_i = \{0, 1\}$, $\beta_j = \{0, 1\}$, $\gamma_k = \{0, 1\}$ and where $n = \{\alpha_1, \alpha_2, ..., \alpha_N, \beta_1, \beta_2, ..., \beta_{N_{tr}}, \gamma_1, \gamma_2, ..., \gamma_{N\text{bus}}\}$ with $|\tilde{\chi}_0\rangle = |\tilde{0}_L\rangle$, $|\tilde{\chi}_1\rangle = |\tilde{1}_L\rangle$, $|\chi_0\rangle = |0_{\text{Bus}}\rangle$, and $|\chi_1\rangle = |1_{\text{Bus}}\rangle$. Because the MS gates return the buses and (along with resetting) the transfer qubits to the initial state they were in before the $\tilde{U}_{CN}^{(j,k)}$ gate, and the logical qubits do not evolve in time, it should be possible to maintain quantum coherence across the set of coupled buses. This allows open branching networks of coupled buses to be formed, enlarging the possible number of interacting qubits in a single quantum computer.

17.5 Conclusions

We have shown that it is possible to design a quantum computer comprised of Josephson junction prism qubits that are coupled via the flux of a resonant LC bus that is capable of carrying out quantum computations in a decoherence free subspace. The effective Hamiltonian is zero between gates, so there is no dynamical evolution of the system. The $|0\rangle$ and $|1\rangle$ physical qubit states are characterized by circulating currents that follow the perimeter of a two cell qubit, and are thus symmetric with respect to the flux threading the cells of the qubit. The logical qubit states are made up of linear combinations of $|01\rangle \equiv |0\rangle \otimes |1\rangle$ and $|10\rangle \equiv |1\rangle \otimes |0\rangle$, with the result that no antisymmetric currents or fluxes are generated by the qubits when Mølmer–Sørensen gates are employed. A coupled SQUID gradiometer that is antisymmetric with respect to the flux threading the two cells of a qubit will not register any coupled flux until a Hadamard transform is carried out on the physical qubits, immediately prior to a measurement. The two basic states are then transformed into $\pm$ combination states that have antisymmetric 'figure 8' circulating currents whose flux is then measurable by the SQUID gradiometer. The coupling of each qubit to an LC resonant bus is made by means of a loop that encircles both cells of the qubit symmetrically. This prevents the direct mutual inductance coupling of the bus to both antisymmetric microwave gate pulses and the measuring SQUID gradiometer.

Pairing the qubits into logical qubits also enables Mølmer-Sørensen gates to be used, since the phase errors that would ensue if QC were based on physical qubits are cancelled out in each pair of qubits. The requirement that the LC resonant loop be smaller in extent than one tenth of a substrate wavelength to remain within the quasistatic limit, constrains the number of qubits that can interact on a single LC resonant bus to around $N = \sim 50 - 100$. To enlarge the system, two resonant loop buses are coupled by means of a 'transfer' logical qubit that couples concurrently to both buses. Direct interaction of the loops is prevented by coupling them antisymmetrically. With more than one transfer qubit per bus, the buses may be coupled into a scalable open network. The use of Mølmer-Sørensen gates insures that the quantum states of the buses and transfer qubits will be returned to their pre gate status. Since the logical qubits do not evolve in time, it should be possible for the system of coupled buses and qubits to maintain quantum coherence.

Acknowledgments. The author appreciates helpful conversations with Alexey Ustinov, Farrukh Abdumalikov, Will Oliver, Al Drehman, and Jianke Yang. This work was carried out with support from Air Force Office of Scientific Research AFOSR.

References

1. J. I. Cirac and P. Zoller,*Phys. Rev. Lett.* 7**4**, 4091 (1995).
2. 2. Y. Makhlin, G. Schön, and A. Shnirman, *Rev. Mod. Phys.* **73**, 357 (2001)
3. O. Buisson and F. W. Hekking, in *Macroscopic Quantum Coherence and Quantum Computing* (Kluwer Academic, Dordrecht), p. 137, (2001).
4. S. Yukon, *Physica C* **368**, 320 (2002).
5. M. J. Feldman and X. Zhou, in *Quantum Computing and Quantum Bits in Macroscopic Systems* (Kluwer Academic, Dordrecht), p. 219, (2004).
6. D. Bacon, J. Kempe, D. A. Lidar, and K. B. Whaley, *Phys. Rev. Lett.* **85**, 1758 (2000)
7. D. Bacon, Decoherence, Control and Symmetry in Quantum Computers, Phd. Thesis, Cal. Inst. Tech. (1997).
8. A. Sørensen and K. Mølmer, *Phys Rev. Lett.* **82**, 1971 (1999).
9. K. Mølmer and A. Sørensen, *Phys Rev. Lett.* **82**, 1835 (1999).
10. A. Sørensen and K. Mølmer, *Phys A* **62**, 1022311 (2000).
11. J. E. Mooij, T.P. Orlando, L. Levitov, L. Tian, C. H. van der Wal, and S. Lloyd, *Science* **285**, 1036 (1999); *Science* **290**, 773 (2000).
12. D. S. Crankshaw and T. P. Orlando, *IEEE Trans. on Appl. Sup*. **11**, 1006 (2001).
13. K. M. Unwin, F. M. Arscott, *Proc. Roy Soc. Edinb.* **71**, 28, (1971).

18

Spatial Bose-Einstein Condensation in Josephson Junction Arrays

M. CIRILLO[1], V. MERLO[1], R. RUSSO[1], M. G. CASTELLANO[2], C. COSMELLI[3], A. TROMBETTONI[4], G. GIUSIANO[5], F. P. MANCINI[5] AND P. SODANO[5]

[1] Russo, Dipartimento di Fisica and INFM, Università di Roma Tor Vergata, 00173 Roma, Italy.
[2] IFN-CNR, 00156 Roma, Italy.
[3] Dipartimento di Fisica and INFN, Università di Roma La Sapienza, 00185 Roma, Italy.
[4] Dipartimento di Fisica and INFM, Università di Parma, 43100 Parma, Italy.
[5] Dipartimento di Fisica e Sezione INFN, Università di Perugia, 06123, Perugia, Italy.

Abstract

Very recent studies hint to the new and exciting possibility that the topology of Josephson Junction Networks (JJN) may be crucial for inducing novel and unexpected macroscopic quantum phenonema. After discussing the theoretical possibility of a spatial Bose-Einstein Condensation (BEC) of Cooper pairs for a JJN built on a comb-graph, we shall describe a simple experiment enabling to observe BEC in these systems. We report also on the progress made by our collaboration in the fabrication of graph-shaped networks and we shall provide preliminary experimental evidences of BEC on a comb shaped JJN.

18.1 Introduction

Theoretical studies seem now to evidence that Josephson Junctions Networks (JJN) fabricated on pertinent graphs may support the emergence of a spatial Bose Einstein Condensation (BEC) of Cooper pairs.[1–4] We shall report here not only on the engineering of an experiment aimed at observing the signatures of BEC in pertinently graph-shaped JJNs,[3,4] but also on the progress made so far in fabricating and testing comb-shaped JJNs.

In this study the focus will be on experimentally exploiting the recently discovered possibility that Josephson junction networks (JJN) built on suitable graphs may support new and interesting quantum macroscopic states.[1,2] In particular it has been evidenced the occurrence of a spatial BEC of Cooper pairs for a JJN built on a comb-graph[1] simple experiment enabling to observe BEC in this system is described.[3]

A comb-shaped array of classical Josephson junctions is usually described by the Hamiltonian

$$H_{\mathrm{JJ}} = -J_0 \sum_{(x,y;x',y')} \cos(\phi_{x,y} - \phi_{x',y'}). \tag{18.1}$$

On each of the L^2 sites (x, y) of the network there is a superconducting grain and the junctions are located between neighbouring sites with Josephson energy J_0; $\phi_{x,y}$ is the phase of the superconducting order parameter (x labels the different fingers and y represents the distance from the backbone). The parameter J_0 is connected to the critical Josephson current I_c of the junctions by the relation $I_c = 2eJ_0/\hbar$.[5] The summation $(x, y; x'y')$ runs over all the distinct nearest-neighbours pairs. Cooper pairs are allowed to condense in the spatially localized ground-state of the system.

If one defines T_{BEC} as the critical temperature at which BEC occurs, for any $T < T_{\mathrm{BEC}}$, the ground-state should be macroscopically filled. The equation relating T_{BEC} to the Josephson energy and of the filling of the system f is given by[1]

$$f \approx \int \mathrm{d}E \frac{1}{\pi\sqrt{4t^2 - E^2}} \frac{k_B T_{\mathrm{BEC}}}{E + \sqrt{8}(J_0/2f)} = \frac{k_B T_{\mathrm{BEC}}}{J_0/f}. \tag{18.2}$$

Equation (18.2) holds for $f \gg 1$. The critical temperature T_{BEC} is then a linear function of J_0

$$T_{\mathrm{BEC}} \approx \frac{J_0}{k_B}. \tag{18.3}$$

Due to the spatial condensation of the Cooper pairs one expects an inhomogeneous distribution of the bosons along the fingers. In fact, below the critical temperature, the average number of bosons $N_{\mathrm{B}}(y)$ on a site (x, y) is a function of the distance y from the backbone and of the temperature T.

In the thermodynamic limit, for $y \gg 1$, one has[3]

$$\frac{N_{\mathrm{B}}(y; T/T_{\mathrm{BEC}})}{f} \approx \frac{T}{T_{\mathrm{BEC}}}. \tag{18.4}$$

To detect the signature of BEC for a JJN built on a comb graph one should perform a measurement of the I–V characteristic of a single finger of the JJN; if one feeds, in fact, an external dc-current I_{ext} at the top of the finger one should expect to observe no voltage unless I_{ext} is larger than the smallest of the critical currents of the Josephson junctions along the finger. Since below T_{BEC} the critical Josephson current of the finger is given by the smallest of the critical currents along the fingers, the measurement of the I–V characteristic of the finger should provide a measurement of the critical current of the junctions at the top of the fingers, which, as pointed out by Eq. (18.4), is the fraction of *noncondensate* Cooper pairs.[3] To make a definite prediction one needs to estimate the value of the Josephson critical current of a single junction of the network as a function of both the temperature and the distance from the backbone. For a large array ($L \gg 1$), far from the backbone $y \gg 1$ and above T_{BEC}, the Cooper pairs are uniformly distributed and the Josephson critical current $I_c^{(0)}$ is the same at each junction. Below T_c the number of bosons exponentially decreases with the distance y from the backbone. The Josephson critical current depends only on the position and on the population of the sites. The relation between the Josephson critical current at site y far from the backbone $y \gg 1$ above and below the critical temperature T_{BEC} is then given by[3]

$$\frac{I_c^{T<T_{\mathrm{BEC}}}(y)}{I_c^{(0)}(y)} \approx \frac{N_{\mathrm{B}}(y, T/T_{\mathrm{BEC}})}{f} \approx \frac{T}{T_{\mathrm{BEC}}} \tag{18.5}$$

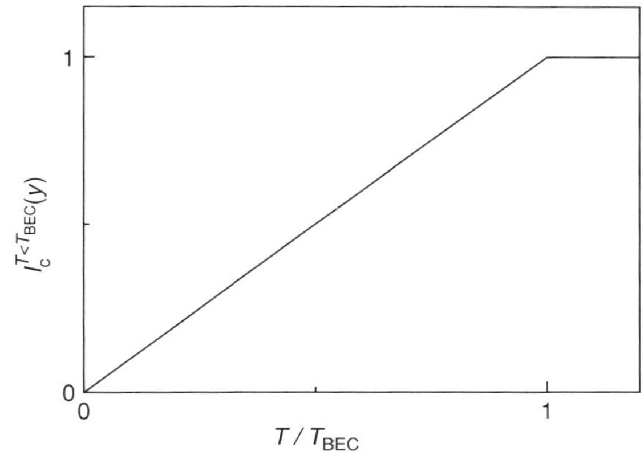

FIGURE 18.1. Josephson critical current as a function of $T/T_{\rm BEC}$ for $y \gg 1$. $I_{\rm c}^{T<T_{\rm BEC}}(y)$ is in units of $I_{\rm c}^{(0)}(y)$, and is therefore equal to 1 for $T \geq T_{\rm BEC}$.

The result in Eq. (18.5) is independent of the total number of bosons in the system. In Fig. 18.1 Eq. (18.5) is plotted as a function of $T/T_{\rm BEC}$. In describing the properties of the comb arrays fabricated by authors' collaboration, numerical simulations show that a straight line behavior is achieved already for $y = 20$.

BEC in a JJN built on a comb predicts then a rather sharp decrease of the Josephson critical current for a junction located away from the backbone and this behavior should affect the measurement of the I–V characteristic along a given finger of the JJN. The linear dependence exhibited by the solid line in Fig. 18.1 is consistent with the fact that in this system, the condensate has dimension 1.

The slope of the linear plot in Fig. 18.1 provides a direct estimate for the fraction of condensate $N_0/N_T \approx 1 - k_B T/J_0$ for a JJN on a comb graph. In fact, far away from the backbone $y \gg 1$, from Eq. (18.5) one has that indeed $1 - I_{\rm c}^{T<T_{\rm BEC}}(y)/I_{\rm c}^{(0)}(y)$ is the fraction of condensate.

18.2 Preliminary Studies for Engineering Graph-Shaped Networks

In order to meet the rather specific requirements related to the transition temperature to the superconducting state and to the physical parameters fixing the number of Cooper pairs per unit of volume, the first task is to determine the best material for the realization of graph-shaped networks.

In Table 18.1 a table of the pertinent parameters for the most used materials: Aluminium, Niobium, Lead, and Titanium is presented. Taking into account that working with critical temperatures $T_{\rm BEC}$ lower than $0.4K$ may become impractical, it results that the Titanium should be avoided. In the following, only the data referring to Nb, Pb or Al have to be retained.

From Table 18.1 one may infer the number of Cooper pairs per junction referred to a standard junction of 1 micron by 1 micron, and for $\delta^* = \delta - 2\lambda_{\rm L}$, where δ is the thickness and $\lambda_{\rm L}$ is the London penetration depth. By assuming for the Cooper pair density the values: Nb $= 129$, Al $= 3.1$, Pb $= 24$ in units of $10^{24}/m^3$, in Table 18.2 the number of Cooper pairs for two typical material thickness is reported.

TABLE 18.1. Physical characteristics of the most common materials in the construction of Josephson junctions. The critical temperature, the Fermi energy, the Fermi velocity, the coherence length, the London penetration depth, the gap at $0K$, the density of quasi particles, the density referred to the atomic density and the atomic density is shown. Where two values are presented, it refers to evaluations from different researchers.

	T_c (K)	E_F (eV)	V_F (10^6 m/s)	ξ_c (μ m)	λ_L (m)	$\Delta(0)$ (meV)	N_{cp} ($10^{24}/m^3$)	$N_{cp} \times 100$ Atom density	N ($10^{28}/m^3$)
Al	1.17	11.7	1.6	1.28	19	0.178	2.75	0.511	18.1
			2.03				3.06		
Nb	9.25		0.6	0.039	32.8	1.403	129	23.2	
Ti	0.40		≈ 0.3	0.44	70	0.061	2.42	0.425	
Pb	7.20	9.47	1.4	0.103	35.1	1.092	15.2	7.18	13.2
			1.83		34.4		23.7		

TABLE 18.2. Number of Cooper pairs ($\times 10^{-6}$) for two typical thickness of 80 and 200 nm.

	δ (80nm)	δ^* (80nm)	δ (200nm)	δ^* (200nm)
Al	0.24	0.13	0.61	0.49
Pb	1.9	0.24	4.7	3.8
Nb	10.3	1.8	26	21

18.3 Fabrication of Comb-Shaped JJNs

The fabrication of Josephson junction arrays suitable for the observation of the condensation predicted by the theory cannot follow, for several reasons, our standard lithographic techniques based on Nb –AlOx –Nb tunnel junctions trilayer.[5] In this fabrication procedure the definition of the windows of the junctions is obtained by selective anodization. In general, when a large number of junctions has to be fabricated, it is difficult to define the lithographic processes for directing the anodizing currents to the Nb trilayer, especially when the geometry of the samples is complex. Therefore, our standards had to be modified to set up a fabrication procedure based on the definition of the junction's windows through an isolating SiO_2 layer. Moreover, it is known that in the Nb-trilayer process it is not trivial to obtain junctions of good quality with a low current density (10 A/cm^2 or below is required). On the basis of the theoretical predictions, we have estimated that in order to observe a condensation temperature in the range (4.2–9.0) K the maximum Josephson current should be of the order of 1 μA.

Although the fabrication process set up by our collaboration needs improvements for what concerns the uniformity of the junction critical currents (the problem arises mainly from the definition of the geometry of the junctions) the results obtained in terms of current densities and current–voltage characteristics are so far very encouraging. The design of the samples closely follows the geometry studied in the theoretical model. An example of the final result of the fabrication process is shown in Fig. 18.2(a) and (b). The areas of the junctions defined by the SiO_2 dielectric are not clearly visible in the Fig. 18.2(b) because they are covered by the top electrode. The size of the junctions, however, vary between 9 and 25 square *microns*. All the superconducting islands have the same area/volume so that the Josephson supercurrent would be distributed evenly between the junctions. The cross-like islands providing the connections for

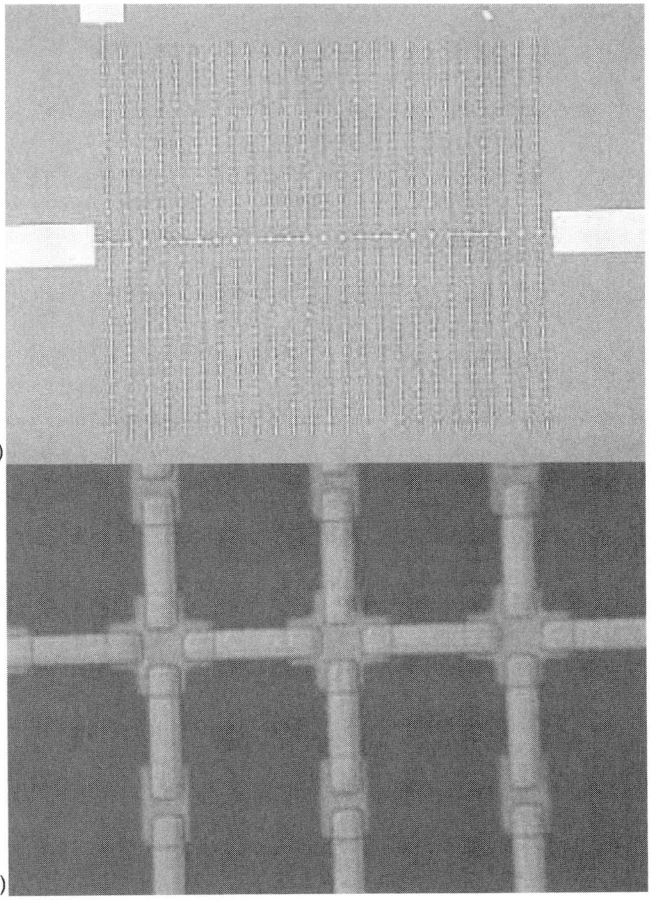

(a)

(b)

FIGURE 18.2. Optical microscope pictures of the comb-shaped Josephson junction arrays samples fabricated within authors' collaboration, (a) an overall view of the chip showing also the superconducting pads connecting the end points of the arrays to current generators and voltmeters, (b) a zoom of the central part showing the backbone cross island along with the Nb-contact electrode connecting them to form Josephson junctions.

the backbone (Fig. 18.2(b)) have an area/volume two times larger than the island containing only two junctions. The backbone, as well as the fingers, contains 50 junctions connected in series. As it is clear from Fig. 18.2, the ends of the arrays are connected directly to the (superconducting) contact pads; a situation which is not strictly the condition proposed in the theoretical models available.[1–4] However, numerical simulations show that the influence of a large superconducting electrode at the end junctions of the arrays does not substantially affect the predicted phenomena. In these numerical approaches the effect of the isolation of the normal metal films contact pads (in place of the superconducting ones) has been evaluated.

As mentioned in the introduction, one expects condensation along the horizontal central part of the graph while a decrease of the Josephson current is expected along the fingers (vertical lines). This phenomena is discussed by providing the current–voltage characteristics of the arrays.

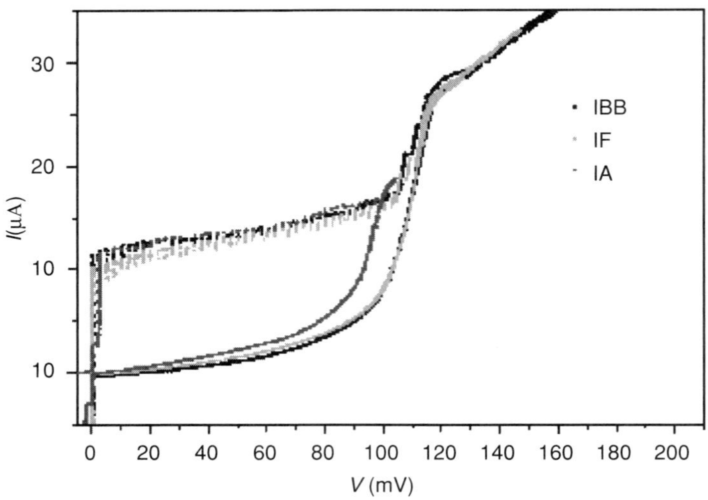

FIGURE 18.3. Current–Voltage characteristics of the Array (IA,red), backbone (IBB, black) and first finger (IF, green), measured at 60mK.

All the chips fabricated within our collaboration contain, along with the comb-shaped arrays, an isolated array with exactly the same shape of the backbone of the comb. The purpose of this extra array is to compare, at every temperature, the behavior of an isolated array with the one embedded in the graph. The extra array is not shown in Fig. 18.2.

18.4 Measurements

The chips fabricated within our collaboration and described in detail in the previous section have been tested at 4.2 K and 60 mK by measuring the *I–V* characteristics of the Josephson isolated array, of the central backbone and of one of the fingers of the comb array. Well distinguishable differences between the current–voltage characteristics of the arrays have been recorded at 60 mK.

Results of our measurements are shown in Fig. 18.3, where we report in black color the current–voltage characteristics of the backbone (label IBB), in green the one of a finger (IF), and in red the characteristics (IA) of the isolated array.

In the design all the arrays have the same number of junctions i.e., 50. However, one can clearly see in Fig. 18.3 that the finger and the backbone have the same number of junctions while in the isolated array some junctions are missing. A scale-expanded version of Fig. 18.3 in which the quasi-particle background has been subtracted from the three curves, evidences more clearly an effect that is also visible in Fig. 18.3. The junctions of the finger arrays have currents which are few microamperes below those of the backbone array. Also, the isolated array maximum currents (red curve) are in between the two.

We are now considering carefully all the possibilities that could generate the effect clearly observed in Fig. 18.4, including, naturally, geometrical problems due to technological fabrication process. We note, however, that our analysis leads to a difference in current between backbone junctions and finger junctions of the order of 2 μA (a rough 20% over the total currents of the

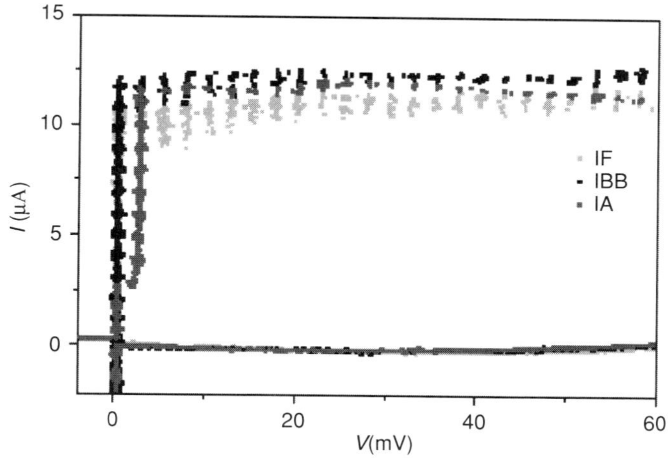

FIGURE 18.4. A scale expansion of Fig. 18.3. One can see more clearly the effects of the difference in the currents. The quasi-particle background has been subtracted from all the curves.

junctions having an area of 10 μm^2). At the same time we are trying to improve our fabrication process in order to obtain more "ideal" current–voltage characteristics and more convincing evidence that a spatial Bose–Einstein condensation in graph-shaped arrays occurs.

References

1. R. Burioni, D. Cassi, I. Meccoli, M. Rasetti, S. Regina, P. Sodano, and A. Vezzani, *Europhys. Lett.* **52,** 251 (2000).
2. R. Burioni, D. Cassi, M. Rasetti, P. Sodano, and A. Vezzani, *J. Phys. B: At. Mol. Opt. Phys* **34,** 4697 (2001).
3. G. Giusiano, F. P. Mancini, P. Sodano, and A. Trombettoni, *Int. J. Mod. Phys. B* **18,** 691 (2004).
4. I. Brunelli, G. Giusiano, F. P. Mancini, P. Sodano, and A. Trombettoni, *J. Phys. B: At. Mol. Opt. Phys* **37,** S275 (2004).
5. A. Barone and G. Paternò, *Physics and applications of the Josephson effect*, (Wiley and sons, New York, 1982).
6. S. Morohashi and S. Hasuo, *J. Appl. Phys.* **61,** 4835 (1987).

19

Cooper Pair Shuttle: A Josephson Quantum Kicked Rotator

ALESSANDRO ROMITO, SIMONE MONTANGERO AND ROSARIO FAZIO

NEST-INFM & Scuola Normale Superiore, Piazza dei Cavalieri, 7 I-56126 Pisa, Italy

Abstract

An experimental setup to realize the quantum kicked rotator by means of a periodically driven superconducting single electron transistor is presented here. The fidelity of such system that displays characteristic features of chaotic dynamics: instability of quantum motion within the Ehrenfest time scale, chaotic diffusion and dynamical localization is studied. The noise effects are taken into account.

19.1 Introduction and System Description

The kicked rotator is the paradigm model to study the effects of classical chaos in the quantum world.[1,2] The exponential localization of the wave function due to interference effects is one of the main characteristic features of quantum chaotic systems. The dynamics of the Quantum Kicked Rotator (QKR) follows the classical exponential instability (characterized by a positive Lyapunov exponent λ) only up to the Ehrenfest time t_E needed for a minimal quantum wave packet to spread in the angle coordinate of the action-angle phase space.[3,4] This sets the time scale at which quantum interference effects starts to be important leading to weak localization correction to the classical behavior.[5] After a localization time $t^\star$ the classical-like diffusive behavior stopped by quantum effects.[1,2] Since typically $t^\star \gg t_E$, the diffusive behavior is possible also in the absence of exponential instability.

Despite the long-standing interest in the QKR, only few proposals have been put forward and the only experimental implementation so far has been realized with cold atoms exposed to time-dependent standing waves of light.[6] Here, we suggest to realize a QKR by means of a Superconducting Single Electron Transistor (SSET).[7*] We analyze its dynamics that in certain limits it reduces to a "generalized" QKR. We also quantify the stability of quantum motion under a perturbation by means of the fidelity[8–15] and following[16] we present a setup through which it is possible to measure it for our proposed Josephson Quantum Kicked Rotator.

*Periodically driven Josephson junction has been already suggested to study quantum chaos in Ref 8

The system we consider, illustrated in Fig. 19.1, is very closely related to the Cooper pair shuttle[17,18] but it operates in the regime where the Josephson coupling is much larger than the charging energy. The perturbation needed to compute the fidelity is realized by coupling the system capacitively to a Cooper pair box. The flexibility in the design allows us to consider several different situations. In the semiclassical regime and for strong enough perturbations the fidelity decay, exponentially with rate given by the Lyapunov exponent or power-law, follows the classical one up to the localization time $t^{\star}$. For $t > t^{\star}$, the fidelity oscillates around a value given by the inverse of the localization length. An additional feature of this implementation is the possibility to tune the dynamics by changing the phases of the superconducting electrodes.

The Cooper pair shuttle is a superconducting device composed by a small superconducting island (grain) coupled to two macroscopic leads.[17] The couplings to left (L) and right (R) electrodes are time dependent with period $2T$ and the grain is never connected to both leads simultaneously. The two leads are macroscopic and have definite phases $\phi_{\mathrm{L,R}}$, while the superconducting island is described by the number n of excess Cooper pairs present on it. In the Cooper pair box, only states consisting of zero and one Cooper pairs are allowed ($|0\rangle$ and $|1\rangle$, respectively). The Hamiltonian describing the system is

$$\hat{H} = \hat{H}_0 \otimes |0\rangle\langle 0| + \hat{H}_1 \otimes |1\rangle\langle 1| , \tag{19.1}$$

$$\hat{H}_0 = \frac{E_{\mathrm{C}}}{2}\left(\hat{n} - \frac{n_{\mathrm{g}}}{2} - \mu\frac{n_{\mathrm{G}}}{2}\right)^2 - \sum_{b=\mathrm{L,R}} E_{\mathrm{J}}^{(b)}(t)\cos(\hat{\varphi} - \phi_b) + \mathrm{const.}, \tag{19.2}$$

$$\hat{H}_1 = \hat{H}_0 + E_{\mathrm{C}}\,\mu\,\hat{n} + \mathrm{const.}, \tag{19.3}$$

where $\hat{\varphi}$ is the conjugate phase of $\hat{n}$ ($[\hat{n}, \hat{\varphi}] = -i$). $E_{\mathrm{C}} = (2e)^2 C_\sigma/(C_\Sigma C_\sigma - C_{\mathrm{int}}^2)$ is the charging energy; all capacitances involved in the definition of the total capacitance of the Cooper pair shuttle, $C_\Sigma = C_{\mathrm{int}} + C_{\mathrm{L}} + C_{\mathrm{R}} + C_{\mathrm{g}}$, and the Cooper pair box, $C_\sigma = C_{\mathrm{int}} + C + C_{\mathrm{G}}$, are shown in Fig 19.1; $n_{\mathrm{g}} = V_{\mathrm{g}}C_{\mathrm{g}}/(2e)$, $n_{\mathrm{G}} = V_{\mathrm{G}}C_{\mathrm{G}}/(2e)$ are dimensionless gate charges. In the Cooper pair box, only states consisting of zero and one Cooper pairs are relevant ($|0\rangle$ and $|1\rangle$). This is guaranteed by the condition $C < C_\sigma \ll C_\Sigma$. We set to zero the Josephson energy of the Cooper pair box and $0 \le \mu = C_{\mathrm{int}}/C_\sigma < 1$. The time dependence of the Josephson energies $E_{\mathrm{J}}^{(\mathrm{L,R})}(t)$ are plotted in Fig. 19.1. When the grain is coupled to one of the leads ("Josephson kick") the corresponding Josephson coupling has value E_J, otherwise $E_{\mathrm{J}}^{(\mathrm{L})}(t) = E_{\mathrm{J}}^{(\mathrm{R})}(t) = 0$. We employ a sudden approximation (switching time $\Delta t \ll 1/E_{\mathrm{J}}$) so that $E_{\mathrm{J}}^{(\mathrm{L,R})}(t)$ can be approximated to step functions.

19.2 Classical and Quantum Dynamics

We first set $\mu = 0$ and study the chaotic dynamics of the Cooper pair shuttle. For the sake of simplicity we assume $V_{\mathrm{g}} = V_{\mathrm{G}} = 0$. The Hamiltonian in Eq. (19.2) is then

$$\hat{H}_0 = \frac{E_{\mathrm{C}}}{2}\hat{n}^2 - E_{\mathrm{J}} \sum_{n\in\mathbb{N}} [\cos(\varphi - \phi_L)\Theta(t - 2nT) + \cos(\varphi - \phi_R)\Theta(t - (2n+1)T)] , \tag{19.4}$$

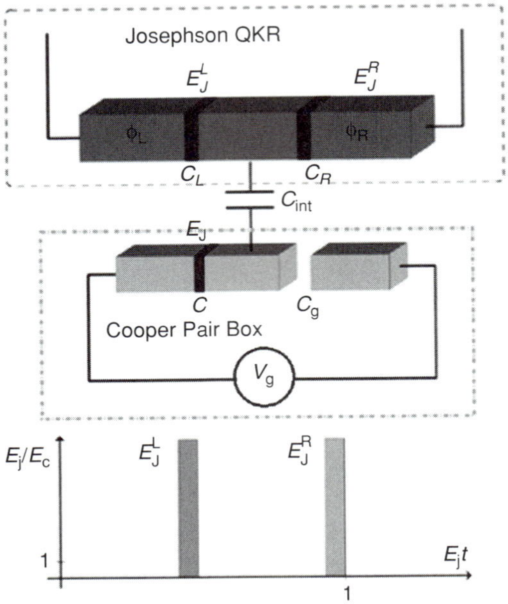

FIGURE 19.1. Upper panel: Schematic drawing of a Cooper pair shuttle (dashed red box) capacitively coupled to a Cooper pair box (dot-dashed green box). Lower panel: Time dependence of the left and right Josephson energies within a single period $2T = 2(t_J + t_C)$.

where $\Theta(t) = \theta(t)\theta(t_J - t)$ are the squared Josephson energy pulses presented in Fig. 19.1. Under the assumption that $E_J \gg E_C$, if E_C cannot induce a significant change of φ during the kick,[19] the dynamics of the Cooper pair shuttle mimics that of a QKR[†] with the additional free parameter $\phi = \phi_R - \phi_L$. This condition reads $nE_C t_J/\hbar \lesssim 1$, thus establishing a condition on the maximum number of allowed charge states involved in the dynamics. The evolution operator is therefore

$$\hat{U}(2T) = e^{ik\cos(\hat{\varphi}-\phi_R)} e^{-i\frac{K}{2k}\hat{n}^2} e^{ik\cos(\hat{\varphi}-\phi_L)} e^{-i\frac{K}{2k}\hat{n}^2}, \tag{19.5}$$

with $k = E_J t_J/\hbar$, $K = (E_C t_C/\hbar)(E_J t_J/\hbar)$. As the parameters k, K are varied, the dynamics of the QKR exhibits several important phenomena, including quantum ergodicity, quantum resonances and dynamical localization.[1] The classical limit is obtained for $k \to \infty$, keeping $K = \text{const}$. The classical dynamics corresponding to [5] depends only on the parameters K and $\phi = \phi_R - \phi_L$, as can be seen by the redefinition $p = (K/k)n$, $\theta = \varphi - \phi_L$, leading to the slightly modified Chirikov map

$$\begin{cases} p_t &= p_{t-1} - K\sin\{\theta_{t-1} - [(t+1) \bmod 2]\phi\} \\ \theta_t &= \theta_{t-1} + p_t, \end{cases} \tag{19.6}$$

where subscripts refer to time measured in units of T. The role of ϕ in the classical map can be investigated by following.[20] We substitute the deterministic description in Eq. (19.6) with a probabilistic one where a random term $\delta\varphi_t$ is added to the second equality of Eq. (19.6). This

[†] Such limit is reached independently on the particular shape of the Josephson kicks.

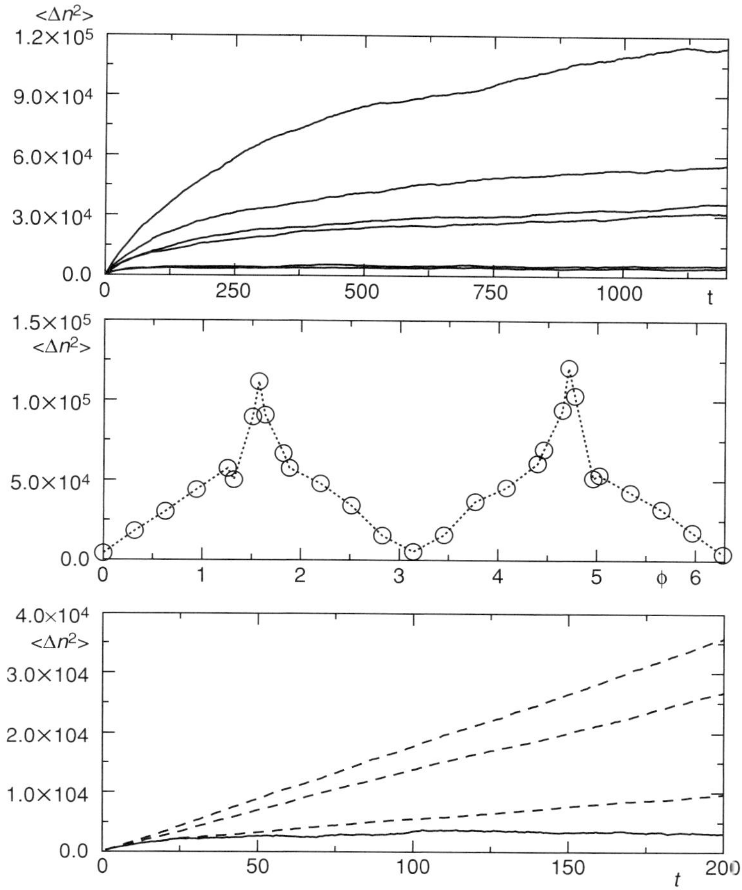

FIGURE 19.2. Upper figure: $\langle(\Delta n)^2\rangle = \langle(n - \langle n\rangle)^2\rangle$ as a function of time for $K = 10$, $k = 15$, $\mu = 0$, and phase difference (from bottom to top) $\phi = 0, 0.05, 0.1, 0.4, 0.8, 0.25$ in units of 2π. Middle figure: Saturation value of $\langle(\Delta n)^2\rangle \propto \ell^2$ as a function of ϕ for the same parameter values as in the upper figure. Lower figure: $\langle(\Delta n)^2\rangle$ as a function of time for $K = 10$, $k = 15$ and from bottom to top $\mu = 0, 10^{-5}, 10^{-3}, 10^{-1}$.

leads to a diffusive dynamics of the charge on the central island for $K > 1$: $\left\langle(n_2 - n_0)^2\right\rangle \stackrel{t\to\infty}{\longrightarrow} D(2t)$, where D is the diffusion coefficient

$$D = \frac{k^2}{2}\left[1 - 2\cos(2\phi)J_2(K) + \mathcal{O}\left(\frac{1}{K}\right)\right]. \qquad (19.7)$$

The QKR follows the classical diffusive behavior up to the localization time $t^\star$. Quantum interference effects, as shown in Fig. 19.2 (upper panel), for $t > t^\star$, suppress this chaotic diffusion: The wave function is exponentially localized in the charge basis, over a localization length ℓ, and we have $\ell \sim t^\star \sim D$ [1]. The charge fluctuations of the central island freeze in time. The localization length can be further tuned by changing the phase difference as demonstrated in the middle panel of Fig. 19.2. The lower panel shows the effect of noise: the localization is destroyed and a slow diffusion appears.[22] In the limit of strong noise, the signature of dynamical localization is completely wiped out: the quantum coherences are destroyed and the wave function diffuses following the classical chaotic behavior.

19.3 Fidelity

We now turn to the discussion of the fidelity defined as $F(t) = |f(t)|^2 = |\langle\psi(0)|\,\hat{f}(t)\,|\psi(0)\rangle\,|^2$, where $\hat{f} = \exp(\frac{i}{\hbar}\hat{H}_1 t)\exp(-\frac{i}{\hbar}\hat{H}_0 t)$ is the echo operator ($\hat{H}_0$ and $\hat{H}_1$ defined in Eqs.(19.2) and (19.3). In this case the fidelity has a simple interpretation. The state evolves with the unperturbed Hamiltonian H_0, at half evolution it is shifted by an amount μ along the coordinate n and finally the state evolves back in time with *the same* Hamiltonian H_0. $F(t)$ measures the overlap of the final and initial state. Note also that the specific form of the perturbation implies that $f(t)$ is a 2π-periodic function of $\mu K/k$. The fidelity is a function of the specific perturbation, but its dependence on time shows rather general features.[11–15] The fidelity decays as a function of time until it reaches a saturation value $1/\ell$ for $t > t^*$, ℓ being the localization length, i.e., the total number of quantum levels involved in the dynamics. This saturation is shown in Fig. 19.3 and the saturation value provides an indirect measurement of the localization length. At very short times $t \lesssim t_p$, perturbation theory holds and the fidelity decay is quadratic: $1 - F(t) \propto (\mu t)^2$.[9] The functional dependence of $F(t)$ for $t_p \lesssim t \lesssim t^*$ is related to the strength of the perturbation. General arguments[11,13] allow us to define a perturbative border such that for $\mu < \mu_p \sim 1/\sqrt{\ell}$ the fidelity decay is Gaussian. Beyond such boundary the decay is exponential: $F(t) \sim \exp(-\Gamma t)$. For $\mu_p < \mu < \mu_c \sim 1$ the decay rate is $\Gamma \sim \mu^2$, in agreement with the Fermi golden rule.[11] For $\mu > \mu_c \sim 1$, the quantum fidelity decay follows the classical one up to the localization time scale $t^\star$.[13] For times small than the Ehrenfest time scale, the decay is exponential, with a perturbation-independent rate $\Gamma = \lambda$,[10] where λ is the Lyapunov exponent characterizing the exponential instability of classical chaotic dynamics. This behavior can be observed in Fig. 19.4. The Lyapunov decay takes place up to time $\sim (1/\lambda)\ln(k/\mu K)$ and is followed by a square root decay: $F(t) \propto 1/\sqrt{Dt}$ (see Fig. 19.4). As can be observed in Fig. 19.3 and 19.4, the fidelity decay follows the classical one up to the localization time scale for experimentally accessible values of $\mu \sim 0.8$.

19.4 The Effect of Noise

Let's now consider the effect of the external environment on our system, and focus on effect of noise due to gate voltage fluctuations or (in some limits) to fluctuating background charges. It amounts in adding a term to the Hamiltonian of the form

$$\hat{H}_0 \to \hat{H}_0 + \xi(t)\hat{n}\,, \tag{19.8}$$

$\xi(t)$ being Gaussian distribute with $\langle\xi(t)\rangle_{\text{stoc}} = 0$ and $\langle\xi(t)\xi(t')\rangle_{\text{stoc}} = \hbar^2\gamma\,\delta(t-t')/T$. Due to the condition $E_J \gg E_C$, this type of noise is relevant only between kicks‡. The effects of such noise on the classical map (19.6) is to add a Gaussian stochastic step to the phase at every kick.

‡ During the kick one should include the effect of critical current fluctuations. As long as fluctuations are slow on the dimensionless time scale t_J/T this amount in considering fluctuations in kick strength.[21]

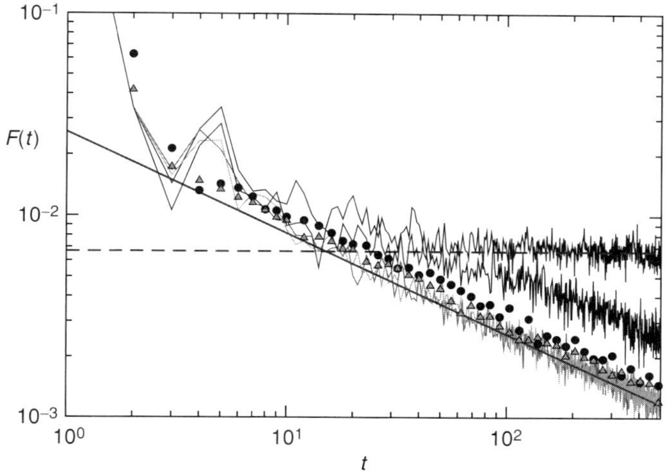

FIGURE 19.3. Fidelity dependence on time for $K = 10$, $k = 15$, $\phi = 0$, $\mu = 0.5$ and different noise strengths: $\gamma t_C = 0, 10^{-5}$ (upper and lower black curves), 10^{-2} (green), 10^{-1} (red). Circles and triangles show the decay of fidelity in the classical limit (with no noise) and in the limit of strong noise respectively. Dashed and full blue lines show the saturation of fidelity in the localized regime ($F(t) = 1/\ell$) and the power law decay $F(t) \propto 1/\sqrt{Dt}$.

The phase randomization is already obtained by the chaotic dynamics, so the presence of $\xi(t)$ does not significantly affect the classical diffusion coefficient (see Fig. 19.3). In the quantum case, both the averaged density matrix and the averaged fidelity amplitude can be determined by a master equation. The evolution of the density matrix is given by

$$\begin{cases} \dot{\hat{\rho}}(t)/T &= \frac{\iota}{\hbar}\left[\hat{H},\hat{\rho}\right], \text{ at kicks} \\ \dot{\hat{\rho}}(t)/T &= \frac{\iota}{\hbar}\left[\hat{H},\hat{\rho}\right] - \frac{\gamma}{2}\left(\hat{n}^2\hat{\rho} + 2\hat{n}\hat{\rho}\hat{n} + \hat{\rho}\hat{n}^2\right), \text{ otherwise} \end{cases} . \tag{19.9}$$

If noise is weak ($\gamma t_C \ll 1$), it gives small corrections to the results discussed so far. The modifications of the fidelity at a generic value of γ are shown in Figs. 19.3 and 19.4 resulting in the destruction of the dynamical localization. In the limit of $\gamma t_C \gg 1$, the system reaches the classical behavior characterized by an exponential decay of fidelity with the Lyapunov exponent at short time scales (Fig. 19.4) and by a behavior $F(t) \propto 1/\sqrt{Dt}$ at arbitrary long time (Fig. 19.3). This non trivial limit allows an analytical treatment: The diagonal terms of the density matrix $\hat{\rho}$, $\rho_n \equiv \langle n|\,\hat{\rho}\,|n\rangle$, are not affected by the noise, while the off-diagonal ones are exponentially suppressed on a time scale $\sim \gamma^{-1}$. Therefore, as $\gamma \to \infty$, only diagonal elements ρ_n of $\hat{\rho}$ survive, and they are determined, at integers multiples of periods, by the map

$$\rho_m(t+1) = \sum_n J_{m-n}^2(k)\rho_n(t)\,. \tag{19.10}$$

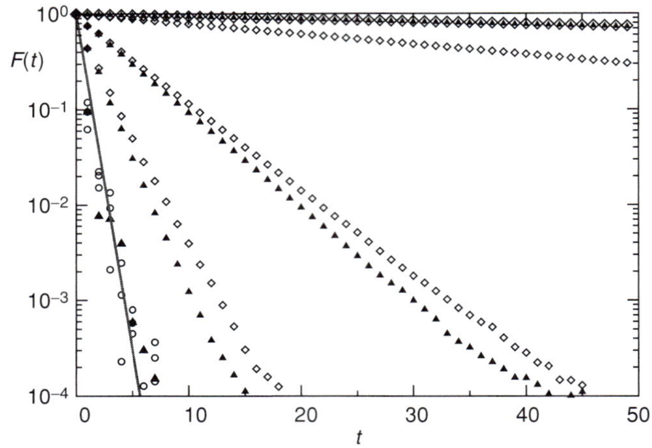

FIGURE 19.4. Fidelity decay for $K = 10$, $k = 2.6 \times 10^3$, $\phi_L = \phi_R = 2\pi/3$. Different curves correspond to different values of of μ and γt_C, in the Fermi golden rule regime (diamonds, from top to bottom $\mu = 5 \times 10^{-2}, 10^{-1}, 3 \times 10^{-1}, 5 \times 10^{-1}$) and in the Lyapunov decay Regime (circles, $\mu = 0.8, 1, 1.5$). Black triangles are obtained in the presence of noise ($\gamma t_C = 1$), for $\mu = 5 \times 10^{-2}, 3 \times 10^{-1}, 5 \times 10^{-1}, 1$. The red line shows the exponential decay $f(t) = \exp(-\lambda t)$ where λ is the Lyapunov exponent of the kicked rotator for $K = 10$: $\lambda = 1.62 \approx \ln(K/2)$.

The Fourier transform of this map has a simple form: every step consist in a multiplication by the Fourier transform of the squared Bessel function:[23]

$$\tilde{p}(x,t) = \left[\sum_l J_l^2(k) e^{-\imath l x}\right]^t = \left[J_0\left(2k \sin\left(\frac{x}{2}\right)\right)\right]^t, \tag{19.11}$$

where $\tilde{p}(x,t)$ is the Fourier transform of $\rho_m(t)$. The map[11] gives a diffusive behavior: $\langle\langle \hat{n}^2(t) \rangle\rangle_{\text{stoc}} = \sum_n \rho_n(t) n^2 = \left(\imath \partial_x^2\right) \tilde{p}(x,t)\big|_{x=0} = (k^2/2)t$, and thus the decay of the fidelity $F(t) \propto 1/\sqrt{Dt}$ follows the classical one. Note also that in the strong damping limit the diffusion coefficient is independent of ϕ. This is because noise destroys time correlations $\langle \sin(\theta_t) \sin(\theta_{t+2}) \rangle$, from where the dependence on ϕ arises.

In presence of noise the averaged fidelity amplitude, which is directly accessible by our experimental setup, display rather different behavior. The evolution of the fidelity amplitude can be determined analytically as $f(t) = Tr[\hat{f}]$ and $\hat{f}$ fulfils the same differential equation Eq. (19.9), once the replacement $[\hat{H}, \hat{\rho}] \longrightarrow (\hat{f}\hat{H}_0 - \hat{H}_1\hat{f})$ has been performed. The same argument used for the density matrix leads to the conclusion that in the $\gamma \to \infty$ limit $\hat{f}$ is diagonal and its evolution is described by the map

$$f_m(t+1) = e^{-i\frac{K}{k}\mu m} \sum_n J_{m-n}^2(k) f_n(t), \tag{19.12}$$

where $f_n \equiv \langle n| \hat{f} |n\rangle$. Using this map, the asymptotic decay of $|\langle f(t) \rangle_{\text{stoc}}|$ can be computed analytically. In fact, in Fourier space, the term $e^{-i\frac{K}{k}\mu m}$ corresponds to a translation $\tilde{p}(x,t) \longrightarrow \tilde{p}(x + \frac{K}{k}\mu, t)$ at each step. As $\tilde{p}$ is a 2π-periodic function of x, the asymptotic behavior of f can be determined as an ergodic average of $\ln|f|$. Finally one obtains $|\langle f(t)\rangle_{\text{stoc}}| = \exp(-ct)$, with $c = -\frac{1}{2\pi}\int_0^{2\pi} \ln|J_0[2k\sin(\theta/2)|d\theta$. It should be noticed that the exponential decay of the fidelity amplitude in the noisy regime is not strictly dependent on the particular form of the perturbation. For example, consider a different echo experiment: the perturbation is a shift of an amount μ along the coordinate φ at half evolution.[§] Again, the fidelity amplitude decay is exponential $\left|\langle f(t)\rangle_{\text{stoc}}\right| = \exp(-ct)$, with $c = |\ln J_0\,(2k\sin(\mu/2))|\,t$.

19.5 Fidelity Measurement

All the features of the fidelity discussed so far can be measured by adapting the protocol presented in[16] as sketched in Fig.19.1. By preparing the system in the initial state $|\psi(0)\rangle\otimes(|1\rangle + |2\rangle)/\sqrt{2}$ and by applying a $\pi/2$-pulse to the Cooper pair box at time t, the fidelity can be extracted by measuring the probability $P_1 = [1 - \mathrm{Re}\,[f(t)]]/2$ of the Cooper pair box being in the state $|1\rangle$. By repeating the procedure for the initial state $|\psi(0)\rangle \otimes (|0\rangle + i\,|1\rangle)/\sqrt{2}$, one can measure $P_1' = [1 - \mathrm{Im}\,[f(t)]]/2$ and therefore the fidelity amplitude. In measuring the fidelity amplitude also noise effects due to fluctuations of the Cooper pair box's gate voltage have to be taken into account. These last fluctuations are uncorrelated to the the previous one, and are treated in the same way by adding the term $\Xi(t)\,|1\rangle\langle 1|$ to the Hamiltonian. One gets $\langle P_g\rangle_{\text{stoc}} + i\langle P_{g'}\rangle_{\text{stoc}} = (1+i)/2 - \exp(-\Gamma t)\langle f(t)\rangle_{\text{stoc}}/2$, where Γ is defined through $\langle \Xi(t)\Xi(t')\rangle_{\text{stoc}} = \hbar^2\Gamma\delta(t-t')/T^2$.

Finally the experimental feasibility of the proposal was commented upon. Due to physical constraints, in the proposed setup the whole parameter space could not be explored. By choosing $t_J \sim 10^{-10}$s and $E_C \sim 10^{-8}$eV, we can access parameter values corresponding to interesting physical regimes. For instance, as shown in Figs. 19.2–19.3, dynamical localization for $K = 10$, $k = 15$ (corresponding to $t_C \sim 5\times 10^{-8}$sec, $E_J \sim 10^{-4}$eV) is observed. In this case the maximum number of levels for which the QKR correctly describes the physics of the system is $\hbar/(E_C t_J) \sim 6\times 10^2$.

Acknowledgments. We acknowledge fruitful discussions with Giuliano Benenti, Italo Guarneri and Valentin Sokolov. This work was supported by EU (SQUBIT2 and EDIQIP).

References

1. F. M. Izrailev, *Phys. Rep.* **196**, 299 (1990).
2. G. Casati, B. V. Chirikov, J Ford, and F. M. Izrailev, Stochastic behavior of classical and quantum Hamiltonian systems, Casati G. and J. Ford, (eds), *Lecture Notes in Physics* **93** Springer, New York, 1979.
3. B.V. Chirikov, *Phys. Rep.* **52**, 263 (1979).
4. G.P. Berman and G.M. Zaslavsky, *Physica* A **91**, 450 (1978).
5. C. Tian, A. Kameneev, and A. Larkin, arXiv:cond-mat/0403482.

[§]The shift in the φ direction can, in principle, be experimentally achieved by applying a voltage bias δ-pulse between the two macroscopic superconductors of the SSET.

6. F. L. Moore, J. C. Robinson, C. F. Bharucha, B. Sundaram, and M. G. Raizen, *Phys. Rev. Lett.* **75**, 4598 (1995); H. Ammann, R. Gray, I. Shvarchuck, and N. Christensen, *Phys. Rev. Lett.* **80**, 4111 (1998); C. Zhang, J. Liu, M.G. Raizen, and Q. Niu, Phys. Rev. Lett. **92**, 054101 (2004).
7. S. Montangero, A. Romito, G. Benenti, R. Fazio, arXiv:cond-mat/0407274
8. R. Graham, M. Schlautmann, and D.L. Shepelyansky, *Phys. Rev. Lett.* **71**, 255 (1991).
9. A. Peres, *Phys. Rev.* A **30** 1610 (1984).
10. R.M. Jalabert and H.M. Pastawski, *Phys. Rev. Lett.* **86**, 2490 (2001).
11. P. Jacquod, P. G. Silvestrov, and C. W. J. Beenakker. *Phys. Rev. E* **64**, 055203(R) (2001).
12. N.R. Cerruti and N.R. Tomsovic, *Phys. Rev. Lett.* **88**, 054103 (2002).
13. G. Benenti and G. Casati, *Phys. Rev. E* **65**, 066205 (2002).
14. T. Prosen and M. Znidaric, *J.Phys. A* **35**, 1455 (2002)
15. F. M. Cucchietti, D. A. R. Dalvit, J. P. Paz, and W. H. Zurek, *Phys. Rev. Lett.* **91**, 210403 (2003).
16. S. A. Gardiner, J. I. Cirac, and P. Zoller, *Phys. Rev. Lett.* **79**, 4790 (1997).
17. L. Y. Gorelik, A. Isacsson, Y. M. Galperin, R. I. Shekhter, and M. Jonson, *Nature* **411**, 454 (2001).
18. A. Romito, F. Plastina, and R. Fazio, *Phys. Rev. B* **68**, 140502 (2003).
19. B. J. Klappauf, D. A. Oskay, D. A. Steck, and M. G. Raizen, *Physica D* 131, 78 (1999).
20. A. B. Rechester, M. N. Rosenbluth, and R. B. White, *Phys. Rev. A* **23**, 2664 (1981).
21. E. Ott, T. M. Antonsen, Jr., and J. D. Hanson, *Phys. Rev. Lett.* **53** 2187 (1984).
22. D.Cohen, *Phys. Rev. A* **44**, 2292 (1991).
23. M. Abramovitz and I. Stegun, *Handbook of mathematical functions*, (Dover, New York, 1972).

20

Size Dependence of the Superconductor-Insulator Transition in Josephson Junction Arrays

S. G. CHUNG

S.G. Chung, Department of Physics and Nanotechnology Research and Computation Center, Western Michigan University, Kalamazoo, MI 49008-5252, USA and Max-Planck-Institut fur Physik komplexer Systeme, Nethnitzer Strasse 38, D-01187 Dresden, Germany; chung@wmich.edu

Abstract

The superconductor-insulator (SI) transition in a single Josephson junction (JJ) and JJ arrays in one and two dimensional geometry have been extensively studied. In such a structure, E_J, E_C and $g = R_Q/R_S$ are, respectively, the Josephson energy, capacitive Coulomb energy of the superconducting islands, and the Ohmic shunt resistance in units of the quantum resistance. $R_Q = h/4e^2$. Experimentally observed phase diagrams in the E_J/E_C - g plane show two interesting features. One is dissipative transition controlled by g and the other is quantum phase transition controlled by E_J/E_C. Here our focus is on the size transition of the Superconductor-Insulator Transition in JJ arrays.

20.1 Introduction

The superconductor-insulator (SI) transition in a single Josephson junction (JJ) and JJ arrays in one and two dimensional geometry have been extensively studied both theoretically[2,5,7,8,13,14] and experimentally.[3,4,6,9,10,17,18] Experimentally observed phase diagrams for a single JJ and for the 48×40 JJ arrays are characterized by parameters E_J, E_C and $g = R_Q/R_S$ which are, respectively, the Josephson energy, capacitive Coulomb energy of the superconducting islands, and the Ohmic shunt resistance in units of the quantum resistance, $R_Q = h/4e^2 = 6.45$ kΩ. Quantum dissipative transition at $g \sim 0.5$ was reviewed by Sondhi et al. Here our focus is on the E_J/E_C axis at $g = 0$ where the critical $E_J/E_C \sim 3$ for a single JJ and ~ 0.3 for the 48×40 JJ arrays. A Monte Carlo study of the thermodynamic limit due to Capriotti et al. reported ~ 0.4 for the latter number. This order of magnitude difference was argued to be an evidence of cooperative phenomena and the phase boundary a true phase boundary. In the case of JJ chain,[4] the opposite size dependence of the resistance which decreases with the size in the superconducting phase and increases in the insulator phase was measured and used as a signature to identify the critical coupling constant where the resistance becomes independent of the chain size. An important remark here is that this experiment still primarily looked at larger sizes close to the thermodynamic limit rather than the finite size effect of the SI transition, which will be discussed in this chapter.

20.2 Finite-size Effect in the 2DXY Model

The theoretically irrefutable treatment of the fully 2D quantum mechanical and strongly correlated systems near the phase boundary is an ever-lasting challenge. A rather robust observation both experimentally and theoretically, however, is that the parameter in the classical 2D XY model E_J/T where T is temperature roughly plays the role of the parameter E_J/E_C at $T = 0$, except, as we will see below, a major difference between the classical phase transition and the quantum phase transition; the critical E_J/T in the former increases with size whereas the critical E_J/E_C decreases. Therefore, the fundamental physics may be in the 2D XY model.

While the 2D XY model has been extensively studied in the past, there are two noteworthy development recently. One is in the context of the planar magnets. It is known that in spite of the fact that the Berezinskii–Kosterlitz–Thouless (BKT) transition describing the SI transition is infinite order, in a dozen of planar magnets, the magnetization appears like the 2nd order transition and a critical magnetization exponent is found to be rather universal $\sim$ 0.23. Note that the intra-layer exchange coupling J and inter-layer J' is $J/J' = 10^3$–10^4, therefore only fluctuation of length $(J/J') =$ 30–100 in lattice units is of 2D character, and thus the 2D XY model is realized in these planar magnets with their sizes 10^3–10^4. Bramwell and Holdsworth[16] used renormalization group, Monte Carlo and finite-size scaling to show that the transition indeed looks like 2-nd order with $T_c \sim 1.08$ for 10^3 spins and ~ 1.02 for 10^4 spins, which are fairly modified from $T_c = 0.892$ believed to be exact for the infinite system. The size dependence is significant, reflecting the essential singularity of the correlation length,

$$\xi \sim \exp(B/\sqrt{T_{\mathrm{BKT}}^{-1} - T^{-1}}) \tag{20.1}$$

where B is a positive constant [C.Itzykson and J.-M.Drouffe, 1989]. The other development is the transfer-matrix density-matrix-renormalization-group (TM-DMRG) method by the present author [S.G.Chung, 1999]. See, for DMRG, [S.R. White, 1993]. Using the Roomay-Wyld finite-size scaling, without any fitting parameters, nor relying on the BKT-RG, we found that $T_c = 1.07 \pm 0.01$ for 2000-3000 spins in agreement with.[16] The TM-DMRG allows us to calculate the partition function and the correlation function precisely as a function of the system size. These recent theories point to the strong size effect in the BKT bearing systems arising from the singular behavior of the correlation length, Eq. (20.1).

20.3 The Effective Hamiltonian

To account for the size dependence of the critical $E_J/T \equiv \beta$, we need to go one step deeper and look for a signature of the SI transition as a function of the system size. The 2D XY model is described by the action

$$L_{XY} = -\beta \sum_{ab} \cos(\phi_a - \phi_b) \tag{20.2}$$

where (ab) denotes the nearest neighbor pair in the square lattice. First we transform this action using the Villain approximation, and the duality transformation to arrive at the 2D sine-Gordon (SG) model[1]

$$L_{SG} = \frac{1}{8\pi^2\beta}\sum_{ab}(\psi_a - \psi_b)^2 - \sum_a 2\cos\psi_a \tag{20.3}$$

It should be noted that ϕ and ψ are dual to each other; ϕ denotes phase whereas ψ particle. Therefore the ordered phase and the disordered phase are interchanged. Next we use the transfer matrix method to evaluate the partition function

$$Z = \int \prod_{i=1}^{N} \mathrm{d}\psi_{\mathrm{i}} \exp(-L_{\mathrm{SG}}) \int \mathrm{d}\psi_{N+1}\delta(\psi_{N+1} - \psi_1) \tag{20.4}$$

where $\psi_{\mathrm{i}} = \mathrm{tr}(\psi_{1i}, \psi_{2i}, \ldots, \psi_{Ni})$ (tr means transpose) is the column vector (we consider $N \times N$ lattice) and we have inserted an identity

$$\delta(\psi_{N+1} - \psi_1) = \sum_n \Psi_n^*(\psi_{N+1})\Psi_n(\psi_1) \tag{20.5}$$

with $\{\Psi_n\}$ being an orthonormal complete set. Due to the delta function, the action in the first column can be written as $\tilde{K}(\psi_{N+1})$,

$$\tilde{K}(\psi) = \exp\left[-\frac{1}{8\pi^2\beta}\sum_{i=1}^{N-1}(\psi_1 - \psi_{N-1})^2 + \sum_{i=1}^{N} 2\cos\psi_i\right] \tag{20.6}$$

We can choose $\{\Psi_n\}$ to be the eigenstate of the transfer matrix equation

$$\int_{-\infty}^{\infty} \mathrm{d}\psi\, K(\psi, \psi')\Psi_n(\psi) = \exp(-\lambda_n)\Psi_n(\psi') \tag{20.7}$$

with the kernel

$$K(\psi, \psi') = \tilde{K}(\psi')\exp\left[-\frac{1}{8\pi^2\beta}\sum_{i=1}^{N}(\psi_i - \psi_i')^2 + \sum_{i=1}^{N} 2\cos\psi_i'\right] \tag{20.8}$$

Repeatedly using the TM equation, one reaches

$$Z = \sum_n \exp\{-\lambda_n(N-1)\}S_n \tag{20.9}$$

with a boundary-effect factor

$$S_n = \int_{-\infty}^{\infty} \mathrm{d}\psi\, \tilde{K}(\psi)\Psi_n(\psi) \int_{-\infty}^{\infty} \mathrm{d}\psi\, \Psi_n(\psi) \tag{20.10}$$

Note that the kernel is real symmetric and therefore all the quantities are real. Finally, we transform the integral Eq. (20.7) to a Schrodinger equation. Inserting the identity

$$\Psi_n(\psi) = \exp\{(\psi - \psi')\frac{\mathrm{d}}{\mathrm{d}\psi'}\}\Psi_n(\psi') \tag{20.11}$$

into Eq. (20.7) and performing the Gaussian integration, we reach a N-body Schriodinger problem, $H_{\mathrm{eff}}\Psi_n = \lambda_n\Psi_n$,

$$H_{\mathrm{eff}} = \sum_{i=1}^{N}\{-\frac{\tau}{2}\frac{\mathrm{d}^2}{\mathrm{d}\psi_i^2} + 2(1 + \cos\psi_i)\} + \frac{1}{2\tau}\sum_{i=1}^{N-1}(\psi_i - \psi_{i+1})^2 \tag{20.12}$$

where we have replaced $-2\cos\psi$ by $2(1+\cos\psi)$ and $\tau \equiv 4\pi^2\beta$. Note that, due to Eq. (20.9), the ground state contribution exponentially dominates with the sistem size, so we will hereafter concentrate on the ground state. A brief summary up to this point will be (1) the Villain approximation, (2) the duality transformation, and (3) only consider the ground state of H_{eff}. The H_{eff} is the well-known sine-Gordon Hamiltonian.

20.4 DMRG

The ground state and the first excited states of the SG Hamiltonian for L up to 200 can be analyzed by DMRG as follows.[12] First determine the basis states at each lattice site by solving the one-body problem, the Mathieu equation

$$\{-\frac{\tau}{2}\frac{\mathrm{d}^2}{\mathrm{d}\psi^2} + 2(1+\cos\psi)\}\Psi(\psi) = \lambda\Psi(\psi) \tag{20.13}$$

To solve this, we limit the ψ space to be $(-M\pi,\ M\pi)$ and take M to be an even integer. Then in the Floquet's solution

$$\Psi_{n\nu} = e^{i\nu\psi}P_{n\nu}(\psi), \tag{20.14}$$

where n is the band index and ν is crystal momentum, ν is determined from the periodic boundary condition $\exp(2\pi M\nu i) = 1. P_{n\nu}(\psi)$ is 2π periodic and can be expanded with a sufficiently large integer J as

$$P_{n\nu}(\psi) = \sum_{k=-J}^{J} C_{n\nu}^{k} e^{ik\psi} \tag{20.15}$$

It is convenient to work on the Wannier functions

$$W_n(\psi - 2\pi m) = \frac{1}{\sqrt{M}}\sum_{\nu} e^{-2\pi m\nu i}\Psi_{n\nu}(\psi) \tag{20.16}$$

where $m = -(M-1)/2, -(M-1)/2+1, \ldots.., (M-1)/2$. The Wannier function is localized at each cosine potential well. Including up to n bands and for fixed number of ν states, M, the dimension of the local basis states is $q = n \times M$ and all the local variables are expressed by $q \times q$ matrices. The infinite algorithm, open boundary condition, and the ground state target have been used.[15] From the previous study, we know that $q = 16$.

20.5 Spontaneous Symmetry Breaking

The SG model is nothing but the arrays of quantum pendula coupled by springs, and one can see that there is a Spontaneous Symmetry Breaking (SSB). In fact at $\tau \to \infty$ in the limit $L \to \infty$, the mass of the pendulum vanishes and the ground state is nondegenerate and translationally symmetric with respect to the phase ψ, whereas at $\tau \to 0$, mass is infinite, the ground state is symmetry broken and infinitely degenerate modulo 2π, i.e., with each ground state describing a zero-point motion near one of the potential minima $\psi = \pi \pm 2\pi$ x integer. Figure 20.1 shows a contour graph for the probability distribution of the phase ψ at the center site in the ground state

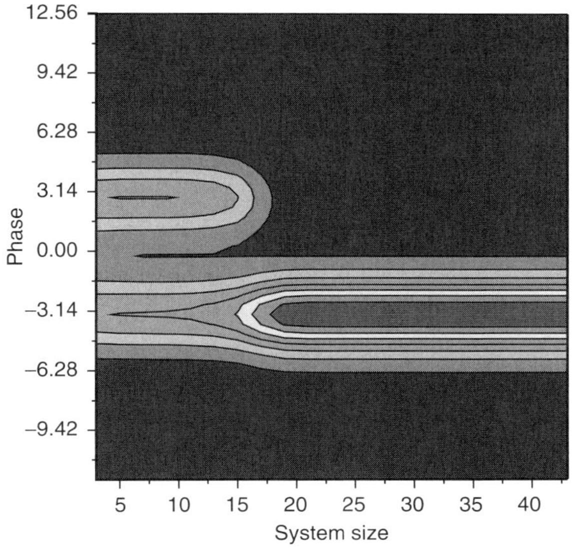

FIGURE 20.1. The contour graph for the probability distribution of the phase as a function of the size. The ground state for $\beta = 0.4$. The symmetry changes from unbroken to broken around system size 17.

vs the system size for $\beta = 0.4$. It is seen that the translational symmetry is broken around $L = 17$ from equal distribution over $-\pi$ and π potential minima to localization around π. Figure 20.2 shows the same for the first excited state indicating localization at the other minimum $-\pi$. At the same time, these states have energy splitting of 10^{-5} in the unit of the Coulomb energy, indicating the ground state degeneracy. It is noted that the distribution of ψ is essentially the same for all the lattice sites, only a few percent fluctuation at the edges. Repeating the calculation with varying β, Fig. 20.3 shows the phase average vs the system size for $\beta = 0.1 \sim 0.6$. The SSB occurs rather abruptly and at small size for small β while it gets more gradual and at larger sizes for increasing β. One can thus draw a phase diagram in the β vs L plane, see Fig. 20.4.

20.6 BKT as a SSB

The variation of critical β with the system size is significant. Two previous theories on the finite-size effect.[11,16] found that T_c in the 2D XY model varies like $T_c = 1.02$ for 10^4 arrays and $T_c = 1.07$ for 2000–3000 arrays, compared with 0.892 for the thermodynamic limit. In terms of the inverse temperatute β, this means an increase of β_c with the array size. Fig. 20.4 gives a whole picture of the size dependence of the critical β. It is interesting to see where the curve is heading. A simple extrapolation at $L \to \infty$ gives $\beta_c = 0.57$ which is different from $1/0.892 = 1.12$. Note, however, that we made the Villain approximation which by RG is known to give $\beta_c = 2/\pi = 0.637$ which is close to 0.57. Still, how do we convince ourselves that the SSB = BKT? For this purpose, a Roomany–Wyld RG scaling was carried out, which tells us that the BKT transition point is a fixed point in the $L \times \text{Gap}(L)$ vs β plane where Gap(L) is the energy gap between the ground state and the first excited state at L. We have found that β_c of BKT under our treatment is 0.57 which is exactly that of the SSB, identifying the BKT as the SSB.

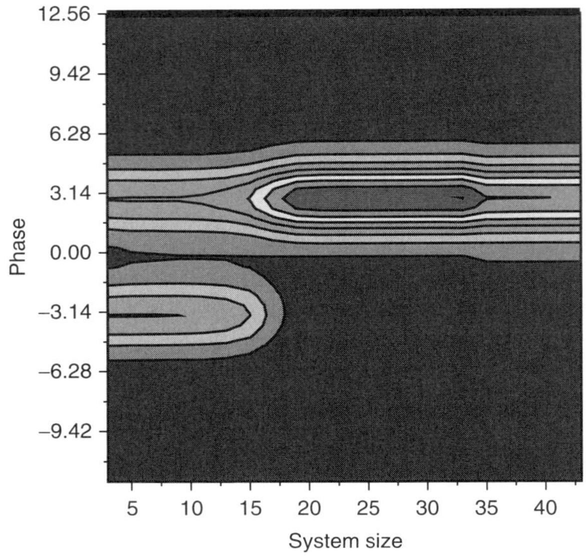

FIGURE 20.2. The same as Fig. 20.1 for the first excited state.

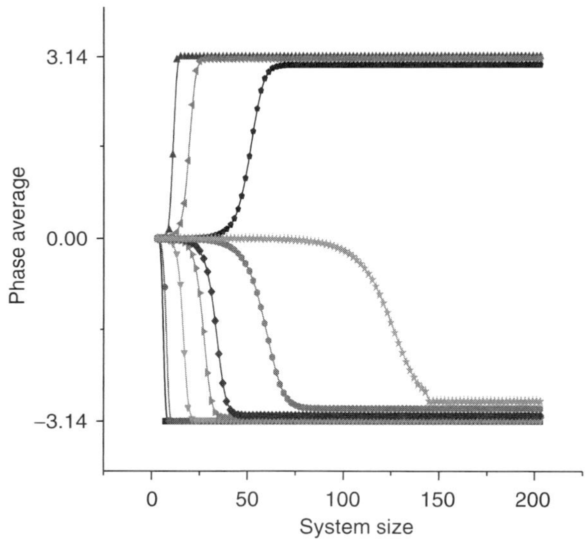

FIGURE 20.3. The phase average vs the system size for $\beta = 0.1, 0.2, 0.3, 0.4, 0.43, 0.46, 0.48, 0.5, 0.52, 0.54$. The smaller the β, the shorter the critical size over which SSB occures.

20.7 Experimental Implications

Let us now discuss the experimental implications of the present theory. First of all, we need to remember that we did a duality transformation which exchanges the ordered phase and the

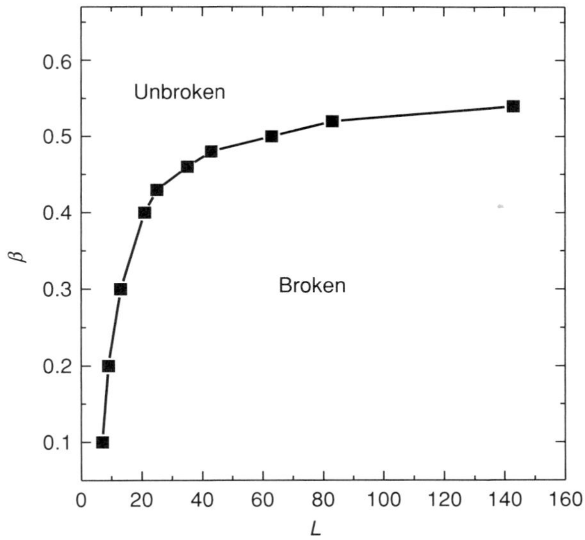

FIGURE 20.4. The spontaneous symmetry breaking in β–L plane.

disordered phase. Thus in Fig. 20.4, the broken state corresponds to the insulator phase and the unbroken states to the superconducting phase. Fig. 20.4 therefore shows the superconductor to insulator transition with the increase of the array size, and that the critical size varies significantly as a function of β. If one fixes the array size conversely, then Fig. 20.4 shows that the critical $\beta \equiv E_J/T$ is smaller for smaller arrays. This is intuitively correct; because of the singular behavior of the correlation length Eq. (20.1), the critical temperature as defined the one where the correlation length reaches the system size can be much higher than the thermodynamic one 0.892. Turn to the experimental findings of Pentilla et al. and Yamaguchi et al. on the large size dependence of the critical E_J/E_C which motivated the present work. In fact, the critical E_J/E_C *decreases* from $\sim$ 3 of the single JJ to $\sim$ 0.3 of the 48 $\times$ 40 JJ arrays, opposite to the E_J/T behavior in the 2DXY model. The key is in the fact that the 2DXY model undergoes the classical phase transition while the BKT transition along the E_J/E_C axis, $T = 0$, is the quantum phase transition. More specifically, the effective SG Hamiltonian, $H_{\rm eff}$ should be compared with that of the JJ chain.

$$H_{\rm JJ} = -4E_C \sum_{i=1}^{N} \frac{d^2}{d\psi_i^2} - E_J \sum_{i=1}^{N-1} \cos(\psi_i - \psi_{i+1}) \tag{20.17}$$

giving the correspondence $\pi^2 E_J/T \sim 4E_C/E_J$, and leading to a speculation that the critical E_J/E_C in the quantum BKT transition would occur at ~ 4 for $L \sim 3$, and ~ 0.8 for $L \sim 50$, that is, the critical E_J/E_C *decreases* with the array size, in accordance with the experiments.

20.8 Conclusions

To conclude, the 2D XY model has been analyzed using the Villain approximation, the dual transformation and the transfer matrix method augmented by DMRG. The effective Hamiltonian

H_{eff} undergoes a spontaneous symmetry breaking which is identified as the BKT transition. The resulting large size dependence of the BKT transition offers an explanation of the large size dependence of the SI transition observed in Josephson junction arrays. The present theory points out an opposite tendency for the classical BKT transition in the 2D XY model, namely the critical $\beta \equiv E_J/T$ should decrease significantly with the decrease of the system size. Turning to the experiment on the JJ chain by E. Chow et al.,[4] these authors pointed out that the resistance decreases in the superconducting phase with the size, while it increases in the insulator phase, and using this phenomena, extracted the critical E_J/E_C where the resistance does not depend on the size. Their focus was, however, still on relatively large sizes near the thermodynamic limit. Our finding indicates a rather strong size dependence at smaller sizes. A systematic experiment encompassing the single JJ to larger arrays looking for both the classical BKT and the quantum BKT would prove the present theory. Finally, this theory of SSB may be regarded as a many-body extension of the single particle band theory of Likharev and Zorin.[7]

Acknowledgments. This work was partially supported by the Visitor Program of the MPI-PKS, Dresden, Germany. The work was also partially supported by the NSF under grant No. DMR990002N and utilized the IBM P690 at the National Center for Supercomputing Applications at the University of Illinois at Urbana-Champaign.

References

1. C. Itzykson and J.-M. Drouffe, *Statistical Field Theory:* Vol.1, (Cambridge University Press, New York, 1989).
2. C. Rojas and V. Jose, Jorge, *Phys. Rev. B*, **54**, 12361 (1996).
3. D.B. Haviland, K. Anderson, and P. Argen, *J. Low Temp. Phys.*, **118**, 733 (2000).
4. E. Chow, P. Delsing, and D.B. Haviland, *Phys. Rev. Lett*, **1998** 204 (1998).
5. G.-L. Ingold and H. Grabert, *Phys. Rev. Lett.*, **83**, 3721 (1999).
6. J.S. Penttila, U. Parts, P.J. Hakonen, M.A. Paalanen, and E.B. Sonin, *Phys. Rev. Lett.*, **82**, 1004 (1999).
7. K.K. Likharev and A.B. Zorin, *J. Low Temp. Phys.*, **59**, 347 (1985).
8. L. Capriotti, A. Cuccoli, A. Fubini, V. Tognetti, and R. Vaia, *Phys. Rev. Lett.*, **91**, 247004 (2003).
9. M. Watanabe and D. Haviland, *Phys. Rev. Lett.*, **86**, 5120 (2001).
10. R. Yagi, S. Kobayashi, and Y. Ootsuka, *J. Phys. Soc. Japan*, **66**, 3722 (1997).
11. S.G. Chung, *Phys. Rev. B*, **60**, 11761 (1999).
12. S.G. Chung, *Phys. Rev. E*, **62**, 3262 (2000).
13. S.G. Chung, *Phys. Rev. B*, **66**, 12503 (2002).
14. S.L. Sondhi, S.M. Girvin, J.P. Carini, and D. Shahar, *Rev. Mod. Phys.*, **69**, 315 (1996).
15. S.R. White, *Phys. Rev. B*, **48**, 10345 (1993).
16. S.T. Bramwell and P.C.W. Holdsworth, *J. Phys.: Condens. Matt.*, **5**, L53 (1993).
17. T. Yamaguchi, R. Yagi, A. Kanda, Y. Ootuka, and S. Kobayashi, *Phys. Rev. Lett.*, **85**, 1974 (2000).
18. W. Kuo and C. Chen, *Phys. Rev. Lett.*, **87**, 186804 (2001).

21

Monte Carlo Method for a Superconducting Cooper-pair-box Charge Qubit Measured by a Single-electron Transistor

HSI-SHENG GOAN

Center for Quantum Computer Technology, University of New South Wales, Sydney, NSW 2052 Australia
Email: goan@physics.uq.edu.au; Mailing address: Center for Quantum Computer Technology, Department of Physics, University of Queensland, Brisbane 4072 Australia

Abstract

The quantum trajectory or stochastic (conditional) master equation for a single Superconducting Cooper-pair Box (SCB) charge qubit measured by a Single-Electron Transistor (SET) detector is presented. This stochastic master equation describes the random evolution of the measured SCB qubit density matrix and is conditioned on a particular realization of the measured electron tunneling events through the SET junctions. Hence it can be regarded as a Monte Carlo method (or a quantum trajectory approach) that allows us to simulate the continuous quantum measurement process. We illustrate the connection between the quantum trajectory approach and the "partially" reduced density matrix approach [Y. Makhlin et al., *Phys. Rev. Lett.* **85**, 4578 (2000)]. We also present Monte Carlo simulation results for the SCB/SET measurement process.

21.1 Introduction

The Single-Electron Transistor (SET) is a highly charge-sensitive electro-meter and has been suggested as a readout device for solid-state charge qubits[1] or spin qubits[2] (through a measurement of a spin-dependent charge transfer). A Monte Carlo method[3] which allows one to follow each electron tunneling event has been successfully applied to simulate transport properties of a SET or more complicated single electronics circuits. This method gives physical insight into the processes taking place in the simulated system. But to our knowledge, it has not yet been formally applied to quantum measurement problems by a SET detector. In this study, such an investigation is provided and derive the *quantum-jump* stochastic master equation (or quantum trajectory

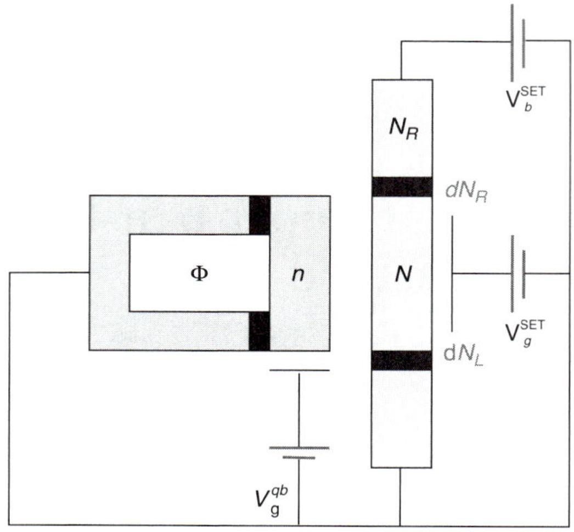

FIGURE 21.1. Schematic illustration of a Cooper-pair-box qubit measured by a SET.

equation) for a single superconducting Cooper-pair box (SCB) charge qubit (generalization to other charge qubit case is simple) continuously measured by a SET (see Fig. 21.1) is derived. This stochastic master equation describes the random evolution of the measured SCB qubit density matrix, which both conditions and is conditioned on a particular realization of the measured electron tunneling events through the SET junctions. We can regard it as a Monte Carlo method that allows us to simulate the continuous quantum measurement process of a charge qubit by a SET. This problem has been extensively studied in Ref.[1] We show that the master equation for the "partially" reduced density matrix presented in Ref.[1] (referred to here as a "partial" course-grain description) can be obtained by taking a "partial" average on the stochastic master equation over the fine grained measurement records of the tunneling events in the SET. The quantum trajectory approach (or Bayesian formalism) was introduced recently[4–6] to describe a charge qubit measured by a low-transparency point contact detector.

21.2 Model Hamiltonian

The Hamiltonian of the SCB/SET system is described[1] as: $\mathcal{H} = \mathcal{H}_{\rm SET} + \mathcal{H}_{\rm L} + \mathcal{H}_{\rm R} + \mathcal{H}_{\rm I} + \mathcal{H}_{\rm T} + \mathcal{H}_{\rm qb} + \mathcal{H}_{\rm int}$. Briefly, $\mathcal{H}_{\rm SET} = E_{\rm SET}(N - N_g)^2$ describe the charging energy of the SET. The charge on the middle island is eN, and the induced charge eN_g is determined by the gate voltage $V_{\rm g}$ and other voltages in the circuit. $\mathcal{H}_{\rm r} = \sum_{ks} \epsilon^r_{ks} c^{r\dagger}_{ks} c^r_{ks}$, where $r = L, R, I$, describes microscopic degrees of freedom of noninteracting electrons in the two leads (left and right) and the middle island of the SET, respectively. To make the charge transfer explicit, two "macroscopic" operators, $e^{\pm i\phi}$ and $e^{\pm i\psi}$ are included in the tunneling Hamiltonian[1] in the SET: $\mathcal{H}_{\rm T} = \sum_{kk's} T^{\rm L}_{kk's} c^{\rm L\dagger}_{ks} c^{I}_{k's} e^{-i\phi} + \sum_{k'k''s} T^{\rm R}_{k'k''s} c^{\rm R\dagger}_{k''s} c^{\rm I}_{k's} e^{-i\phi} e^{i\psi} + {\rm H.c.}$. The SCB qubit consists of a superconducting island (box) weakly linked to a superconducting reservoir by a Josephson tunnel junction. The tunnel junction of the SCB qubit shown in Fig. 21.1 is split into two to form a SQUID loop which allows us to control its effective Josephson energy $E_{\rm J}$ with a small external magnetic field or magnetic flux Φ. The effective Hamiltonian of the decoupled SCB qubit, written in the charge eigen basis of the number n of extra Cooper pair on the island of SCB, is $H_{\rm qb} = \frac{1}{2}(E_{\rm ch}\hat{\sigma}_z - E_{\rm J}\hat{\sigma}_x)$, where $\hat{n} = (1 - \hat{\sigma}_z)/2$ with eigenvalues $n = 0$ or 1. The capacitive Coulomb coupling between the charge on the SET island and that on the SCB qubit is represented by $\mathcal{H}_{\rm int} = 2E_{\rm int}N\hat{n}$. We will consider the case that the leading tunneling process in the SET are sequential transitions between two adjacent charge states N and $N+1$ (say, $N = 0$ and $N+1 = 1$ states to represent the extra charge on the SET island). This would be the case if the applied transport voltage across the SET is not too high and the temperature is low (for simplicity, consider the zero temperature case). Since, effectively, only two adjacent charge states $N = 0$ and $N+1 = 1$ are considered, the charge transfer operators $e^{\pm i\phi}$, in this case, satisfy $e^{-i\phi}|N\rangle = 0 = e^{i\phi}|N+1\rangle$, $e^{i\phi}|N\rangle = |N+1\rangle$ and $e^{-i\phi}|N+1\rangle = |N\rangle$. The other set of charge transfer operators satisfy $e^{\pm i\psi}|m\rangle = |m \pm 1\rangle$, where m represents the number of electrons that have tunneled into the right lead (drain) of the SET.

21.3 Measurement records and conditional density matrix

To be able to describe the measured qubit in a pure state continuously, one needs to have the maximum knowledge about the change of its state. When the qubit interacts with (is measured by) the SET, this information is lost to the SET. For example, each time when an electron tunnels onto or off the SET island, it will cause a change (e.g., a phase shift) of the qubit state. One can recover this information lost, provided that a detailed measurement record from the SET is available. The transport of electrons through the SET occurs via real states of the central island, from $N \to N+1 \to N$. The information of detecting the mth electron just tunneling into the drain only tells us that the island state now is in $|N\rangle$ state. Thus knowing the "partially" reduced density matrix $\rho_N(m,t)$ at every time[1] does not provide us with the full information.

One can imagine that in the transport process, electrons may spend different times in the intermediate $|N+1\rangle$ state, causing different phase shifts to the qubit. If the record of the times when electrons tunneling onto and off the island is not available from the measurement results of the SET, our knowledge about the precise qubit state decreases. When this happens, averaging the random dwelling times of electrons on the island over a period of time or over an ensemble of systems will then lead to the decoherence of the qubit. Hence, one needs to have a measurement record which records whether or not an electron tunnels onto or off the central island of the SET at each time interval $\mathrm{d}t$. This time interval $\mathrm{d}t$ should be much smaller than the typical qubit system evolution or response time so that no information is lost as far as the qubit system evolution is concerned. In this sense, effectively the qubit is continuously monitored or measured. For this purpose, we introduce $\mathrm{d}N_{\rm Lc}(t)$ and $\mathrm{d}N_{\rm Rc}(t)$ to represent, in the *quantum-jump* case,[5] the number (either zero or one) of tunneling events seen in infinitesimal time $\mathrm{d}t$ through the left and right junctions of the SET, respectively. Throughout the paper, the subscript or superscript c indicates that the quantity to which it is attached is conditioned on previous measurement results. If no tunneling electron is detected, the result is *null*, i.e., $\mathrm{d}N_{\rm Lc}(t) = 0$ and $\mathrm{d}N_{\rm Rc}(t) = 0$. If there is *detection* of a tunneling electron in time $\mathrm{d}t$, then $\mathrm{d}N_{\rm Lc}(t) = 1$ or $\mathrm{d}N_{\rm Rc}(t) = 1$. We can think of $\mathrm{d}N_{\rm Rc}(t)$ as the increment in the number of electrons $N_{\rm Rc}(t) = \sum \mathrm{d}N_{\rm Rc}(t)$ passing through the right junction of the SET into the drain in the infinitesimal time $\mathrm{d}t$. It is the variable $N_{\rm R}(t) = m(t)$, the accumulated electron number transmitted through the SET into the drain, which is used in Ref.[1] Since the nature of detection results is classical and that of electrons tunneling through the SET is stochastic, $dN_{Lc/Rc}(t)$ should represent a classical random process. The measurement record in each single run of experiment is the set of times $\{t_L^{(i)}\}$ and $\{t_R^{(i)}\}$ when electrons tunnel onto or off the SET island, respectively (i.e., ones of $dN_{Lc}(t)$ and $dN_{Rc}(t)$ over the entire detection time).

At first, one may expect that at the end of each time interval dt, there are four possible measurement outcomes, $dN_{Lc}(t)\,[1-dN_{Rc}(t)]$, $dN_{Rc}(t)\,[1-dN_{Lc}(t)]$, $[1-dN_{Lc}(t)]\,[1-dN_{Rc}(t)]$, and $dN_{Lc}(t)\,dN_{Rc}(t)$. However, let us consider the case in the sequential tunneling dominated regime that the probability of electrons tunneling onto and off the SET island within the same infinitesimal time interval dt is rather small. In fact, the respective probability of $dN_{Lc}(t)$ or $dN_{Rc}(t)$ equal to unity is proportional to dt (see Eqs. (21.10) and (21.11)). Thus the product of $dN_{Lc}(t)\,dN_{Rc}(t) = 1$ occurs with probability proportional to dt^2. Since only terms to order dt in the master equations, we can neglect the case that $dN_{Lc}(t)$ and $dN_{Rc}(t)$ both equal one within the same infinitesimal time interval. The possible measurement outcomes then become: $dN_{Lc}(t)$, $dN_{Rc}(t)$ and $[1-dN_{Lc}(t)-dN_{Rc}(t)]$. The first two terms, in this case, represent that an electron tunneling event through, respectively, the left and right junctions of SET is detected at the end of the time interval $[t, t+dt)$. While the last term $[1-dN_{Lc}(t)-dN_{Rc}(t)]$ represents that no tunneling event is observed in $[t, t+dt)$. Thus the conditioned density matrix $W_c(t+dt)$ to order dt at the end of the time interval $[t, t+dt)$ can be written as

$$
\begin{aligned}
W_c(t+dt) = {} & dN_{Lc}(t)\{W_{L1c}(t+dt)/\mathrm{Tr}[W_{L1c}(t+dt)]\} \\
& + dN_{Rc}(t)\{W_{R1c}(t+dt)/\mathrm{Tr}[W_{R1c}(t+dt)]\} \\
& + [1-dN_{Lc}(t)-dN_{Rc}(t)]\{W_{0c}(t+dt)/\mathrm{Tr}[W_{0c}(t+dt)]\},
\end{aligned}
\tag{21.1}
$$

where $W_{L1c}(t+dt)$, $W_{R1c}(t+dt)$, and $W_{0c}(t+dt)$ are the unnormalized density matrices, given that an electron tunneling event through left or right junction of the SET island, or no tunneling event is measured at the end of the time interval $[t, t+dt)$. Equation (21.1) simply states that when $dN_{Lc} = 1$ and $dN_{Rc} = 0$, the normalized conditioned density matrix is $W_{L1c}(t+dt)/\mathrm{Tr}[W_{L1c}(t+dt)]$, and so on. Self-consistently, the ensemble averages $E[dN_{Lc}(t)]$ and $E[dN_{Rc}(t)]$ of the classical stochastic processes $dN_{Lc}(t)$ and $dN_{Rc}(t)$ should equal respectively the probabilities (quantum average) of electrons tunneling through the left and right junctions of the SET in time dt, i.e., $E[dN_{Lc}(t)] = \mathrm{Tr}[W_{L1c}(t+dt)]$ and $E[dN_{Rc}(t)] = \mathrm{Tr}[W_{R1c}(t+dt)]$. Formally, we can write the currents through the junctions as $I_{Lc}(t) = e\,dN_{Lc}(t)/dt$ and $I_{Rc}(t) = e\,dN_{Lc}(t)/dt$.

21.4 Stochastic master equation

The question now is to find expressions for $W_{L1c}(t+dt)$, $W_{R1c}(t+dt)$, and $W_{0c}(t+dt)$ in the model. This has been done in Ref.[7] by using similar derivations in Ref.[5] and the same assumptions and approximations in Ref.[1] The conditional master equation can then be obtained:[7]

$$
\begin{aligned}
d\rho_N^c(t+dt) = {} & -[dN_{Lc}(t)+dN_{Rc}(t)]\rho_N^c(t) \\
& + dN_{Rc}(t)\,[\check{\Gamma}_R\rho_{N+1}^c(t)/\mathcal{P}_{R1c}(t)] \\
& - dt\Big\{(i/\hbar)[\mathcal{H}_{qb}, \rho_N^c(t)] + \check{\Gamma}_L\rho_N^c(t) \\
& - [\mathcal{P}_{L1c}(t)+\mathcal{P}_{R1c}(t)]\rho_N^c(t)\Big\}\,,
\end{aligned}
\tag{21.2}
$$

$$\begin{aligned}
d\rho_{N+1}^{c}(t+dt) = & -[dN_{Lc}(t) + dN_{Rc}(t)]\rho_{N+1}^{c}(t) \\
& + dN_{Lc}(t)\,[\check{\Gamma}_{L}\rho_{N}^{c}(t)/\mathcal{P}_{L1c}(t)] \\
& - dt\,\Big\{(i/\hbar)[\mathcal{H}_{qb} + 2E_{int}\hat{n}, \rho_{N+1}^{c}(t)] + \check{\Gamma}_{R}\rho_{N+1}^{c}(t) \\
& - [\mathcal{P}_{L1c}(t) + \mathcal{P}_{R1c}(t)]\rho_{N+1}^{c}(t)\Big\}\ ,
\end{aligned} \tag{21.3}$$

where the effective rates $\check{\Gamma}_L$ and $\check{\Gamma}_R$ are defined as

$$\check{\Gamma}_L\rho_N^c = \Gamma_L\rho_N^c + (\Gamma_L - \Gamma_L')\{\hat{n}, \rho_N^c\}/2, \tag{21.4}$$

$$\check{\Gamma}_R\rho_{N+1}^c = \Gamma_R\rho_{N+1}^c + (\Gamma_R - \Gamma_R')\{\hat{n}, \rho_{N+1}^c\}/2. \tag{21.5}$$

The rates $\Gamma_{L/R}$ and $\Gamma'_{L/R}$ represent the tunneling rates (in the left or right junction) with and without the presence of the extra Cooper pair on the island of the SCB (i.e., $n = 1$ or $n = 0$), respectively. They are determined by the chemical potentials $\mu_{L/R}$ of the leads and the induced charge N_g on the SET's island:

$$\Gamma_L = (2\pi\alpha_L/\hbar)[\mu_L - (1 - 2N_g)E_{SET}], \tag{21.6}$$

$$\Gamma_R = (2\pi\alpha_R/\hbar)[(1 - 2N_g)E_{SET} - \mu_R], \tag{21.7}$$

$$\Gamma_L' = \Gamma_L - (4\pi\alpha_L E_{int}/\hbar), \tag{21.8}$$

$$\Gamma_R' = \Gamma_R + (4\pi\alpha_R E_{int}/\hbar), \tag{21.9}$$

where $\alpha_{L/R} = R_Q/(4\pi^2 R_{L/R})$, $R_Q = h/e^2$ is the resistance quantum, and $R_{L/R}$ represents the resistance of the left or right junction. Self-consistently, $E[dN_{Lc}(t)]$ and $E[dN_{Rc}(t)]$ should equal their respective quantum averages, and can be written as

$$E[dN_{Lc}(t)] = \mathrm{Tr}[W_{L1c}(t + dt)] = \mathcal{P}_{L1c}(t)dt, \tag{21.10}$$

$$E[dN_{Rc}(t)] = \mathrm{Tr}[W_{R1c}(t + dt)] = \mathcal{P}_{R1c}(t)dt, \tag{21.11}$$

where

$$\mathcal{P}_{L1c}(t) = \Gamma_L\mathrm{Tr}[\rho_N^c(t)] + (\Gamma_L' - \Gamma_L)\mathrm{Tr}[\hat{n}\,\rho_N^c(t)] \tag{21.12}$$

$$\mathcal{P}_{R1c}(t) = \Gamma_R\mathrm{Tr}[\rho_{N+1}^c(t)] + (\Gamma_R' - \Gamma_R)\mathrm{Tr}[\hat{n}\,\rho_{N+1}^c(t)]. \tag{21.13}$$

Note that $\mathcal{P}_{Lc}(t)$ and $\mathcal{P}_{Rc}(t)$ appearing in Eqs. (21.2) and (21.3) are due to the normalization requirement for the density matrix after each detection interval dt as in Eq. (21.1). One can then use Eqs. (21.2)–(21.13) to simulate the conditional (stochastic) system dynamics under continuous quantum measurements by the SET.

A set of typical quantum trajectories (or stochastic time evolutions) is shown in Figs. 21.2(a)–(h) and its corresponding randomly distributed moments of detections are presented in Fig. 21.2(i) and (j). Note that, due to Coulomb blockade, the SET measurement record in Figs. 21.2(i) and (j) is in an order of exactly alternating $dN_{LC} = 1$ and $dN_{Rc} = 1$ time sequence. The conditional evolutions of the qubit alone shown in Figs. 21.2(e) and (f) can be obtained from the sum of the joint state evolutions of Figs. 21.2(a) and (b), or Figs. 21.2(c) and (d), respectively. The probabilities, $P_{0/1,c} = \mathrm{Tr}_{qb}[\rho_{0/1,c}] = \rho_{0/1,c}^{00} + \rho_{0/1,c}^{11}$, of the SET island state alone in Figs. 21.2(g) and (h) can be obtained by summing the evolutions in Figs. 21.2(a) and (c), or Figs. 21.2(b) and (d), respectively.

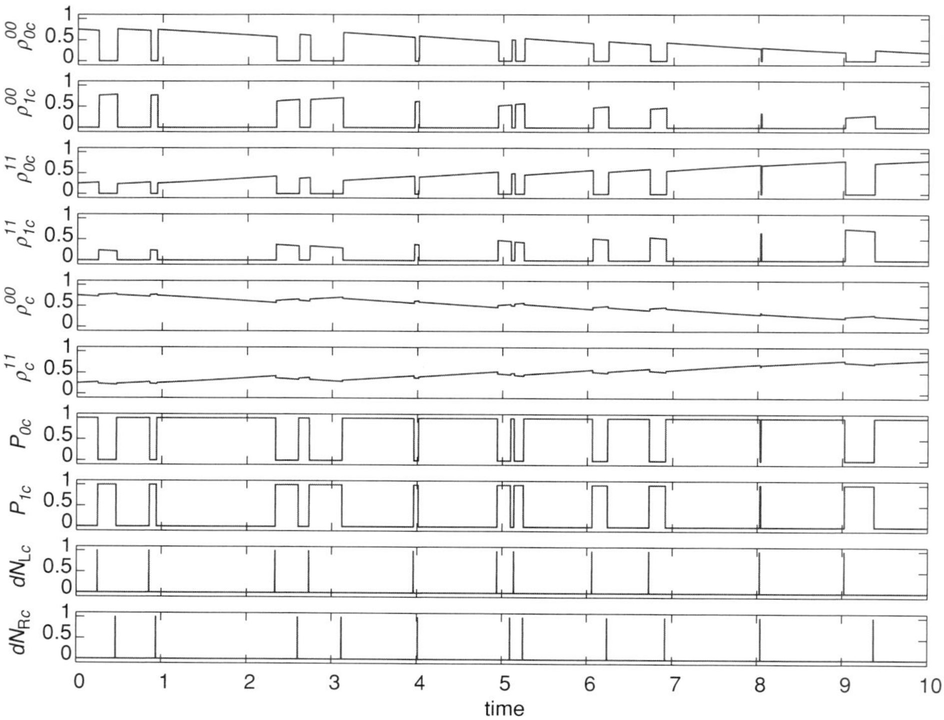

FIGURE 21.2. Set of typical quantum trajectories and corresponding detection record for the initial conditions: qubit state in $(\sqrt{3}\,|0\rangle + |1\rangle)/2$ and the SET island in $N = 0$ state. Other parameters are $E_{\rm int} = 1.5E_{\rm ch} = 1500E_{\rm J}$, $\Gamma_{\rm L} = (3/4)\Gamma_{\rm R} = 3E_{\rm ch}/\hbar$, $\alpha_{\rm L} = \alpha_{\rm R} = 0.03$ and the time is in unit of $\hbar/E_{\rm ch}$.

The conditional evolutions in Figs. 21.2(a)-(h) differ considerably from their ensemble average counterparts shown in Fig 21.3. Each quantum trajectory (stochastic state evolution), mimics a single history of the system state in a single run of the continuous measurement experiment. The stochastic element in the quantum trajectory corresponds exactly to the consequence of the random outcomes of the detection record of the tunneling events in the SET. The macroscopic ensemble measurement properties can be calculated by using large ensembles of single electron tunneling events (fine grained measurement records). The conditional time evolutions in Fig. 21.2 are not smooth, but exhibit jumps, and they do not tend towards a steady state. Although little similarity can be observed between the time evolutions in Figs. 21.2 and Fig. 21.3, averaging over many different individual realizations shown in Fig. 21.2 leads to a closer and closer approximation to their ensemble average in Fig. 21.3.

21.5 Connection to "Partially" Reduced Density Matrix

It has been shown in Ref.[7] that the master equation of the "partially" reduced density matrix, in Ref.[1] can be obtained by taking a "partial" average on Eqs. (21.2) and (21.3) over the fine grained

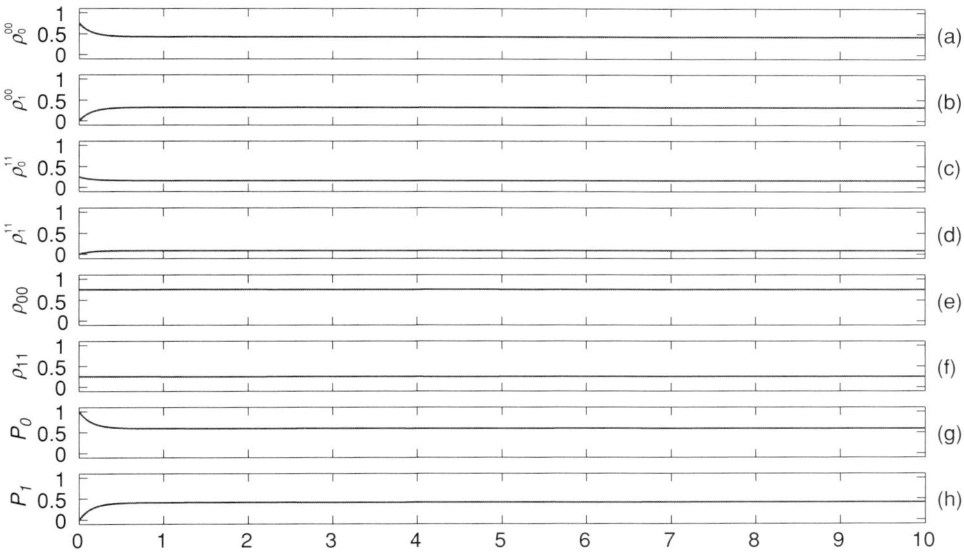

FIGURE 21.3. The unconditional, ensemble-averaged time evolutions. The initial conditions and parameters are the same as those in Fig. 21.2.

measurement records of the tunneling events in the SET. If the sum is taken over all possible values of electrons m that have tunneled into the drain on the "partially" reduced density matrix [i.e., tracing out the detector states completely, the master equation of the "partially" reduced density matrix Ref.[1] then reduces to the master equation of the reduced density matrix. This procedure of reducing Eqs. (21.2) and (21.3) to the "partially" reduced density matrix master equation and then to the reduced density matrix equation, by successively disregarding information that distinguishes different states of the detector, provides a connection between the approach of Ref.[1] and the more detailed stochastic master equation used here.

The probability distribution $P(m, t)$ that $m(= N_{\rm R})$ electrons have tunneled through the right junction into the drain during time t, discussed in Ref.,[1] can be explicitly simulated through constructing the histogram of the accumulated number of electrons $N_{\rm Rc} = \sum {\rm d}N_{\rm Rc}$ up to time t for many realizations of the detection records (generated together with their corresponding quantum trajectories), and then normalizing the distribution to one. The plot $P(m, t)$ in solid line in Fig. 21.4 is obtained using the method of Fourier analysis[1] of the partially reduced density matrix. This distribution splits into two and their weights correspond closely to the initial qubit diagonal elements of $\rho^{11}(0) = 0.25$ and $\rho^{00}(0) = 0.75$. The simulation of the normalized histogram in Fig. 21.4 using 2000 quantum trajectories and their corresponding detection records is already in good agreement with the plot in solid line. However, the possible individual realizations of measurement records and their corresponding quantum trajectories (e,g., in Fig. 21.2) do provide insight into, and aid in the interpretation of the ensemble average properties. This is one of the appealing features of the quantum trajectory approach.

For a charge qubit measured by a low-transparency point contact detector, this appealing feature of the quantum trajectories is illustrated in Ref.[6] Another advantage of the quantum trajectory approach (or Bayesian formalism) is that it may describe a quantum feedback process.

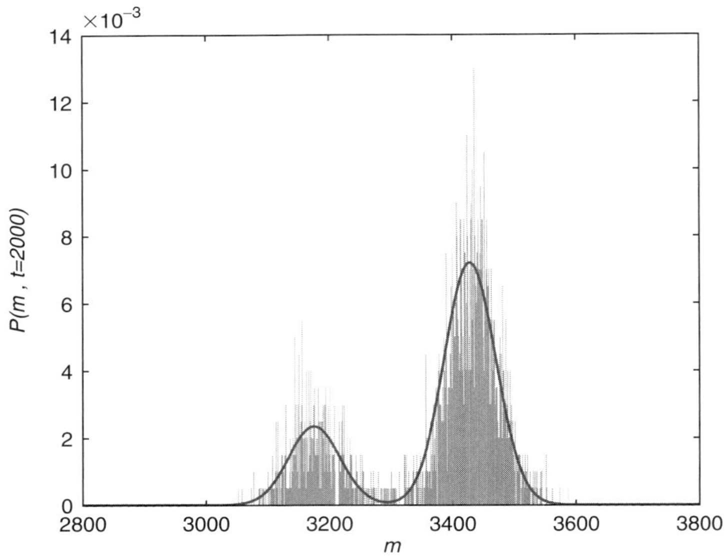

FIGURE 21.4. Probability distribution $P(m, t = 2000)$. The initial conditions and parameters are the same as those in Fig. 21.2.

It has been shown[8] that one may utilize the measurement output for the feedback control and manipulation of a qubit state.

21.6 Conclusion

To summarize, the stochastic master equation for the SCB/SET system has been derived, which can be regarded as a Monte Carlo method that allows to simulate the continuous quantum measurement process of the SCB qubit by the SET. Following each electron tunneling event through the SET junctions gives insight into the processes taking place in the SCB/SET system. It is also illustrated that by taking a "partial" average over the fine grained measurement records of the tunneling events in the SET, this stochastic master equation reduces to the master equation presented in Ref.[1]

Acknowledgments. The author would like to thank G. J. Milburn for useful discussions. This work was partly supported by the Australian Research Council, the Australian government and by the US National Security Agency (NSA), Advanced Research and Development Activity (ARDA) and the Army Research Office (ARO) under contract number DAAD19-01-1-0653. The author would also like to acknowledge the support from a Hewlett–Packard Fellowship.

References

1. Y. Makhlin, G. Schön and A. Shnirman, *Phys. Rev. Lett.* **85**, 4578 (2000); *Rev. Mod. Phys.* **73**, 357 (2001).

2. D. Loss and D. P. DiVincenzo, *Phys. Rev. A* **57**, 120 (1998); B. E. Kane, *Nature*, **393**, 133 (1998).
3. N. S. Bakhvalov et al., *Sov. Phys. JETP* **68**, 581 (1989).
4. A. N. Korotkov, *Phys. Rev. B* **60**, 5737 (1999); *Phys. Rev. B* **63**, 115403 (2001).
5. H.-S. Goan et.al, *Phys. Rev. B* **63**, 125326 (2001); H.-S. Goan and G. J. Milburn, *Phys. Rev. B* **64**, 235307 (2001).
6. H.-S. Goan, *Quantum Inf. Comput.* **3**, 121 (2003).
7. H.-S. Goan, to appear in *Phys. Rev. B* (2004).
8. R. Ruskov and A. N. Korotkov, *Phys. Rev. B* **66**, 041401(R) (2002).

22

On the Conversion of Ultracold Fermionic Atoms to Bosonic Molecules via Feshbach Resonances

EHOUD PAZY, AMI VARDI, AND YEHUDA B. BAND

Chemistry department, Ben-Gurion University of the Negev, Israel

Abstract

It is demonstrated that symmetry plays an important roll in the production, through Feshbach sweep experiments, of diatomic bosonic molecules from fermionic atomic mixtures of two hyperfine spin-states by constructing the many-body state. It is shown that the experimental procedure employed to prepare these atomic mixtures, leads to a mixture of even and odd many body parity pair-states. The symmetry of the many body state is shown to be the reason behind the 0.5 limited transfer experimental efficiency observed independently by two experimental groups who studied the conversion of ultracold Fermi gases of ^{40}K and ^{6}Li atoms into diatomic Bose gases. The higher than 50% transfer efficiency, as is currently experimentally obtained is achieved by a magnetic sweep rate which is much slower than the "the collision rate" in the gas, which is here related to a pair decorrelation rate.

22.1 Introduction

Recent experiments with ultracold Fermi gases have demonstrated macroscopic coherence[1–8] i.e., superfluid phases. Such experiments involve modifying the effective interaction between fermionic atoms of two hyperfine levels, via a Feshbach resonance thereby producing diatomic molecules. In employing such magnetically tuned scattering resonances between two-component Fermi gases experiments have been able to produce, Bose–Einstein condensates of molecules formed by Fermionic atoms of ^{6}Li and ^{40}K, as well as to probe the theoretically predicted crossover from a Bose–Einstein condensate (BEC) to a Bardeen-Cooper-Schrieffer (BCS) superfluid.

The first stage in such experiments is the sympathetic cooling of the Fermi gas allowing the experimentalists to achieve the fermionic degenerate regime, in which the temperature T is less than the Fermi temperature T_F. After reaching the degenerate regime, experiments have produced mixtures of two hyperfine levels in the vicinity of a Feshbach resonance, which conveniently allows adjustment of the effective interaction between atoms, characterized only by their s-wave scattering length at low temperatures. By slowly varying the magnetic field one can sweep through the magnetic Feshbach resonance, starting at a given magnetic field value with a weak attractive interaction between atoms of different hyperfine levels, i.e., small negative scattering length, and lowering the magnetic field as a function of time thereby making it more and more negative, and finally have it jump to a positive value at the other side of the resonance corresponding to a repulsive interaction between atoms leading to a formation of diatomic molecules. In four such recent experiments a Fermi gas of atoms was converted into an ultracold Bose gas of molecules by adiabatic passage through a Feshbach resonance.[1–4] In two of these experiments[1,4] ultracold $^{40}K_2$ molecules were produced from a quantum degenerate Fermi gas of ^{40}K atoms, whereas in Refs. [2, 3] ^{6}Li atoms were converted to diatomic molecules. An interesting feature of these studies is that when the sweep rate through the Feshbach resonance was comparable to the background elastic scattering rate, yet slow with respect to the atom–molecule coupling rate, a maximum atom–molecule transfer efficiency of 50% was reported.[1,2] However, when the sweep rate was much slower, approaching "close to thermal equilibrium" conditions, higher conversion efficiency was attained.[3,4]

In the following Section the experimental procedure for establishing a two hyperfine atomic mixture is described in detail. By constructing the many-body state employed in these experiments involving two-component Fermi gases it is shown by general symmetry arguments that the initial state is a mixture of even and odd parity pair-states in Section 22.3. In Section 22.4 the resulting 50% transfer efficiency as resulting from the symmetry of the system is explained. In Section 22.5 the dependence of the transfer efficiency on a further rate the decorrelation rate and how a transfer efficiency of above 0.5 are obtained is described.

22.2 Initial State Preparation

The first stage of the experimental procedure to establish the initial state for the Feshbach experiments is the production of a gas of atoms in a single hyperfine spin state defined by $|fm_f\rangle$, which is cooled to a very low temperature by sympathetic cooling. Polarized fermions cannot undergo s-wave collisions, since the colliding atoms have a symmetric spin-state and therefore any pair is only allowed to interact through odd-parity partial waves (e.g., p-wave). Thus in order to enable the s-wave scattering of the fermionic atoms in the experiments of Refs. [2–4] statistical mixtures of two spin-states are prepared. The method employed to produce this statistical mixture of two Fermi gases in the above experiments is the rapid driving of the Zeeman transition via an RF pulse in the presence of a magnetic field having a gradient. As different atoms are driven by the RF pulse between the lowest two $|f\ m_f\rangle$ states, the magnetic field gradient results in an inhomogeneous broadening of the transition. During and after the RF pulse elastic-scattering collisions between atoms in the resulting different spin-states commence and result in evaporative cooling of the trapped gas as hot atoms leave the trap. This procedure effectively splits the many-body system into two subsystems corresponding to two different spinstates, as depicted in Fig. 22.1. Both the ^{6}Li and the ^{40}K Feshbach experiments involve such two atomic spin-states $|\chi\rangle \equiv |f_\chi m_\chi\rangle$ with $\chi = \{a,\ b\}$.

22.3 General symmetry arguments

Since all coherence between the two spin-states is destroyed due to the inhomogeneous broadening of the transition the initial state of the system, after cooling, is best described by a density

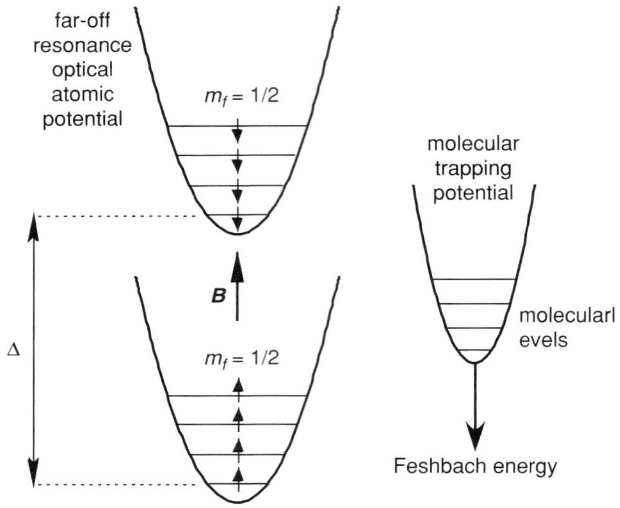

FIGURE 22.1. Schematic drawing of the trapping potentials for atoms with different spin projections, m_a and m_b (explicitly labeled $\uparrow$ and $\downarrow$ as appropriate for the ^{6}Li case), and for diatomic molecules (on the right side of the figure) whose potential is swept downward in energy as a function of time. Δ is the Zeeman energy splitting. The molecular potential force constant is twice as large as that of the atoms, but the energy level spacings of the atoms and molecules are equal.

matrix composed of an *incoherent mixture* of two decoupled Fermi gases, one for each spin-state:

$$\rho_0^N = \mathcal{A}\Big[\Big(\rho_{N/2}(a,\ T)|a\rangle^{(N/2)}\ {}^{(N/2)}\langle a|\Big) \otimes \Big(\rho_{N/2}(b,\ T)|b\rangle^{(N/2)}\ {}^{(N/2)}\langle b|\Big)\Big]\mathcal{A}^{-1}, \qquad (22.1)$$

where $\rho_{N/2}(\chi,\ T)$ denotes the motional part of the density matrix for $N/2$ atoms with spin projection $\chi = \{a, b\}$ at temperature T, and $\mathcal{A}$ is the anti-symmetrization operator of all the particles.

To illustrate the above claim and to determine the symmetries of the system the many-body state of the cooled system is constructed explicitly in the limit as $T \to 0$. To do so we start by anti-symmetrizing each of the wave functions for the two Fermi seas separately. Since atoms in each Fermi sea have the same spin projection, their many-body wave function is a tensor product of an odd parity Slater determinant for the spatial degrees of freedom and a totally symmetric spin configuration. It is evident from the construction of the pertinent Young tableaux (see Fig. 22.2) that the anti-symmetrization of the two Fermi seas can be done in only two ways: a totally anti-symmetric spatial wave function times a spin-symmetric configuration $|f = N/2,\ m_f = 0\rangle$ (the row Young tableau on the right hand side of the equation in Fig. 22.2), and a state which is spin-antisymmetric upon exchange of particles of different spin corresponding to $|f = 0,\ m_f = 0\rangle$ (the second tableau on the right hand side of the equation in Fig. 22.2). This is the many-body analogy of the two spin 1/2 case in which the two body state naturally falls into two classes: symmetric or antisymmetric in the spatial coordinates corresponding to spin antisymmetric (singlet) and spin symmetric (triplet) states respectively, such that the total fermionic wave function is antisymmetric with respect to exchange of the two particles.

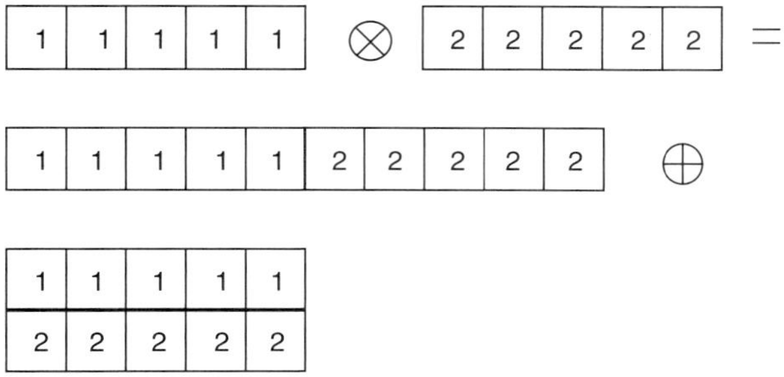

FIGURE 22.2. Young Tableaux spin degrees of freedom corresponding to a totally spin-symmetric configuration for each of the two Fermi seas (each row tableau corresponds to a different spin projection m_a denoted by 1 and m_b denoted by 2) and its decomposition into a totally symmetric $F = N/2$ and a $F = 0$ many-body configuration.

22.4 Explaining the Experimental Limited Transfer Efficiency

Considering the symmetry of the Feshbach Hamiltonian showing that it couples to molecules only pairs of atoms in spin antisymmetric (singlet) states, then it is illustrated that the reduced two-particle density matrix obtained by tracing out all but one particle of one spin-state and another particle of the other spin-state contains an equal weighted sum of spin-antisymmetric (spin-singlet for ^{6}Li), spin-symmetric (e.g., the triplet spin-symmetric superposition of the form $|\uparrow\downarrow\rangle + |\downarrow\uparrow\rangle$ for the ^{6}Li case). It therefore becomes evident that when sweeping through the resonance sufficiently slowly so that the Landau-Zener[10] transition is traversed adiabatically,[11] since only even parity spin-singlet pairs, can produce molecules through the Feshbach sweep that the conversion of ultracold Fermi gases of ^{40}K and ^{6}Li atoms into diatomic Bose gases is limited to 0.5.

Consider the Feshbach atom–molecule coupling Hamiltonian

$$\hat{H}_{\mathrm{F}} = g \sum_{\mathbf{q},\mathbf{k}} \hat{c}^{\dagger}_{\mathbf{q}/2-\mathbf{k},a} \hat{c}_{\mathbf{q}/2+\mathbf{k},b} \hat{b}_{\mathbf{q}} + \mathrm{h.c.}, \tag{22.2}$$

where g is the coupling coefficient, $\hat{c}^{\dagger}_{\mathbf{q}/2-\mathbf{k},\sigma}$ is the usual atomic creation operator for an atom in spin-state $\sigma = \{|a\rangle, |b\rangle\}$ respectively, and $\hat{b}_{\mathbf{q}}$ is the molecular annihilation operator. Written explicitly for fermionic operators $\hat{c}_{\mathbf{k},a}$, $\hat{c}_{\mathbf{k},b}$, and bosonic operators $\hat{b}_{\mathbf{q}}$, the Hamiltonian in Eq. (22.2) takes the form

$$\hat{H}_{\mathrm{F}} = \frac{g}{2} \sum_{\mathbf{q},\mathbf{k}} \left(\hat{c}^{\dagger}_{\mathbf{q}/2-\mathbf{k},a} \hat{c}^{\dagger}_{\mathbf{q}/2+\mathbf{k},b} - \hat{c}^{\dagger}_{\mathbf{q}/2-\mathbf{k},b} \hat{c}^{\dagger}_{\mathbf{q}/2+\mathbf{k},a} \right) \hat{b}_{\mathbf{q}} + \mathrm{h.c.}, \tag{22.3}$$

where we have used the anti-commutation relation $\{\hat{c}_{\mathbf{k}^{\dagger},a}, \hat{c}_{\mathbf{k}',b}\} = 0$.

It is evident that the fermionic Feshbach coupling Hamiltonian in Eq. (22.3) is anti-symmetric under spin permutation and that it has even parity. It only allows for the association of atom pairs colliding with even parity partial waves (effectively restricted at ultracold temperatures, to s-wave scattering) with an antisymmetric spin-state (a spin-singlet in the ^{6}Li case). Defining the commuting operators

$$\hat{S}_{\mathbf{q}} = \sum_{\mathbf{k}} \left(\hat{c}_{\mathbf{q}/2-\mathbf{k},a} \hat{c}_{\mathbf{q}/2+\mathbf{k},b} - \hat{c}_{\mathbf{q}/2-\mathbf{k},b} \hat{c}_{\mathbf{q}/2+\mathbf{k},a} \right), \tag{22.4}$$

$$\hat{T}_{\mathbf{q}} = \sum_{\mathbf{k}} \left(\hat{c}_{\mathbf{q}/2-\mathbf{k},a} \hat{c}_{\mathbf{q}/2+\mathbf{k},b} + \hat{c}_{\mathbf{q}/2-\mathbf{k},b} \hat{c}_{\mathbf{q}/2+\mathbf{k},a} \right), \tag{22.5}$$

the Feshbach coupling Hamiltonian for fermionic atoms simply reads

$$\hat{H}_{\mathrm{F}} = \frac{g}{2} \sum_{\mathbf{q}} \hat{S}_{\mathbf{q}}^{\dagger} \hat{b}_{\mathbf{q}} + \text{h.c.} \tag{22.6}$$

To determine the two-particle collision probability we now consider the reduced two-particle density matrix $\rho_0^{(2)} = \mathrm{Tr}_{(N-2)} \left(\rho_0^N \right)$ obtained by tracing out all but one particle with spin projection a and another particle with spin projection b. It is readily seen from the Young tableaux in Fig. 22.2 that tracing out all but two of the particles, one with spin up and the other with spin down, will leave a two particle reduced density matrix which is an equally weighted sum of a spin symmetric (two box row tableau) and a spin antisymmetric (two box column tableau) atomic pair. Therefore, if a given atom pair is spatially antisymmetric, then each of the atoms comprising this pair will be spatially antisymmetric with any other atom of opposite spin due to the symmetry of the many-body state. The spin part of $\rho_0^{(2)}$ is identical for all atomic pair-states, and is obtained by projection on to the momentum states of any given colliding pair,

$$\begin{aligned} \langle \mathbf{q}/2 - \mathbf{k} | \langle \mathbf{q}/2 + \mathbf{k} | \rho_0^{(2)} | \mathbf{q}/2 + \mathbf{k} \rangle | \mathbf{q}/2 - \mathbf{k} \rangle &= \frac{1}{2} (|a\rangle|b\rangle\langle b|\langle a| + |b\rangle|a\rangle\langle a|\langle b|) \\ &= \langle \mathbf{q} | \frac{1}{2} (|S\rangle\langle S| + |T\rangle\langle T|) | \mathbf{q} \rangle. \end{aligned} \tag{22.7}$$

where $|\mathbf{q}/2 \pm \mathbf{k}\rangle$ is the single-particle spatial wave function with momentum $\mathbf{q}/2 \pm \mathbf{k}$, $|T\rangle = (|a\rangle|b\rangle + |b\rangle|a\rangle)/\sqrt{2}$ is a spin symmetric triplet state, and $|S\rangle = (|a\rangle|b\rangle - |b\rangle|a\rangle)/\sqrt{2}$ is a spin antisymmetric (singlet) state. In the middle line of Eq. (22.7) the projection onto momentum states to simplify the notation is omitted. Equation (22.7) demonstrates the simple result that in the absence of coherence between colliding particles in different spin-states, collisions with a spin-triplet (via a p-wave) are just as probable as collisions with a spin-singlet (via an s-wave). In the above expressions, an explicit orbital-spin form of the triplet state is given by

$$|T\rangle = [|\phi_a(1)\phi_b(2)\rangle - |\phi_b(1)\phi_a(2)\rangle] \times [|f_a m_a, f_b m_b\rangle + |f_b m_b, f_a m_a\rangle], \tag{22.8}$$

where $\phi_a(i)$ is the orbital for the ith atom of species a. For example, for the ^{6}Li case,

$$|T\rangle = [|\phi_a(1)\phi_b(2)\rangle - |\phi_b(1)\phi_b(2)\rangle][|\uparrow\downarrow\rangle + |\downarrow\uparrow\rangle]. \tag{22.9}$$

Similarly for $|S\rangle$, but with the signs reversed.

The above considerations show that any given particle pair has an equal probability to be in a spatially symmetric or anti-symmetric state and since it has been shown that the Feshbach resonance transfers only spatially symmetric pairs into molecules a 0.5 probability is expected for a given collision to produce a molecule. As opposed to a thermal gas, in which collisions can be viewed as independent, the probability for forming a molecule in any single Feshbach collision is fixed by the symmetry of the many-body state. This can be demonstrated by considering the state formed by the tensor product of a spin-symmetric pair with another fermion and anti-symmetrizing:

$$\begin{aligned}&\mathcal{A}\{[|\mathbf{k}\rangle_1|\mathbf{k}'\rangle_2 - |\mathbf{k}'\rangle_1|\mathbf{k}\rangle_2](|a\rangle_1|b\rangle_2 + |b\rangle_1|a\rangle_2) \otimes |\mathbf{k}''\rangle_3|a\rangle_3\}\\ &\quad= [|\mathbf{k}\rangle_1|\mathbf{k}'\rangle_2 - |\mathbf{k}'\rangle_1|\mathbf{k}\rangle_2](|a\rangle_1|b\rangle_2 + |b\rangle_1|a\rangle_2) \otimes |\mathbf{k}''\rangle_3|a\rangle_3\\ &\qquad-[|\mathbf{k}\rangle_3|\mathbf{k}'\rangle_2 - |\mathbf{k}'\rangle_3|\mathbf{k}\rangle_2](|a\rangle_3|b\rangle_2 + |b\rangle_3|a\rangle_2) \otimes |\mathbf{k}''\rangle_1|a\rangle_1\\ &\qquad-[|\mathbf{k}\rangle_1|\mathbf{k}'\rangle_3 - |\mathbf{k}'\rangle_1|\mathbf{k}\rangle_3](|a\rangle_1|b\rangle_3 + |b\rangle_1|a\rangle_3) \otimes |\mathbf{k}''\rangle_2|a\rangle_2.\end{aligned} \tag{22.10}$$

Tracing out one of the particles results in only spin symmetric pair states, implying that no spin antisymmetric collisions are allowed until the spin symmetric pair decoheres. Using similar argumentation, it is also evident that pairs of spin symmetric pairs can not interact in such a way that two of the four atoms scatter as a spin antisymmetric, even parity, pair. Eq. (22.10), is just a specific three particle example of the general case which is apparent by considering the Young tableaux in Fig. 22.2 from which one can infer that all two particle states are either odd parity spin-symmetric for all atoms, or even parity spin-antisymmetric for all atoms with different spin projections. It explicitly depicts the case when one has to combine a two box row tableau with a further box tableau, and since the states with the same spin projection need to be symmetric this can only be done in one way.

It is therefore clear that as long as pair correlations are maintained, constituent atoms of the spin symmetric pairs remaining after Feshbach coupling cannot interact with another atom to produce a spin-antisymmetric state via elastic scattering collisions. It is therefore shown that if the process of adiabatically scanning the magnetic field from high to low field can be described in terms of a Landau–Zener transition, since only the spin antisymmetric, even parity, pairs are coupled to molecules by the Feshbach Hamiltonian in Eq (22.6), the maximal anticipated conversion efficiency is 0.5. The spin-symmetric pairs cannot interact via the Feshbach resonance and therefore cannot be converted to molecules. Moreover even though the Feshbach sweep produces holes in the Fermionic atomic sea, since it can couple any two atoms attempting to create a molecule, these holes can not be filled by the remaining atoms, i.e., atoms which have not been converted to molecules, atomic population. Such a hole filling process has been predicted to cause rapid heating of the fermionic sea[9] which is not apparent in the Feshbach sweep experiments. The reason as demonstrated above is that these atoms are non-interacting at low temperatures which explains the experimental observation that the remaining atoms do not heat.

22.5 Transfer efficiencies above 0.5

According to the symmetry of the many body state to obtain a transfer efficiency which is above 50%, as is commonly achieved in current experiments, the odd parity spin-symmetric states must decorrelate before the constituent atoms can interact again via the Feshbach resonance. Therefore in achieving a higher than 50% efficiency the sweep rate was much slower, slower than the approaching "close to thermal equilibrium" conditions. Consequently, further molecule-producing s-wave/spin-antisymmetric collisions can take place and higher than 50% efficiency may be obtained as reported by Cubizolles *et. al.*[3] and Greiner *et. al.*[4] in all these experiments the sweep rate through the Feshbach resonance was three orders of magnitude slower than the 0.5 efficiency experiments.[1,2]

To demonstrate how introducing a decorrelation rate for spin-triplets (triplet to singlet transitions) effects the transition probability we employ the simplest possible model which involves a single parameter: the ratio of the Feshbach sweep rate γ_f to the decorolation rate, γ_{st},

$$\frac{\partial n_s}{\partial t} = -\gamma_f n_s + \gamma_{st} n_t, \tag{22.11}$$

$$\frac{\partial n_t}{\partial t} = -\gamma_{st} n_t, \tag{22.12}$$

$$\frac{\partial n_m}{\partial t} = \gamma_f n_s, \tag{22.13}$$

n_s and n_t is the density of the correlated singlet and triplet pairs of atoms, and n_m is the density of molecules. Using initial conditions $n_s(0) = n_t(0) = n_a(0)/2,\ n_m(0) = 0$, it is easy to verify that $n_s(t) + n_t(t) + n_m(t)$ is a conserved quantify independent of time and equal to $n_a(0)$, the initial density of atoms in spin-state a (and b). The analytic solution to Eqs. (22.11)–(22.13) given the above initial conditions is:

$$n_s(t) = \frac{n_a(0)}{2(\gamma_f - \gamma_{st})}[\gamma_{st} e^{-\gamma_{st} t} + (\gamma_f - 2\gamma_{st}) e^{-\gamma_f^t}], \tag{22.14}$$

$$n_t(t) = \frac{n_a(0)}{2} e^{-\gamma_{st} t}, \tag{22.15}$$

$$n_m(t) = \frac{n_a(0)}{2(\gamma_f - \gamma_{st})}[2\gamma_f - 2\gamma_{st} - \gamma_f e^{-\gamma_{st}^t} - (\gamma_f - 2\gamma_{st}) e^{-\gamma_f^t}]. \tag{22.16}$$

The inset of Fig. 22.4 shows the concentrations normalized to $n_a(0)$ as a function of time in units of γ_f^{-1} for $\gamma_{st} = 0.001\gamma_f$. If the measurement of the number of molecules is made at time $t = 10\gamma_f^{-1}$, only about half of the atoms will have been converted to molecules, but if the measurement is made at $t = 1000\gamma_f^{-1}$, the inset of Fig. 22.4 shows that 82% of the atoms will have been converted. Fig. 22.4 shows the normalized molecule concentration versus time for three different decorrelation rates. The number of atoms converted to molecules is about 50% over an extended range of times for the very small ratio $\gamma_{st}/\gamma_f = 0.001$, but for $\gamma_{st} = 0.01\gamma_f$, the number of atoms converted to molecules is already 82% at $t = 100\gamma_f^{-1}$ and increases thereafter. For $\gamma_{st} = 0.1\gamma_f$, nearly 100% conversion is achieved beyond about $t = 50\gamma_f^{-1}$.

This model is too simplistic. It does not account for the initially filled finite-temperature Fermi sea of a and b, spatial inhomogeneity of the two gases, moreover additional terms in the rate Eqs (22.11)–(22.13) to account for destruction of molecules by atom–molecule and molecule–molecule collisions can be easily added. It does however emphasize how only the spin antisymmetric (spatially even-parity) component couples to the Feshbach resonance and how the spin symmetric (spatially odd-parity) component can hang the conversion efficiency at around 50% when the ratio of the decorrelation rate and the Feshbach sweep rate is small.

Measurement of the conversion efficiency as a function of sweep rate in the whole range from very fast to very slow should serve to verify or contradict the picture of the conversion efficiency painted here.

22.6 Summary

Constructing the many-body state initially prepared by experimentalists in Feshbach sweep experiments, it has been shown that if we consider pairs of atoms, one from each spin-state, half the pairs of atoms are antisymmetric spinstates and half are symmetric spin-states. More explicitly, the reduced two-particle density matrix obtained by tracing out all but one particle of one spin-state and another particle of the other spin-state contains 50% spin-antisymmetric atom-pairs, interacting via an s-wave (even parity spatial state), and 50% spin-symmetric atom-pairs, interacting via a p-wave (odd parity spatial state). Explicitly demonstrating that the Feshbach

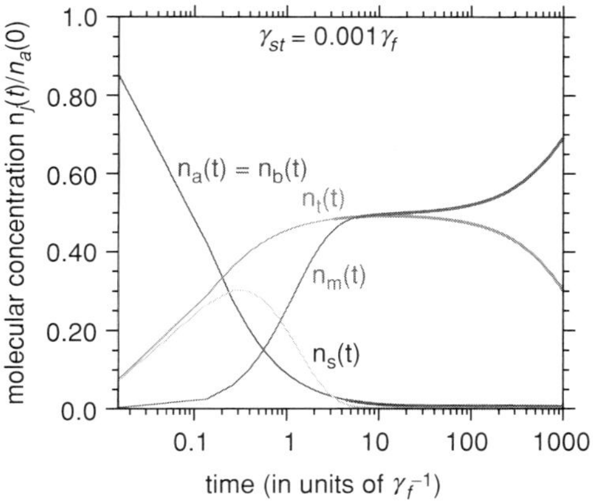

FIGURE 22.3. Concentrations of spin-singlet and spin-triplet pairs of atoms, n_s and n_t, and molecules, n_m, versus time.

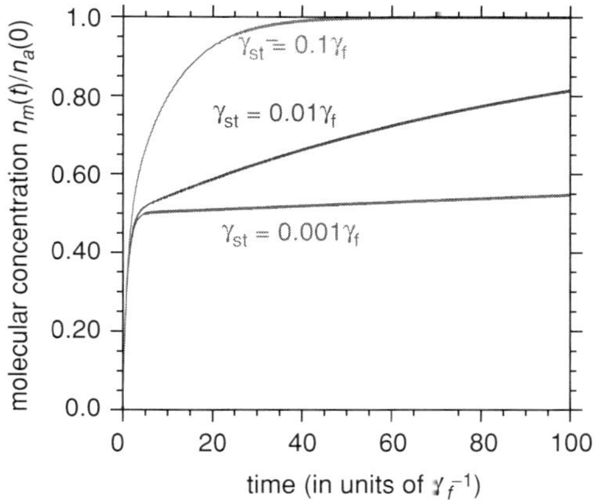

FIGURE 22.4. Molecular concentration n_m versus time for several values of the parameter γ_{st}. The curve for $\gamma_{st} = 0.001\gamma_f$ is identical to the molecule curve labeled $n_m(t)$ in figure on the left.

Hamiltonian is an s-wave resonance, i.e., spin-triplet pairs cannot Feshbach scatter, it is established that only half of the two particle collisions will result in molecule creation. By noting that for a cold atomic gas collisions are not independent it is explained that the observed 50% saturation of the atom–molecule conversion efficiency in the fast (yet adiabatic) Feshbach sweep regime is a result of the initial state preparation. By introducing the spin-triplet decorrelation rate it is demonstrated that employing a simplistic model how the transfer efficiency critically depends on this rate. It has been shown that for experiments in which the ratio of the decorrelation rate and the Feshbach sweep rate is small the conversion efficiency is pinned around 50% and when the sweep rate is slower than the time required for spin-triplet states to decorrelate, the conversion efficiency can grow beyond 50%.

Acknowledgments. This work was supported in part by grants from the U.S.-Israel Binational Science Foundation (grant Nos. 2002214, 2002147), the Minerva Foundation through a grant for a Minerva Junior Research Group, and the Israel Science Foundation for a Center of Excellence (grant No. 8006/03).

References

1. C. A. Regal, C. Ticknor, J. L. Bohn and D. S. Jin, *Nature* **424**, 47 (2003).
2. K. E. Strecker, G. B. Partridge and R. G. Hulet, *Phys. Rev. Lett.* **91**, 080406 (2003).
3. J. Cubizolles *et al.*, *Phys. Rev. Lett.* **91**, 240401 (2003).
4. M. Greiner, C. A. Regal, and D. S. Jin, *Nature* **426**, 537 (2003).
5. S. Jochim *et al.*, *Phys. Rev. Lett.* **91**, 240402 (2003).
6. C. A. Regal, M. Greiner, and D. S. Jin, *Phys. Rev. Lett.* **92**, 040403 (2004).
7. S. Jochim *et al.*, *Science* **302**, 2101, (2003).
8. W. Ketterle and H. J. Miesner, *Phys. Rev. A* **56**, 3291 (1997).
9. E. Timmermans, *Phys. Rev. Lett.* **87**, 240403 (2001).
10. L. D. Landau, *Z. Phys. Sowjetunion* **2**, 46 (1932); C. Zener, *Proc. R. Soc. London A* **137**, 696 (1932).
11. F. H. Mies, E. Tiesinga, and P. S. Julienne, *Phys. Rev. A* **61**, 022721 (2000).

23

Revealing Anisotropy in a Paul Trap Through Berry Phase

M. Scala, B. Militello and A. Messina

INFM, MIUR and Dipartimento di Scienze Fisiche ed Astronomiche, Via Archirafi 36, I-90123 Palermo, Italy

Abstract

When an ion confined in an anisotropic bidimensional Paul trap is subjected to a laser beam oriented along an arbitrary direction, the interaction between its electronic and vibrational degrees of freedom is described by a time-dependent Hamiltonian model as a consequence of the lack of symmetry. Appropriately choosing the laser frequency, the Hamiltonian model turns out to be sinusoidally oscillating at the difference between the proper frequencies of the center of mass of the ion. Thus, if the anisotropy of the trap is sufficiently small, the evolution of the system can be considered as adiabatic. In the context of this physical situation we have calculated the Berry phase acquired in a cycle by the instantaneous eigenstates of the trapped ion Hamiltonian. Suitably choosing the initial condition and a physical observable we succeed to forecast physical effects directly traceable back to the accumulated Berry phase. In particular we indeed bring to light that the mean value of the chosen observable after a cycle is the negative of that calculated at the same instant of time in the case of isotropic traps. This effect demonstrates that and how the Berry phase can be exploited to evidence the existence of a weak anisotropy in a Paul trap.

23.1 Introduction

Over the last years we have witnessed a continuous development of sophisticated techniques, which allow cooling and traping single atoms, thus providing a physical system that is ideal to make experiments to test the bases of quantum mechanics. An ion confined in a Paul trap is subjected to a particular configuration of oscillating electromagnetic fields responsible for a centre of mass (c.m.) motion describable as that of a quantum harmonic oscillator [1,2,3]. Driving the ion by a classical laser beam causes to induce couplings between the vibrational and electronic degrees of freedom of the ion well described by simple spin-boson Hamiltonian models, provided the ion can be treated as a two-level system. In particular, when the amplitude of the oscillations of the ion c.m. are much smaller than the laser wavelength (Lamb–Dicke limit) and the driving field is tuned on one of the vibrational sidebands of the atomic transition, the interaction is described by Jaynes–Cummings-like models [4]. The advantage of using ion traps instead of cavities to test the dynamics induced by Hamiltonian models of this kind is that the coupling between the ion and its environment can be neglected in a lot of experimental situations.

In this study we focus our attention on *anisotropic* bidimensional Paul traps, defined by the feature that the oscillation frequencies along the two axes of the trap are different. Our aim is to bring to light effects directly attributable to the presence of anisotropy in the trap. To this end we show that, driving the ion by a laser beam oriented along an arbitrary direction different from the axes of the trap, there is no way to make the interaction picture Hamiltonian of the system time-independent. We find that a particular resonance condition exists under which this Hamiltonian is sinusoidally oscillating at the difference between the two oscillation frequencies of the c.m. of the ion. This implies that when the anisotropy is small enough the system dynamics can be evaluated using the adiabatic approximation approach. The system we are studying proves then to be an ideal candidate to seek effects stemming from the Berry or geometric phase accumulation [5,6].

The main result reported in this paper is twofold. On the one hand we demonstrate that anisotropy may be exploited as an effective resource to test the theory of non-dynamical phases. On the other hand we show that the circumstance that experiments for measuring the Berry phase are feasible provides an effective way to reveal and quantitatively appreciate the presence of anisotropy in the trap.

This paper is organized as follows: in the next section we describe the physical system under study and obtain its time-dependent effective Hamiltonian under the action of a suitable laser field. In the third section we analyze the physical system adiabatic evolution calculating the Berry phase acquired by the eigenstates of the Hamiltonian, while in the fourth section we search for physical effects directly traceable back to geometric phase accumulation. Finally in the fifth section some conclusive remarks are given.

23.2 Physical System and Hamiltonian Model

If we drive a two level ion with mass m in a bidimensional Paul trap by a laser beam oriented along a direction that makes an angle α with the x-axis, the Hamiltonian of the system in a frame rotating at the laser frequency is:

$$H = H_{\text{trap}} + H_{\text{int}} \tag{1}$$

$$H_{\text{trap}} = \hbar\left(\nu_a a^\dagger a + \nu_b b^\dagger b\right) + \hbar\frac{\delta}{2}\sigma_z \tag{2}$$

$$H_{\text{int}} = \hbar\Omega\left\{e^{-ik(x\cos\alpha + y\sin\alpha)}\sigma_- + e^{ik(x\cos\alpha + y\sin\alpha)}\sigma_+\right\} \tag{3}$$

where $\delta = \omega_0 - \omega$, ω_0 is the Bohr frequency associated to the two electronic levels of the ion andω and k are respectively the frequency and the modulus of the wave vector of the laser. Expanding the exponentials appearing in H_{int} yields:

$$H_{\text{int}} = \hbar\Omega e^{k^2(\Delta x^2 \cos^2\alpha + \Delta y^2 \sin^2\alpha)} \times \left\{ \sum_{l=0}^{\infty} \frac{(-ik\Delta x \cos\alpha)^l}{l!} a^l \sum_{m=0}^{\infty} \frac{(-ik\Delta x \cos\alpha)^m}{m!} a^{\dagger m} \right. \times \sum_{n=0}^{\infty} \frac{(-ik\Delta y \sin\alpha)^n}{n!} b^n \sum_{p=0}^{\infty} \frac{(-ik\Delta y \sin\alpha)^p}{p!} b^{\dagger p} \tag{4}$$

where

$$\Delta x = \sqrt{\frac{\hbar}{2m\nu_a}}, \qquad \Delta y = \sqrt{\frac{\hbar}{2m\nu_b}} \tag{4a}$$

In the interaction picture with respect to the Hamiltonian H_{trap} given by Eq. (2), H_{int} becomes the sum of terms oscillating at frequencies which are linear combinations of ν_a, ν_b and ω with integer coefficients. In the Lamb–Dicke limit

$$k\Delta x \cos\alpha, k\Delta y \sin\alpha \ll 1 \tag{5}$$

most of the terms in Eq. (4) turn out to be negligible. Moreover introducing the resonance condition

$$\delta = \nu_a + \nu_b \tag{6}$$

the remaining non-rapidly oscillating terms are those proportional to $a^{\dagger 2}\sigma_-$, $b^{\dagger 2}\sigma_-$, $a^\dagger b^\dagger \sigma_-$ and their hermitian conjugates. Performing the rotating wave approximation the interaction picture Hamiltonian is then given by

$$\begin{aligned} H_{\text{int}}^I(t) &\approx \hbar\Omega e^{k^2\left(\Delta x^2 \cos^2\alpha + \Delta y^2 \sin^2\alpha\right)} \left[\left(-\frac{k^2\Delta x^2}{2} \cos^2\alpha\, a^{\dagger 2} e^{-i\Delta\nu t}\right.\right. \\ &\left.\left. -\frac{k^2\Delta b^2}{2}\sin^2\alpha\, b^{\dagger 2} e^{i\Delta\nu t} - k^2\Delta x\Delta y \cos\alpha \sin\alpha\, a^\dagger b^\dagger\right)\sigma_- + h.c.\right] \\ &= -\hbar\lambda\left[\left(\cos\theta\, a^\dagger e^{-i\frac{\Delta\nu}{2}t} + \sin\theta\, b^\dagger e^{i\frac{\Delta\nu}{2}t}\right)\sigma_- + h.c.\right] \end{aligned} \tag{7}$$

where

$$\cos\theta = \frac{\Delta x \cos\alpha}{\sqrt{(\Delta x \cos\alpha)^2 + (\Delta y \sin\alpha)^2}}, \tag{7a}$$

$$\sin\theta = \frac{\Delta y \sin\alpha}{\sqrt{(\Delta x \cos\alpha)^2 + (\Delta y \sin\alpha)^2}}, \tag{7b}$$

$$\lambda = \frac{\Omega}{2} e^{k^2(\Delta x^2 \cos^2\alpha + \Delta y^2 \sin^2\alpha)} \left\{(\Delta x \cos\alpha)^2 + (\Delta y \sin\alpha)^2\right\} \tag{7c}$$

and $\Delta\nu = \nu_b - \nu_a$.

At the end of this calculation, the clean result we obtain is a periodically time-dependent Hamiltonian. This fact suggests that the system under scrutiny could be useful to investigate effects related to Berry phase. It is worth noting that even with a different choice of δ, we would anyhow obtain a time-dependent Hamiltonian. Therefore the fact that the interaction picture Hamiltonian is intrinsically time-dependent due to the lack of symmetry of the trap establishes a deep connection between Berry phase and the presence of anisotropy.

23.3 Calculation of Berry Phase

We know that when the Hamiltonian $H(t)$ of a physical system is slowly varying with time, the so-called *adiabatic approximation* is valid. Assuming $H(t)$ possesses a non-degenerate discrete spectrum at any arbitrarily fixed t, the time evolution of the nth eigenstate $|n(0)\rangle$ of $H(0)$ generates at the time instant t the nth eigenstate $|n(t)\rangle$ of $H(t)$ [7]. Berry showed that there is a simple way to express the phase acquired by this state when, after some time T, the Hamiltonian comes back to its initial form, i.e., $H(T) = H(0)$. Assume that the Hamiltonian depends on a set of parameters $\mathbf{R}(t) = (R_1(t), R_2(t), \ldots)$, i.e., $H(t) \equiv H(\mathbf{R}(t))$, and call $E_n(\mathbf{R})$ the eigenvalues of $H(\mathbf{R})$ and $|n(R)\rangle$ the corresponding eigenstates: the evolution of the Hamiltonian $H(t)$ corresponds to a curve in the space of parameters and the condition $H(T) = H(0)$ becomes $\mathbf{R}(T) = \mathbf{R}(0)$. If the initial state of the system is:

$$|\Psi(0)\rangle = |n(0)\rangle \tag{8}$$

the state of the system at instant T can be written down as [5]:

$$|\Psi(T)\rangle = e^{i\gamma_n^C} e^{-\frac{i}{\hbar}\int_0^T \mathrm{d}t' E_n(\mathbf{R}(t'))} |\Psi(0)\rangle \tag{9}$$

where the second exponential is called *dynamical phase factor*, which reduces to the usual phase factor in the case of time-independent Hamiltonian, and the quantity γ_n^C is given by:

$$\gamma_n^C = i \oint_C \langle n(\mathrm{R}) \mid \nabla_{\mathrm{R}} n(\mathrm{R})\rangle \cdot \mathrm{dR} \tag{10}$$

where the curve is given by the path followed by $\mathbf{R}(t)$ in the space of parameters. The quantity in Eq (10) is called *Berry phase* or *geometrical phase*, because its value depends only on the curve C.

In the case under investigation in this paper, the evolution can be considered as adiabatic if the anisotropy of the system is sufficiently small. More quantitatively, a condition for the validity of the adiabatic approximation is that the square modulus of the ratio between the maximum angular velocity of $|n(t)\rangle$ and the minimum Bohr frequency of $|n(t)\rangle$ has to be much smaller than unity [7]. In our case this condition becomes:

$$\left|\frac{\Delta\nu}{\lambda}\right|^2 \ll 1 \tag{11}$$

Putting $\phi = \Delta\nu t/2$, the Hamiltonian in Eq (7) may be cast in the following form:

$$\begin{aligned} H_\phi &= -\hbar\lambda\left[\left(\cos\theta\, e^{-i\phi}a^\dagger + \sin\theta\, e^{i\phi}b^\dagger\right)^2 \sigma_- + h.c.\right] \\ &= -\hbar\lambda\left[A_\phi^{\dagger 2}\sigma_- + A_\phi^2\,\sigma_+\right] \end{aligned} \tag{12}$$

where

$$A_{\phi}^{\dagger} \equiv \cos\theta\, e^{-i\phi} a^{\dagger} + \sin\theta\, e^{i\phi} b^{\dagger} \tag{13}$$

Following Berry's formalism with

$$\mathbf{R} = (\cos\phi, \sin\phi) \tag{14}$$

and introducing the creation operator

$$B_{\phi}^{\dagger} \equiv \sin\theta\, e^{i\phi} a^{\dagger} + \cos\theta\, e^{-i\phi} b^{\dagger} \tag{15}$$

which corresponds to a mode orthogonal to that of A_{ϕ} and finally using the Fock states

$$|N\rangle_{\phi} \equiv |N\rangle_{A_{\phi}} |0\rangle_{B_{\phi}} = \frac{A_{\phi}^{\dagger N}}{\sqrt{N!}} |0\rangle_{a} |0\rangle_{b} \tag{16}$$

the eigenstates of Eq. (12) can be written down as

$$|\Psi_{N\pm}(\phi)\rangle = \frac{1}{\sqrt{2}} \left(|N\rangle_{\phi} |-\rangle \mp |N-2\rangle_{\phi} |+\rangle \right) \tag{17}$$

whose corresponding eigenvalues are

$$E_{N\pm} = \pm\hbar\lambda\sqrt{N(N-1)} \tag{18}$$

Denoting by $U(t)$ the evolution operator, the dynamics of the eigenstates Eq. (17) after a cycle is given by

$$U(T)\, |\Psi_{N\pm}(\phi = 0)\rangle = e^{-i\frac{E_{N\pm}}{\hbar}t} e^{i\gamma_{N\pm}(T)} |\Psi_{N-}(\phi = 0)\rangle \tag{19}$$

Using Eqs. (10) and (14) and varying ϕ from 0 to 2π we obtain the Berry phase:

$$\gamma_{N-}(T) = \gamma_{N+}(T) \equiv \gamma_{N}(T) = -2\pi \left(2\sin^{2}\theta - 1\right) N \tag{20}$$

where terms independent of N have been omitted since they do not give rise to observable effects.

23.4 Anisotropy vs. Berry Phase Effects

We now wish to look for some simple physical effects that can be directly traced back to Berry phase and then to the presence of anisotropy in a bidimensional Paul trap. First we choose an initial state quite simple in structure, experimentally feasible and appropriate to reveal Berry phase accumulation, being the quantum superposition of eigenstates of $H(0)$ accumulating different Berry phases. As a consequence we shall be able to find a suitable physically transparent observable sensitive to the non-dynamical phase acquired by the eigenstates of $H(0)$ appearing in the initial state expansion.

To this end let us then start from the following superposition

$$|\Psi(t = 0)\rangle = \frac{1}{\sqrt{2}} \left(|N\rangle_{\phi=0} + |N+1\rangle_{\phi=0} \right) |-\rangle \tag{21}$$

In view of Eq. (17)–(20) the evolution of this state after a cycle may be cast into the following form:

$$
\begin{aligned}
|\Psi(t=T)\rangle &= \frac{1}{\sqrt{2}}\left(U(T)\,|N\rangle_{\phi=0}\,|-\rangle + U(T)\,|N+1\rangle_{\phi=0}\,|-\rangle\right) \\
&= \frac{1}{\sqrt{2}}\left\{e^{i\gamma_N(T)}\left[\cos\left(\lambda\sqrt{N(N-1)T}\right)\,|N\rangle_{\phi=0}\,|-\rangle\right.\right. \\
&\quad \left. + i\sin\left(\lambda\sqrt{N(N-1)T}\right)|N-2\rangle_{\phi=0}\,|+\rangle\right] \\
&\quad \times e^{i\gamma_{N+1}(T)}\left[\cos\left(\lambda\sqrt{(N+1)\,NT}\right)\,|N+1\rangle_{\phi=0}\,|-\rangle\right. \\
&\quad \left.\left. + i\sin\left(\lambda\sqrt{(N+1)\,NT}\right)|N-1\rangle_{\phi=0}\,|+\rangle\right]\right\}
\end{aligned} \tag{22}
$$

In accordance with our strategy, we consider as a "candidate" observable sensitive to Berry phase accumulation, the operator

$$
\hat{O} = \frac{A^{\dagger}_{\phi=0} + A_{\phi=0}}{2} \tag{23}
$$

which has a clear physical meaning, being proportional to the position operator of the vibrational mode corresponding to $\phi = 0$.

If we choose the direction α of the laser such that $\theta = \pi/6$, we obtain the following mean value after a cycle

$$
\begin{aligned}
\langle\Psi(t=T)|\,\hat{O}\,|\Psi(t=T)\rangle &= \frac{1}{2}\cos\left\{\,\left[\gamma_N(T) - \gamma_{N+1}(T)\right] + \nu T\,\right\} \\
&\quad \times\left[\cos\left(\lambda\sqrt{N(N-1)T}\right)\,\cos\left(\lambda\sqrt{(N+1)\,NT}\right)\sqrt{N+1}\right. \\
&\quad \left. + \sin\left(\lambda\sqrt{(N-1)\,NT}\right)\,\sin\left(\lambda\sqrt{(N+1)\,NT}\right)\sqrt{N-1}\right]
\end{aligned} \tag{24}
$$

Since for the chosen value of θ it turns out that

$$
\gamma_N - \gamma_{N+1} = \pi \tag{25}
$$

in view of Eq. (20), such a quantity is the negative of what we would obtain if the only phase factor acquired by an eigenstate of the Hamiltonian were the dynamical one. If we take N sufficiently large, we can consider this difference in sign as *macroscopic*. Moreover, if we consider that the eigenvalues in Eq. (18) are independent of time and then that the dynamical phase in our anisotropic trap is the same that we would have in an isotropic trap, where the geometrical phase is absent, we obtain the following result: *the macroscopic difference in sign we have found is a direct manifestation of anisotropy which allows us to distinguish at a certain instant T between an isotropic trap and an anisotropic one with oscillation frequencies such that* $\Delta\nu = 4\pi/T$.

23.5 Conclusions

In this paper we have presented, through a very specific example, the connection between the anisotropy of a Paul trap and Berry phase established by laser fields. Indeed, the difference between the oscillation frequencies of the trap is responsible for the impossibility of finding exact resonance conditions when a laser field propagates along directions different from the axes of the trap. This leads, in a natural way, to the implementation of time-dependent vibronic couplings turning out to be slowly varying in the case of weak anisotropy, causing geometric phase accumulation. Such a behavior has been brought to light by the evaluation of a suitably chosen vibrational observable which connects vibronic states acquiring different Berry phases. In fact, at a specific instant of time, its mean value exhibits a change in sign completely determined by the non-dynamical phase. For a large enough number of excitations, such an effect may be considered as *macroscopic*.

References

1. P. K. Ghosh, *Ion Traps*, Clarendon Press, Oxford (1995).
2. D. Wineland et al., *J. Res. Natl. Inst. Stand. Technol.* **103**, 258 (1998).
3. Leibfried et al., *Rev. Mod. Phys.* **75**, 281 (2003).
4. W. Vogel and R.L. de Matos Filho, *Phys. Rev. A* **52**, 4214 (1995).
5. M.V. Berry, *Proc. R. Soc. Lond. A* **392**, 45–57 (1984).
6. M.V. Berry, *J. Mod. Opt.* **34**, 1401–1407 (1987).
7. A. Messiah, *Quantum Mechanics*, John Wiley & Sons, New York (1958).

24

Distilling Angular Momentum Schrödinger Cats in Trapped Ions

B. MILITELLO, AND A. MESSINA

INFM, MIUR and Dipartimento di Scienze Fisiche ed Astronomiche dell'Università di Palermo, Via Archirafi, 36, I-90123 Palermo. Italy

Abstract

A quantum system (S) in interaction with a repeatedly measured one (M) undergoes a non-unitary time evolution pushing it into some specific subspaces. It is assumed that the measured system is found in the initial state at each step, and that both the M–S interaction and its duration between two measurement acts are the same during all the process. Such three ingredients (measured state, interaction, interaction duration) give raise to a "selection rule" determining what states are S decay (the "residual" states) and what states are decay-preserved (the "distilled" states). On the basis of this strategy, we report a method for distilling the vibrational state of a trapped ion center of mass in order to obtain angular momentum Schrödinger Cat superposition.

24.1 Introduction

Trapped ions provide a fruitful physical scenario wherein fundamental aspects of quantum mechanics may be tested and applications dealing with quantum technology are realizable.[1,2] In fact, a trapped ion is nothing but a charged particle confined into a finite space region through inhomogeneous time-dependent electric fields generated in a quadrupolar trap, i.e., a disposal whose hyperboloid-like geometry is responsible for an electric field spatial dependence similar to that of a quadrupole. The center of mass of the confined ion behaves approximately as a harmonic oscillator.[3] Moreover, in most of the cases, the internal degrees of freedom of the ion, associated with the motion of the electrons around the nucleus, are describable as those of a two-level system. Therefore, a compound spin 1/2-boson physical system is indeed provided

In this physical scenario a wide variety of Hamiltonian models may be implemented,[4] and moreover, a huge quantity of both classical and non-classical states have been realized: coherent states, squeezed states, Fock number states and Schrödinger Cat states.[5] Here we focus on the generation of coherent superposition of macroscopically well distinguishable harmonic oscillator quantum states possessing the same energy but opposite angular momentum. In other words, we propose a scheme for obtaining the superposition of states performing clockwise and counter-clockwise motion. Generation strategies of such states have been already proposed,[6] even if they require a very high degree of temporal resolution and a long experiment duration.

In this contribution, in the spirit of quantum non-demolition measurements,[7] we propose a distillation process based upon the idea that a physical system in interaction with a repeatedly measured one undergoes a non-unitary time evolution provoking the decay of most of the states and as a consequence preserving few of them (the 'filtered' or 'distilled' states).[8] The relevant selection rule discriminating between distilled and non-distilled states is proved to be related with the survival probabilities subjected to the unitary evolution.[9]

Under certain conditions concerning both the initial state and the specific vibronic interaction, it is possible to generate coherent 50% superposition of angular momentum eigenstates.[10]

The chapter is organized as follows. In the next section the trapped ion system is described and the strategy proposed to distill angular momentum quantum states is sketched. In the Section 24.3 the generation (distillation) scheme of Schrödinger Cats is reported. Finally, in Section 24.4, some conclusive remarks are provided.

24.2 Distilling Angular Momentum 'cats' Through Vibronic Couplings

24.2.1 Vibronic Couplings in trapped ions

A quadrupolar sinusoidally time-dependent electric field can force the center of mass of the charged particle to move approximately as if it was confined into a three-dimensional well characterized by quadratic attractive potentials. When the particle is an ion possessing just two atomic levels involved in the dynamics, the result is a compound fermionic-bosonic system describable via the following Hamiltonian ($\hbar = 1$)

$$H_{\text{trap}} = \sum_{j=x,y,z} \nu_j \hat{a}_j^+ \hat{a}_j + \frac{\omega_0}{2} \hat{\sigma}_3$$

where ν_j are the center of mass harmonic oscillator frequencies, $\hat{a}_j$ $\left(\hat{a}_j^+\right)$ the related annihilation (creation) operators, $\omega_0$is the Bohr frequency between the two electronic levels considered, and $\hat{\sigma}_3$ is the third Pauli operator. In this chapter we concentrate on the degenerate two-dimensional case, that is the motion along $z$is never excited and then factorized and $\nu_x = \nu_y \equiv \nu$ is assumed.

Once the ion is trapped, a wide variety of vibronic couplings may be realized through the action of laser fields. For instance, using a laser field directed along x and tuned to the kth red sideband, i.e., $\omega_0 - k\nu$, in the so called Lamb–Dicke limit (i.e., assuming that the ion centre of mass oscillations have amplitudes much smaller than the laser wavelengths), and making the rotating wave approximation, one obtains, in the interaction picture with respect to $\hat{H}_{\text{trap}}$, the k-bosons Jaynes–Cummings model,

$$\hat{H}_{\text{JC}} = \gamma \left(a_x^k \hat{\sigma}_+ + \text{h.c.}\right)$$

where $\hat{\sigma}_+$ are the lowering and raising Pauli operators, and γ measures the coupling strength.

Acting upon the system through two laser fields one directed along $x$and the other one directed along y, both tuned to the second red sideband, and in the Lamb–Dicke limit, the interaction picture Hamiltonian turns out to be

$$\hat{H}_{\text{AM}} = \gamma \left[\left(a_x^2 + a_y^2\right) \hat{\sigma}_+ + \text{h.c.}\right]$$

Before closing this section it can be mentioned that a huge class of experimental setup, i.e., a set of suitably directed, tuned and polarized laser fields, is responsible for an interaction picture vibronic coupling which may be cast in the form

$$\hat{H}_{\Omega} = \gamma\left(\hat{\Omega}\hat{\sigma}_{+} + \hat{\Omega}^{+}\hat{\sigma}_{-}\right)$$

$\hat{\Omega}$ being a bosonic operator individuated as the specific laser configuration is given.

24.2.2 *Distilling angular momentum eigenstates*

The time evolution associated with $\hat{H}_{\mathrm{AM}}$ is straightforwardly evaluated as

$$\hat{U}(t) \equiv e^{-i\hat{H}_{\mathrm{AM}}t} = \cos\left(t\sqrt{\hat{H}^2_{\mathrm{AM}}}\right) - i\sin\left(t\sqrt{\hat{H}^2_{\mathrm{AM}}}\right)\frac{\hat{H}_{\mathrm{AM}}}{\sqrt{\hat{H}^2_{\mathrm{AM}}}}$$

where

$$\hat{H}^2_{\mathrm{AM}} = \gamma^2\left[\left(\hat{a}_x^+ + \hat{a}_y^+\right)\left(\hat{a}_x + \hat{a}_y\right)|-\rangle\langle-| + \left(\hat{a}_x + \hat{a}_y\right)\left(\hat{a}_x^+ + \hat{a}_y^+\right)|+\rangle\langle+|\right]$$

from which it immediately follows

$$\hat{H}^{2n}_{\mathrm{AM}} = \gamma^{2n}\left[\left[\left(\hat{a}_x^+ + \hat{a}_y^+\right)\left(\hat{a}_x + \hat{a}_y\right)\right]^n|-\rangle\langle-| + \left[\left(\hat{a}_x + \hat{a}_y\right)\left(\hat{a}_x^+ + \hat{a}_y^+\right)\right]^n|+\rangle\langle+|\right]$$

providing a very high degree of manageability of the series of $\sqrt{H^2_{\mathrm{AM}}}$.

It deserves to be observed that in correspondence to an eventually vanishing eigenvalue of $\hat{H}^2_{\mathrm{AM}}$ the action of the operator $\frac{\sin\left(t\sqrt{\hat{H}^2_{\mathrm{AM}}}\right)}{\sqrt{\hat{H}^2_{\mathrm{AM}}}}$ on the relevant eigenstate approaches identity times t, hence providing a meaningful explicit expression for the evolution operator is in any case.

Assume now that the system, initially prepared into the upper electronic level, is subjected to such an evolution for a time τ, and then subjected to an electronic state measurement returning $|+\rangle$ as a result. Moreover, let the system again undergo the unitary evolution $\hat{U}(\tau)$, and again its electronic state is measured and found to be $|+\rangle$, and so on N times. The complete action on the vibronic system is given by the non-unitary operator

$$\hat{W}^{(N)}(\tau) \equiv \Xi_N\left[|+\rangle\langle+|\hat{U}(\tau)\right]^N|+\rangle\langle+| = \Xi_N\left[\hat{V}(\tau)\right]^N|+\rangle\langle+|$$

with

$$\hat{V}(\tau) \equiv \langle+|\hat{U}(\tau)|+\rangle = \langle+|\cos\left(\tau\sqrt{\hat{H}^2_{\mathrm{AM}}}\right)|+\rangle = \cos\left(\tau\langle+|\sqrt{\hat{H}^2_{\mathrm{AM}}}|+\rangle\right)$$

$$\Xi_N \equiv \left[\prod_{k=1}^{N}\sqrt{\wp_k}\right]^{-1}$$

$\wp_k$ being the probability of finding the ion into the state $|+\rangle$ at the kth step. Few simple algebraic manipulations lead to the expression

$$\hat{V}(\tau) = \cos\left(\gamma\tau\sqrt{\left(\hat{a}_x + \hat{a}_y\right)\left(\hat{a}_x^+ + \hat{a}_y^+\right)}\right)$$

Taking into account that

$$(\hat{a}_x + \hat{a}_y)\left(\hat{a}_x^+ + \hat{a}_y^+\right) = \hat{N}_T^2 + 4\hat{N}_T - \hat{L}_z^2 + 4$$

with $\hat{N}_T = \hat{a}_x^+\hat{a}_x + \hat{a}_y^+\hat{a}_y$ and $\hat{L}_z = i\left(\hat{a}_x\hat{a}_y^+ - \hat{a}_x^+\hat{a}_y\right)$, one obtains the explicit final form

$$\hat{V}(\tau) = \cos\left[\gamma\,\tau\sqrt{\hat{N}_T^2 + 4\hat{N}_T - \hat{L}_z^2 + 4}\right]$$

The operators $\hat{N}_T$ and $\hat{L}_z$ have transparent physical meaning. Indeed the first one is the total vibrational excitation number, while the second one is the z-component of the angular momentum operator of the trapped oscillator.

Since $\hat{N}_T$ and $\hat{L}_z$ commute, they possess a common basis. Let us denote by $|n_T, m\rangle$ the common eigenstate pertaining to the eigenvalues $n_T = 0, 1, 2, \ldots$ and m ($m = 0, \pm 2, \pm 4, \ldots, \pm n_T$ if n_T is even, and $m = \pm 1, \pm 3, \ldots, \pm n_T$ if n_T is odd) of $\hat{N}_T$ and $\hat{L}_z$ respectively. It is straightforward to see that such eigenstates are eigenstates of $\hat{V}(\tau)$too. Observe now that for large enough N, the Nth power of a cosine approaches zero when the argument is different from a multiple of π, while it is ± 1 otherwise. This circumstance leads to the following statement. For large enough N, the non-unitary action $\hat{W}^{(N)}(\tau)$, projects the initial state into the vibrational subspace generated by

$$\left\{|n_T, m\rangle : \gamma\,\tau\sqrt{n_T^2 + 4n_T - m^2 + 4} = l_{n_T,m}\pi, \quad l_{n_T,m} \in Z\right\}$$

eventually including the phase factor -1, if $l_{n_T,m}N$ is odd.

In this sense the procedure described is the realization of a projection operator. The vibrational subspace preserved by $\hat{W}^{(N)}(\tau)$may be adjusted with a suitable choice of the pulse area $\gamma\,\tau$. In fact, the incommensurability of the different square roots corresponding to different n_T and m in most of the cases guarantees, as we will see in the following example, the possibility of selecting a single subspace with prefixed excitation number and angular momentum component along z.

Summarizing, repeatedly measuring the electronic state of a trapped ion subjected to the action of a laser field configuration responsible for the suitable vibronic coupling $\hat{H}_{\text{AM}}$, it is possible to force the system toward a finite-dimensional subspace characterized by well defined excitation number and angular momentum along z.

It is worth noting that this procedure is a conditional one. Indeed its success is subordinated to the result of the electronic measurements. In particular, the probability of distillation success $\wp$ is given by the probability of finding the electronic system into the state $|+\rangle$ at each step $\prod \wp_k$. It is possible to prove that, such a success probability is equal to the norm of the initial vibrational state projected into the distilled subspace. More precisely, denoting by $\hat{P}_{\text{d}}$ the projector onto the distilled subspace, and by $\hat{\rho}_i$ the initial (eventually non-pure) vibrational state, it turns out that

$$\wp \equiv \prod_{k=1}^{N} \wp_k = \left\| \hat{P}_{\text{d}}\hat{\rho}_i \right\|^2$$

This result gives us a very readable and easy to evaluate expression for the success of the conditional distillation process. Moreover it is perfectly in accordance with the statement that what we have realized is a projection operator, and implies that the process fails if the initial state has no component in the target subspace (indeed in this case $\wp = 0$).

24.3 Controllable Cat-Like Superposition

The selection rule which discriminates between suppressed and preserved angular momentum subspaces, involves just the square of m, hence preserving at the same time both $-m$ and $-m$. This is the crucial point of the angular momentum Schrödinger Cat generation scheme that is presented here.

Consider a Fock state directed along a direction forming an angle θ with respect to the $x$axis. Such a state is nothing but the state

$$|n_T\rangle_\theta = \frac{1}{\sqrt{n_T!}}(a_x^+ \cos\theta + a_y^+ \sin\theta)^{n_T} \, |0,0\rangle$$

which is easily classified as a two-mode SU(2)state,[6,10,11] whose general formula

$$|\mu, j\rangle = \frac{1}{\left(1+|\mu|^2\right)^j} \sum_{k=0}^{n_T} \begin{pmatrix} 2j \\ k \end{pmatrix}^{1/2} \mu^k \, |2j-k, k\rangle$$

shows that

$$|\mu = \tan\theta, j = n_T/2\rangle = |n_T\rangle_\theta \, .$$

Moreover, if we consider the common eigenstates of $\hat{N}_T$ and $\hat{L}_z$, possessing maximum (minimum) angular momentum projection compatible with the number of excitations, $|n_T, m = \pm n_T\rangle$, it is easy to verify that it is a two-mode SU(2)state too, being $|\mu = \pm i, j = n_T/2\rangle = |n_T, m = \pm n_T\rangle$ [6, 10]. Each one of them describes the ion performing a circular motion, which is clockwise or counter-clockwise depending on the sign of m.

The overlap between two SU(2)states may be easily evaluated, and in particular it turns out

$$\langle \mu = \pm i, j \mid \mu = \tan\theta, j\rangle = \frac{e^{\mp i2j\theta}}{2^j}$$

which is a non-vanishing quantity. Moreover, since the weights of such overlaps are equal, it is possible to extract, through our methods, a 50% superposition of $|n_T, m = n_T\rangle$ and $|n_T, m = -n_T\rangle$ just choosing a pulse area $\gamma\tau$ preserving $m = \pm n_T$ and suppressing all other components of the initial state $|n_T\rangle_\theta$. The relative phase between the coefficients of the distilled superposition will be the same as in the initial state, and hence equal to $4j\theta = 2n_T\theta$.

As a very specific example of generation, consider the Fock state $|n_T = 4\rangle_\theta$ as initial condition. Since it possesses a well-defined number of excitations it is expressible as the superposition of the five states $|n_T = 4, m\rangle$ with $m = 0, \pm 2, \pm 4$ (The eigenvalues of the angular momentum $\hat{L}_z$, canonically equivalent to $\hat{a}_x^+\hat{a}_x - \hat{a}_y^+\hat{a}_y$, span the subspace of fixed $\hat{N}_T$ with step of 2).[12] The corresponding 'square roots' $\sqrt{4^2 + 4\cdot 4 - m^2 + 4}$ are 6, $4\sqrt{2}$ and $2\sqrt{5}$ respectively, which are incommensurable.

Therefore, choosing the pulse area such that $2\sqrt{5}\gamma\tau = 2\pi$, the state

$$|\psi_d\rangle = \frac{1}{\sqrt{2}}\left[|n_T = 4, m = 4\rangle + e^{i8\theta}\,|n_T = 4, m = -4\rangle\right]$$

is distilled up to a global phase factor.

It is worth noting that the relative phase between the two terms of the resulting superposition is well controllable through the choice of the initial linear motion direction, i.e., through the choice of θ.

The efficiency of the generation scheme is easily evaluated as

$$\left\| \hat{P}_{\mathrm{d}} \left| n_T = 4 \right\rangle_{\theta\,\theta} \left\langle n_T = 4 \right| \right\| = \frac{1}{4^2} + \frac{1}{4^2} = \frac{1}{8}$$

being

$$\hat{P}_{\mathrm{d}} = |n_T = 4, m = 4\rangle \langle n_T = 4, m = 4| + |n_T = 4, m = -4\rangle \langle n_T = 4, m = -4|$$

It deserves to be remarked that the distillation is in principle 'perfect' in the limiting case $N \to \infty$. On the contrary, for finite N, the discrepancy between the actual result of the process and the expected one may be evaluated considering the weights of the residual contributions $\cos^N(\gamma\,\tau\sqrt{n_T^2 + 4n_T - m^2 + 4})$ for non-distilled m. In the given example, after $N = 5$steps, the agreement between the ideal and the actual distilled state is higher than 95%.

24.4 Conclusive Remarks

In this chapter a new method for generating, in the context of trapped ions, Schrödinger Cat-like states that are coherent superposition of a single two-dimensional harmonic oscillator state of motion with well-defined number of excitations and angular momentum along z has been presented. In fact what one obtains is the superposition of clockwise and counter clockwise circular states of motion.

The proposed strategy, in addition to the virtue of higher accuracy and efficiency, possesses the feature of a well controllable phase between the two terms of the superposition. In particular, such a phase is proportional to the direction (i.e., the corresponding angle) of the initial linear motion of the ion center of mass.

As a conclusive remark we mention that, since generally few steps are required to complete the distillation process, the experiment duration results to be short enough to neglect decoherence incoming.

24.5 Acknowledgments

We thank Hiromichi Nakazato and Sabrina Maniscalco for stimulating discussions.

References

1. D. Leibfried, R. Blatt, C. Monroe, and D. Wineland, *Rev. Mod. Phys.* **75**, 281 (2003).
2. D. J. Wineland, C. Monroe, W. M. Itano, D. Leibfried, B. E. King, and D. M. Meekhof, *J. Res. Natl. Inst. Stand. Technol.* **103**, 259 (1998).
3. P. K. Ghosh, *Ion Traps* (Clarendon Press, Oxford, 1995).
4. W. Vogel and S. Wallentowitz, Manipulation of the quantum state of a trapped ion, in *Coherence and Statistics of Photons and Atoms*, ed. by Jan Perina, (Wiley, New York, 2001).
5. D. M. Meekhof et al., *Phys. Rev. Lett.* **76**, 1796 (1996); C. Monroe, et al., *Science* **272**, 1131 (1996).
6. S. Maniscalco, A. Messina, and A. Napoli, *J. Mod. Opt.* **47**, 2113 (2000); S.Maniscalco et al., *Phys. Rev. A* **61** 053806 (2000); S. Maniscalco et al., *J. Mod. Opt.* **48**, 2065 (2001).
7. V. B. Braginsky and F. Ya. Khalili, *Quantum Measurement* (Cambridge University Press, 1992); D. F. Walls and G. J. Milburn, *Quantum Optics*, (Springer–Verlag, Berlin, 1994); R. L. de Matos Filho and

W. Vogel, *Phys. Rev. Lett.* **76** 4520 (1996); L. Davidovich, M. Orszag, and N. Zagury, *Phys. Rev. A* **54**, 5118 (1996).
8. H. Nakazato, T. Takazawa and K. Yuasa, *Phys. Rev. Lett.* **90**, 060401 (2003).
9. B. Militello, H. Nakazato, and A. Messina, *Steering Distillation Processes through Zeno Effect*, quant-ph/0404058
10. B. Militello and A. Messina, *Distilling Angular Momentum nonclassical States in Trapped Ions*, quant-ph/0405079
11. A. M. Perelomov, *Commun. Math. Phys.* **26**, 222–236 (1972).
12. C. Cohen-Tannoudji, B. Diu, F. Laloe., *Quantum Mechanics* (New York, Wiley-Interscience, 1998).

25

Linear-response conductance of the normal conducting single-electron pump

R. Schäfer, B. Limbach, P. vom Stein, and C. Wallisser

Forschungszentrum Karlsruhe, Institut für Festkörperphysik, D-76121 Karlsruhe, Germany.

Abstract

For single-electron devices the model Hamiltonian matches its experimental realization very closely. This has been proven most impressively for the normal conducting single-electron transistor (SET) with the aid of Quantum-Monte-Carlo methods. However, the lowest order perturbation theory, the so called sequential model, gives only a qualitative description in most practical cases. It ignores quantum fluctuations due to the coupling of charge states by the finite conductance of the tunneling contacts, which are fundamental constituents of the devices. Perturbation expansion (PE) has been suggested do deal with the impact of the coupling. To get more insight into the applicability of PE high quality conductance data on single-electron devices other than the SET are desirable. Here we present measurements of the linear response conductance of single-electron pumps (SEP).

25.1 Introduction

In this contribution an experimental study of the linear-response conductance of two single electron islands in series, an arrangement nicknamed Single Electron Pump (SEP)[1–3] is presented and sketched in Fig. 25.1. The linear-response conductance varies with the gate voltages and is bound between temperature dependent values $G_{\min}(T)$ and $G_{\max}(T)$. The tunnel conductances of the outer contacts (G_l and G_r) are varied and the inner contact (G_m) independently increasing it up to values of the order of $G_K = e^2/h$ in either case. The samples are carefully designed to allow the most complete characterization. The only relevant parameter which needs to be determined indirectly is an overall charging energy, which is estimated from the high temperature behavior of the conductance. Distinct corrections beyond the lowest order perturbation treatment in the tunneling Hamiltonian, the so-called "sequential model", manifest themselves most clearly by a logarithmic correction to $G_{\max}(T)$.

From intensive studies of the single-electron transistor[4–6] *we gained confidence that we know how* to model single-electron devices with great precision even if the coupling of the island to its leads is large. However, as is exemplified in Section. 25.2, it is not clear at present whether the exceptional good agreement found for the transistor by comparing experiments and sophisticated

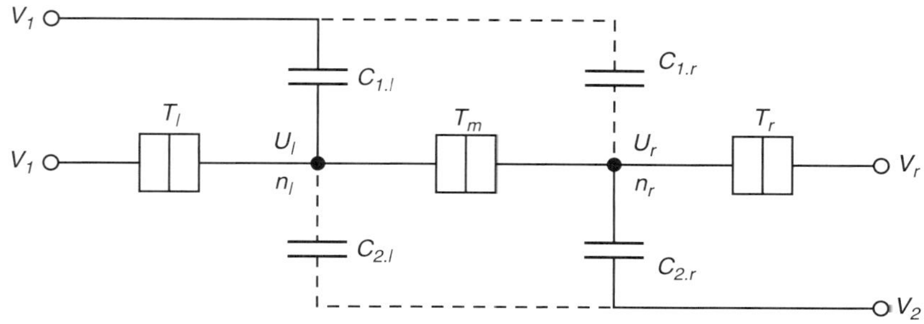

FIGURE 25.1. Schematic representation of the single-electron pump. The three tunneling contacts T_l, T_m, and T_r are laid out in a row forming two islands (each contact $i \in \{l, m, r\}$ is characterized by its capacitance C_i and conductance $G_i \equiv g_i G_K$). n_l and n_r count the number of electrons by which the left and right island charge differs from neutrality. The arrangement is biased by the voltage difference $V_r - V_l$. The electrostatic potentials on the islands (U_l and U_r) can be tuned by V_1 and V_2 which couple directly to the islands by the capacitances C_{1l} and C_{2r}. C_{2l} and C_{1r} represent the experimentally unavoidable stray capacitances.

theoretical work outlasts tests in more complex arrangements of single-electron islands. The experimental data presented here make a sensitive test of theoretical concepts possible. This however requires elaborate calculations, which are out of the scope of the present work. Since a detailed understanding of fluctuation effects and how to model them is a prerequisite for future applications such a test is of considerable importance.

The report here is restricted to the linear-response conductance since only for this quantity low electronic temperatures on the single electron island can be achieved. Finite transport voltages and currents lead to strong heating of the islands due to the weak coupling between the electron and phonon system.[7] The coupling enters into the description of the experiments as a fitting parameter making an unambiguous comparison with theoretical calculations almost impossible. Furthermore the comparison is less sensitive at elevated temperatures since the higher order corrections are most important at the lowest achievable temperatures.

Depending on the ratio G_s/G_m, where $G_s^{-1} = G_l^{-1} + G_r^{-1}$, non-monotonic behavior of $G_{\max}(T)$. This *non-trivial* phenomenon is naturally explained by the sequential model, which is presented for the system at hand in Section. 25.3. Experimental details and the results are given in Section. 25.4 and discussed in Section. 25.5.

25.2 Motivation for the Measurements

It is generally accepted that normal conducting single-electron devices can be modeled by a Hamiltonian made up of three parts, $H_{\text{se}} = H_{\text{qp}} + H_{\text{ch}} + H_T$. The first part, H_{qp}, describes the quasiparticles on the various leads and islands which constitute the device. It is treated in a free Fermion approximation. The second part, H_{ch}, describes the electrostatic energy E_{ch} of the capacitively coupled islands. It can be written as $E_{\text{ch}} = \sum_{i,j} \Delta n_i E_{i,j} \Delta n_j$, where i and j index the single-electron islands. We use the notation $\Delta n_i = n_i - n_{0,i}$, with n_i counting the number of excess electrons on island i and $en_{0,i}$ denoting the polarization charge induced by gate voltages to that particular island. $E_{i,j}$ is the inverse of the capacity matrix of the islands in units of $e^2/2$.

Finally, the third part of the Hamiltonian describes tunneling between islands and between islands and leads. It can be further broken up into a sum over individual tunneling contacts. A tunneling contact between island i and j is accounted for by the term $H^T_{i,j} = \sum_{kq} T_{kq} a^\dagger_k a_q + \text{H. c.}$, where k and q label the quasiparticle states in island i and j, respectively. It is common practice, justified by the success of the theory, to ignore state and energy dependence of the tunneling matrix elements T_{kq} and to write $G^T_{i,j} = (2e^2/h)\Omega_i N_i(0)\Omega_j N_j(0)|T_{kq}|^2$. Here $G^T_{i,j}$ is the Ohmic conductance of the voltage biased junction and $\Omega_i N_i(0)$ quantifies the level density at the Fermi edge of junction i. The Hamiltonian H_{se} in the form presented here ignores the influence of lead impedances in the environment of the device. Such impedances are of experimental relevance only if they are large (of the order of $1/G_K$) and in very close vicinity to tunnel junctions or at high voltage bias.[8] Their influence is further reduced in multi junction devices[8] and is thus completely ignored throughout this chapter.

The Hamiltonian H_{se} has been treated by several methods, usually starting with the Hamiltonian $H_0 = H_{\text{qp}} + H_{\text{ch}}$ and taking H_T as a perturbation into account. The states of H_0 are characterized by the number of excess electrons on each island. H_T leads to fluctuation of these numbers. The probably simplest way to describe the perturbation is worked out by the sequential model, which treats tunnel rates in golden rule approximation and estimates the occupation probabilities of charge states from the steady state solution of a master equation.[8,9] The qualitative performance of single-electron devices is reliably described by this lowest order perturbation theory. However, in most practical cases a quantitative description requires the inclusion of higher order contribution. Conventional perturbation expansion in the tunneling conductance lead to convincing results only if a *single* charge state is occupied at zero temperature.[10] Degeneracy leads to divergence. e.g., single-island circuits (box and transistor) at the degeneracy point can be mapped onto a multi channel Kondo model[11] and need accordingly sophisticated treatments. Here renormalization group procedures[12] as well as quantum Monte Carlo calculations[13] have been very successful. At high temperatures and high conductances semiclassical calculation can be used.[5,13]

Experimental work on the single-electron transistor have been carried out up to $G = 20e^2/h$, were G is the parallel conductance of the transistor island through its tunneling contacts. Careful comparisons with theoretical predictions show that the model Hamiltonian H_{se} describes single-electron devices with amazingly high accuracy.[4–6,14]

Despite the inherent obstacles Schoeller, König, and Schön succeeded to develop a systematic perturbation expansion based on a real-time diagrammatic approach[15] and applied it successfully to the transistor.[16] They included the virtual occupation of all relevant charge states (the number of which exceeds the number of occupied states in the sequential model considerably) and got, up to second order in g, regular expression for important quantities like the current and the mean island charge even at the degeneracy point. Their result agrees with the experiments and the path integral Monte Carlo results[6,13] up to $g \approx 1.5$. Despite this success doubts on the applicability of PE in more general cases has occurred lately. Even for the SET, the validity range away from degeneracy is not clear. e.g., in the blockaded regime deviation between PE and experimental results are stronger than at the degeneracy point while the quantum Monte Carlo data are in full accordance with the experiments.[6] This counterintuitive finding is not understood at present. Furthermore, intermediate expressions of the PE calculation are burdened by a cut-off energy U, which has to be taken larger than all relevant energy scales. In many expressions for the SET – including the mean charge on the transistor island and the conductance – these cut-offs cancel. But this happens to be somewhat fortunate, since no general rule is known which guarantees the cancellation. e.g., the occupation probabilities of the individual charge states, which are certainly relevant quantities, contain the cut-off parameter in the form $\log(\beta U/2\pi)$ explicitly.[16] But this means that either perturbation theory fails to calculate certain quantities or that H_{se} is the wrong starting point for the modeling of these quantities and needs to be replaced by a more realistic model. e.g., including the energy dependence of the tunneling matrix elements in H_T would naturally introduce an energy cut-off. However, this would be a surprising result indeed: from numerical solutions by quantum Monte Carlo methods we know that at least the conductance of the single-electron transistor is perfectly described by H_{se}.

More insight into this problem can be gained by analyzing more complex single-electron devices than the transistor. A first step in this direction was taken by Pojohla et al.[17] who studied the SEP using RG procedures as well as the systematic perturbation-expansion methods developed by König and Schoeller.[14,16] Experimental results presented below, however, can not be directly compared to the results of Ref. [17]. The final results are given as graphs representing the outcome of elaborate numerical calculations. Pojohla et al.[17] treated only special cases of general interest featured by high symmetrical sample parameters. State of the art fabrication techniques do not allow to reproduce samples with properties corresponding to these special situations. Today we have to live with considerable statistical spreading of sample parameters. A detailed comparison has to start with the given sample parameters and requires a reevaluation of the numerics implied by Ref. [17].

25.3 Sequential Model

The capacitance matrix of the system depicted in Fig. 25.1 is given by $\mathbf{C} = \left(\begin{smallmatrix} C_{\Sigma l} & -C_m \\ -C_m & C_{\Sigma r} \end{smallmatrix}\right)$, where $C_{\Sigma i} = C_i + C_{1i} + C_{2i} + C_m$. With $\left(\begin{smallmatrix} E_l & E_{co} \\ E_{co} & E_r \end{smallmatrix}\right) = (e^2/2)\mathbf{C}^{-1}$ the charging energy reads $E_{ch} = E_l(n_{0l} - n_l)^2 + E_r(n_{0r} - n_r)^2 + E_{co}(n_{0l} - n_l)(n_{0r} - n_r)$. We have used $n_{0i} = (C_i V_i + C_{1i} V_1 + C_{2i} V_2)/e$, and n_i counts the number of electrons by which the charge on island $i \in \{l, r\}$ differs from neutrality. Note that $V_i, i \in \{l, r\}$ is infinitesimal small in this chapter. For the sake of simplicity we shift from (n_{0l}, n_{0r}) to a different coordinate system by the relations $n_x = n_{0l} + n_{0r}, n_y = n_{0r} - n_{0l} - \kappa n_x$, where $\kappa = (C_{sr} - C_{sl})/C_s, C_{si} = C_{\Sigma i} - C_m$, and $C_s = C_{sr} + C_{sl}$. For the charging energy we get $E_{ch} = E_{cx}(n_x - n_s)^2 + E_m(n_y + \Delta n + \kappa n_s)^2$ with the energy parameters $E_{cx} = e^2/(2C_s)$ and $E_m = (e^2/2)C_s/(C_m C_s + C_{sl} C_{sr})$. The coordinate n_x is associated with the change of the total charge number $n_s = n_l + n_r$, while n_y redistributes the charge between both islands ($\Delta n = n_l - n_r$).

Each point in the plane spanned by n_x and n_y can be mapped onto a charge ground state (n_l, n_r) which gives the lowest possible $E_{ch}(n_l, n_r)$. This procedure divides the (n_x, n_y) plane into the grid of hexagonal cells depicted in Fig. 25.2(a). The length E_{co}/E_{cx} of the horizontal cell boundaries is a measure of the coupling strength ($E_{co} \propto C_m$). Due to the periodicity it suffices to study the linear-response conductance in a small exemplary portion of the (n_x, n_y) plane. Within the sequential model, the linear-response conductance of the system vanishes exponentially at low temperatures except close to the triple points in the (n_x, n_y) plane, where the ground-state energy of all three adjacent states is degenerate. The conductance peaks near these points. It is worthwhile to mention a peculiar behavior of the SEP: exactly at the triple points the low temperature conductance is constant and given by $G_0/3$ where $G_0^{-1} = G_l^{-1} + G_r^{-1} + G_m^{-1}$. Although for $T \ll E_{co}/k_B$ the maximal conductance G_{max} is temperature independent as well, the system assumes G_{max} at a slightly different position in the n_x direction.[17] Taking only three

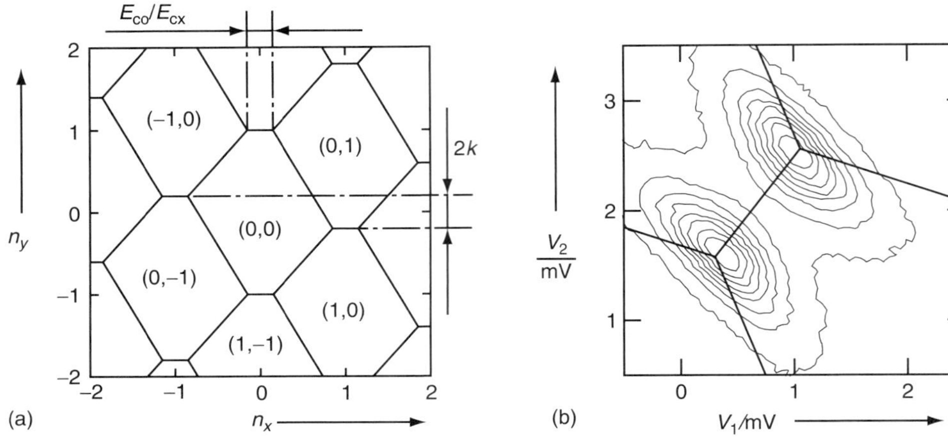

FIGURE 25.2. (a) Stability diagram for the SEP sketched in Fig. 25.1. The hexagonal cells mark regions where the indicated charge state (n_l, n_r) possesses the lowest energy $E_{\text{ch}}(n_l, n_r)$. (b) Measurement of the linear-response conductance of sample 2 as a function of both gate voltages V_1 and V_2 at 150 mK. The outermost contour line indicates a conductance of 0.01 μS. The following lines range from 0.05 μS to 0.4 μS with a 0.05 μS spacing. The stability diagram (thick lines) is deformed in the coordinates of this figure.

states into account (the occupation probability of all other states is exponentially small for $T < E_{\text{co}}/k_B$) we get

$$G(n_x)|_{n_y=-\kappa} = \frac{G_s}{2+e^{-\beta\Delta E}}\left(\frac{g_s}{g_m} + \frac{e^{\beta\Delta E}-1}{\beta\Delta E}\right)^{-1}, \tag{25.1}$$

with $\beta = 1/(k_B T)$, $g_s = G_s/G_K$, and $\Delta E = 2E_{cx}(n_x - 1) + E_{\text{co}}$. Depending on the ratio g_s/g_m the peak position deviates from the triple point at $\Delta E = 0$ for $g_m \neq 2g_s$.

In a measurement as a function of n_x and n_y the conductance displays a periodic grid of peaks (grouped as pairs), marking the endpoints of the horizontal boundaries in Fig. 25.2(a). With rising temperature the peaks broaden and shift towards the center of the horizontal boundaries. Fig. 25.2(b) gives an example at $T = 150\,\text{mK} \sim 0.09E_{cx}/k_B$, where thermal broadening already is effective in merging the two separate peaks. Finally the two peaks merge completely and the conductance takes its maximal value at the mid-points, e. g., at $n_x = 1$ and $n_y = -\kappa$.

The techniques for solving the sequential model are well documented in the literature (e. g., Ref. [8]) so we do not comment on this calculation in detail. It requires the solution of a master equation which one may set up using golden rule rates for the inelastic tunneling events. For $T \ll E_c/k_B$ at most four (n_l, n_r) states are occupied with reasonable probability. Restricting the master equation to these four states allows for an analytical solution. At higher temperatures numerical relaxation methods are used.

25.4 Experimental Details and Results

Three samples in two different layouts are studied. All samples have been produced by standard shadow-evaporation technique from aluminum with aluminum–oxide barriers. The barriers

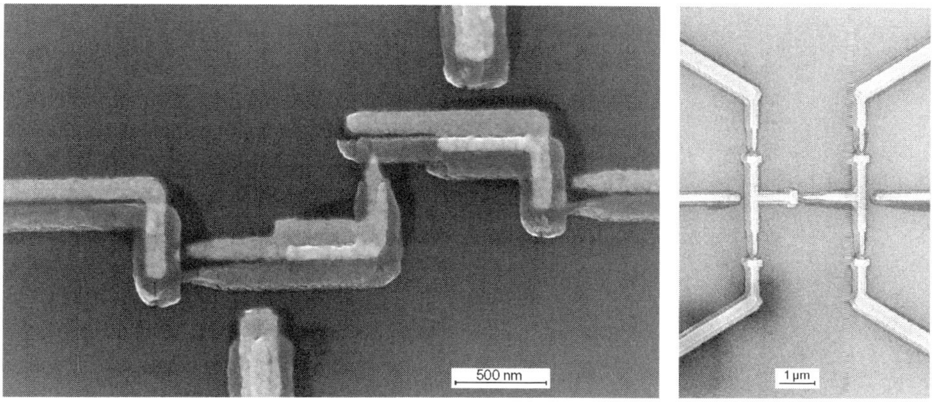

FIGURE 25.3. Scanning electron microscope pictures of sample 1 (left) and sample 2 (right).

for the middle and outer contacts are fabricated in different oxidation steps, making different barrier thicknesses for internal and external contacts possible. Sample 1 has a simple layout (see Fig. 25.3 left). In this layout the serial conductance is accessible, but the conductance of the individual contacts remains unknown. In the slightly more complex structure of sample 2 and 3 (see Fig. 25.3 right) each island is connected to independent leads via two contacts. This permits measuring the conductance of different serial combinations of the contacts, and thus the individual contact conductances can be determined. In the final experiments both external contacts of each island are operated in parallel (connected to the same voltage source), and the two contacts then act exactly as a single contact. Any degradation (considering noise performance or sensitivity to external disturbances) for the latter samples compared to sample 1 was not detected. It was noted however, that in general the charging energy E_c is lower in the complex design compared to the simple one so as to reduce the experimentally accessible value of βE_c. For this reason a sample with the simple layout was included in the investigation.

From the positions of the conductance peaks in the (n_x, n_y) plane at low temperatures (see Fig. 25.2) we obtain the parameters κ and E_{co}/E_{cx}. In addition we measure the conductance in the high temperature region where it does not depend on either n_x nor on n_y. In close analogy to the high temperature expansion for the SET, it behaves as $G(T) \approx G_0\,((1 - E_c/(3k_BT)))$. It can be shown[18, 19] that the relation

$$E_{cx} = \frac{E_c}{G_0}\left(\frac{e_+}{G_l} + \frac{e_-}{G_r} + \frac{e_+ + e_- - 2E_{co}/E_{cx}}{G_m}\right)^{-1}$$

holds, where $e_\pm = ((E_{co}/E_{cx})\,(\kappa \pm 1) \mp 2)\,/\,(\kappa \mp 1)$ and E_c is an experimentally determined fitting parameter. Table 25.1 gives all relevant sample parameters.

To simplify the analysis of our data and facilitate the comparison with theoretical considerations focus was on the temperature dependence of three special values of the conductance in the (n_x, n_y) plane, namely $G_{\min}$, $G_{\max}$, and G_m, the latter defined as the conductance at the center of the horizontal boundaries in Fig. 25.2(a) (i. e., at $n_x = 1, n_y = -\kappa$). In Fig. 22.4–22.6 main findings has been displayed. In addition to the experimental results the outcome of a calculation in the framework of the sequential tunneling model using the parameters from Table 25.1 is presented.

TABLE 25.1. Parameters of the three samples.

samp.	$\frac{E_{cx}}{k_B}$ (K)	$\frac{E_m}{k_B}$ (K)	$\frac{E_{co}}{k_B}$ (K)	$\frac{E_c}{k_B}$ (K)	κ	g_l	g_r	g_m	G_0 (μS)
1	2.8	5.8	1.3	5.6	−0.018	0.44	0.44	0.04	1.42
2	1.6	3.0	0.9	2.5	0.10	0.52	0.83	1.32	10.0
3	1.5	4.7	0.3	4.5	0.0013	0.73	0.57	0.03	0.95

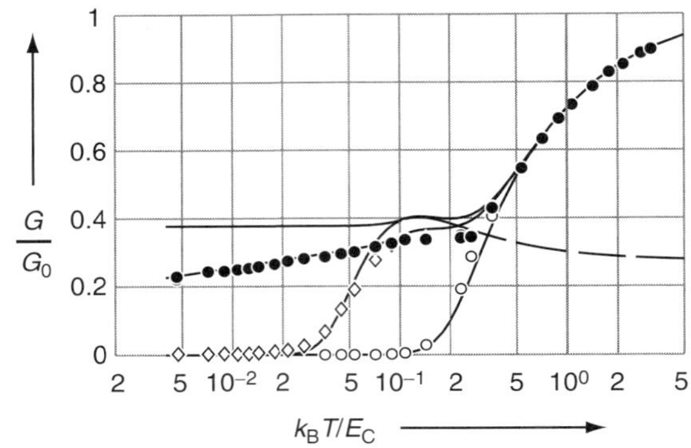

FIGURE 25.4. Linear response conductance of sample 1 as a function of temperature. Shown are G_{min} (○), G_{max} (●), and G_m (◇, the conductance at $n_x = 0.5$). The dashed lines are the result of an analytical solution of a four state model (see text). As solid lines the outcome of the sequential model with parameters from Table 25.1 is shown. For the thin solid line a term of the form $\alpha \log(k_B T/(2E_{co}))$ has been added to the maximal conductance as calculated from the sequential model. Here α is a fitting parameter.

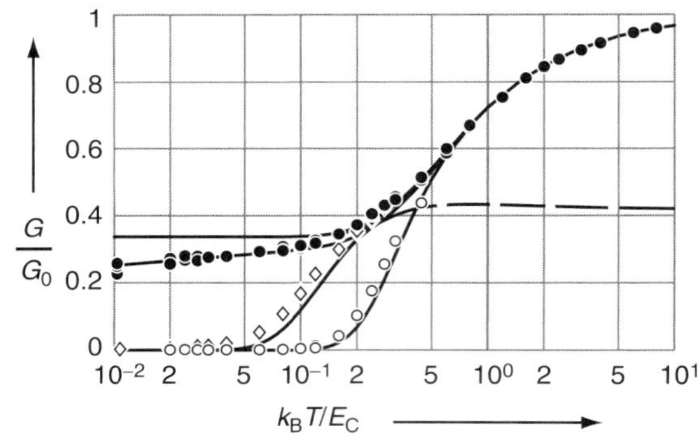

FIGURE 25.5. Sample 2. Symbols and line styles are described in the caption of Fig. 25.4.

25.5 Discussions

The sequential tunneling model is in good agreement with the experimental data with the exception of G_{max} at low temperature. The latter deviations are discussed at the end of this section. The overall behavior (see Fig. 25.4–25.6) is governed by two scales (E_{co}/k_B and E_c/k_B) at which $G_m(T)$ and $G_{min}(T)$, respectively, start to increase and finally merge with $G_{max}(T)$. To get finite

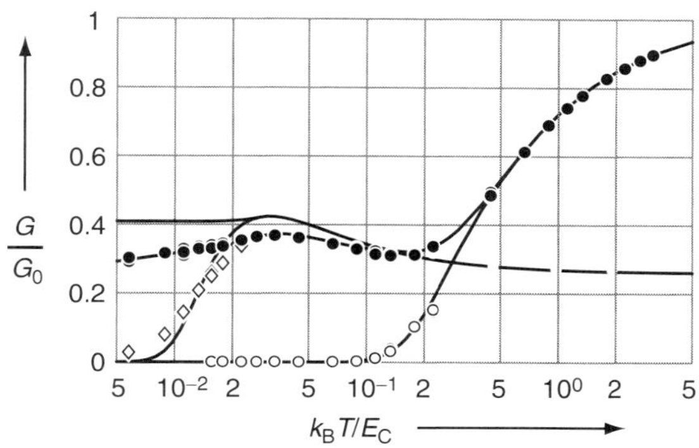

FIGURE 25.6. Sample 3. Symbols and line styles are described in the caption of Fig. 25.4.

conductance through the SEP at least three charge states have to be occupied in thermal equilibrium. At $T \ll E_{co}/k_B$ this is only possible near the triple points where three adjacent states are occupied. Here e. g., the sequence $(0, 0) \rightarrow (1, 0) \rightarrow (0, 1) \rightarrow (0, 0)$, corresponding to a charge transfer from left to right, occurs with finite probability. The inverse process is equiprobable, but under voltage bias a net current occurs. At $n_x = 1, n_y = -\kappa$ where G_m is measured, the charging energy of states $(0, 0)$ and $(1, 1)$ lie $\Delta E = E_{co}$ above the two fold degenerate ground state $((0, 1), (1, 0))$. As a result the linear response conductance at $T \ll E_{co}/k_B$, as calculated from the sequential model, is exponentially suppressed since no charge transfer is possible using only two states. At $T \sim E_{co}/k_B$ the states $(0, 0)$ and $(1, 1)$ are thermally occupied with increasing probability leading first to an exponential increase of G_m and finally to the merging of G_m and G_{max}. For $n_x = n_y = 0$ states besides $(0, 0)$ are occupied for $T \gtrsim E_c/k_B$ only. Thus G_{min} is exponentially small for $T \ll E_c/k_B$.

The most striking feature of our measurements is the non-monotonic dependence of G_{max} on the temperature found for sample 3. It is clear from Fig. 25.6 that the phenomenology is correctly described by the sequential model. To get more insight into the nature of the drop of G_{max} at $T > E_{co}/k_B$ $G_m(T)$ is analyzed, which coincides with G_{max} in the relevant temperature range. The four-state approximation mentioned above yields an analytical solution

$$G_m^{(4)} = \frac{G_s}{2} \frac{\beta E_{co}}{\sinh(\beta E_{co})} \left(1 - \frac{2g_s}{g_m} \frac{\beta E_{co}}{1 - e^{\beta E_{co}}}\right)^{-1}. \tag{25.2}$$

This function for our sample parameters in Fig. 25.4 as dashed lines. It has a distinct maximum for $g_m < g_s$, the position and strength of which depends on the ratio g_s/g_m. The above approximation breaks down at $T \sim E_c/k_B$ where more than four states are occupied significantly resulting in a rapid rise of the conductance. For an experimental observation of a local minimum two relations ($E_{co} < E_c$ and $g_m < g_s$) have to be fulfilled.

In Fig. 25.7 the n_x coordinate of the conductance maximum n_{max} as a function of temperature is plotted. Again, reasonable agreement with the sequential model is found. For $T \rightarrow 0$ the position of the maximum approaches the location of the triple point. At low temperature the conductance peak is described by Eq. (25.1). The maximum can shift in either direction depending on the ratio g_s/g_m. This can be used to determine g_s/g_m for sample 1 ($g_s/g_m = 5$) where this ratio can not be measured directly due to the simple layout (see above).

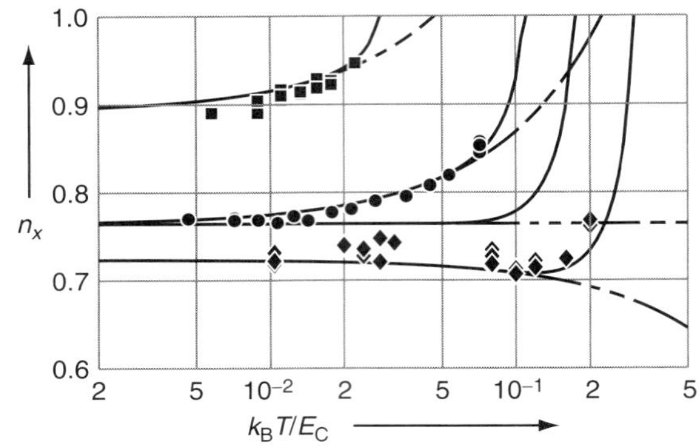

FIGURE 25.7. Position of the conductance maxima as a function of temperature. •: sample 1, ♦: sample 2, ■: sample 3. Lines: Predictions of the full sequential model (solid) and Eq. 25.1 (three-state approximation, dash dotted). For sample 1 the peak position has been used to adjust $g_s/g_m = 5$. Assuming $g_l = g_r = g_m$ gives the thin lines failing to fit the data.

The sequential model fails to predict the temperature dependence of $G_{\max}$ at low temperature. This deviation can be described phenomenologically by adding a term of the form $\Delta G_{\mathrm{qf}} = \alpha \log(k_B T/(2E_{\mathrm{co}}))$. The thin solid lines in Fig. 25.4–25.6 display the outcome of a fitting procedure in α that minimizes the mean square deviation between the measured values of $G_{\max}$ and $G_{\mathrm{seq}} + \Delta G_{\mathrm{qf}}(\alpha)$, showing very good agreement. $\alpha_1 = 45$ nS, $\alpha_2 = 200$ nS, and $\alpha_3 = 33$ nS for sample 1, 2, and 3, respectively was found . For the SET much the same behavior is observed and attributed to quantum fluctuations of the charge states. Pohjola et al.[17] analyzed the linear response of the SEP by renormalization group methods. They also found a logarithmic behavior of the low temperature conductance, in qualitative agreement with our experimental result. For a detailed comparison with our experiment a calculation using our sample parameters is highly desirable.

A careful look at Fig. 25.4–25.6 reveals considerable deviations between G_m and the sequential predictions, too. An evaluation of G_m taking higher order contributions into account can yield additional information on the validity range of the perturbation expansion. The deviations between $G_{\min}$ and the sequential model are less pronounced. This is expected, since in the double island structure, cotunneling contributions are strongly suppressed as compared to the single-electron transistor.[10]

25.6 Summary

In summary an experimental study of the linear response conductance of the SEP in a regime is presented where quantum fluctuations of the charge eigenstates can not be ignored. Depending on the ratio g_s/g_m the SEP shows a remarkably rich phenomenology even within the framework of lowest-order perturbation theory, the so-called sequential tunneling model. Most strikingly, in the easily accessible regime $E_{\mathrm{co}} < E_{cx}$ and $g_m \ll 2g_s$ a pronounced non-monotonic temperature dependence of the conductance has been observed. At low temperature deviations from the sequential behavior due to quantum fluctuations become clearly visible. They are described in close analogy to the SET by a logarithmic correction term of the form $\Delta G_{\mathrm{qf}} = \alpha \log(k_B T/(2E_{\mathrm{co}}))$, in qualitative agreement with the findings of Ref. [1]. A reevaluation of the formulae of Ref. [1] with parameters in accordance with our experiment so as to check the applicability of perturbation expansion for devices more complex than the SET is proposed.

Acknowledgments. We acknowledge useful discussions with G. Göppert, G. Johannson, P. Joyez, J. König, H. Pothier, and H. v. Löhneysen. This work has been carried out as part of SFB 195.

References

1. H. Pothier, P. Lafarge, C. Urbina, D. Esteve, M. H. Devoret, Single-electron pump based on charging effects, *Europhysics Lett.* **17**(3), 249–254 (1992).
2. H. Pothier, P. Lafarge, P. F. Orfila, C. Urbina, D. Esteve, M. H. Devoret, Single electron pump fabricated with ultrasmall normal tunnel-junctions, *Physica B* **169**(1–4), 573–574 (1991).
3. S. V. Lotkhov, S. A. Bogoslovsky, A. B. Zorin, J. Niemeyer, Operation of a three-junction single-electron pump with on-chip resistors, *Appl. Phys. Lett.* **78**(7), 946–948 (2001).
4. P. Joyez, V. Bouchiat, D. Esteve, C. Urbina, and M. H. Devoret, Strong tunneling in the single-electron transistor, *Phys. Rev. Lett.* **79**(7), 1349–1352 (1997).
5. D. Chouvaev, L. S. Kuzmin, D. S. Golubev, and A. D. Zaikin, Strong tunneling and Coulomb blockade in a single-electron transistor, *Phys. Rev. B* **59**(16), 10599–10602 (1999).
6. C. Wallisser, B. Limbach, P. vom Stein, R. Schfer, C. Theis, G. Göppert, and H. Grabert, Conductance of the single-electron transistor: A comparison of experimental data with Monte Carlo calculations, *Phys. Rev. B* **66**(12), 125314-1–8 (2002).
7. R. L. Kautz, G. Zimmerli, and J. M. Martinis, Self-heating in the coulomb-blockade electrometer, *J. Appl. Phys.* **73** (5), 2386–2396 (1993).
8. G.-L. Ingold and Y. V. Nazarov, in *Single Charge Tunneling*, ed. by H. Grabert and M. H. Devoret (Plenum Press, New York, 1992), pp. 21–107.
9. D. V. Likharev and K. K. Averin, in *Mesoscopiv Phenomena in Solids*, ed. by B. L. Altshuler, P. A. Lee, and R. A. Webb (Elsevier Science Publishers B. V., Amsterdam, 1991), pp. 173–271.
10. D. V. Averin and Y. V. Nazarov, in *Single Charge Tunneling*, ed. by H. Grabert and M. H. Devoret (Plenum Press, New York, 1992), pp. 217–247.
11. K. A. Matveev, Quantum fluctuations of the charge of a metal particle under the Coulomb blockade conditions, *Sov. Phys. JETP* **72** (5), 892–899 (1991), [*Zh. Eksp. Teor. Fiz.* **99**, 1598 (1191)].
12. J. König and H. Schoeller, Strong Tunneling in the Single-Electron Box, *Phys. Rev. Lett.* **81** (16), 3511–3514 (1998).
13. G. Göppert, B. Hpper, and H. Grabert, Conductance of the single-electron transistor for arbitrary tunneling strength, *Phys. Rev. B* **63**(15), 9955–9958 (2000).
14. J. König, H. Schoeller, and G. Schön, Cotunneling at Resonance for the Single-Electron Transistor, *Phys. Rev. Lett.* **78**(23), 4482–4485 (1997).
15. H. Schoeller and G. Schön, Mesoscopic quantum transport - resonant-tunneling in the presence of a strong Coulomb interaction, *Phys. Rev. B* **50** (24), 18436–18452 (1994).
16. J. König, H. Schoeller, and G. Schön, Cotunneling and renormalization effects for the single-electron transistor, *Phys. Rev. B* **58**(12), 7882–7892 (1998).
17. T. Pohjola, J. König, H. Schoeller, G. Schön, Strong tunneling in double-island structures, *Phys. Rev. B* **59**(11), 7579–7589 (1999).
18. K. P. Hirvi, J. P. Kauppinen,A. N. Korotkov, M. A. Paalanen, J. P. Pekola, Arrays of normal-metal tunnel-junctions in weak coulomb-blockade regime, *Appl. Phys. Lett.* **67**(14), 2096–2098 (1995).
19. B. Limbach, Forschungszentrum Karlsruhe Technical Report No. FZKA 6791, 2002 (unpublished); http://bibliothek.fzk.de/zb/berichte/FZKA6791.pdf.

26

Transmission Eigenvalues' Statistics for a Quantum Point Contact

G. CAMPAGNANO, YA. M. BLANTER, AND YU. V. NAZAROV

Kavli Institute of NanoScience, Delft University of Technology, Lorentzweg 1, 2628 CJ Delft, The Netherlands

Abstract

The transport properties of a quantum point contact in presence of nearby impurities within the semiclassical approach is studied here. We find the distribution function of transmission eigenvalues is function only of the average reflection eigenvalue and of the number of open transport channels is shown.

26.1 Introduction

The Quantum Point Contact (QPC) is one of the reference systems of mesoscopic physics. This device is usually realized by confining a 2DEG by electric gates. The width of the obtained constriction is tuned by varying the voltage applied to the gates. The experiments show that the conductance is quantized when the size of the constriction and the wavelength of the electrons at the Fermi energy are of the same order.[1] If the width of the constriction change slowly in comparison with the wavelength of an electron the theoretical description is easily obtained within the adiabatic approximation.[2] The two-dimensional motion of an electron confined between the gates is equivalent to *one-dimensional* scattering of an electron at a potential barrier. The height of the barrier is different for different transport channels. Semi-classically, the electron is fully transmitted if its energy exceeds the top of the barrier in a given channel, and is fully reflected otherwise. Thus, semi-classical transmission eigenvalues of a QPC are strongly degenerate: one has a finite number of transmission eigenvalues equal to one and an infinite number of transmission eigenvalues equal to zero. This picture would manifest experimentally in the precise quantization of conductance as a function of gate voltage.

In real experiments, this degeneracy is lifted, all experimental studies show that conductance does not rise in ideal steps, this is in general due to the non complete adiabaticity of the system. The question whether the transmission eigenvalues are degenerate is also important for other reasons. Let us consider a QPC prepared in a superconducting material, discrete subgap (Andreev) states develop.[4] These states describe quasiparticles localized at the QPC. The number of these states equals the number of transport channels, and their energies are expressed via transmission eigenvalues T_n, $E_n = \Delta\sqrt{1 - T_n \sin^2\varphi/2}$, with Δ and φ being the superconducting gap and the phase difference across the QPC. If the transmission eigenvalues are degenerate, the Andreev levels are also degenerate. Thus, any small perturbation would lift this degeneracy and produce a number of states with very close energies. Such a perturbation would then drastically affect properties of the system. In this chapter, how the degeneracy of transmission eigenvalues is lifted by the scattering on impurities which are always present in and around the QPC is shown. Properties of a disordered QPC have been investigated (see Refs. [5–9]), mostly in relation to the disorder smearing of conductance steps or evolution of conductance fluctuations in ballistic regime. In contrast to the previous study, the case when the conductance of the QPC is only slightly modified by the impurities, or, in other words, the impurity-related splitting of transmission eigenvalues is much less than one is investigated. This regime is realized for low concentration of impurities.

The main result is that in this regime, reflection amplitudes are Gaussian distributed with zero average and second-order correlation function which does not depend on the channel index. This provides us with a new class of random matrix theory. The results for the distribution function of transmission eigenvalues are universal — they only depend on the number of transport channels and on the average reflection eigenvalue. All other information can be extracted from these two parameters.

26.2 Model of QPC with Impurities

Following Ref. [2] we describe the QPC as a constriction between infinitely high walls separated by the distance $d(x)$ (Fig. 26.1). Let us consider the Schrödinger equation,

$$\left[-\frac{\hbar^2}{2m}\nabla^2 + V(x, y)\right]\psi(x, y) = E\psi(x, y), \tag{26.1}$$

with the boundary conditions,

$$\psi(x, y = \pm d(x)/2) = 0.$$

were

$$V(x, y) = \sum_i v(x - x_i, y - y_i),$$

with v being the single impurity potential, and the sum is taken over impurity positions.

We assume that the width of the constriction $d(x)$ changes smoothly. In this case we can employ the adiabatic approximation and separate the transverse motion,

$$\psi(x, y) = \sum_n \phi_n(x)\varphi_n^{(x)}(y).$$

The transverse wave functions $\varphi_n^{(x)}(y)$ that satisfy the boundary conditions are

$$\varphi_n^{(x)}(y) = \sqrt{\frac{2}{d(x)}}\sin\left[\frac{n\pi}{d(x)}\left(y + \frac{d(x)}{2}\right)\right].$$

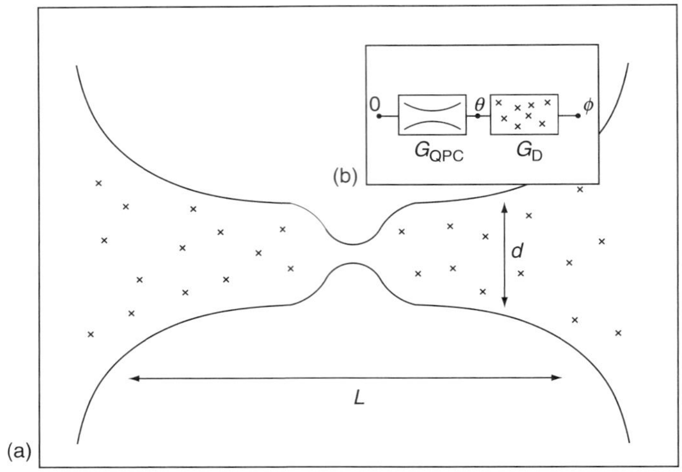

FIGURE 26.1. (a) Layout of a disordered QPC. (b) An equivalent circuit representing a disordered QPC as a clean QPC and a diffusive resistor in series.

Substituting this into Eq. (26.1) and disregarding the terms containing the derivatives of $d(x)$, we obtain a *one-dimensional* equation for the longitudinal wave function,

$$\left[-\frac{\hbar^2}{2m}\frac{d^2}{dx^2} + \epsilon_n(x) - E\right]\phi_n(x) = -\sum_m V_{nm}(x)\phi_m(x), \tag{26.2}$$

with the channel-dependent effective potential barrier

$$\epsilon_n(x) = \frac{\hbar^2\pi^2 n^2}{2md^2(x)},$$

and the matrix element of the disorder potential,

$$V_{nm}(x) = \int_{-d(x)/2}^{d(x)/2} \mathrm{d}y\, \varphi_n^{(x)}(y)V(x,y)\varphi_m^{(x)}(y).$$

Eq. (26.2) is the generalization of the equations previously written in Refs. [2,3] to the case of disordered QPC.

In the semi-classical (WKB) approximation, in the absence of disorder, for each transport channel, electrons with the energies above (below) the top of the barrier are perfectly transmitted (reflected). This approximation breaks down if the energy of an electron coincides with the top of the barrier. Here, we do not consider this case. When an electron is ideally transmitted its wave function is

$$\phi_n^{(0)}(x) = \sqrt{\frac{p_n(\infty)}{p_n(x)}}\exp\left[\frac{i}{\hbar}\int^x p_n(z)\mathrm{d}z\right], \tag{26.3}$$

with the channel-dependent momentum

$$p_n(x) = [2m(E - \epsilon_n(x))]^{1/2}.$$

26.3 Scattering Matrix Elements and Correction to the Conductance

Let us consider the scattering matrix $\hat{S}$,

$$\hat{S} = \begin{pmatrix} \hat{r} & \hat{t} \\ \hat{t}^T & \hat{r}' \end{pmatrix},$$

which is unitary due to the current conservation requirement. The conductance of the system is expressed via Landauer formula

$$G = G_{\mathrm{Q}} \mathrm{Tr}\, \hat{t}^\dagger \hat{t} = G_{\mathrm{Q}} \sum_n T_n,$$

where T_n are the eigenvalues of the matrix $\hat{t}^\dagger \hat{t}$, and $G_{\mathrm{Q}} = e^2/\pi\hbar$ is the conductance quantum. Without impurities, the matrix $\hat{t}^\dagger \hat{t}$ is diagonal, with the elements describing the transmission of an electron in the same open transport channel equals one and all others equal zero In this case, the conductance is $G_0 = G_{\mathrm{Q}} N$, with N being the number of open transport channels.

To treat the effect of disorder, it is more convenient to investigate the matrix $\hat{r}^\dagger \hat{r}$, with the eigenvalues (*reflection eigenvalues*) $R_n = 1 - T_n$. In the following, we calculate the correction to the transmission eigenvalue R_n due to disorder.

It is possible to consider the perturbative expansion of the reflection matrix $\hat{r}$ in the disorder potential and show that

$$r_{nm} = \int_{-\infty}^{\infty} \mathrm{d}x' \frac{i\hbar}{m\sqrt{p_n(x')p_m(x')}} \exp\left[\frac{i}{\hbar}\int^{x'} \mathrm{d}z\, (p_n(z) + p_m(z))\right] V_{nm}(x'), \tag{26.4}$$

and that the only non zero correlators are

$$\langle |r_{nm}|^2 \rangle = \frac{n_i}{\hbar^2 v_F^2} \frac{L}{d} |\tilde{v}(\pi)|^2 \left(1 + \frac{\delta_{nm}}{2}\right), \tag{26.5}$$

where $\tilde{v}(\pi)$ is the Fourier transform of the single impurity potential with the momentum transfer $2k_F$, and n_i is the concentration of impurities per unit area. The cannel index n and m refer to open channels (see Ref. [10] for all the details.)

The correction to the conductance reads

$$\langle \delta G \rangle = -G_{\mathrm{Q}} \sum_{n,m=0}^{N} \langle |r_{nm}|^2 \rangle,$$

and in the case of large number of open channels N can be written as

$$\langle \delta G \rangle \simeq -\frac{e^2 n_i}{\pi\hbar^3 v_F^2} \frac{L}{d} |\tilde{v}(\pi)|^2 N^2. \tag{26.6}$$

For further reference, we identify the average reflection eigenvalue $\langle R \rangle$ by means of Landauer formula, $\langle \delta G \rangle = -G_{\mathrm{Q}} N \langle R \rangle$,

$$\langle R \rangle = \frac{n_i}{\hbar^2 v_F^2} \frac{L}{d} |\tilde{v}(\pi)|^2 N.$$

The correlation function of the reflection amplitudes can then be expressed via only one parameter $\langle R \rangle$,

$$\langle |r_{nm}|^2 \rangle = \frac{1}{N} \langle R \rangle \left(1 + \frac{\delta_{nm}}{2}\right). \tag{26.7}$$

26.4 Distribution function

Now, we turn to the calculation of the distribution function of transmission eigenvalues,

$$\rho(T) = \left\langle \sum_n \delta(T - T_n) \right\rangle,$$

where the sum is taken over all open channels, and disorder averaging is performed. The distribution function is normalized so that its integral is the number of open channels N. For a clean QPC, $\rho(T) = N\delta(1 - T)$.

We first consider the case of one open channel and perform the disorder averaging directly. Then, we obtain analytical results in the limiting case of very large number of channels based on the circuit theory: a disordered QPC is presented as a clean QPC connected in series with a diffusive resistor. For intermediate values of N, analytical calculation was not performed. Results obtained by a random matrix simulation are reported in Ref. [10].

26.4.1 *One open channel*

For one open channel, a "brute force" disorder averaging is performed. The reflection probability is $R = |r|^2$, where the channel indices are suppressed. In its turn, the reflection amplitude r is function to the potential V. It is a complex quantity, and first the distribution function of the transmission amplitude is calculated, which is defined as

$$\mathcal{P}(r) = \langle \delta\left(\mathrm{Re}(r - r[V])\right) \delta\left(\mathrm{Im}(r - r[V])\right)\rangle .$$

Representing delta-functions as integrals, we write explicitly

$$\mathcal{P}(r) = \left\langle \int_{-\infty}^{\infty} \frac{\mathrm{d}\omega_1 \mathrm{d}\omega_2}{(2\pi)^2} \exp\left\{i\omega_1 \mathrm{Re}\left[r[V] - r\right] + i\omega_2 \mathrm{Im}\left[r[V] - r\right]\right\} \right\rangle .$$

Performing the averaging with the Gaussian distribution $\tilde{\mathcal{P}}[V]$, and introducing the short-hand notation

$$r[V] = \int \mathrm{d}\mathbf{r} F(\mathbf{r}) V(\mathbf{r}),$$

we obtain

$$\mathcal{P}(r) = \int_{-\infty}^{\infty} \frac{\mathrm{d}\omega_1 \mathrm{d}\omega_2}{(2\pi)^2} \exp\left\{ -\frac{1}{2} \int \mathrm{d}\mathbf{r_1} \mathrm{d}\mathbf{r_2} \left[\omega_1 \mathrm{Re} F(\mathbf{r_1}) + \omega_2 \mathrm{Im} F(\mathbf{r_1})\right] \langle V(\mathbf{r_1}) V(\mathbf{r_2})\rangle \right.$$

$$\left. \left[\omega_1 \mathrm{Re} F(\mathbf{r_2}) + \omega_2 \mathrm{Im} F(\mathbf{r_2})\right] - i\omega_1 \mathrm{Re}\, r - i\omega_2 \mathrm{Im}\, r \right\} .$$

Now we use Eq. (26.4) and disregard the terms containing rapidly oscillating functions. Calculating the integrals over $\mathrm{d}\omega_1$ and $\mathrm{d}\omega_2$, we finally obtain

$$\mathcal{P}(r) = \frac{1}{\pi \langle |r|^2 \rangle} \exp\left(-\frac{|r|^2}{\langle |r|^2 \rangle} \right). \tag{26.8}$$

Since $\mathrm{d}R = 2|r|\mathrm{d}|r| = \pi^{-1}\mathrm{d}\,\mathrm{Re}r\ \mathrm{d}\,\mathrm{Im}r$, we can rewrite Eq. (26.8) in terms of the reflection eigenvalue,

$$\rho(R) = \frac{1}{\langle R \rangle} \exp\left(-\frac{R}{\langle R \rangle} \right); \quad \rho(T) = \rho(1 - R). \tag{26.9}$$

Thus, for one channel the reflection eigenvalue is Poisson distributed. This result is actually not surprising. Indeed, both real and imaginary parts of r are linearly related to the potential V. This means they are both Gaussian with zero average. Moreover, from Eq. (26.4) it follows that they have the same dispersion, and the Gaussian distribution for r Eq. (26.8) and Poisson distribution for R Eq. (26.9) is arrived. These distributions are *universal*: all the information about the type and amplitude of disorder is encoded in only one number, which is the average reflection eigenvalue $\langle R \rangle$.

26.4.2 Circuit theory

If the number N of open channels is large ($G \gg G_Q$), the distribution function $\rho(T)$ can be calculated analytically by means of the circuit theory developed by one of the authors.[11] To this purpose, a disordered QPC is represented as a clean QPC connected in series to a diffusive conductor (Fig. 26.1). Such a point of view, to authors' knowledge, was first adopted in Ref. 8 for investigation of conductance fluctuations. The input parameters are the conductances of both circuit elements (connectors), G_{QPC} and G_D. Each connector is subject to a phase difference ϕ, which generates the pseudo-current $I(\phi)$. The relation $I(\phi)$ is determined by the distribution function of transmission eigenvalues,

$$\rho(T) = \frac{1}{2\pi G_Q} \frac{1}{T\sqrt{1-T}} \mathrm{Re}\left[I\left(\pi + 2i \operatorname{arccosh} \frac{1}{\sqrt{T}}\right)\right].$$

In our case, the current–phase relations are

- Diffusive conductor: $I(\phi) = G_D \phi$
- Quantum point contact $I(\phi) = 2G_{QPC} \tan(\phi/2)$

The circuit theory shows how these two elements can be combined to get the distribution function of the entire circuit. To this purpose, we introduce the phases: ϕ in the left reservoir; zero in the right reservoir, and θ (to be calculated) in the point (node) separating the QPC and diffusive conductor. Thus, the QPC is subject to the phase difference $\phi - \theta$, and the diffusive conductor to the phase difference θ. The pseudo-current must be conserved, from which we get the following equation for θ,

$$G_D \theta = 2G_{QPC} \tan((\phi - \theta)/2). \tag{26.10}$$

After solving this equation, we find the pseudo-current $I(\phi) = G_D \theta(\phi)$, and eventually the distribution function.

Eq. (26.10) can not be solved analytically for an arbitrary relation between G_D and G_{QPC}, and we restrict ourselves to the case of low impurity concentration, $G_D \gg G_{QPC}$. We obtain

$$I(\phi) = G_D \left[\frac{\phi - \pi}{2} + \sqrt{\frac{(\phi - \pi)^2}{4} + \frac{4G_{QPC}}{G_D}} \right],$$

and the distribution function follows,

$$\rho(T) = \frac{G_D}{2\pi G_Q} \frac{1}{T\sqrt{1-T}} \sqrt{T - T_c}, \tag{26.11}$$

for $T > T_c$, $T_c = 1 - 4G_{QPC}/G_D$, and zero for $T < T_c$. For further comparison with the simulation results, we calculate the average reflection coefficient $\langle R \rangle = G_{QPC}/G_D \ll 1$ and rewrite the distribution function (26.11) as

$$\rho(R) = \frac{N}{2\pi \langle R \rangle} \frac{1}{1-R} \sqrt{\frac{4\langle R \rangle}{R} - 1}. \tag{26.12}$$

We see that in the limiting case $N \gg 1$ only channels with the transmission close to perfect exist: reflection eigenvalues can only be lower than $4\langle R \rangle$, which is a small number. This is in contrast to the one-channel case, where all values of the reflection coefficient are permitted.

26.5 Conclusions

The statistical properties of a quantum point contact with nearby impurities was investigated. The results were obtained within the adiabatic approximation, i.e., assuming a smooth profile for the constriction. The main result is that the correction to the conductance and the distribution function of transmission eigenvalues can be expressed as function of only two parameters, the number of open channel and the average reflection coefficient. Further application of these results will come in the field of SNS junction. There is, indeed, a one to one correspondence between transmission eigenvalues and energy of localized Andreev's states. This correspondence is remarkable because, as suggested in Ref. [12], the Andreev's states can be employed to realize a solid state quantum-bit.

Acknowledgments. This work was supported by the Netherlands Foundation for Fundamental Research on Matter (FOM).

References

1. B. J. van Wees, H. van Houten, C. W. J. Beenakker, J. G. Williamson, L. P. Kouwenhoven, D. van der Marel, and C. T. Foxon, *Phys. Rev. Lett.* **60**, 848 (1988); D. A. Wharam, T. J. Thornton, R. Newbury, M. Pepper, H. Ahmed, J. E. F. Frost, D. G. Hasko, D. C. Peacock, D. A. Ritchie, and G. A. C. Jones, *J. Phys. C* **21**, L209 (1988).
2. L. I. Glazman, G. B. Lesovik, D. E. Khmel'nitskii, and R. I. Shekhter, *Pis'ma Zh. Éksp. Teor. Fiz.* **48**, 218 (1988); *JETP Lett.* **48**, 238 (1988).
3. M. Büttiker, *Phys. Rev. B* **41**, 7906 (1990).
4. A. Furusaki and M. Tsukada, *Physica B* **165 & 166**, 967 (1990); C. W. J. Beenakker, *Phys. Rev. Lett.* **67**, 3836 (1991).
5. I. Kander, Y. Imry, and U. Sivan, *Phys. Rev. B* **41**, 12941 (1990).
6. J. A. Nixon, J. H. Davies, and d H. U. Baranger, *Phys. Rev. B* **43**, 12638 (1991).
7. L. I. Glazman and M. Jonson, *Phys. Rev. B* **44**, 3810 (1991).
8. D. L. Maslov, C. Barnes, and G. Kirczenow, *Phys. Rev. Lett.* **70**, 1984 (1993); *Phys. Rev. B* **48**, 2543 (1993).
9. C. W. J. Beenakker and J. A. Melsen, *Phys. Rev. B* **50**, 2450 (1991); C. W. J. Beenakker, *Rev. Mod. Phys.* **69** 731 (1997).
10. G. Campagnano, O. N. Jouravlev, Ya. M. Blanter, and Yu. V. Nazarov, *Phys. Rev. B* **69**, 235219 (2004).
11. Yu. V. Nazarov, in *Quantum Dynamics of Submicron Structures*, ed. by H. A. Cerdeira, B. Kramer, and G. Schön, (Kluwer Academic Publishers, Dordrecht 1995), p.687.
12. N. M. Chtchelkatchev and Yu. V. Nazarov, *Phys. Rev. Lett.* **90**, 226806 (2003)

27

Creating Entangled States between SQUID Rings and Electromagnetic Fields

M. J. EVERITT[1], T. D. CLARK[1], P. B. STIFFELL[1], J. F. RALPH[2], AND A. VOURDAS[3]

[1] Centre for Physical Electronics and Quantum Technology, University of Sussex, Falmer, Brighton, BN3 9QT, UK.
m.j.everitt@physics.org
t.d.clark@sussex.ac.uk
[2] Department of Electrical Engineering and Electronics, The University of Liverpool, L69 3GJ, UK.
[3] Department of Computing, Bradford University, Bradford, West Yorkshire, BD7 1DP, UK

Keywords: SQUID, Entangled states

Abstract

The problem of entangling a SQUID ring inductively coupled to an electromagnetic field mode is considered here. It is demonstrated by solving the time dependent Schrödinger equation that a system with varying superpositions of states and levels of entanglement between the components of this system can be created. These calculations can be extended to include the effects of dissipative environments coupled to the SQUID ring-field mode system. It is also demonstrated that even including such decohering environments, it is possible to maintain a sufficient degree of entanglement over time scales that would be required for quantum technologies.

27.1 Introduction

Over the last few years it is seen that quantum technologies become established as a very active field of study encompassing the physics, mathematics, engineering and computer science communities. For example, quantum mechanically encrypted communication is now a reality with off the shelf kits commercially available (for example see website: http://www.magiqtech.com/). However, quantum information processing, and specifically quantum computation, have still to be implemented with any degree of really significant success. To take an example, Shor's factorisation algorithm has very significant implications as regards future developments in these quantum technologies. However, even now it appears that the most sucessful implementation of this algorithm has only managed to compute the prime factors of 15. In order to move beyond such limits we clearly need to pursue scalable devices so as to increase the computational

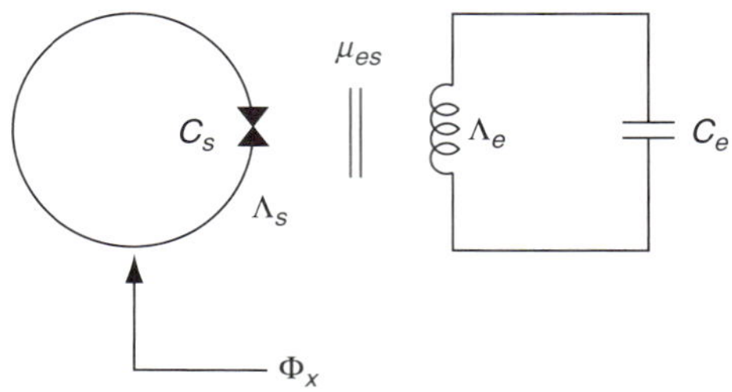

FIGURE 27.1. A schematic for the coupled SQUID ring–electromagnetic field (here represented by an *LC* circuit).

capacity available for quantum technologies. Among the prime contenders for such technologies are circuit systems based on superconducting devices[1–8]. It is our opinion that within this field (Superconducting Quantum Interference Device) (SQUID) rings will play a major role in any future developments. Where SQUID rings (in this work a thick superconducting ring enclosing a single Josephson weak link) are utilised, not only does the weak link enable the ring to become tunable device (via an applied magnetic flux), but it also generates the non-perturbative properties that can be used in the manipulation of quantum information.[9–11]

In using SQUID rings for quantum technologies the first step is to be able to create any desired superposition of states in such systems. In recent years there has been much progress, both theoretically and experimentally, in this direction[8,11] and there seems little doubt that, in the near future, the control over creating these superpositions will improve greatly. The next step is to understand how such macroscopic quantum objects may couple to each other, and how one may control fundamentally important properties such as the level of entanglement between them.

In this study we take the specific example of a SQUID ring, inductively coupled to an electromagnetic field mode, as illustrated in Fig. 27.1. It should be noted that we need not restrict ourselves to em field modes; instead we could replace the field mode by another SQUID ring with very deep wells.[12] The control of entanglement has already been discussed in such systems in a previous publication[11]. In particular, that it is indeed possible to create a state of maximal entanglement between SQUID rings and em field modes have been shown. However, in order to achieve this the author's had to be very precise in the control of the external, classical bias flux applied to the ring. In this chapter the same example system is taken and the issue of the creation of highly entangled states by a suitable time dependent magnetic bias flux (Φ_x) applied to the SQUID ring as addressed. To be specific this external flux is sweeped over a relatively large range using a ramp, linearly increasing in time.

27.2 Background

In this description of the coupled SQUID ring-field mode system the well known, lumped component, ring Hamiltonian is used, which is given by

$$H_s = \frac{Q_s^2}{2C_s} + \frac{(\Phi_s - \Phi_x)^2}{2L_s} - \hbar\nu \cos\left(\frac{2\pi\,\Phi_s}{\Phi_0}\right) \tag{27.1}$$

where, Φ_s is the magnetic flux threading the ring and Q_s is the conjugate electric displacement flux across the weak link, with $[\Phi_s,\ Q_s] = i\hbar$. Here, C_s and L_s are, respectively, the weak link capacitance and the SQUID ring inductance such that the characteristic ring oscillator frequency is $\hbar\omega_s = \hbar/\sqrt{L_s C_s} \gg k_B T$, where T is the operating temperature, and $\Phi_0 = h/2e$ is the superconducting flux quantum. In our calculations the circuit parameters for the ring is taken as $C_s = 5 \times 10^{-15}$ F, $L_s = 3 \times 10^{-10}$ H and the matrix coupling element for Josephson pair tunnelling through the weak link $\hbar\nu/2 = 0.015\Phi_0^2/L_s$. These correspond realistically to the current state of the art in the SQUID rings for the experimental investigation of macroscopic superposition of states. Moreover, devices with these parameters are perfectly accessible with modern micro-fabrication techniques.

The em-field mode is moduled via the usual Hamiltonian

$$H_f = \frac{Q_f^2}{2C_f} + \frac{\Phi_f^2}{2L_f} \tag{27.2}$$

where, for simplicity of computation, consider $C_f = C_s$ and $L_f = L_s$ and the field mode oscillator frequency is $\omega_f = 1/\sqrt{L_f C_f}$. Again, it is assume for the quantum regime that $\hbar\omega_f \gg k_B T$.

If it is assume that these circuits are only inductively coupled together, the energy for this is given by

$$H_i = -\frac{\mu}{2L_s}\Phi_f(\Phi_s - \Phi_x). \tag{27.3}$$

where the coupling constant $\mu = M/L_t$ for a mutual inductance M between the ring and the field mode. For the calculations this coupling constant is set to $\mu = 0.01$. The total Hamiltonian for the system is then

$$H = H_s + H_f + H_i \tag{27.4}$$

In this chapter the structure of the state vector is considered rather than the expectation values of any particular observables. In order to do this in a physically transparent manner the state vector in terms of a natural, complete, othonormal, basis comprising of the eigenstates of H_s and H_f will be described. Explicitly, this takes the form

$$|\psi\rangle = \sum_{i,j\in\mathbb{N}} c_{ij}|i_s\rangle \otimes |j_f\rangle \tag{27.5}$$

where $c_{ij} \in \mathbb{C}$, $H_s|i_s\rangle = \sigma_s|i_s\rangle$ and $H_f|j_f\rangle = \lambda_f|j_f\rangle$. Thoughout this chapter the following notation $|i_s j_f\rangle = |i_s\rangle \otimes |j_f\rangle$ is used.

As the focus in the level of entanglement in the system the standard entropic measure of $S = S(\rho_s) + S(\rho_f) - S(\rho)$ will also be used when there is no dissipation[13] while $I_i = S(\rho) - S(\rho_i), i = s, f$ is used when dissipation is present.[14,15] Here ρ is the density operator for the system and ρ_s and ρ_f are the density operators for the ring and the field respectively.

27.3 The Hamiltonian's Spectrum

Considering some properties of this system that the spectrum of the Hamiltonian Eq. (27.4) yields. This spectrum is shown in Fig. 27.2(a) It should be noted that within these energy eigenvalues there are two kinds of lifted degeneracies. The first kind are those caused by the weak link and are splittings for the ring only, an example of which is shown by the circle 1 in Fig. 27.2(a). This is termed as a type 1 splitting. The second form is caused by the lifting of degenerate eigenvalues of H by the interaction term H_i. An example of this, type 2, splitting (or exchange region) is shown within circle 2 of Fig. 27.2(a).

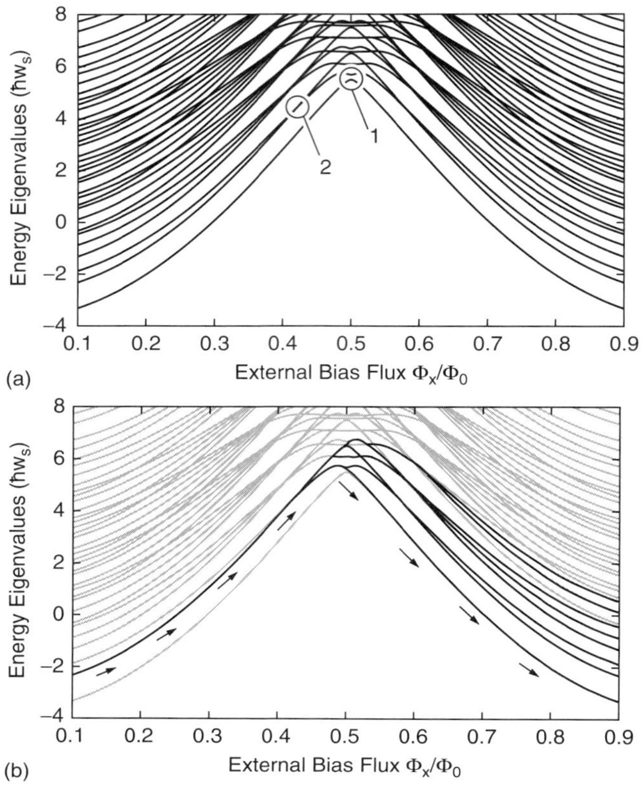

FIGURE 27.2. Energy eigen values for the SQUID ring-em field system as described in the text. (a) The circles show the splitting between (1) the state $|0_s 0_f\rangle$ and $|1_s 0_f\rangle$ (2) the state $|0_s 1_f\rangle$ and $|1_s 0_f\rangle$. (b) Illustration of the energy levels that may be included in the system state vector by ramping in external flux (ramp direction is indicted by the arrows).

Type 1 splittings will result in changes in the SQUID ring component of the wavefunction and are the lifted degeneracies found in the uncoupled SQUID ring spectrum. In the coupled system they function exactly as would be expected of them. For example, the transformation $|0_s 1_f\rangle \rightarrow [A|0_s\rangle + B|1_s\rangle] \otimes |1_f\rangle$ could be achieved via rapid passage though one such splitting.

However passing through a type 2 splitting will result in the creation of superpositions within the tensor product space of the system as a whole. This type of splitting, shown in circle 2 in Fig. 27.2(a), generates transformations of the form $|0_s 1_f\rangle \rightarrow A|0_s 1_f\rangle + B|1_s 0_f\rangle$. This will happen even if the external flux is held statically at such a point. By direct comparison with Bell states it can be seen that it is just such splittings that will cause entanglement between these two objects (ring and field mode).

In this chapter we consider ramping the external bias flux linearly over the interval $[0.1\Phi_0,\ 0.9\Phi_0]$. It can be seen from Fig. 27.2(a) many type 1 and type 2 splitings will be crossed. The resulting state will, therefore, depend on the speed at which type 1 regions is crossed and the amount of time we stay near type 2 regions. For an arbitrary initial state, it can be seen that the situation will very quickly become complicated. In order to provide a specific example, for the initial state $|0_s 1_f\rangle$ it is shown in Fig. 27.2(b) all the possible states that could be included in the final state vector after ramping over the interval $[0.1\Phi_0,\ 0.9\Phi_0]$. In practice, not all of these states will contribute to the final state vector of the system and the exact proportion of each state present in the superposition will be dependent on the speed of the ramp applied. In this chapter the velocity of this ramp is characterized by the time t_r taken to traverse the interval $[0.1\Phi_0,\ 0.9\Phi_0]$.

27.4 Rapid Passage

Following on from the results presented in the previous section, now the case of rapidly traversing this $[0.1\Phi_0,\ 0.9\Phi_0]$ region of external flux, taking $t_r = 200\omega_s^{-1}$ is considered. Again we consider the initial state of the system to be $|0_s 1_f\rangle$. By solving the Time Dependent Schrödinger Equation (TDSE) it is seen in Fig. 27.3(a) that, as we expected, using such a ramp does indeed create additional components in the state vector. Specifically, these extra states are $|0_s 0_e\rangle$ and $|1_s 1_e\rangle$.

Furthermore, because the form of the final state is governed mainly through rapid passage across the $0.5\Phi_0$ point in external flux (circle 1 in Fig. 27.2(a)) a significant level of entanglement between the ring and the field is not expected. This is quantified by computing the entanglement entropy S for the system. As can be seen from the results plotted in Fig. 27.3(b), using such a fast ramp does not result in a significant level of entanglement between the field and the ring.

27.5 Towards the Adiabatic Limit

The lack of entanglement observed in the results presented in the previous section can be understood by the observation that sufficient time is not spent within type 2 exchange regions. Hence, there is little opportunity for the ring-field system to form a state with a strong level of entanglement. If we now make use of a slower ramp rate, not only will this facilitate type 2 transitions, but will also reduce the effect of the type 1, rapid passage, transitions.

Considering again the initial state $|0_s 1_f\rangle$, the TDSE is solved once more, but now utilising a much slower ramp rate with $t_r = 1600\omega_s^{-1}$. As we can see in Fig. 27.4(a) the final state of the ring-field system is dominated by a superposition of the original state $|0_s 1_e\rangle$ and $|1_s 0_e\rangle$. By comparison with Bell states, this superposition is expected to be a reasonably well entangled state. This is confirmed by the computed entanglement entropy shown in Fig. 27.4(b).

It should be noted that there is a small contribution of $|11\rangle$ within this final state, which is cause by some degree of rapid passage at the half integer.

27.6 The Effect of Dissipation

In order to model the effects of dissipation the master equation is now solved for an open quantum system coupled to a thermal bath. This takes the usual form

$$\frac{\partial}{\partial t}\rho = -\frac{i}{\hbar}[H,\ \rho] + \sum_{i=f,s} \frac{\gamma i}{2\hbar}(M_i + 1)\left(2a_i\rho a_i^\dagger - a_i^\dagger a_i\rho - \rho a_i^\dagger a_i\right)$$
$$+\frac{\gamma i}{2\hbar}M_i\left(2a_i^\dagger\rho a_i - a_i a_i^\dagger\rho - \rho a_i a_i^\dagger\right). \qquad (27.6)$$

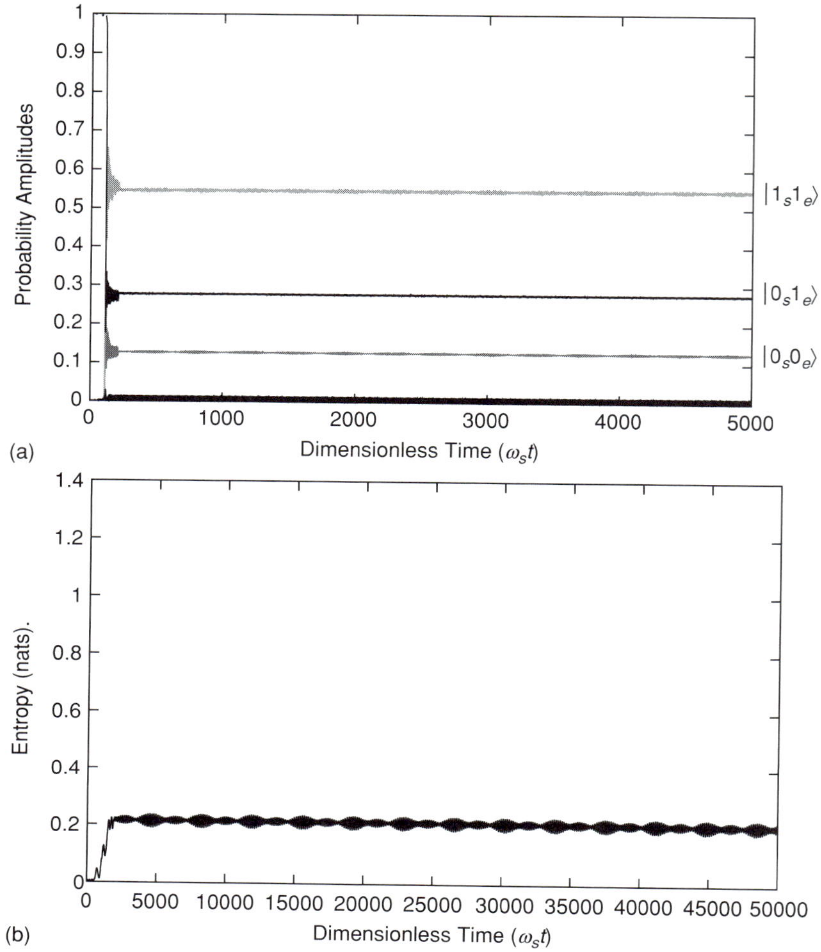

FIGURE 27.3. The affect of appling a rapid flux ramp on the SQUID ring. (a) Probabilities amplitude for being in the basis state shown. (b) Entanglement entry for the SQUID ring-field mode system.

where the mean photon number $M_i = (\exp(\hbar\omega_b/k_B T) - 1)^{-1}$ and γ_i is the damping rate between each of the components in the coupled system to its respective thermal bath at temperature T. In this the temperature is set as chapter $T = 40$mK with $\omega_b = \omega_{s,\ f}$ and $\gamma_i = 2 \times 10^{-5}\omega_{s,\ f}$.

In Fig. 27.5(a) the probability densities for the coupled system corresponding to those used to compute the results of Fig. 27.4. is shown. Here, however, decoherence to the thermal bath is accounted for using the master Eq. (27.6). For the chosen parameter values it is seen that the state $|0_s 0_f\rangle$ takes on a prominent role (and $|2_s 0_f\rangle$ ceases to be completely negligible). This is to be expected, since although $|0_s 0_f\rangle$ is not the vacuum state of the system (i.e., the lowest eigenstate of H), it is a close approximation to it.

In order to quantify the effect of dissipation on this system in Fig. 27.5(b) the entanglement entropy is shown, as introduced by Adami and Cerf.[14,15] Using this measure, where negative

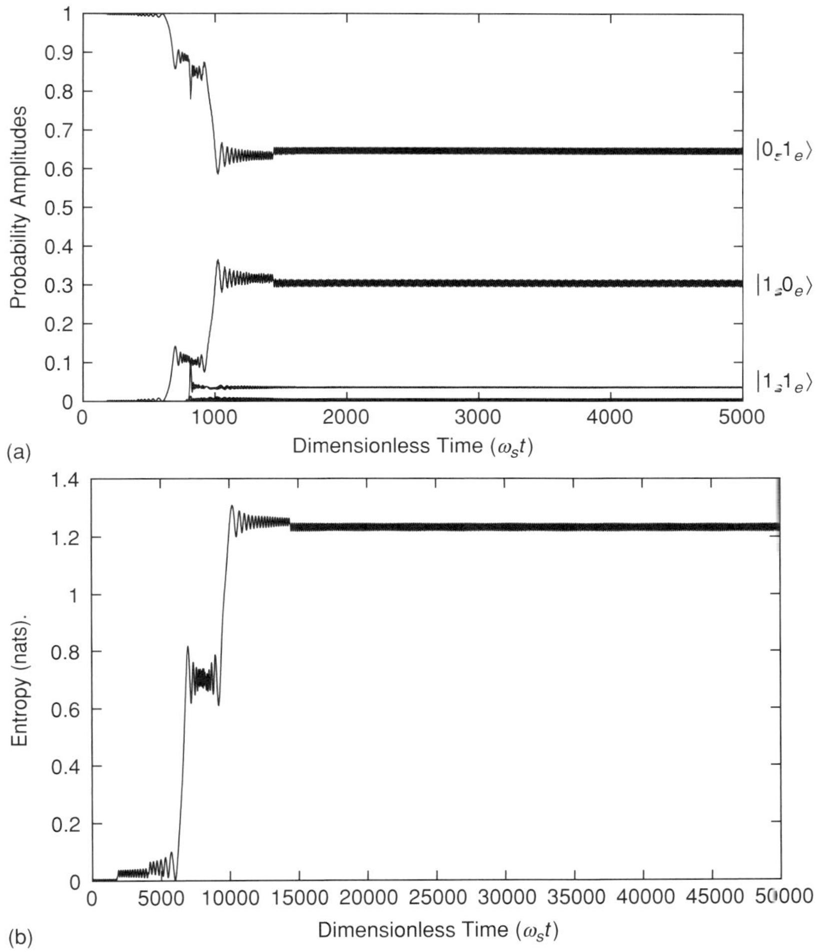

FIGURE 27.4. The affect of applying a slow flux ramp on the SQUID ring. (a) Probability amplitude for being in the basis state shown. (b) Entanglement entry for the SQUID ring-field mode system.

values imply that the system is entangled, it can be seen that we maintain a strong level of entanglement over the period of time considered.

27.7 Conclusions

In this Chapter it was shown that it is possible to create highly entangled states between SQUID rings and em-field modes, both treated quantum mechanically. Moreover, we have also demonstrated that it is possible to create superposition states in SQUID rings without necessarily entangling with the coupled field mode. It may prove possible that these results can be extended to encompass controlled creation of sophisticated, multi-level (qdit) entangled quantum states for quantum information processing. In addition, similar results may be obtainable for networks of

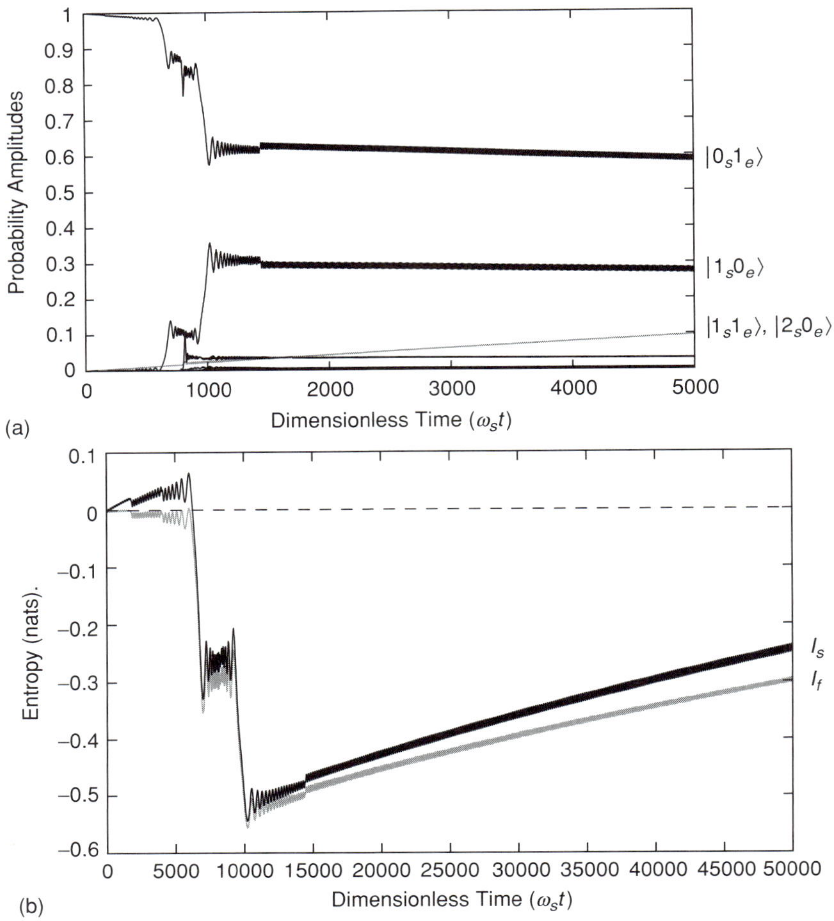

FIGURE 27.5. The effect of applying, in the presence of dissipation, a slow flux ramp on the SQUID ring. (a) Probability amplitudes for being in the basis state shown. (b) Entanglement entropy for the SQUID ring-field mode system. Note here we use the measure of Adami and Cerf where a negative value implies quantum entanglement.

coupled SQUID rings. This could yield a higly configurable tensor product space where each component is a multi-level (qdit) flux controllable quantum system.

Acknowledgments. We would like to thank to the Engineering and Physical Sciences Research Council for the generous funding of this work through its support of the U.K. Quantum Circuits Network.

References

1. I. Chiorescu, Y. Nakamura, C. J. P. M. Harmans, and J. E. Mooji, Coherent quantum dynamics of a superconducting flux qubit, *Science*, **299** (5614), 1869–1871 (March 2003).

2. J. M. Martinis, S. Nam, and J. Aumentado, Rabi oscillations in a large josephson-junction qubit, *Phys. Rev. Lett.*, **89**(11), 117901 (Sep 2002).
3. P. Silvestrini, B. B. Ruggiero, C. Granata, and E. Esposito, Supercurrent decay of josephson junctions in non-stationary conditions: Experimental evidence of macroscopic quantum effects, *Phys. Lett. A*, **267**, 45–51 (2000).
4. Y. Nakamura, C. D. Chen, and J. S. Tsai, Spectroscopy of energy-level splitting between two macroscopic quantum states of charge coherently superposed by josephson coupling, *Phys. Rev. Lett.*, **79**, 2328–2331 (1997).
5. C. H. van der Wal, A. C. J. ter Haar, F. K. Wilhelm, R. N. Schouten, C. J. P. M. Harmans, T. P. Orlando, S. Lloyd, and J. E. Mooij, Quantum superposition of macroscopic persistent-current states, *Science*, **290**, 773–777 (2000).
6. T. P. Orlando, J. E. Mooji, L. Tian, C. H. van der Wal, L. S. Levitov, S. Lloyd, and J. J. Mazo, Superconducting persistent-current qubit, *Phys. Rev. B-Condens Matter*, **60**, 15398–15413 (1999).
7. Y. Makhlin, G. Schon, and A. Shnirman, Josephson-junction qubits with controlled couplings, *Nature*, **398**, 305–307 (1999).
8. J. R. Friedman, V. Patel, W. Chen, S. K. Tolpygo, and J. E. Lukens, Quantum superposition of distinct macroscopic states, *Nature*, **406**, 43–46 (2000).
9. M. J. Everitt, P. Stiffell, T. D. Clark, A. Vourdas, J. F. Ralph, H. Prance, and R. J. Prance, Fully quantum-mechanical model of a squid ring coupled to an electromagnetic field, *Phys. Rev. B*, **6314** 144530 (2001).
10. M. J. Everitt, T. D. Clark, P. Stiffell, H. Prance, R. J. Prance, A. Vourdas, and J. F. Ralph, Quantum Statistics and entanglement of two electromagnetic field modes coupled via a mesoscopic squid ring, *Phys. Rev. B*, **6418** 184517 (2001).
11. M. J. Everitt, T. D. Clark, P. B. Stiffell, H. Prance, R. J. Prance, and J. F. Ralph, Controlled creation of maximally entangled states using a squid ring coupled to an electromagnetic field, quant-ph/0307181 (2003).
12. C. Harmans, Qubits in superconducting circuits, Condensed Matter and Materials Physics Conference (CMMP04) (2004).
13. S. M. Barnett and S. J. D. Phoenix, Information-theory, squeezing, and quantum correlations, *Phys. Rev. A*, **44**, 535–545 (1991).
14. N. J. Cerf and C. Adami, Negative entropy and information in quantum mechanics, *Phys. Rev. Lett.*, **79**, 5194–5197 (1997).
15. N. J. Cerf and C. Adami, Information theory of quantum entanglement and measurement, *Physica D*, **120**, 62–81 (1998).

28

Frequency Down Conversion and Entanglement between Electromagnetic Field Modes via a Mesoscopic SQUID Ring

P. B. STIFFELL,[1] M. J. EVERITT,[1], T. D. CLARK,[1], J. F. RALPH[2], AND A. VOURDAS[3]

[1] Centre for Physical Electronics and Quantum Technology, University of Sussex, Falmer, Brighton, BN3 9QT, UK.
[2] Department of Electrical Engineering and Electronics, The University of Liverpool, L69 3GJ, UK.
[3] Department of Computing, Bradford University, Bradford, West Yorkshire, BD7 1DP, UK.
p.b.stiffell@physics.org
t.d.clark@sussex.ac.uk

Abstract

In this chapter, and by analogy with Quantum Optics, the properties of SQUID rings (here, a Josephson weak link enclosed by a thick superconducting ring) in the quantum regime are discussed. In particular, the possibility of frequency down conversion is investigated, via a quantum mechanical SQUID ring, between quantised electromagnetic field oscillator modes. It is shown that because of the non-perturbative properties of the SQUID ring (essentially due to the cosinusoidal Josephson coupling term in the ring potential) multi-photon down conversion can take place between an input photon and output photons at sub-integers of the input frequency. In addition, these output photons are created in an entangled state with a degree of entanglement that could be modified through the static magnetic flux applied to the SQUID ring.

28.1 Introduction

In the review published in Ref.[1] it was found that, by analogy with the field of quantum optics, Superconducting Quantum Interference Devices (SQUID) rings can act as a highly non-linear medium, inducing a whole range of manifestly quantum phenomena, including frequency conversion, entanglement with other quantum systems and squeezing. With the example of Quantum Optics in mind, this could have a very significant impact on the development of quantum technologies such as quantum computing and quantum information processing.[2,3] In the quantum regime this strong (or, in the limit, extreme) non-linear behavior arises because of the Josephson cosine coupling energy in the SQUID ring Hamiltonian, which as we have demonstrated, can generate remarkable nonperturbative effects.[4] Furthermore, as we shall show, the coupling between a SQUID ring and external electromagnetic (em) fields can be made strong, if required, which is generally not the case for optical media. In addition, this interaction between em fields (or field modes) can be adjusted through the static (or quasi-static) classical magnetic bias flux (Φ_x) applied to the ring.[5] This would appear to allow for much greater dynamical control than is available in quantum optics. In this chapter a theoretical viewpoint is considered for the way the non-perturbative behavior affects the interaction of a quantum SQUID ring with (external) em field oscillator modes. In particular, the manner in which entanglement and frequency down conversion between input and output field modes can be generated is investigated through the intermediary of such a SQUID ring.

In order to demonstrate that this behavior can exist, the possibility of energy transfer between an initial (input) em field mode and two other (output) field modes linked via a quantum regime SQUID ring is first considered. The schematic for this four mode system is shown in Fig. 28.1(a). As a first example, it is assumed that the characteristic oscillator frequencies of the input and output field oscillator modes and the SQUID ring are close, but not identical, thus eliminating symmetry issues, i.e., $\omega_{e_1} \approx \omega_{e_{2,3}} \approx \omega_s$, where the subscript 1 refers to the input em mode, 2 and 3 to the output modes and s to the SQUID ring. That energy can be transferred between these input and output modes via the SQUID ring will be shown. Furthermore, we shall demonstrate that these (output) modes exhibit entanglement through (as it were) the SQUID ring acting as a highly nonperturbative quantum machine is demonstrated. This entanglement is oscillatory, as the energy in the SQUID ring-oscillator mode(s) system washes back and forth between the elements of this system. However, it is noted (as above) that this transfer of energy between the em field modes is dependent on the external bias flux incident on the ring.[6] Within the model description this means that energy transfer can be turned on or turned off by appropriate adjustment of the external flux bias, as can the degree of entanglement between the SQUID ring and the field oscillator modes.

As an extension of this study, and that authors' first results were not an artifact of the particular frequencies chosen, a situation is then considered where the output em modes were set at approximately half the frequency of the output modes, i.e., $0.5\omega_{e_1} \approx \omega_{e_2} \approx \omega_{e_3}$. Again, it will be demonstrated, quantum down conversion and quantum entanglement are still seen in this coupled system at these parameter values. Of course, in such a non-perturbative system there exist major

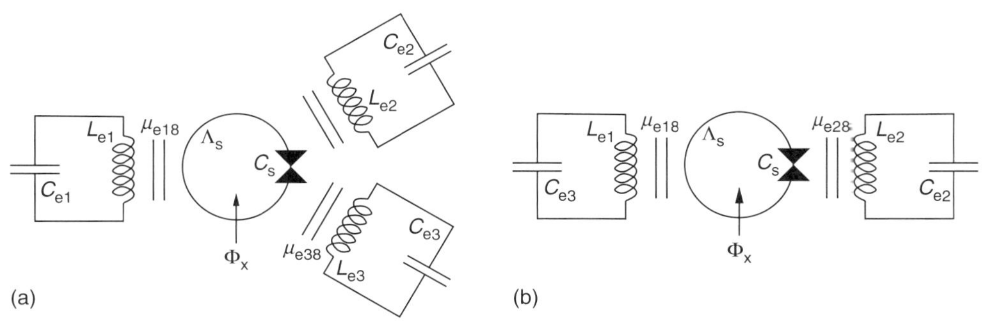

FIGURE 28.1. Schematics for the two systems under investigation. (a) Schematic of a four-mode system. (b) Schematic of a three-mode system.

problems concerning the computational power required to perform accurate theoretical simulations. However, with some constraints, we have been able to compute rather large ratio quantum frequency down conversions in the input-output system shown schematically in Fig. 28.1(b). Thus, within this configuration it is demonstrated that energy is transferred between the input and output modes where there are frequency differences of factors 5 and 10 ($\omega_{e_1} = 5\omega_{e_2}$ and $\omega_{e_1} = 10\omega_{e_2}$). It will be shown that down conversion is present in both these cases, although it is less efficient for the larger frequency difference ratio. Given this, it seems reasonable to suggest that with sufficient computing power and time available, very much higher frequency down conversion ratios could be computed together with the concomitant entanglement together of the lower frequency output em modes. In each cased considered it is assumed that both the SQUID ring and the em oscillators are operating in the quantum regime such that $\hbar\omega_s, \hbar\omega_{1,2,3} \gg k_B T$ for operating temperature T.

28.2 Entanglement of em Fields via a SQUID Ring

With reference to the SQUID ring-field oscillator modes schematic of Fig. 28.1(a), the quantum description required to demonstrate that entanglement can occur between the various elements of this coupled system is considered. Starting with the SQUID ring, this can be described by a translated Hamiltonian of the form[7]

$$H_s = \frac{Q_s^2}{2C_s} + \frac{\Phi_s^2}{2\Lambda_s} - \hbar\nu \cos\left(2\pi\frac{\Phi_s + \Phi_x}{\Phi_0}\right) \tag{28.1}$$

where Φ_s and Q_s are, respectively, the magnetic flux and displacement flux operators for the ring, $\hbar\nu/2$ is the matrix element for Josephson pair tunnelling through the weak link in the ring, C_s and Λ_s are, respectively, the weak link capacitance and the ring inductance (with the characteristic ring oscillator frequency given by $\omega_s = 1/\sqrt{\Lambda_s C_s}$) and $\Phi_0 = h/2e$ is the superconducting flux quantum. As depicted in Fig. 28.1(a), each field mode is coupled inductively to the SQUID ring with individual coupling terms given by $H_{e_i s}, i = 1, 2, 3$

$$H_{e_i s} = \frac{\mu_{e_i s}}{\Lambda_s}\Phi_{e_i}\Phi_s \tag{28.2}$$

where Φ_{e_i} is the flux operator for the em field mode in question and $\mu_{e_i s}$ is the coupling strength between the field mode and the SQUID ring. This, along with the Hamiltonian for the em field modes $H_{e_i} = \hbar\omega_{e_i}(n_{e_i} + 1/2)$, leads to a total system Hamiltonian of the form

$$H = H_s + \sum_{i=1}^{3} H_{e_i} - H_{e_i s} \tag{28.3}$$

where, in terms of the lumped component inductances and capacitances (L_i and C_i, respectively) of Fig. 28.1(a), the oscillator mode frequencies are given by $\omega_{e_i} = 1/\sqrt{L_i C_i}$. Through this Hamiltonian in Eq. (28.3), and using the initial state $|\Psi_0\rangle = |n\rangle_{e_1} \otimes |0\rangle_s \otimes |0\rangle_{e_2} \otimes |0\rangle_{e_3}$, we form an evolution operator solution[8] to the Schrödinger equation to investigate the properties of the ringfield modes system. By adjusting the relative frequencies between the modes of this coupled system, and through the appropriate setting of the external flux Φ_x applied to the SQUID ring, it was demonstrated that this system can display a wide variety of quantum behavior, which should be applicable in the new quantum technologies. In order to lessen computational complexities this chapter is restricted to the case where the oscillator frequency of the input mode, e_1 is equal to that of the SQUID ring (ω_s). Nevertheless, even with this restriction, the behavior of the coupled system when the input and output em modes differ in frequency could be investigated. To start with the case where the two output modes in the four mode (input field mode, SQUID ring and two output modes) system were close together in frequency was considered.

28.3 Entanglement of field modes at similar frequencies

In this situation as an initial approach, it was assumed that the two output modes (e_2 and e_3) were of approximately the same frequency as the input (e_1) and chose realistic circuit parameter values for the SQUID ring based quantum system, namely $\hbar\omega_s = 0.00604\Phi_0^2/\Lambda_s$, $\hbar\nu = 0.035\Phi_0^2/\Lambda_s$, $C_s = 5 \times 10^{-15}$F and $\Lambda_s = 3 \times 10^{-10}$H, with the (inductive) flux coupling strength between the ring and the field modes set at $\mu_{e_1s} = 0.005$ and $\mu_{e_2s} = \mu_{e_3s} = 0.01$ and the initial photon occupancy of the input mode set at $n = 2$. It is also important to note that while we restricted ourselves only to similar frequencies between input and output modes are considered there was a difference between them, the selected values being $\omega_{e_2} = 0.99\omega_{e_1}$ and $\omega_{e_3} = 1.01\omega_{e_1}$.

Once the system parameter values and coupling strengths were chosen, it was needed to establish the optimal bias flux to be applied to the SQUID ring. Using a combination of authors' previous experience,[5,6] and trial and error, it was found that the best energy transfer occurred for an external bias flux of $\Phi_x = 0.43795\Phi_0$. In Fig. 28.2(a) is shown as a function of dimensionless time ($\tau = \omega_s t$) the way in which the energy in the input mode is transferred to the output modes. This demonstrates that the photon occupancy of the output modes is directly linked to the transfer of energy from the input mode.

A primary purpose of this study was to characterize the level of entanglement present in a system and in order to do this it was required to define the measure by which authors' intend to quantify this entanglement. The entanglement between two modes[9] is defined as

$$I(AB) = S(\rho_A) + S(\rho_B) - S(\rho_{AB}) \tag{28.4}$$

where $S(\rho_i) = -\mathrm{Tr}[\rho_i \ln(\rho_i)]$ is the entropy of each system, or subsystem. In Fig. 28.2(b) this entanglement entropy for the two output modes is shown. As can be seen, the output modes are, by this measure, significantly entangled over almost all of the time evolution of the system. This result was, of course, predicated by the assumption that the input mode and the output modes are very close in frequency. Even though, in authors' opinion, the result obtained here is important it would be better if this restriction could be lifted. With this encouraging outcome as a guide, was then removed this limitation.

28.3.1 *Entanglement of dissimilar input/output frequency field modes*

With the same SQUID ring parameters that were used to calculate the results of Fig. 28.2, but using a uniform coupling strength of $\mu_{e_1s} = \mu_{e_2s} = \mu_{e_3s} = 0.01$ and an input mode occupancy of $n = 1$, the change in behavior was then investigated that could occur if the output modes were set at around one half the frequency of the input mode, specifically $\omega_{e_2} = 0.499\omega_{e_1}$ and $\omega_{e_3} = 0.501\omega_{e_1}$. In order to proceed, it was again needed to determine the optimal value of the external bias flux Φ_x that would generate the best frequency down conversion and entanglement. For the system of Fig. 28.1(a) this was found to be $\Phi_x = 0.4498\Phi_0$.

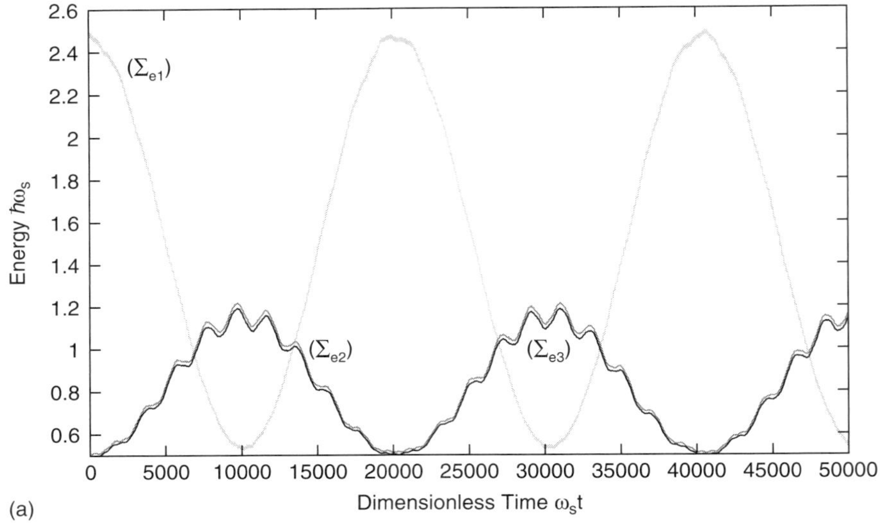

(a) Energy expectation values for the input mode e_1 (light grey) and the output modes e_2 (dark grey) and e_3 (black) versus dimensionless time $\tau = \omega_s t$.

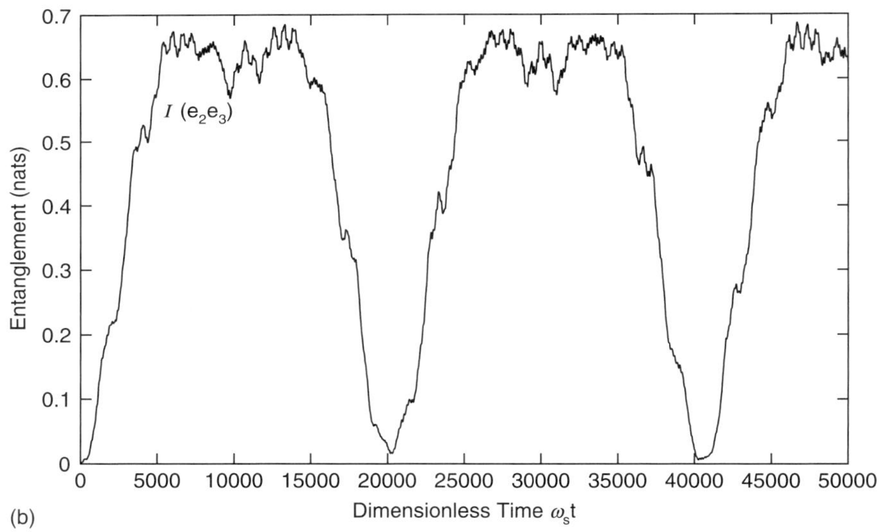

(b) Entanglement entropy between the output modes e_2 and e_3 versus dimensionless time $\tau = \omega_s t$.

FIGURE 28.2. Responses of the four-mode system with similar frequencies for all modes.

Having identified the Φ_x bias point to use with these new mode frequency values, we then looked to reproduce the characteristic features presented in the results of the previous section. Solving the evolution of the system in Eq. (28.3) the energy transfer between the input and output modes via the SQUID ring was computed; the results are shown in Fig. 28.3(a). As before,

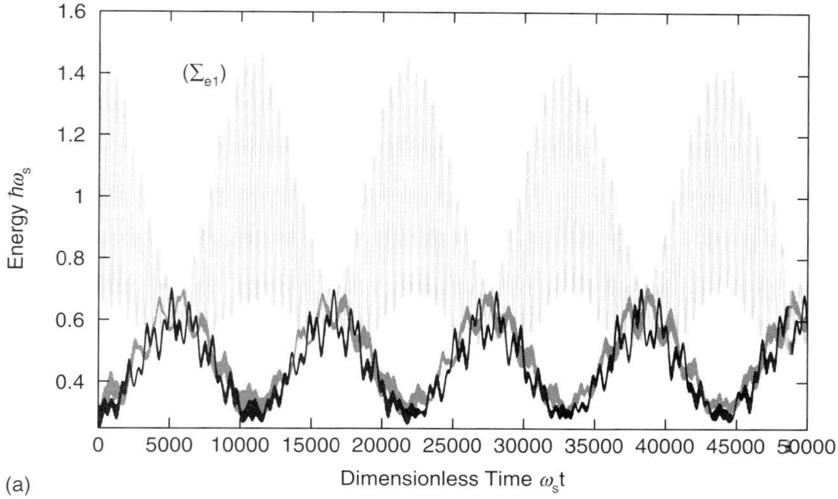

(a)

(a) Energy expectation values for the input mode e_1 (light grey) and the output modes e_2 (dark grey) and e_3 (black) versus dimensionless time $\tau = \omega_s t$.

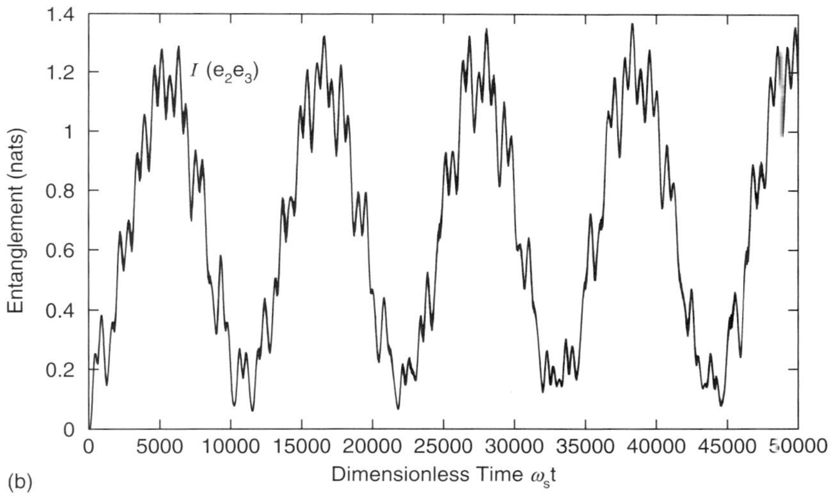

(b)

(b) Entanglement entropy between the output modes e_2 and e_3 versus dimensionless time $\tau = \omega_s t$.

FIGURE 28.3. Responses of the four-mode system with output frequencies approximately half that of the input mode.

the correspondence between the energy in the input mode and that in the two output modes is seen. However, unlike the case discussed in the previous subsection, the presence of very high frequency oscillations is seen in this result. These appear to arise because some of the system energy is oscillating between the SQUID ring and the input mode instead of being tranferred through to the two output modes.

Despite this rather inefficient transfer of energy between the input and output field modes, there exists a very significant level of entanglement entropy between the two output modes, considerably more than in Fig. 28.2(b). This is shown in Fig. 28.3(b). This increase in the entanglement entropy is most likely due to the way in which the energy of the single photon within the input mode is divided between the two output modes, both being very close to half the frequency of the input mode. Conversely, in the first case it was considered that there were two photons in the input mode the energy of which was transferred via the SQUID ring to generate two photons, one in each of the two output modes with the energy distributed almost equally between these two modes.

An important point highlighted by these calculations is the way in which a SQUID ring may be used to generate frequency down converted photons between em field modes as well as the manner in which the transfer of energy, and the entanglement, within the system can be affected by the frequency differences between the input and output field modes. In order to extend the discussion consider some higher frequency differences between input and output modes. However, in order to accomplish this due to limitations on the computational power available the investigation had to be constrained to simpler three-mode systems.

28.4 Energy down Conversion at Higher Frequency Ratios

Using the simpler three mode system of Fig. 28.1(b), and the SQUID ring circuit parameters of the previous section, energy transfer was then investigated where a significant frequency difference existed between the input and output modes. Since the simulation of this system is much less demanding computationally it was possible to investigate higher order down conversion processes as well as any entanglement between the input and output field modes. Clearly, being able to control photons in field modes in a given state of entanglement is of crucial importance in the fields of quantum computing and quantum information processing.[2,3]

28.4.1 Factor 5 down conversion

For our first look at down conversion with larger frequency ratios we used an output mode of one fifth the frequency of the input, $\omega_{e_2} = 0.2\omega_{e_1}$. The coupling strengths set were $\mu_{e_1 s} = 0.01$ and $\mu_{e_2 s} = 0.05$. The reason for making the output coupling much larger than the input coupling was to ensure that the energy transfer occurred mainly from input to output and did not oscillate between the two. With this established, it was noted that as the frequency differences between input and output in our system increased it becomes progressively more difficult to predict the dependence of the energy transfer between these modes on the external bias flux Φ_x applied to the SQUID ring. To deal with this it was therefore needed to determine the way in which the time averaged energy expectation values of the output mode varied with Φ_x. In this factor 5 down conversion example the best energy transfer to the output mode was found for $\Phi_x = 0.49738\Phi_0$.

Using this value of external flux, it could then be shown the way in which the energy in the input and output modes varied with dimensionless time ($\tau = \omega_s t$). In Fig. 28.4(a) it is seen that energy is indeed being down converted from the high frequency input mode to the output mode of much lower frequency. It is also shown, in grey, in the same figure the energy in the output mode when the input mode is empty (zero photons). It is noted here whatever energy is there in the output mode arises due to the coupling energy in the system. However, the very clear difference between this and the energy content in the output mode when a photon is

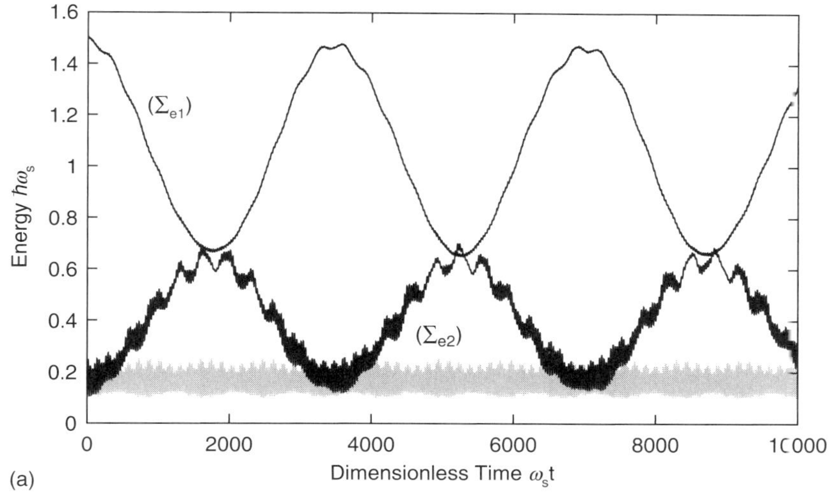

(a) Energy expectation values of the em field modes for a factor 5 down conversion.

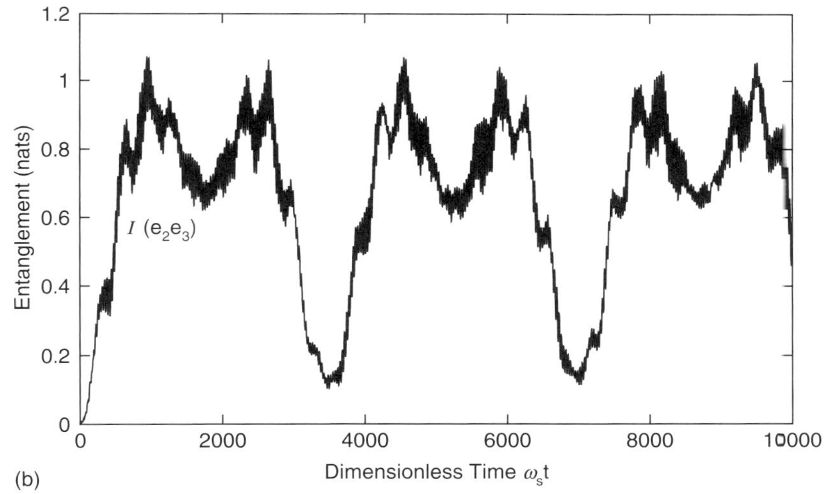

(b) Entanglement entropy between the two em field modes for a factor 5 down conversion.

FIGURE 28.4. Responses of the three-mode system for factor 5 down conversion.

present in the input mode establishes that we are observing quantum down conversion through the SQUID ring.

Concomitant with this down conversion of energy is the idea that the modes in the system may become entangled. Using the entanglement measure of Eq. (28.4) it is found, unsurprisingly, that the strength of entanglement between the input and output modes in the main oscillates in step with the level of energy transfer into the output mode. This is shown in Fig. 28.4(b). Although the form of this entanglement clearly oscillates with the energy in the output mode, it is also significant that as the energy within the input mode diminishes so does the entanglement. This is understandable if we consider the fact that if the input mode were able to transfer all of its energy to the output mode it would be forced into its ground state, and hence would become separable from the rest of the system.

In order to demonstrate the general nature of this down conversion process it is useful to compute the behavior of our coupled three mode system for an even larger frequency ratio, namely a factor of 10.

28.5 Larger frequency ratio (factor of 10) down conversion

Here again a system simulated, which is of the form depicted in Fig. 28.1(b), this time using a factor of 10 frequency difference between the input and output modes, i.e. $\omega_{e2} = 0.1\omega_{e1}$ with coupling strengths $\mu_{e1s} = \mu_{e2s} = 0.01$. The reason for reducing the coupling strength to the output mode, relative to the previous example, was to keep the energy transition regions [5, 6], for both the input and output modes, aligned in external bias flux. We found that the best energy transfer to the output mode occurred for $\Phi_x = 0.486\Phi_0$.

Following the same calculational path as for the factor 5 case, above, in Fig. 28.5(a) the variation in the energy in the output mode as the energy in the input mode is changed is shown. For comparison, the energy in the output mode when no input photons are present in the system is shown in grey.

In line with the previous examples entanglement between the input and output modes of this system was looked for. The result is shown in Fig. 28.5(b). Again, a similar form was seen as in Fig. 28.4(b), although the reduced coupling energy has made the response smoother. As before, the level of entanglement still oscillates with the energy in the output mode, and still displays a reduction when the occupancy of the input mode reaches its minimum.

28.6 Conclusions

In this study it is shown that a highly non-perturbative SQUID ring coupled to em field modes, all operating in the quantum regime, can lead to the transfer of energy between input and output field modes even when these differ very substantially in frequency. As we have emphasized, this kind of (quantum) behavior has its analogue in the non-linear media utilised in the field of quantum optics.[10] However, by comparison, the SQUID ring, acting as a highly non-perturbative system, can generate very much stronger non-linear effects than its quantum optics analogues. Furthermore, this non-perturbative/non-linear behavior can be easily controlled through the application of an external magnetic bias flux. However, space constraints have made it impracticable to present a full survey of this dependence on bias flux for each of the systems of Fig. 28.1, Nevertheless, together with the strong coupling that can exist between a SQUID ring and em field modes, this control mechanism would appear to make SQUIDs favored candidate for applications in future quantum technologies.

In this chapter it is also demonstrated that a quantum regime SQUID ring can generate energy down conversion between (quantum) input and output em field modes with concomitant entanglement of photons in these modes. Furthermore, even though we were limited for computational reasons to rather small frequency ratios in calculating these processes, our results would appear to be quite general. Thus, with more computer power available we would expect much higher

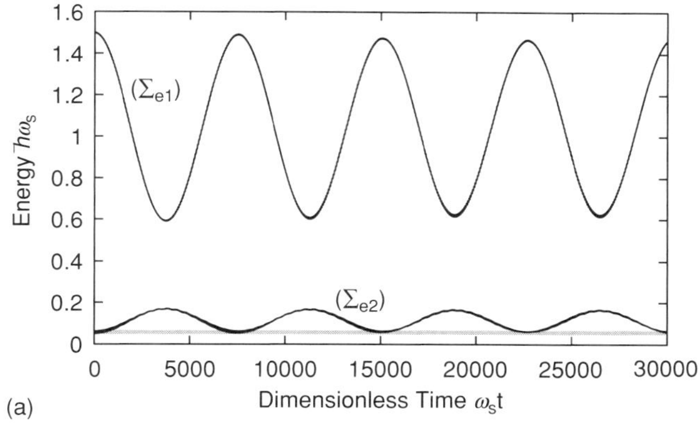

(a) Energy expectation values of the em field modes for a factor 10 down conversion.

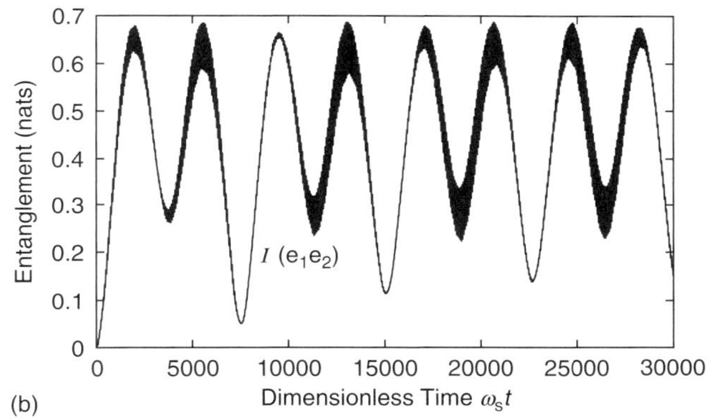

(b) Entanglement entropy between the two em field modes for a factor 5 down conversion.

FIGURE 28.5. Responses of the three-mode system for factor 10 down conversion.

(orders of magnitude) frequency down conversion ratios would appear as solutions of the evolution of Eq. (28.3). This appears to be borne out in the consistent behavior displayed in Figs. 28.4 and 28.5. Nevertheless, according to our computed results, there seems to be a reduction in the efficiency of energy conversion as the frequency differences increase; this may in turn affect the level of entanglement that is attainable. However, the strong variation due to external flux may indicate that we have, thus far, failed to identify the optimum value of Φ_x at which to bias our SQUID ring. It is also noteworthy that the entanglements shown in Fig. 28.3(b) indicate that it may prove practicable to entangle a large number of photons together in the output field modes, having been generated by down conversion from a much higher energy (frequency) photon in the input mode. If the entanglement to the input mode can be found to fall away, as indicated in Figs. 28.5(b) and 28.4(b), whilst still maintaining this entanglement between the output modes, this could have real significance in the preparation of states for quantum computing.[2,11]

That a quantum regime SQUID ring can be used in an analogous way to the non-linear media used in quantum optics is very encouraging since it implies that many, if not all of the effects and results established in this field may be translated directly into solid state circuits based on superconductors and the Josephson effect. It is clear that if this turns out to be correct it will have a very strong impact on the development of quantum technologies based on electronic circuit systems.[11,12] However, in any realistic system we will have to find a way to prevent the constant oscillation of energy between the output and inputs of our system. In the future we hope it proves possible to achieve this goal through the dynamical control of the bias flux applied to the SQUID ring in the system.

Acknowledgments. We would like to thank the Engineering and Physical Sciences Research Council for its funding of the work presented in this paper through its sponsorship of the Quantum Circuits Network programme.

References

1. M. J. Everitt, T. D. Clark, P. B. Stiffell, A. Vourdas, J. F. Ralph, R. J. Prance, and H. Prance, Superconducting analogs of quantum optical phenomena: Macroscopic quantum superpositions and squeezing in a superconducting quantum-interference device ring, *Phys. Rev. A*, **69**, 043804 (2004).
2. T. P. Orlando, J. E. Mooij, L. Tian, C. H. van der Wal, L. S. Levitov, S. Lloyd, and J. J. Mazo, Superconducting persistent-current qubit, *Phys. Rev. B-Condens Matter*, **60**, 15398–15413 (1999).
3. H. K. Lo, S. Popescu, and T. P. Spiller, (eds), *Introduction to Quantum Computation and Information*, (World Scientific, New Jersey, 1998).
4. T. P. Spiller, T. D. Clark, R. J. Prance, and A. Widom, *Progress in Low Temperature Physics*, vol. XIII, (Elsevier Science Publishers, 1992).
5. M. J. Everitt, T. D. Clark, P. Stiffell, H. Prance, R. J. Prance, A. Vourdas, and J. F. Ralph, Quantum statistics and entanglement of two electromagnetic field modes coupled via a mesoscopic squid ring, *Phys. Rev. B*, **6418**, 184517 (2001).
6. M. J. Everitt, P. Stiffell, T. D. Clark, A. Vourdas, J. F. Ralph, H. Prance, and R. J. Prance, Fully quantum-mechanical model of a squid ring coupled to an electromagnetic field, *Phys. Rev. B*, **6314**, 144530 (2001).
7. T. D. Clark, J. Diggins, J. F. Ralph, M. Everitt, R. J. Prance, R. Whiteman, A. Widom, and Y. N. Srivastava, Coherent evolution and quantum transitions in a two level model of a squid ring, *Ann. Phys.*, **268**, 1–30 (1998).
8. J. J. Sakurai, *Modern Quantum Mechanics*, (Addison-Wesley Publishing Company, revised edn, 1994).
9. G. Lindbland, *Commun. Math. Phys.*, **33**, 305 (1973).
10. F. Dell'Anno, S. De Siena, and F. Illuminati, Structure of multiphoton quantum optics. i. canonical formalism and homodyne squeezed states, *Phys. Rev. A.*, **69**, 033812, (2004).
11. Y. Makhlin, G. Schon, and A. Shnirman, Josephson-junction qubits with controlled couplings, *Nature*, **398**, 305–307 (1999).
12. P. Silvestrini, B. B. Ruggiero, C. Granata, and E. Esposito, Supercurrent decay of josephson junctions in non-stationary conditions: Experimental evidence of macroscopic quantum effects, *Phys. Lett. A*, **267**, 45–51 (2000).

29

Photon-induced entanglement of distant mesoscopic SQUID rings

A. VOURDAS[1], D. I. TSOMOKOS[1], AND C. C. CHONG[2]

[1] Department of Computing, University of Bradford, West Yorkshire, BD7 1DP, England.
[2] Institute of High Performance Computing, 1 Science Park Road, 117 528, Singapore.

Keywords: SQUID, Entangled states

Abstract

An experiment that involves two distant mesoscopic SQUID rings is studied. The superconducting rings are irradiated with correlated photons, which are produced by a single microwave source. Classically correlated (separable) and quantum mechanically correlated (entangled) microwaves are considered, and their effect on the Josephson currents is quantified. It is shown that the currents tunnelling through the Josephson junctions in the distant rings, are correlated.

29.1 Introduction

A fundamental property of Superconducting Quantum Interference devices (SQUIDs) is that they exhibit quantum coherence at the macroscopic level.[1] This property may be used for the purposes of quantum information processing.[2,3]

A lot of research on superconducting devices investigates their interaction with classical microwaves. On the other hand the use of nonclassical microwaves makes the system fully quantum mechanical and interesting quantum phenomena arise. For example, in this chapter we show that entangled two-mode microwaves produce correlated currents in distant SQUID rings.

Nonclassical electromagnetic fields at low temperatures ($k_B T \ll \hbar\omega$) have been studied for more than twenty years both theoretically and experimentally.[4] The interaction of SQUID rings with nonclassical microwaves has been studied in the literature.[5,6]

In previous publications[7] the effects of entangled electromagnetic fields on distant electron interference experiments have been studied. In this chapter we review and extend further this study in the context of SQUID rings. Two mesoscopic SQUID rings are considered, which are far from each other and are irradiated with entangled microwaves, produced by a single source (Fig. 29.1). It is shown that the Josephson currents in the distant SQUID rings are correlated. The photon-induced correlations between the currents are quantified. It is shown that the current

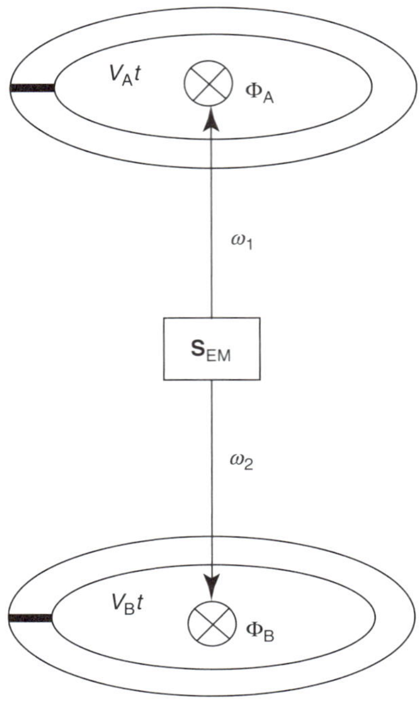

FIGURE 29.1. Two distant mesoscopic SQUID rings A and B are irradiated with nonclassical microwaves of frequencies ω_1 and ω_2, correspondingly. The microwaves are produced by the source S_{EM} and are correlated. Classical magnetic fluxes $V_A t$ and $V_B t$ are also threading the two rings A and B, correspondingly.

correlations depend on whether the photons are classically correlated (separable) or quantum mechanically correlated (entangled). The difference between separable and entangled microwave density matrices[8] is in the nondiagonal elements; and the effect of these nondiagonal elements on the Josephson currents is explicitly calculated.

29.2 Interaction of a Single SQUID Ring with Nonclassical Microwaves

In this section a single SQUID ring is considered and its interaction with both classical and nonclassical microwaves is studied.

For irradiation with classical microwaves, the Josephson current is $I_A = I_1 \sin\theta_A$, where $\theta_A = 2e\Phi_A$ is the phase difference across the junction due to the total flux Φ_A through the ring. The external field approximation is assumed, where the back reaction (the additional flux induced by the SQUID ring current) is neglected; i.e., the flux $\mathcal{L}I_A$, where $\mathcal{L}$ is the self-inductance of the ring, is negligible in comparison to Φ_A. The magnetic flux has a linear and a sinusoidal component:

$$\Phi_A = V_A t + \phi_A; \qquad \phi_A = A\sin(\omega_1 t). \tag{29.1}$$

Consequently the observed current is

$$I_A = I_1 \sin[\omega_A t + 2eA\sin(\omega_1 t)]; \qquad \omega_A = 2eV_A. \tag{29.2}$$

Consider now the interaction of a SQUID ring with nonclassical microwaves, which are carefully prepared in a particular quantum state and are described by a density matrix ρ. The dual quantum variables of the nonclassical field are the vector potential A_i and the electric field E_i. Integrating these over the SQUID ring the magnetic flux and the electromotive force operators $\hat{\phi} = \oint_C A_i dx_i$, $\hat{V}_{\text{EMF}} = \oint_C E_i dx_i$ are obtained.

In the external field approximation the flux operator evolves as

$$\hat{\phi}(t) = \xi 2^{-1/2}[\hat{a}^\dagger \exp(i\omega t) + \hat{a}\exp(-i\omega t)], \tag{29.3}$$

where ξ is a parameter proportional to the area of the SQUID ring and the $\hat{a}^\dagger, \hat{a}$ are the photon creation and annihilation operators. Consequently the phase difference θ_{A} is the operator

$$\hat{\theta}_{\text{A}} = \omega_{\text{A}} t + q[\hat{a}^\dagger \exp(i\omega t) + \hat{a}\exp(-i\omega t)], \qquad q = \sqrt{2}e\xi; \tag{29.4}$$

and the current also becomes an operator, $\hat{I}_{\text{A}} = I_1 \sin\hat{\theta}_{\text{A}}$. Expectation values of the current are calculated by taking its trace with respect to the density matrix ρ, which describes the nonclassical electromagnetic fields,

$$\langle \hat{I}_{\text{A}} \rangle = \text{Tr}(\rho \hat{I}_{\text{A}}) = I_1 \text{Im}[\exp(i\omega_{\text{A}} t)\tilde{W}(\lambda_{\text{A}})], \tag{29.5}$$

$$\lambda_{\text{A}} = iq\exp(i\omega_1 t). \tag{29.6}$$

$\tilde{W}(x)$ is the Weyl function[9] given by

$$\tilde{W}(x) = \text{Tr}[\rho D(x)]; \qquad D(x) = \exp(x\hat{a}^\dagger - x^*\hat{a}) \tag{29.7}$$

where $D(x)$ is the displacement operator. Higher moments of the Josephson current quantify the quantum statistics of the electron pairs tunnelling through the junction.

29.3 Interaction of two Distant SQUID Rings with Entangled Microwaves

In this section two mesoscopic SQUID rings far apart from each other are considered, which are refered to as A and B (Fig. 29.1). They are irradiated with correlated microwaves. Let ρ be the density matrix of the microwaves, and $\rho_{\text{A}} = \text{Tr}_{\text{B}}\rho$, $\rho_{\text{B}} = \text{Tr}_{\text{A}}\rho$, the density matrices of the microwaves interacting with the two SQUID rings A, B, correspondingly. When the density matrix ρ is factorizable as $\rho_{\text{fact}} = \rho_{\text{A}} \otimes \rho_{\text{B}}$ the two modes are not correlated. If it can be written as $\rho_{\text{sep}} = \sum_i p_i \rho_{\text{A}i} \otimes \rho_{\text{B}i}$, where p_i are probabilities, it is called separable and the two modes are classically correlated. Density matrices which cannot be written in one of these two forms are entangled (quantum mechanically correlated).[8]

The currents in the two SQUIDs are

$$\langle \hat{I}_{\text{A}} \rangle = I_1 \text{Tr}(\rho_{\text{A}} \sin\hat{\theta}_{\text{A}}), \qquad \langle \hat{I}_{\text{B}} \rangle = I_2 \text{Tr}(\rho_{\text{B}} \sin\hat{\theta}_{\text{B}}). \tag{29.8}$$

The expectation value of the product of the two current operators is given by:

$$\langle \hat{I}_{\text{A}} \hat{I}_{\text{B}} \rangle = I_1 I_2 \text{Tr}(\rho \sin\hat{\theta}_{\text{A}} \sin\hat{\theta}_{\text{B}}). \tag{29.9}$$

In general $\langle \hat{I}_{\text{A}} \hat{I}_{\text{B}} \rangle$ is different from $\langle \hat{I}_{\text{A}} \rangle \langle \hat{I}_{\text{B}} \rangle$ and we calculate the ratio

$$R = \frac{\langle \hat{I}_{\text{A}} \hat{I}_{\text{B}} \rangle}{\langle \hat{I}_{\text{A}} \rangle \langle \hat{I}_{\text{B}} \rangle}. \tag{29.10}$$

For factorizable density matrices $\rho_{\text{fact}} = \rho_A \otimes \rho_B$ it is seen that $R_{\text{fact}} = 1$. For separable density matrices ρ_{sep} the ratio R_{sep} is not necessarily equal to one and numerical results for various examples are shown below.

The second moments are also calculated as

$$\langle \hat{I}_{\mathrm{A}}^2 \rangle = I_1^2 \mathrm{Tr}[\rho_{\mathrm{A}}(\sin\hat{\theta}_{\mathrm{A}})^2], \qquad \langle \hat{I}_{\mathrm{B}}^2 \rangle = I_2^2 \mathrm{Tr}[\rho_{\mathrm{B}}(\sin\hat{\theta}_{\mathrm{B}})^2]. \tag{29.11}$$

The statistics of the photons affects the statistics of the tunnelling electron pairs, which is quantified with the $\langle \hat{I}_{\mathrm{A}} \hat{I}_{\mathrm{B}} \rangle$, $\langle \hat{I}_{\mathrm{A}}^2 \rangle$, $\langle \hat{I}_{\mathrm{B}}^2 \rangle$ (and also with the higher moments).

29.3.1 *Microwaves in number states*

Consider microwaves in the separable (mixed) state

$$\rho_{\text{sep}} = \frac{1}{2}(|N_1 N_2\rangle\langle N_1 N_2| + |N_2 N_1\rangle\langle N_2 N_1|), \tag{29.12}$$

where $N_1 \neq N_2$. Also consider microwaves in the entangled state $|s\rangle = 2^{-1/2}(|N_1 N_2\rangle + |N_2 N_1\rangle)$, which is a pure state. The density matrix of $|s\rangle$ is

$$\rho_{\text{ent}} = \rho_{\text{sep}} + \frac{1}{2}(|N_1 N_2\rangle\langle N_2 N_1| + |N_2 N_1\rangle\langle N_1 N_2|), \tag{29.13}$$

where the ρ_{sep} is given by Eq. (29.12). It is seen that the ρ_{ent} and the ρ_{sep} differ only by the above nondiagonal elements.

In this example, the reduced density matrices are the same for both the separable and entangled states

$$\rho_{\text{sep,A}} = \rho_{\text{ent,A}} = \rho_{\text{sep,B}} = \rho_{\text{ent,B}} = \frac{1}{2}(|N_1\rangle\langle N_1| + |N_2\rangle\langle N_2|). \tag{29.14}$$

Consequently in this example $\langle \hat{I}_{\mathrm{A}} \rangle_{\text{sep}} = \langle \hat{I}_{\mathrm{A}} \rangle_{\text{ent}}$, and also $\langle \hat{I}_{\mathrm{B}} \rangle_{\text{sep}} = \langle \hat{I}_{\mathrm{B}} \rangle_{\text{ent}}$.

For the density matrix ρ_{sep} of Eq. (29.12) we find

$$\langle \hat{I}_{\mathrm{A}} \rangle = \frac{I_1}{2} \exp\left(-\frac{q^2}{2}\right)[L_{N_1}(q^2) + L_{N_2}(q^2)] \sin(\omega_{\mathrm{A}} t), \tag{29.15}$$

$$\langle \hat{I}_{\mathrm{B}} \rangle = \frac{I_2}{2} \exp\left(-\frac{q^2}{2}\right)[L_{N_1}(q^2) + L_{N_2}(q^2)] \sin(\omega_{\mathrm{B}} t), \tag{29.16}$$

where the $L_n^{\alpha}(x)$ are Laguerre polynomials. The currents $\langle \hat{I}_{\mathrm{A}} \rangle$, $\langle \hat{I}_{\mathrm{B}} \rangle$ are in this example independent of the microwave frequencies ω_1, ω_2.

The expectation value of the product of the two currents [Eq. (29.9)] is

$$\langle \hat{I}_{\mathrm{A}} \hat{I}_{\mathrm{B}} \rangle_{\text{sep}} = I_1 I_2 \exp(-q^2) L_{N_1}(q^2) L_{N_2}(q^2) \sin(\omega_{\mathrm{A}} t) \sin(\omega_{\mathrm{B}} t). \tag{29.17}$$

Consequently the ratio R of Eq. (29.10) is

$$R_{\text{sep}} = \frac{4 L_{N_1}(q^2) L_{N_2}(q^2)}{[L_{N_1}(q^2) + L_{N_2}(q^2)]^2}. \tag{29.18}$$

In this example the R_{sep} is time-independent.

The moments of the currents, defined by Eq. (29.11), are also calculated:

$$\langle \hat{I}_A^2 \rangle = \frac{I_1^2}{2}\left\{1 - \frac{1}{2}\exp(-2q^2)[L_{N_1}(4q^2) + L_{N_2}(4q^2)]\cos(2\omega_A t)\right\}, \tag{29.19}$$

$$\langle \hat{I}_B^2 \rangle = \frac{I_2^2}{2}\left\{1 - \frac{1}{2}\exp(-2q^2)[L_{N_1}(4q^2) + L_{N_2}(4q^2)]\cos(2\omega_B t)\right\}. \tag{29.20}$$

For the case of ρ_{ent} the $\langle \hat{I}_A \rangle$, $\langle \hat{I}_B \rangle$ are the same as in Eqs. (29.15), (29.16); and the $\langle \hat{I}_A^2 \rangle$, $\langle \hat{I}_B^2 \rangle$ are the same as in Eqs. (29.19), (29.20). However the $\langle \hat{I}_A \hat{I}_B \rangle$ is

$$\langle \hat{I}_A \hat{I}_B \rangle_{\mathrm{ent}} = \langle \hat{I}_A \hat{I}_B \rangle_{\mathrm{sep}} + I_{\mathrm{cross}}, \tag{29.21}$$

where

$$\begin{aligned} I_{\mathrm{cross}} &= -\frac{I_1 I_2}{2}\exp(-q^2) L_{N_1}^{N_2-N_1}(q^2) L_{N_2}^{N_1-N_2}(q^2)[\cos(\omega_A t + \omega_B t) \\ &\quad -(-1)^{N_1-N_2}\cos(\omega_A t - \omega_B t)]\cos(\Omega t), \end{aligned} \tag{29.22}$$

$$\Omega = (N_1 - N_2)(\omega_1 - \omega_2). \tag{29.23}$$

It is seen that the effect of entangled microwaves on Josephson currents is different from the effect of separable microwaves. In this case the ratio R of Eq. (29.10) is

$$R_{\mathrm{ent}} = R_{\mathrm{sep}} + \frac{I_{\mathrm{cross}}(t)}{\langle \hat{I}_A \rangle \langle \hat{I}_B \rangle}, \tag{29.24}$$

which is a time-dependent quantity oscillating around the R_{sep}.

29.3.2 Microwaves in coherent states

Consider microwaves in the classically correlated state

$$\rho_{\mathrm{sep}} = \frac{1}{2}(|A_1 A_2\rangle\langle A_1 A_2| + |A_2 A_1\rangle\langle A_2 A_1|), \tag{29.25}$$

where $|A_1\rangle$, $|A_2\rangle$ are microwave coherent states. We also consider the entangled state $|u\rangle = \mathcal{N}(|A_1 A_2\rangle + |A_2 A_1\rangle)$, with density matrix

$$\rho_{\mathrm{ent}} = 2\mathcal{N}^2 \rho_{\mathrm{sep}} + \mathcal{N}^2(|A_1 A_2\rangle\langle A_2 A_1| + |A_2 A_1\rangle\langle A_1 A_2|), \tag{29.26}$$

where the normalization constant is given by

$$\mathcal{N} = \left[2 + 2\exp\left(-|A_1 - A_2|^2\right)\right]^{-1/2}. \tag{29.27}$$

For microwaves in the separable state of Eq. (29.25) the reduced density matrices are

$$\rho_{\mathrm{sep,A}} = \rho_{\mathrm{sep,B}} = \frac{1}{2}(|A_1\rangle\langle A_1| + |A_2\rangle\langle A_2|), \tag{29.28}$$

and hence the current in A is

$$\begin{aligned} \langle \hat{I}_A \rangle_{\mathrm{sep}} &= \frac{I_1}{2}\exp(-\frac{q^2}{2})\{\sin[\omega_A t + 2q|A_1|\cos(\omega_1 t - \theta_1)] \\ &\quad + \sin[\omega_A t + 2q|A_2|\cos(\omega_1 t - \theta_2)]\}. \end{aligned} \tag{29.29}$$

where $\theta_1 = \arg(A_1)$, and $\theta_2 = \arg(A_2)$. A similar expression yields the current in B. We have also calculated numerically the ratio R_{sep}.

For microwaves in the entangled state of Eq. (29.26) the reduced density matrices are

$$\rho_{\mathrm{ent,A}} = \rho_{\mathrm{ent,B}} = \mathcal{N}^2(|A_1\rangle\langle A_1| + |A_2\rangle\langle A_2| + \tau|A_1\rangle\langle A_2| + \tau^*|A_2\rangle\langle A_1|), \quad (29.30)$$

where $\tau = \langle A_1|A_2\rangle = \exp(-|A_1|^2/2 - |A_2|^2/2 + A_1^* A_2)$.

The current in A is

$$\langle \hat{I}_{\mathrm{A}}\rangle_{\mathrm{ent}} = 2\mathcal{N}^2\langle \hat{I}_{\mathrm{A}}\rangle_{\mathrm{sep}} + \mathcal{N}^2 E F_1 \exp\left(-\frac{q^2}{2}\right) I_1, \quad (29.31)$$

where $E = \exp[-|A_1|^2 - |A_2|^2 + 2|A_1 A_2|\cos(\theta_1 - \theta_2)]$, and

$$F_1 = [\exp(q|A_1|S_{A,1} - q|A_2|S_{A,2}) + \exp(-q|A_1|S_{A,1} + q|A_2|S_{A,2})]\sin(\omega_A t + q|A_1|C_{A,1} + q|A_2|C_{A,2}), \quad (29.32)$$

with $S_{A,1} = \sin(\omega_1 t - \theta_1)$, $S_{A,2} = \sin(\omega_1 t - \theta_2)$, $C_{A,1} = \cos(\omega_1 t - \theta_1)$, $C_{A,2} = \cos(\omega_1 t - \theta_2)$. A similar expression yields the current in B, and we have also calculated numerically the ratio R_{ent}.

29.3.3 Numerical results

In the numerical results of Figs.29.2 and 29.3 the microwave frequencies are $\omega_1 = 1.2 \times 10^{-4}$, $\omega_2 = 10^{-4}$, in units where $k_B = \hbar = c = 1$. The critical currents are $I_1 = I_2 = 1$. The other parameters are $\xi = 1$, $\omega_{\mathrm{A}} = \omega_1$, $\omega_{\mathrm{B}} = \omega_2$, $N_1 = 1$, $N_2 = 4$, and $\theta_1 = \theta_2 = 0$.

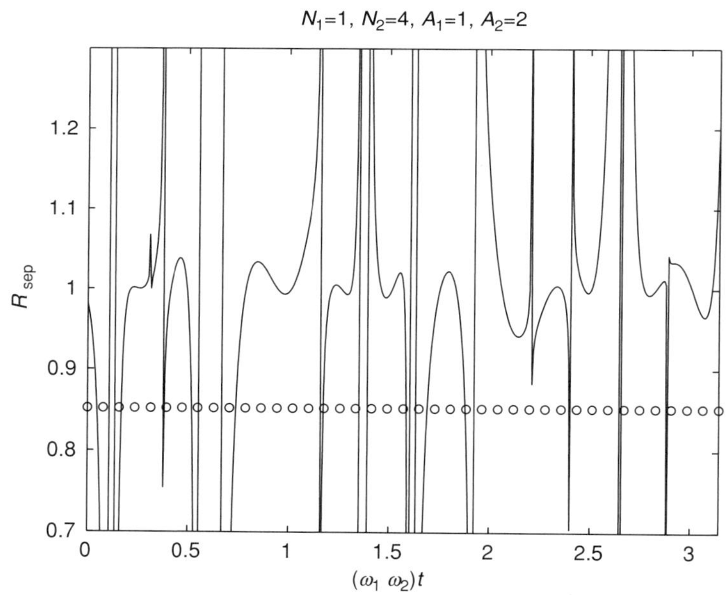

FIGURE 29.2. R_{sep} against $(\omega_1 - \omega_2)t$ for the number state of Eq. (29.12) with $N_1 = 1$, $N_2 = 4$ (line of circles), and the coherent state of Eq. (29.25) with $A_1 = 1$, $A_2 = 2$ (solid line). The photon frequencies are $\omega_1 = 1.2 \times 10^{-4}$ and $\omega_2 = 10^{-4}$, in units where $k_B = \hbar = c = 1$.

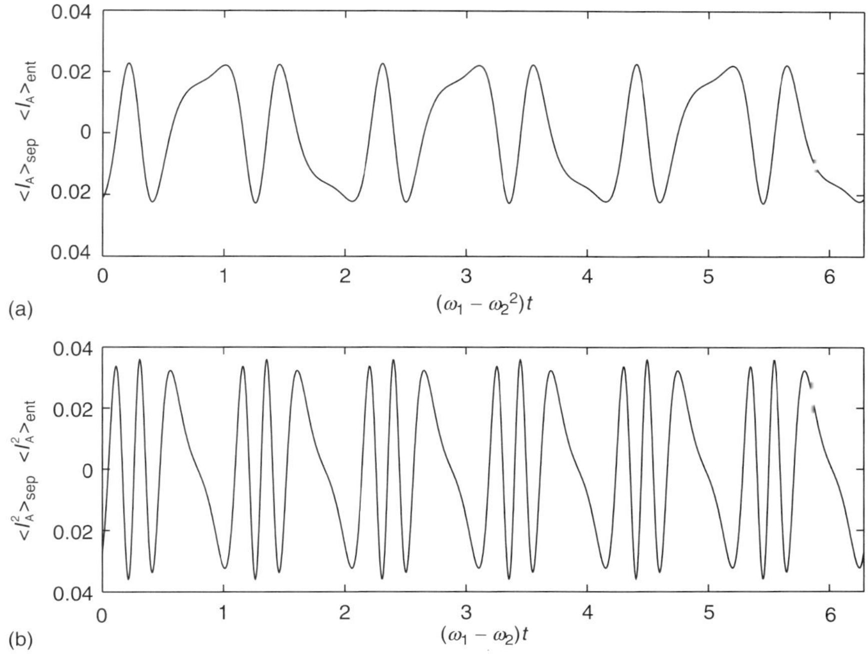

FIGURE 29.3. (a) $\langle \hat{I}_{\mathrm{A}} \rangle_{\mathrm{sep}} - \langle \hat{I}_{\mathrm{A}} \rangle_{\mathrm{ent}}$, and (b) $\langle \hat{I}^2_{\mathrm{A}} \rangle_{\mathrm{sep}} - \langle \hat{I}^2_{\mathrm{A}} \rangle_{\mathrm{ent}}$ against $(\omega_1 - \omega_2)t$ for the coherent state $\rho_{\mathrm{sep,A}}$ of Eq. (29.28) and $\rho_{\mathrm{ent,A}}$ of Eq. (29.30) with $A_1 = 1$, $A_2 = 2$. The photon frequencies are $\omega_1 = 1.2 \times 10^{-4}$ and $\omega_2 = 10^{-4}$, in units where $k_B = \hbar = c = 1$.

For a meaningful comparison between microwaves in number states and microwaves in coherent states, we take them to have the same average number of photons, $|A_1|^2 = N_1$ and $|A_2|^2 = N_2$.

In Fig. 29.2 R_{sep} is plotted against $(\omega_1 - \omega_2)t$ for currents induced by microwaves in the number state of Eq. (29.12) with $N_1 = 1$, $N_2 = 4$ (line of circles), and the coherent state of Eq. (29.25) with $A_1 = 1$, $A_2 = 2$ (solid line). It is seen that two different separable photon states with the same average number of photons give rise to different correlations between the SQUID currents.

In Fig. 29.3 we plot (a) $\langle \hat{I}_{\mathrm{A}} \rangle_{\mathrm{sep}} - \langle \hat{I}_{\mathrm{A}} \rangle_{\mathrm{ent}}$, and (b) $\langle \hat{I}^2_{\mathrm{A}} \rangle_{\mathrm{sep}} - \langle \hat{I}^2_{\mathrm{A}} \rangle_{\mathrm{ent}}$, against $(\omega_1 - \omega_2)t$ for microwaves in the coherent state $\rho_{\mathrm{A,sep}}$ of Eq. (29.28) and $\rho_{\mathrm{A,ent}}$ of Eq. (29.30) with $A_1 = 1$, $A_2 = 2$. In Fig. 29.3(a) it is seen that the Josephson current in SQUID ring A is different for irradiation with separable and entangled microwaves in coherent states. In Fig. 29 3(b) it is seen that irradiation with separable and entangled coherent states leads to different second moments of the current, which implies that the quantum statistics of electron pairs tunnelling the Josephson junction of SQUID ring A are different in these two cases.

29.4 Discussions

The interaction of two distant SQUID rings A and B with two-mode nonclassical microwaves are considered, which are produced by the same source. The flux, the phase difference and the

Josephson currents are operators and their expectation values with the density matrix of the non-classical microwaves give the physically observed quantities. That external field approximation is assumed, where the electromagnetic field created by the Josephson currents (back reaction) is neglected and various quantities are calculated.

It has been shown that the Josephson currents in the two rings are correlated in the sense that $\langle \hat{I}_A \hat{I}_B \rangle$ is different from $\langle \hat{I}_A \rangle \langle \hat{I}_B \rangle$ (for non-factorizable density matrices). Those examples are considered where the photons are classically correlated and quantum mechanically correlated; and it was shown that the non-diagonal terms in the latter case affect the Josephson currents. Further work in this direction could be the formulation of Bell-type inequalities for the Josephson currents, which are obeyed in the case of separable microwaves and violated in the case of entangled microwaves.

References

1. N. Byers and C. N. Yang, *Phys. Rev. Lett.* **7**, 46 (1961); F. Bloch, *Phys. Rev. B* **2**, 109 (1970); A. Barone and G. Paterno, *Physics and Applications of the Josephson Effect* (Wiley, New York, 1982); M. Tinkham, *Introduction to Superconductivity* (McGraw-Hill, New York, 1996).
2. Y. Makhlin, G. Schøn, and A. Shnirman, *Rev. Mod. Phys.* **73**, 357 (2001); M.A. Kastner, *ibid.* **64**, 849 (1992); G. Schøn and A. D. Zaikin, *Phys. Rep.* **198**, 237 (1990).
3. I. Chioreseu, Y. Nakamura, C. Harmans, and J. Mooij, *Science* **299**, 1869 (2003); Y. Nakamura, Y.A. Pashkin, and J. S. Tsai, *Nature* **398**, 786 (1999); C. H. van der Wal, *et al.*, *Science* **290**, 773 (2000).
4. R. Loudon and P. L. Knight, *J. Mod. Optics* **34**, 709 (1987); R. Loudon, *The Quantum Theory of Light* (Oxford University Press, Oxford, 2000); D. F. Walls and G. Milburn, *Quantum Optics* (Springer, Berlin, 1994).
5. A. Vourdas, *Phys. Rev. B* **49**, 12 040 (1994); *Z. Phys. B* **100**, 455 (1996); A. Vourdas and T. P. Spiller, *ibid.* **102**, 43 (1997); A. A. Odintsov and A. Vourdas, *Europhys. Lett.* **34**, 385 (1996).
6. L. M. Kuang, Y. Wang, and M. L. Ge, *Phys. Rev. B* **53**, 11 764 (1996); J. Zou, B. Shao, and X. S. Xing, *ibid.* **56**, 14 116 (1997); *Phys. Lett. A* **231**, 123 (1997); *Z. Phys. B* **104**, 439 (1997); M. J. Everitt, *et al.*, *Phys. Rev. B* **63**, 144 530 (2001); W. Al-Saidi and D. Stroud, *ibid.* **65**, 014512 (2002).
7. D. I. Tsomokos, C. C. Chong, and A. Vourdas, *Phys. Rev. A* **69**, 013810 (2004); *Phys. Rev. B*, to appear (2004).
8. R. F. Werner, *Phys. Rev. A* **40**, 4277 (1989); R. Horodecki and M. Horodecki, *ibid.* **54**, 1838 (1996); A. Peres, *Phys. Rev. Lett.* **77**, 1413 (1996); V. Vedral *et al.*, *ibid.* **78**, 2275 (1997); V. Vedral, *Rev. Mod. Phys.* **74**, 197 (2002).
9. See, for example, S. Chountasis and A. Vourdas, *Phys. Rev. A* **58**, 848 (1998) and references therein.

30

Time Evolution of two distant SQUID rings irradiated with entangled electromagnetic field

A. Konstadopoulou[1], R. Migliore[2], Z. Ahmad Zukarnain[3], A. Vourdas[3], and A. Messina[1]

[1] Department of Computing, University of Bradford, Bradford BD7 1DP, U. K.
akonstad@bradford.ac.uk
[2] Dipartimento di Scienze Fisiche ed Astromoniche University of Palermo, Via Archirafi 36, 90123 Palermo (Italy)
rosanna@fisica.unipa.it
messina@fisica.unipa.it
[3] Department of Computing, University of Bradford, Bradford BD7 1DP, U.K.
Z. Ahmadzukarnain@brad.ac.uk
a.vourdas@brad.ac.uk

Keywords: SQUID, Entangled states

Abstract

Two distant mesoscopic SQUID rings are irradiated with two mode microwaves produced by the same source. The time evolution of the system is studied. The two microwave modes are correlated. It is shown that the currents tunnelling through the Josephson junctions in the distant rings, are also correlated.

30.1 Introduction

Entanglement is an important feature of quantum mechanics which might have important implications for quantum information processing.[1] In the present chapter how entangled two-mode microwaves can be used to induce correlated Josephson currents in two distant SQUID rings is studied.

SQUID rings are potential candidates as quantum gates.[2,3] The creation of entangled currents in distant SQUID rings could lay an important role in the general areas of quantum communications and quantum cryptography.

The work is in the general context of research in the interaction of SQUIDs with nonclassical microwaves. In this case the microwaves are carefully prepared in a particular quantum state described by a density matrix ρ and the quantum noise of the microwaves affects the Josephson currents. For example, in Fig. 30.1 microwaves in coherent and squeezed states are shown. The average value of the electric field $\langle E \rangle$ is the same in both cases, but the quantum noise ΔE is

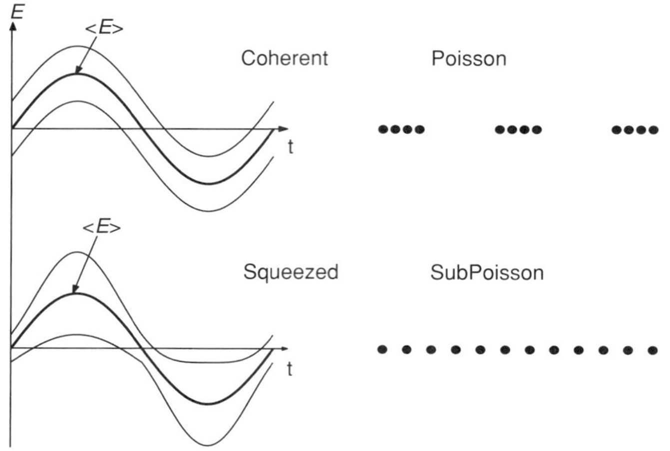

FIGURE 30.1. Coherent (top) and Squeezed (bottom) microwaves.

different; and the corresponding Josephson currents are also different. Another way of explaining this is through the quantum statistics of photons which is Poissonian for coherent states and sub-Poissonian for squeezed states. The statistics of photons threading the SQUID rings affects the Josephson currents. Such phenomena have been discussed in Ref. [4, 5].

In this chapter we study the effect of entangled microwaves on Josephson currents. The experiment depicted in Fig. 30.2 where correlated microwaves produced by the same source irradiate two distant SQUID rings is considered. Two cases are studied: classically correlated (separable) and quantum mechanically (entangled) microwaves.

It is shown that the expectation value of the product of two physically meaningful quantities in the two SQUID rings are different from the products of the expectation values of the same quantities. This shows the occurrence of correlations between physical observables pertaining to the two distant SQUIDs.

In Ref. [7] a similar experiment has been studied using the external field approximation in which the back reaction i.e., the electromagnetic field created by the Josephson currents, is neglected. In this chapter numerically the time evolution of the system is studied and the back reaction is taken into account.

30.2 Interaction of Two SQUID Rings with Nonclassical Microwaves

A system comprising of two SQUID rings (Fig. 30.2) irradiated with entangled electromagnetic fields is considered. The Hamiltonian describing the system is

$$H = \frac{\Delta_A}{2}\sigma_z^A + \frac{\Delta_B}{2}\sigma_z^B + \omega_1(\alpha_1\alpha_1^\dagger + \frac{1}{2}) + \omega_2(\alpha_2\alpha_2^\dagger + \frac{1}{2}) + \lambda_{A}\sigma_x^A(\alpha_1 + \alpha_1^\dagger) + \lambda_{B}\sigma_x^B(\alpha_2 + \alpha_2^\dagger). \quad (30.1)$$

where, α_1, $\alpha_1^\dagger$ and α_2, $\alpha_2^\dagger$ are the annihilation and creation operators of the two microwave modes; ω_1, ω_2 are the frequencies of the two microwave modes; $\lambda_A(\lambda_B)$ is the coupling constant between the first (second) SQUID and the first (second) mode of the microwaves. The SQUID rings are approximated with two-level systems where $\Delta_A(\Delta_B)$ is the energy difference between the excited level and the ground level.

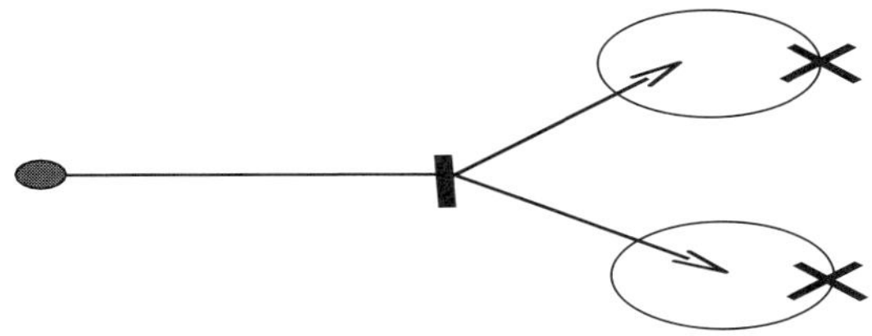

FIGURE 30.2. Correlated microwaves produced by the same source irradiate two distant SQUID rings.

It is assumed that the density matrix of the system at $t = 0$ is given by

$$\rho = \rho_A \otimes \rho_B \otimes \rho_F \tag{30.2}$$

where ρ_A, ρ_B are the density matrices of the two SQUID rings and ρ_F is the density matrix of the two-mode electromagnetic field.

The time evolution of the system have been calculated numerically as

$$\rho(t) = e^{iHt}\rho(0)e^{-iHt} \tag{30.3}$$

The infinite dimensional Hilbert spaces of the two microwave modes are truncated at N_{max} which is taken to be much greater than the average value of photons in the microwaves. As a measure of accuracy of the approximation, at a generic instant time t, the traces of the truncated matrices were calculated. In the limit $N_{max} \to \infty$ they equal to **1**; and in the truncated case they should be very close to **1**. In all our results the traces were greater than 0.98.

The following reduced density matrices were also calculated with partial traces

$$\begin{aligned}
\rho_S(t) &= Tr_{F_1,F_2}\rho(t)\\
\rho_A(t) &= Tr_B\rho_S(t)\\
\rho_B(t) &= Tr_A\rho_S(t)\\
\rho_F(t) &= Tr_{A,B}\rho(t)\\
\rho_{F_1}(t) &= Tr_{F_2}\rho_F(t)\\
\rho_{F_2}(t) &= Tr_{F_1}\rho_F(t)
\end{aligned} \tag{30.4}$$

where $\rho_S(t)$ is the density matrix of the combined system of the two SQUID rings and $\rho_{F_1}(t)$ and $\rho_{F_2}(t)$ are the density matrices of the two electromagnetic modes, correspondingly.

The purpose of this chapter is to demonstrate that although at $t = 0$, $\rho_S(0) = \rho_A(0) \otimes \rho_B(0)$, at a later time t, $\rho_S(t)$ is in general, different than $\rho_A(t) \otimes \rho_B(t)$ and the two SQUID rings are correlated although they are apart from each other.

In order to demonstrate this explicitly, the 'average level' has been calculated

$$\begin{aligned}
\langle N_A N_B\rangle &= Tr[\rho_S(\sigma_z^A \otimes \sigma_z^B)]\\
\langle N_A\rangle &= Tr[\rho_A\sigma_z^A]\\
\langle N_B\rangle &= Tr[\rho_B\sigma_z^B]
\end{aligned} \tag{30.5}$$

For completeness we have also calculated

$$\begin{aligned}
\langle N_{F_1} N_{F_2}\rangle &= Tr[\rho_F(\alpha_1^\dagger\alpha_1 \otimes \alpha_2^\dagger\alpha_2)]\\
\langle N_{F_1}\rangle &= Tr[\rho_{F_1}\alpha_1^\dagger\alpha_1]\\
\langle N_{F_2}\rangle &= Tr[\rho_{F_2}\alpha_2^\dagger\alpha_2]
\end{aligned} \tag{30.6}$$

Furthermore, the expectation value of the Pauli matrices σ_x^A and σ_x^B have been calculated as follows:

$$\begin{aligned}\langle I_A I_B\rangle &= Tr[\rho_S(\sigma_x^A \otimes \sigma_x^B)]\\ \langle I_A\rangle &= Tr[\rho_{S_1}\sigma_x^A]\\ \langle I_B\rangle &= Tr[\rho_{S_2}\sigma_x^B]\end{aligned} \tag{30.7}$$

These quantities are indicative of the currents in the SQUID rings in the adopted two-level model, because the current operator flowing in the SQUID ring A (B) is proportional to the magnetic flux through the loop $\phi_A = \frac{\phi_0}{2}\sigma_x^A$ $(\phi_B = \frac{\phi_0}{2}\sigma_x^B)$.

Furthermore, the expectation value of the magnetic flux $\alpha_i + \alpha_i^\dagger$ (i = 1, 2) induced by the two field modes have been calculated, as follows:

$$\begin{aligned}\langle \Phi_1\Phi_2\rangle &= Tr\{\rho_F[(\alpha_1+\alpha_1^\dagger)\otimes(\alpha_2+\alpha_2^\dagger)]\}\\ \langle \Phi_1\rangle &= Tr[\rho_{F_1}(\alpha_1+\alpha_1^\dagger)]\\ \langle \Phi_2\rangle &= Tr[\rho_{F_2}(\alpha_2+\alpha_2^\dagger)].\end{aligned} \tag{30.8}$$

Values of the parameters compatible with the current experimental setup have been chosen as $\omega_1 = \omega_2 = \Delta_A = \Delta_B = 1.3 \times 10^{-4}$, $\lambda_A = \lambda_B = 4.2 \times 10^{-5}$ in units where $\hbar = c = k_B = 1$[5,6]. In all our examples it is chosen

$$\rho_S(0) = |g\rangle\langle g| \otimes |e\rangle\langle e| \tag{30.9}$$

where $|g\rangle$ and $|e\rangle$ are the ground and excited states of the SQUID rings, respectively.

Microwaves in the separable (mixed) state is considered

$$\rho_{\rm sep} = \frac{1}{2}(|01\rangle\langle 01| + |10\rangle\langle 10|), \tag{30.10}$$

Also microwaves in the entangled state $|s\rangle = 2^{-1/2}(|01\rangle + |10\rangle)$ is considered. The density matrix of $|s\rangle$ is

$$\rho_{\rm ent} = \rho_{\rm sep} + \frac{1}{2}(|01\rangle\langle 10| + |10\rangle\langle 01|), \tag{30.11}$$

where $\rho_{\rm sep}$ is given by Eq. 30.10. It is seen that the $\rho_{\rm ent}$ and the $\rho_{\rm sep}$ differ only by the above nondiagonal elements.

In this example, the reduced density matrices are the same for both the separable and entangled states:

$$\rho_{\rm sep,A} = \rho_{\rm ent,A} = \rho_{\rm sep,B} = \rho_{\rm ent,B} = \frac{1}{2}(|0\rangle\langle 0| + |1\rangle\langle 1|). \tag{30.12}$$

In order to make the distinction between classically correlated and quantum mechanically correlated (entangled) currents, the entropic criterion of Ref. [8] is adopted. The conditional entropies have been calculated

$$E_A = S(\rho_S) - S(\rho_A);\ E_B = S(\rho_S) - S(\rho_B);\ S(\rho) = -Tr(\rho \ln \rho) \tag{30.13}$$

as criterion for entanglement. When $E_i < 0 (i = A, B)$ the system is entangled (although the converse is not true, i.e., an entangled system might have $E_i < 0$).

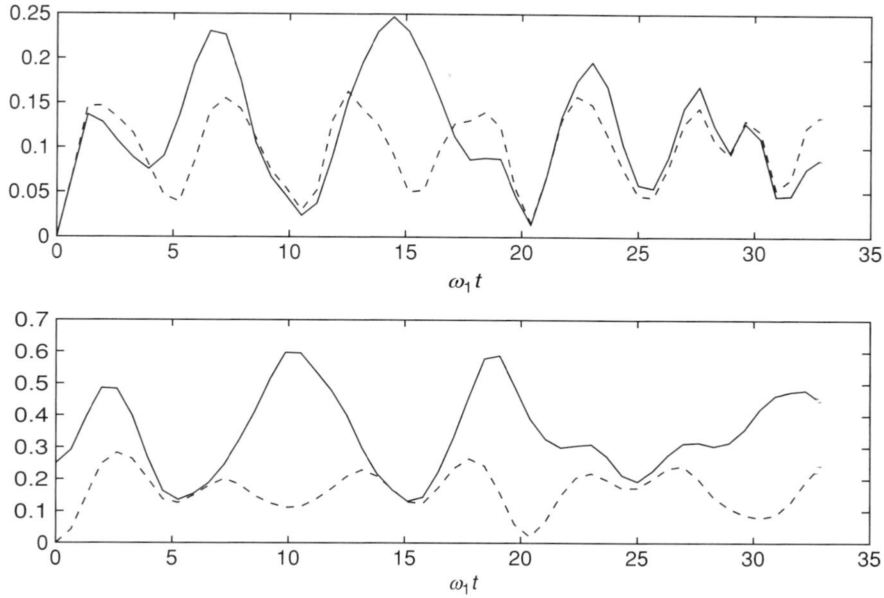

FIGURE 30.3. Time evolution of the system as function of dimensionless time $\omega_1 t$. At $t = 0$ system is in the state of Eq. (30.2) where ρ_F is in the separable state of Eq. (30.10). In the top figure we show $\langle N_A N_B \rangle$ (broken line) and $\langle N_A \rangle \langle N_B \rangle$ (solid line) as functions of dimensionless time; in the bottom figure we show $\langle N_{F_1} N_{F_2} \rangle$ (broken line) and $\langle N_{F_1} \rangle \langle N_{F_2} \rangle$ (solid line) as functions of dimensionless time.

In Fig. 30.3 we assume that at $t = 0$ the system is in the state given by Eq. 30.2 where ρ_F is in the separable state of Eq. 30.10. In the top figure it is shown $\langle N_A N_B \rangle$ (broken line) and $\langle N_A \rangle \langle N_B \rangle$ (solid line) as functions of dimensionless time $\omega_1 t$. The results show that $\langle N_A N_B \rangle$ is different from $\langle N_A \rangle \langle N_B \rangle$ (and in effect that the covariance $\langle N_A N_B \rangle - \langle N_A \rangle \langle N_B \rangle \neq 0$) and therefore average levels of the two SQUID rings are correlated. In the bottom figure $\langle N_{F_1} N_{F_2} \rangle$ (broken line) and $\langle N_{F_1} \rangle \langle N_{F_2} \rangle$ (solid line) as functions of dimensionless time is shown. It is seen that $\langle N_{F_1} N_{F_2} \rangle$ is different from $\langle N_{F_1} \rangle \langle N_{F_2} \rangle$ as expected because the electromagnetic field is produced to be correlated.

In Fig. 30.4 the results relative to the Josephson currents ($\langle I_A I_B \rangle$ top figure) and the electromagnetically induced magnetic fluxes ($\langle \Phi_1 \Phi_2 \rangle$ bottom figure) as functions of the dimensionless time $\omega_1 t$ are plotted. It is assumed that at $t = 0$ the system is in the state of Eq. (30.2) where ρ_F is in the separable state of Eq. (30.10) (solid line); and the entangled state of Eq. (30.11) (broken line). The results show that in the case of entangled electromagnetic fields the Josephson currents are also entangled.

In Fig. 30.5 the entropies E_A and E_B of Eq. (30.13) for the two SQUID rings are shown, as functions of the dimensionless time $\omega_1 t$. It is assumed that at $t = 0$ system is in the state of Eq. (30.2) where ρ_F is in the separable state of Eq. (30.10) (solid line); and the entangled state of Eq. (30.11) (broken line). In the case of entangled electromagnetic fields these quantities are negative and therefore the SQUID rings are entangled.

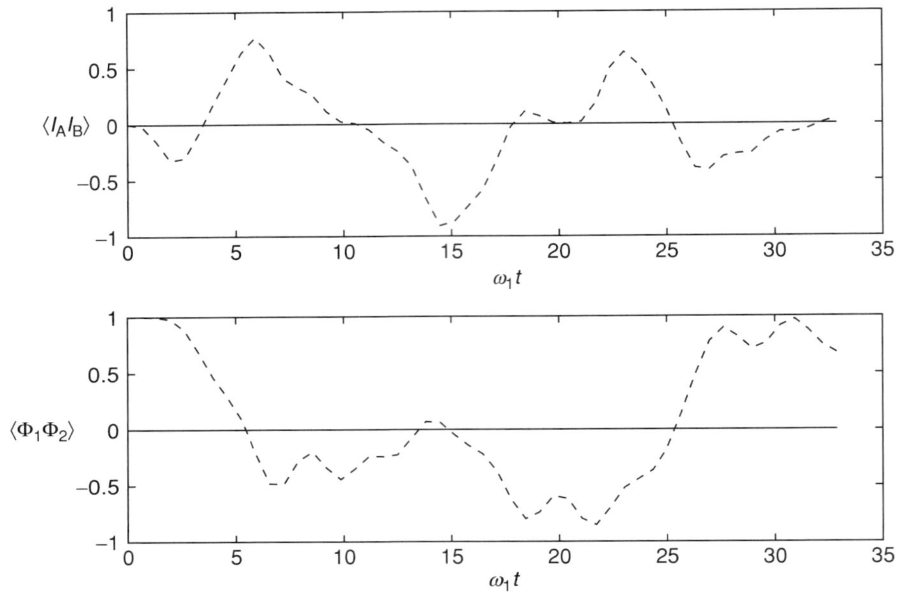

FIGURE 30.4. Time evolution of the system as function of dimensionless time $\omega_1 t$. At $t = 0$ system is in the state of Eq. (2) where ρ_F is in the separable state of Eq. (10) (solid line) and in the entangled state of Eq. (11) (broken line). In the top figure we show $\langle I_A I_B \rangle$ and in the bottom figure we show $\langle \Phi_1 \Phi_2 \rangle$

30.3 Discussions

The time evolution of a system that involves two distant SQUID rings A and B irradiated with two-mode nonclassical microwaves is studied. The system is described with the Hamiltonian of Eq. (30.1). We have considered microwaves which at $t = 0$ are entangled (quantum mechanically correlated); or separable (classically correlated). The corresponding density matrices differ in the off-diagonal elements, and we have shown that they affect some of the measurable quantities in the SQUID rings (Fig. (30.4)). In addition to that, it is shown through the calculation of entropic quantities (Fig. 30.5) that for entangled microwaves the corresponding Josephson currents are also entangled. In the case of classically correlated (separable) microwaves the corresponding Josephson currents are classically or quantum mechanically correlated (the entropies of Eq. (30.13) are positive).

The results are in the general context of the study of the effect of nonclassical microwaves on the operation of Josephson devices. (In authors' previous work it was shown how the quantum noise of the microwaves affects the Josephson currents.) In the present study how the entanglement of two-mode microwaves causes entanglement between Josephson currents in distant SQUID rings is shown. The work could be useful in the general context of quantum communications.

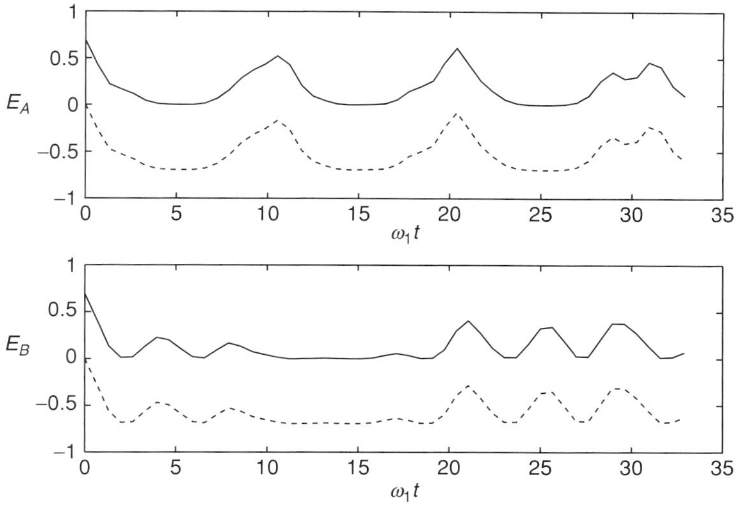

FIGURE 30.5. Entropies E_A (top), E_B (bottom) as functions of dimensionless time $\omega_1 t$. The system at $t = 0$ is in the state of Eq. (30.2) where ρ_F is in the separable state of Eq. (30.10) (solid line) and in the entangled state of Eq. (30.11) (broken line).

Acknowledgments. A. K. gratefully acknowledges financial support from Nuffield Institute under the NUF-NAL04 and Palermo University International Co-operation Division under the project 'Entanglement and decoherence in mesoscopic systems', 2003–2004.

References

1. R. F. Werner, *Phys. Rev. A* **40**, 4277 (1989); R. Horodecki and M. Horodecki, *ibid.* **54**, 1338 (1996); A. Peres, *Phys. Rev. Lett.* **77**, 1413 (1996); V. Vedral *et al.*, *ibid.* **78**, 2275 (1997); V. Vedral, *Rev. Mod. Phys.* **74**, 197 (2002).
2. Y. Makhlin, G. Schön, and A. Shnirman, *Rev. Mod. Phys.* **73**, 357 (2001); M. A. Kastner, *ibid.* **64**, 849 (1992); G. Schön and A.D. Zaikin, *Phys. Rep.* **198**, 237 (1990).
3. I. Chiorescu, Y. Nakamura, C. Harmans, and J. Mooij, *Science* **299**, 1869 (2003); Y. Nakamura, Y. A. Pashkin, and J. S. Tsai, *Nature* **398**, 786 (1999); C. H. van der Wal, *et al.*, *Science* **290**, 773 (2000).
4. A. Vourdas, *Phys. Rev.* **B49** 12040 1994. A. Konstadopoulou, J. M. Hollingworth, M. Everitt, A. Vourdas, T. D. Clark and J. F. Ralph, *IEE Proc. Science, Meas. Tech.*, 148, 229 (2001); A. Konstadopoulou, J. M. Hollingworth, M. Everitt, A. Vourdas, T. D. Clark and J. F. Ralph, *Eur. Phys. J.* **B32**, 279 (2003).
5. R. Migliore, A. Konstadopoulou, A. Vourdas, T.P. Spiller, A. Messina, *Phys. Lett.* **A319**, 67, (2003).
6. R. Migliore and A. Messina, *Phys. Rev.* **B67** (2003) 134505.
7. D. I. Tsomokos, C. C. Chong, and A. Vourdas, *Phys. Rev. A* **69**, 013810 (2004); *Phys. Rev. B*, to appear (2004).
8. N. J. Cerf and C. Adami, *Phys. Rev. Lett.* **77**, 5194 (1997)
9. N. J. Cerf and C. Adami, *Physica* **D120**, 62 (1998).

31

Phase diagram of dissipative two-dimensional Josephson junction arrays

LUCA CAPRIOTTI[1], ALESSANDRO CUCCOLI[2], ANDREA FUBINI[2], VALERIO TOGNETTI[3], AND RUGGERO VAIA[4]

[1] Kavli Institute for Theoretical Physics, University of California, Santa Barbara, CA 93106, USA.
[2] Dipartimento di Fisica dell'Universia di Firenze and Istituto Nazionale per la Fisica della Materia, UdR Firenze, Via G. Sansone 1, I-50019 Sesto Fiorentino (FI), Italy.
[3] Dipartimento di Fisica dell'Universita di Firenze, Istituto Nazionale per la Fisica della Materia, UdR Firenze, and Istituto Nazionale di Fisica Nucleare, Sez. di Firenze, Via G. Sansone 1, I-50019 Sesto Fiorentino (FI), Italy.
[4] Istituto di Fisica Applicata "N. Carrara" del CNR, Via Madonna del Piano, 1, I-50019 Sesto Fiorentino (FI), Italy. and Istituto Nazionale per la Fisica della Materia, UdR Firenze, Via G. Sansone 1, I-50019 Sesto Fiorentino (FI), Italy.

31.1 Introduction

A variety of condensed-matter systems with an intermediate (*mesoscopic*) scale have been recently developed. The characteristic quantum effects involving a macroscopic number of particles cause peculiar properties of all such devices, which make them appealing for applications. However the sizeable dimension of the devices implies that the relevant dynamical variables have to be considered as coupled to a very large (infinite) number of degrees of freedom of the surrounding *environment* (or *dissipation bath*). In these open systems, interaction with the environment leads in general to dissipation and decoherence, which strongly affect the behavior of the system. It is thus essential to study both quantum and dissipative effects. Two-dimensional (2D) arrays of nanosized Josephson junctions, both unshunted[1] and shunted[2] can indeed be produced, giving the opportunity to experimentally test the effects of dissipation.

31.2 Path Integral Monte Carlo for Dissipative Systems

The quantum dissipation is usually addressed within the Caldeira–Leggett (CL) framework.[3] Dissipation is described in terms of a linear coupling of the system with a bath of harmonic oscillators.[4–6] The path integral expression of the partition function of a quantum dissipative system,

$$\mathcal{Z} = \oint \mathcal{D}[\boldsymbol{q}]\, e^{-S[\boldsymbol{q}]}, \tag{31.1}$$

is given by the Euclidean action

$$S[\boldsymbol{q}] = \int_0^{\beta\hbar} \frac{\mathrm{d}u}{\hbar} \left[\frac{1}{2}\dot{\boldsymbol{q}}(u)\, \boldsymbol{A}\dot{q}(u) + V(\boldsymbol{q}(u)) \right] + S_{\mathrm{d}}[\boldsymbol{q}]. \tag{31.2}$$

The effects of dissipation are contained in the influence action,

$$S_{\mathrm{d}}[\boldsymbol{q}] = \frac{1}{2\hbar} \int_0^{\beta\hbar} \mathrm{d}u \int_0^{\beta\hbar} \mathrm{d}u' \boldsymbol{q}(u) \boldsymbol{K}(u-u')\, \boldsymbol{q}(u'). \tag{31.3}$$

Here, $\boldsymbol{q} \equiv \{q_i\}_{i=1,\ldots,M}$ denotes the vector whose components are the M coordinates of the investigated system and $\boldsymbol{A} \equiv \{\boldsymbol{A}_{ij}\}$ is the mass matrix. The kernel $\boldsymbol{K}(u) \equiv \{K_{ij}(u)\}$ is an $M \times M$ matrix, which is in general non local both in space and imaginary time. It depends on the temperature $T = (k_{\mathrm{B}}\beta)^{-1}$, is a symmetric and periodic function of the imaginary time u, $\boldsymbol{K}(u) = \boldsymbol{K}(-u) = \boldsymbol{K}(\beta\hbar - u)$, with vanishing average over a period.

When Josephson Junction Arrays (JJA) are considered, the influence action Eq. (31.2) is a good description of dissipative effects due to currents flowing to the substrate or through shunt resistances.[7] For such systems, an alternative dissipative mechanism due to radiative effects of the electromagnetic field was also suggested.[8,9] This leads to a so-called *anomalous* (or p-coupling) dissipation,[10] where the influence action Eq. (31.2) is replaced with:

$$S_{\mathrm{ad}}[p(u)] \int_0^{\beta\hbar} \frac{\mathrm{d}u}{2\hbar} \int_0^{\beta\hbar} \frac{\mathrm{d}u'}{\beta\hbar} \boldsymbol{p}(u)\boldsymbol{K}(u-u')\boldsymbol{q}(u'). \tag{31.4}$$

where $\boldsymbol{p}$ is the M-component vector of the momenta canonically conjugated to the coordinates $\boldsymbol{q}$. The physical consequences of the two different types of dissipation were investigated and discussed in Refs. [6, 11] by the effective potential method.[12] The relevant difference is that standard dissipation quenches the quantum fluctuations of the coordinates, driving their behavior towards the classical one, while the anomalous dissipation has an opposite effect, enhancing the quantum fluctuations of $\boldsymbol{q}$.

The standard Path-Integral Monte Carlo (PIMC) method, for computing the integral in Eq. (31.1), divides the imaginary-time interval $[0, \beta\hbar]$ into P slices of width $\epsilon = \beta\hbar/P$ where P is called *Trotter number*. The partition function $\mathcal{Z}$ is obtained as the $P \to \infty$ extrapolation of

$$\mathcal{Z}_P = C\beta^{-\frac{MP}{2}} \prod_{i=1}^{M} \int \mathrm{d}q_{i0} \int \prod_{\ell=1}^{P-1} \mathrm{d}q_{i\ell}\, e^{-S_P[\{\boldsymbol{q}_\ell\}]}, \tag{31.5}$$

where the path $\boldsymbol{q}\,(u)$ turns into the P discrete quantities $\boldsymbol{q}_\ell = q(\ell\epsilon)$, and $\dot{\boldsymbol{q}}(u) \to (P/\beta\hbar)(\boldsymbol{q}_\ell - \boldsymbol{q}_{\ell-1})$; $S_P[\boldsymbol{q}_\ell]$ represents the discretized form of the action Eq. (31.2) and C is a temperature-independent normalization.

The application of the standard PIMC approach to dissipative systems is made difficult by the fact that the kernel $\boldsymbol{K}(u-u')$ is a non local (and long ranged) function of the imaginary time.[5] Fourier path-integral approaches,[13] possibly supported by the partial-averaging scheme[14] can perform better: however, the evaluation of the integral over the continuous path they still involve is a serious shortcoming in the case of many-body systems.

It was proposed[15] to start from the finite-P expression in Eq. (31.5) of the standard PIMC for the partition function and make there a lattice (discrete) Fourier transform, changing the integration variables from q_ℓ to q_{ik} by setting:

$$\boldsymbol{q}_\ell = \bar{\boldsymbol{q}} + \sum_{k=1}^{P-1} \boldsymbol{q}_k e^{i2\pi \ell k/P}, \tag{31.6}$$

so that the discretized partition function reads:

$$\begin{aligned} \mathcal{Z}_P &= C\beta^{-\frac{PM}{2}} \prod_{i=1}^{M} \int d\bar{q}_i \int \prod_{\ell=1}^{P-1} dq_{ik} \\ &\times \exp\left\{ -\sum_{k=1}^{P-1} \boldsymbol{q}_k \left[\tfrac{2P^2}{\beta\hbar^2} \sin^2 \tfrac{\pi k}{P} \boldsymbol{A} + \tfrac{\beta}{2} \boldsymbol{K}_k \right] \boldsymbol{q}_k^* - \tfrac{\beta}{P} \sum_\ell V(\boldsymbol{q}_\ell) \right\}, \end{aligned} \tag{31.7}$$

where $\boldsymbol{K}_k \equiv \{K_{ij,k}\}$ is the Matsubara transform of the dissipation kernel matrix at the Matsubara frequency $\nu_k = 2\pi k/\beta\hbar$.

In order to get a reliable estimate of statistical averages in the thermodynamic limit ($M \to \infty$), finite-size effects have to be negligible; as a consequence the number M must be large enough, making the extrapolation to $P \to \infty$ problematic. However, such difficulty can be largely circumvented by making use of our knowledge of both the finite- and infinite-P exact partition function of pure bilinear actions.[16] According to Ref.[16], any PIMC estimate $G(P)$ of a given quantity G obtained at finite P can be corrected by adding the exact ($P \to \infty$) value $G_{\mathrm{HA}}^{(\mathrm{h})}$ and subtracting the finite-P estimate $G_{\mathrm{HA}}^{(\mathrm{h})}(P)$ of the same quantity. This can be done within the self-consistent harmonic approximation, getting:

$$G_{\mathrm{HA}}(P) = G(P) + \left[G_{\mathrm{HA}}^{(\mathrm{h})} - G_{\mathrm{HA}}^{(\mathrm{h})}(P) \right]. \tag{31.8}$$

As it has been clearly shown in the applications,[15] the last step turns out to be essential and truly effective in the investigation of many-body systems.

31.3 Josephson Junction Arrays

Two-dimensional (2D) JJA are one of the best experimental realizations of a model belonging to the XY universality class and permit to check and study a variety of phenomena related to both the thermodynamics and the dynamics of vortices. In these systems a Berezinskii–Kosterlitz–Thouless (BKT) transition[17] separates the low-temperature superconducting (SC) state from the normal (N) state, the latter displaying no phase coherence.[18] At nanoscale size of the junctions, the quantum fluctuations of the superconducting phases cause new interesting features. These appear to be the consequence of the non-negligible energy cost of charge transfer between SC islands. Indeed small capacitances are involved and the phase and charge are canonically conjugate variables. A relevant effect is the progressive reduction of the SC-N transition temperature as evidenced by experimental data[1] and confirmed for small quantum coupling g (see definition below) by a semiclassical investigation.[7]

Recently, arrays of nanosized junctions, both unshunted[1] and shunted[2], have given the opportunity to experimentally approach the quantum (zero-temperature) phase transition. However, the mechanism of suppression of the BKT in the neighborhood of the quantum critical point and its connection with the observed reentrance of the array resistance as function of the temperature is not yet clear.[18,19]

The JJA on the square lattice is described by a quantum XY model action:

$$S[\varphi] = \int_0^{\hbar\beta} \frac{du}{\hbar} \left\{ \sum_{\boldsymbol{ij}} \frac{\hbar^2 C_{\boldsymbol{ij}}}{8e^2} \dot{\varphi}_{\boldsymbol{i}}(u)\dot{\varphi}_{\boldsymbol{j}}(u) - \frac{E_{\mathrm{J}}}{2} \sum_{\boldsymbol{id}} \cos\varphi_{\boldsymbol{id}}(u) \right\}, \tag{31.9}$$

where $\varphi_{\boldsymbol{id}} = \varphi_{\boldsymbol{i}} - \varphi_{\boldsymbol{i}+\boldsymbol{d}}$ is the phase difference between the Josephson phases on the ith and the $(\boldsymbol{i} + \boldsymbol{d})$th neighboring superconducting islands, and the index $\boldsymbol{d}$ labels the four nearest-neighbour displacements. The capacitance matrix reads

$$C_{\boldsymbol{ij}} = C \left[\eta\delta_{\boldsymbol{ij}} + (z\delta_{\boldsymbol{ij}} - \sum_{\boldsymbol{d}} \delta_{\boldsymbol{i},\boldsymbol{j}+\boldsymbol{d}}) \right], \tag{31.10}$$

where $C_0 \equiv \eta C$ and C are, the self- and mutual capacitances of the islands respectively. The standard samples of JJA are well described by the limits $\eta \ll 1$. The quantum character of the array is determined by the Coulomb interaction between the Cooper pairs, described by the kinetic term through Josephson relation $\dot{\varphi}_{\boldsymbol{i}} = 2eV_{\boldsymbol{i}}$. The Josephson coupling term causes the Cooper-pair tunneling across the junctions.

The quantum fluctuations are ruled by the *quantum coupling* parameter $g = \sqrt{E_{\mathrm{C}}/E_{\mathrm{J}}}$, where $E_{\mathrm{C}} = (2e)^2/2C$ is the characteristic charging energy (for $\eta \ll 1$). In the following we use the *dimensionless temperature* $t \equiv k_{\mathrm{B}}T/E_{\mathrm{J}}$. We always assume the presence of at least a very weak Ohmic dissipation due to the currents flowing to the substrate or through shunt resistances[2]: this allows us to consider the phase as an extended variable.[18] Apart from this, dissipative effects are negligible provided that the shunt resistance $R_{\mathrm{S}} \gg R_{\mathrm{Q}}g^2/(2\pi t)$, where $R_{\mathrm{Q}} \equiv h/(2e)^2$ is the quantum resistance. For smaller R_{S} an explicit dissipative contribution must be added to the action Eq. (31.9), e.g., in the form of the Caldeira–Leggett term,[7,18] as given in Eq. (31.3), where $\varphi_{\boldsymbol{i}}(u)$ play the role of $\boldsymbol{q}_{\boldsymbol{i}}(u)$. The two situations are the cases of the experiments in Ref. [1] and Ref. [2], respectively, where an increase of the BKT transition temperature was found for increasing dissipation. This can be easily understood taking into account that the dissipative term in Eq. (31.3) results from the coupling of the phase $\varphi_{\boldsymbol{i}}$ with environmental variables constituting an implicit measurement of $\varphi_{\boldsymbol{i}}$.

31.4 Numerical Simulations and Phase Diagram

Using the PIMC algorithm described in Section 31.2, the dependence on g of the BKT transition temperature of the JJA model described by the action Eq. (31.9) with the additional dissipative term Eq. (31.3) is obtained.

The discretized path can be written as:

$$\varphi_{\boldsymbol{i},l} = \bar{\varphi}_{\boldsymbol{i}} + 2\sum_{k=1}^{N} \mathrm{Re}\left[\varphi_{\boldsymbol{i}k} e^{-i\frac{2\pi}{P}lk}\right] = \bar{\varphi}_{\boldsymbol{i}} + 2\sum_{k=1}^{N} \left[a_{\boldsymbol{i}k} \cos\frac{2\pi lk}{P} + b_{\boldsymbol{i}k} \sin\frac{2\pi lk}{P}\right], \tag{31.11}$$

where $\bar{\varphi}_{\boldsymbol{i}}$ is the zero-frequency component of the Euclidean path, and choosing an odd Trotter number $P = 2N + 1$. The advantage of using the transformed variables $\{\bar{\varphi}_{\boldsymbol{i}}, a_{\boldsymbol{i}k}, a_{\boldsymbol{i}k}\}$ is twofold: the influence action Eq. (31.3) becomes diagonal and PIMC sampling can be performed with an independent move amplitude on each frequency component.

Using Eq. (31.11), the JJA action Eq. (31.9) plus the dissipative term Eq. (31.3) reads

$$S[\varphi] = \sum_{\boldsymbol{ij}} \sum_{k=1}^{N} T_{\boldsymbol{ij},k}(a_{\boldsymbol{i}k}a_{\boldsymbol{j}k} + b_{\boldsymbol{i}k}b_{\boldsymbol{j}k}) - \frac{1}{2Pt} \sum_{\boldsymbol{id}} \sum_{l=1}^{P} \cos \varphi_{\boldsymbol{ij},\boldsymbol{l}}, \tag{31.12}$$

where $\varphi_{\boldsymbol{i}j,l} = \varphi_{\boldsymbol{i},l} - \varphi_{\boldsymbol{j},l}$ is to be expressed as in Eq. (31.11), and the 'kinetic' matrix

$$T_{\boldsymbol{ij},k} = \frac{P^2}{\beta e^2} C_{\boldsymbol{ij}} \sin^2 \frac{\pi k}{P} + \beta K_{\boldsymbol{ij},k}, \tag{31.13}$$

involves $K_{\boldsymbol{ij},k}$, i.e., the discrete FT of the dissipative kernel matrix $K_{\boldsymbol{ij}}(u)$. Any macroscopic thermodynamic quantity is obtained through its estimator as generated from the discretized action Eq. (31.12).

Actual simulations were made on $L \times L$ lattices (up to $L = 96$) with periodic boundary conditions; the move amplitudes were dynamically adjusted for each k-component; this procedure is very effective for reproducing the effect of strong quantum fluctuations in the high-g region, at difference with the standard PIMC algorithm which showed serious ergodicity problems, though eventually giving the same results. Furthermore, an over-relaxation algorithm for the zero-frequency mode proved to effectively reduce the autocorrelation times.[20–22]

In order to determine the transition temperature, a very sensitive method is provided by the scaling law of the helicity modulus Υ (a quantity proportional to the phase stiffness),

$$\Upsilon = \left(\frac{\partial^2 f(k_0)}{\partial k_0^2} \right)_{k_0=0}, \tag{31.14}$$

which measures the response of the dimensionless free energy per lattice site $f(k_0) = F(k_0)/(L^2 E_J)$ when a uniform twist k_0 along a fixed direction $\boldsymbol{u}$ is applied to the boundary conditions (i.e., $\varphi_{\boldsymbol{i}} \to \varphi_{\boldsymbol{i}} + k_0 \boldsymbol{u} \cdot \boldsymbol{i}$, with the unitary vector $\boldsymbol{u}$). The PIMC estimator for Υ is easily obtained, in analogy to that of Ref. [23], by derivation of the discretized path-integral expression of the free energy

$$f(k_0) = \frac{t}{L^2} \ln Z_P(k_0) \tag{31.15}$$

Eventually we get

$$\Upsilon_P = \frac{1}{L^2 P} \sum_{\boldsymbol{id}} \sum_{l=1}^{P} \cos \varphi_{\boldsymbol{id},l} - \frac{1}{2t L^2 P^2} \sum_{\boldsymbol{d}} \left(\sum_{\boldsymbol{i}} \sum_{l=1}^{P} \sin \varphi_{\boldsymbol{id},l} \right)^2. \tag{31.16}$$

Kosterlitz's renormalization group equations provide the critical scaling law for the finite-size helicity modulus Υ_L:

$$\frac{\Upsilon_L(t_{\mathrm{BKT}})}{t_{\mathrm{BKT}}} = \frac{2}{\pi} \left(1 + \frac{1}{2 \ln(L/L_0)} \right), \tag{31.17}$$

where L_0 is a non-universal constant. Following Ref. [24], the critical temperature t_{BKT} can be found by fitting $\Upsilon_L(t)/t$ vs L for several temperatures according to Eq. (31.17) with a further multiplicative fitting parameter $A(t)$. In this way, the critical point can be determined by searching the temperature such that $A(t_{\mathrm{BKT}}) = 1$, getting results like those reported in Fig. 31.1. This method allows for a precise identification of t_{BKT}: at temperature higher (lower) than the critical one the helicity modulus decreases (increases) much faster with L than $\Upsilon_L(t_{\mathrm{BKT}})$. At higher values of the quantum coupling, $g > g^*$, the helicity modulus scales to zero with $L \to \infty$ and $P \to \infty$ at any temperature.[25]

Systematic extrapolations in the Trotter number and in the lattice size have been done, in order to ascertain the good approach to the quantum and the thermodynamic limit. In particular, no anomaly was found in the finite-P behavior: the extrapolations in the Trotter number appear to be well-behaved, in the expected asymptotic regime $\mathcal{O}(1/P^2)$,[26] for $P \sim 60$. Moreover, the extrapolation to infinite lattice-size clearly indicates that Υ_L scales to zero at $t = 0.1$, while it remains finite and *sizeable* at $t = 0.2$. Therefore, we conclude that the reentrant behavior of the helicity modulus appears to be a genuine effect present in the model, rather than a finite-Trotter or finite-size artifact.

When the interaction of the phase variable with a heat bath, as given by Eq. (31.3), is present through a variable shunt resistance, the quantum phase fluctuations are decreased by the dissipation so that the BKT transition temperature rises. This was well reproduced by PQSCHA at low coupling.[7]

Two points must be noticed. The reentrance is present also with dissipation and disappears only with rather significant dissipation strength $R_Q/R_S \sim 0.15$. At the highest dissipation, $R_Q/R_S > 0.5$, a change in curvature is present and the critical temperature asymptotically vanishes.

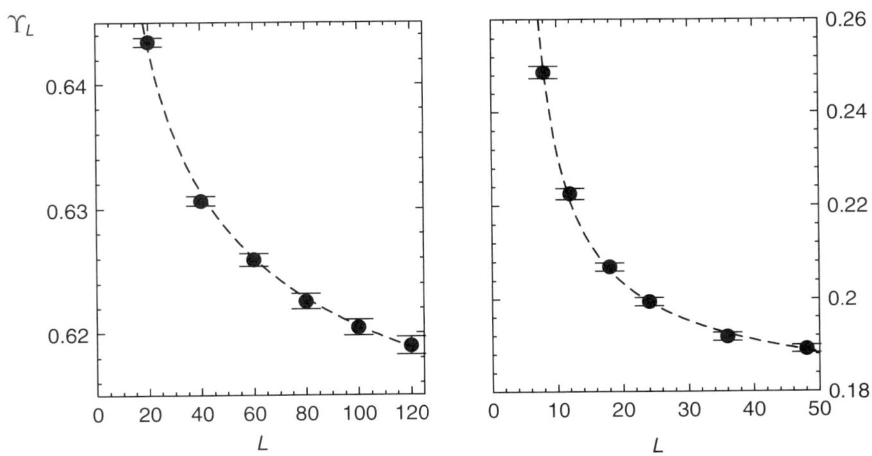

FIGURE 31.1. Size scaling of the helicity modulus Υ_L at the transition temperature. Symbols are PIMC data and the dashed-lines are the one-parameter fit with Eq. (31.17), i.e., with $A(t) = 1$. Left panel: $g = 0$ and $t = 0.892[L_0 = 0.456(6)]$; right panel: $g = 3.4$ and $t = 0.25[L_0 = 3.32(3)]$.

31.5 Discussion of the Results

In order to understand the physical reasons of the reentrance observed in the phase stiffness, the following two quantities have been studied:

$$\langle \cos \varphi_{ij} \rangle, \quad \Delta_{\varphi}^2 = \langle (\varphi_{ij}(u) - \bar{\varphi}_{ij})^2 \rangle, \tag{31.18}$$

where $\bar{\varphi}_{ij}$ is the phase average on the imaginary path, being ij nearest-neighbor sites. The first quantity, $\langle \cos \varphi_{ij} \rangle$, is a measure of the total (thermal plus quantum) short-range fluctuations of the Josephson phase and its maximum occurs where the overall fluctuations are lowest. The second quantity represents instead the pure-quantum spread of the phase difference between two neighboring islands and has been recently studied in the single junction problem;[27] more precisely, Δ_{φ}^2 measures the fluctuations around the static value (i.e., the zero-frequency component of the Euclidean path), it is maximum at $t = 0$ and tends to zero in the classical limit, i.e., $(g/t) \to 0$.

The quantities in Eq. (31.18) on a 8×8 lattice are compared in Fig. 31.3 for two values of the quantum coupling, in the semiclassical ($g = 1.0$) and in the extreme quantum ($g = 3.4$) regime. In the first case $\langle \cos \varphi_{ij} \rangle$ decreases monotonically by increasing t and the pure-quantum phase spread Δ_{φ}^2 shows a semiclassical linear behavior which is correctly described by the PQSCHA. At variance with this, at $g = 3.4$, where the reentrance of $\Upsilon(t)$ is observed, $\langle \cos \varphi_{ij} \rangle$ shows a pronounced maximum at finite temperature. Besides the qualitative agreement with the mean-field prediction of Ref. [28], a much stronger enhancement of the maximum above the $t = 0$ value is found. This single-junction effect, in a definite interval of the quantum coupling ($3.2 < g < 3.4$), is very effective to drive the reentrance of the phase stiffness. As for the transition in the region of high quantum fluctuations and low temperature, the open symbols in Fig. 31.2 represent the approximate location of the points (t, g) where $\Upsilon(t)$ becomes zero within the error bars: in their neighborhood no BKT-like scaling law was found. This fact opens two possible interpretations: (1) the transition does not belong to the *XY* universality class; (2) it does, and

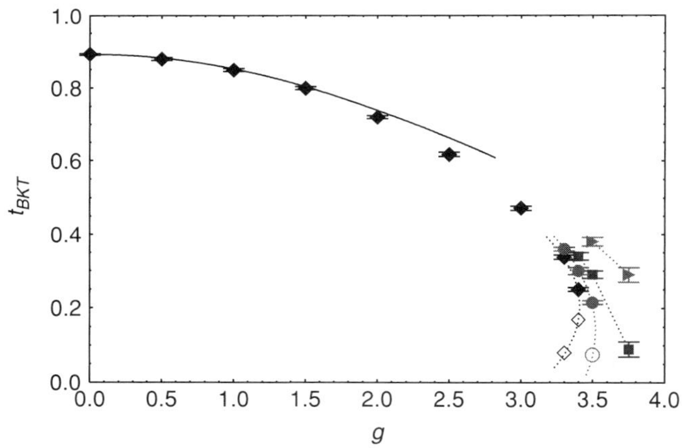

FIGURE 31.2. Phase diagram. BKT transition temperatures versus coupling g. Full rhombs represent the BKT transition temperatures for very low dissipations compared with the semiclassical approach (full line). The full symbols refer different intensities of the dissipation from $R_Q/R_S = 0.15$ (lowest), $R_Q/R_S = 0.25$ and $R_Q/R_S = 0.15$ (highest). Open symbols refer to reentrance phenomena for vanishing and $R_Q/R_S = 0.15$ dissipation.

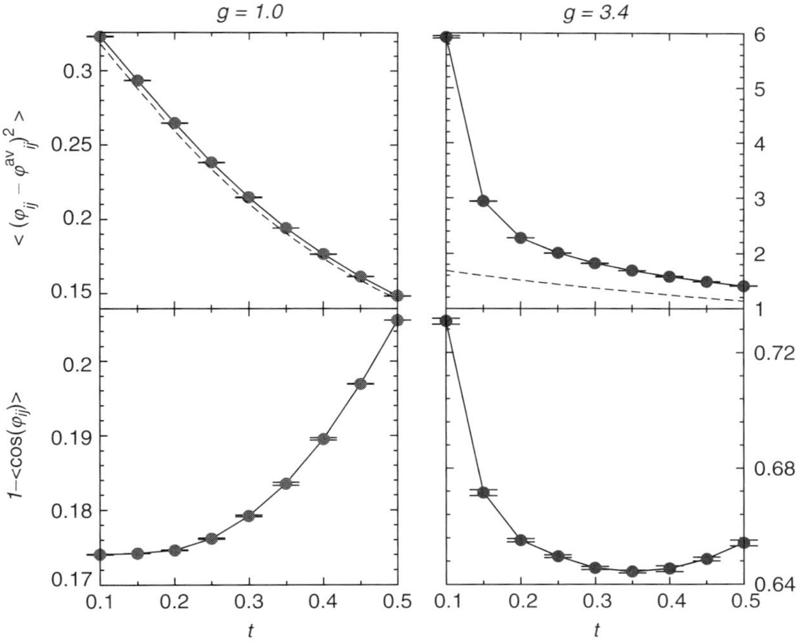

FIGURE 31.3. Behaviour of fluctuations at different values of coupling. Full circles refer to PIMC calculations. Dashed lines are the semiclassic results.

in this case the control parameter is not the (renormalized) temperature, but a more involute function of both t and g.

Acknowledgments. Fruitful discussions with G. Falci, R. Fazio and F. Guinea are gratefully acknowledged. We acknowledge support by INFN, INFM, and MIUR-COFIN2002.

References

1. H. S. J. van der Zant, W. J. Elion, L. J. Geerligs and J. E. Mooij, *Phys. Rev. B* **54**, 10081 (1996).
2. Y. Takahide, R. Yagi, A. Kanda, Y. Ootuka, and S. Kobayashi, *Phys. Rev. Lett.* **85**, 1974 (2000).
3. A. O. Caldeira and A. J. Leggett, *Ann. Phys. (NY)* **149**, 374 (1983).
4. G. W. Ford, M. Kac and P. Mazur, *J. Math. Phys.* **6**, 504 (1965).
5. see for instance: U. Weiss, *Quantum Dissipative Systems*, 2nd edn (World Scientific, Singapore 1999).
6. A. Cuccoli, A. Fubini, V. Tognetti, and R. Vaia, *Phys. Rev. B* **60** 231 (1999).
7. A. Cuccoli, A. Fubini, V. Tognetti, and R. Vaia, *Phys. Rev. B* **61**, 11289 (2000).
8. F. Sols and I. Zapata, in *New Developments on Fundamental Problems in Quantum Physics*, ed. by M. Ferrero and A. van der Merwe (Kluwer, Dordrecht, 1997).
9. Yu. Makhlin, G. Schoen and A. Shnirman, in *New Directions in Mesoscopic Physics*, eds R. Fazio, V.F. Gantmakher, Y. Imry (Kluwer, Dordrecht, 2003).
10. A. Leggett, *Phys. Rev. B* **30**, 1208 (1984).
11. A. Cuccoli, A. Fubini, V. Tognetti, and R. Vaia, *Phys. Rev. E* **64**, 066124-1 (2001).
12. R. Giachetti and V. Tognetti, *Phys. Rev. Lett.* **55**, 912 (1985).

13. J. D. Doll and D. L. Freeman, *J. Chem. Phys.* **80**, 2239, (1984).
14. M. Eleftheriou, J. D. Doll, E. Curotto and D. L. Freeman, *J. Chem. Phys.* **110**, 6657, (1999).
15. L. Capriotti, A. Cuccoli, A. Fubini, V. Tognetti and R. Vaia, *Europhys. Lett.* **58**, 155, (2002).
16. A. Cuccoli, A. Macchi, A. G. Pedrolli, V. Tognetti, and R. Vaia, *Phys. Rev. B* **51**, 12369, (1995).
17. V. L. Berezinskii, *Zh. Eksp. Teor. Fiz.* **59**, 907 (1970) [Sov. Phys. JEPT **32**, 493 (1971)]; J. M. Kosterlitz and D. J. Thouless, *J. Phys. C* **6**, 1181 (1973).
18. R. Fazio and H. S. J. van der Zant, *Phys. Rep.* **355**, 235 (2001).
19. H. M. Jaeger, D. B. Haviland, B. G. Orr, and A. M. Goldman, *Phys. Rev. B* **40**, 182 (1989).
20. L. Capriotti, A. Cuccoli, A. Fubini, V. Tognetti, and R. Vaia, *Physica Status Solidi* **B 237**, 23 (2003).
21. L. Capriotti, A. Cuccoli, A. Fubini, V. Tognetti, and R. Vaia, *Phys. Rev. Lett.* **91**, 247004 (2004).
22. F. R. Brown and T. J. Woch, *Phys. Rev. Lett.* **58**, 2394 (1987).
23. C. Rojas and J. V. Jose, *Phys. Rev. B* **54**, 12361 (1996). (1994).
24. K. Harada and N. Kawashima, *J. Phys. Soc. Jpn.* **67**, 2768 (1998); A. Cuccoli, T. Roscilde, V. Tognetti, R. Vaia, and P. Verruchi, *Phys. Rev. B* **67**, 104414 (2003).
25. P. Olsson, *Phys. Rev. Lett.* **73**, 3339 (1994); M. Hasenbusch and K. Pinn, *J. Phys. A* **30**, 63 (1997); S. G. Chung, *Phys. Rev. B* **60**, 11761 (1999).
26. M. Suzuki, *Quantum Monte Carlo Methods in Equilibrium and Nonequilibrium Systems*, ed. M. Suzuki (Springer-Verlag, Berlin, 1987).
27. C. P. Herrero and A. Zaikin, *Phys. Rev. B* **65** 104516 (2002).
28. P. Fazekas, B. Muhlschlegel, and Schroter, *Z. Phys. B* **57**, 193 (1984).

32

Persistent currents in a superconductor/ normal loop

MUN DAE KIM[1] AND JONGBAE HONG[2]

[1] Center for Theoretical Physics, Seoul National University, Seoul 151-742, Korea
mdkim@phya.snu.ac.kr
[2] School of Physics, Seoul National University, Seoul 151-742, Korea
jbhong@phya.snu.ac.kr

Abstract

The persistent currents of normal/superconducting loop modulated by the phase shift, η, of electron wave functions due to Andreev reflection process is studied. The persistent currents show the periodicity of the superconducting unit flux quantum, $\Phi_0/2$, regardless of the system parameters, when the energy splitting E is smaller than the pair potential Δ. In the limit, $E \ll \Delta$, the persistent currents take the Josephson type formula, $I \propto \sqrt{\Delta}\sin(\pi/2 - 2\eta)$.

32.1 Introduction

The long range proximity effect in the Normal/Superconductor (NS) hybrid junctions[1–3] has been studied in connection with the Andreev reflection process.Solving the Bogoliubov-de Gennes (BdG) equation with boundary conditions, the excitation spectrum and the currents flowing through the junction can be obtained. The electrons in normal sector becomes the Cooper pairs in the superconducting sector via the intermediate quasi particle states. The flux conservation condition, then, requires the currents of the electrons and of the quasi particles should have the same magnitude, which however has been missed in previous studies. For the Superconductor/Normal/Superconductor (SNS) junction the total energy of a pair of electrons has been considered to be fixed to the constant chemical potential. But, for the loop with the Aharonov–Bohm (AB) flux, it is well-known that the total energy is a function of the external flux, Φ_{ext}. Therefore, in calculating the persistent currents of a NS loop, the total energy as a variable is considered and the condition of minimum free energy is taken into account.

First, the case of the SNS junction is examined and that the current conservation condition cause only the change in the amplitude of currents is shown. However, for the NS loop, a quite different behavior is shown; the persistent currents have the periodicity of the superconducting flux quantum, $\Phi_0/2$ with $\Phi_0 \equiv h/e$, regardless of the system parameters, while it is so only for the limiting case of thin normal segment in the previous studies.[4–8] The result obtained here is consistent with the experiments of the Andreev interferometer. Moreover it is demonstrated that the persistent currents are modulated by the phase shift, 2η, in the Andreev reflection process at NS boundaries. In the low energy shift limit, the persistent currents take the Josephson type formula, $I/\sin(\pi/2 - 2\eta) \propto \sqrt{\Delta}$, in contrast to the case with insulating weak link where $I/\sin\phi \propto \Delta$[5,6] with the phase difference ϕ. Finally we point out that, when the external flux is zero, the degenerate states can be a good candidate for a qubit due to the low noise from the external magnetic field source.

32.2 SNS Junction

For SNS junction, the current through the junction has been studied previously,[2] where the wave function, $\Psi(z) = (u(z)\;\; v(z))^T$, is represented as $u(z) \sim Ae^{ik_0 z}(Ce^{i\lambda z})$ and $v(z) \sim Be^{ik_1 z}(De^{i\lambda z})$ in normal (superconducting) sector and the pair potential is $\Delta(z) = \Delta e^{i\chi_1}$ $(\Delta e^{i\chi_2})$ for $z < -d/2$ $(z > d/2)$ in superconducting sector and $\Delta(z) = 0$ for $-d/2 < z < d/2$ in normal sector. The currents can be obtained from the wave vector difference between the particle part and the hole part, $k_0 - k_1$, in normal sector, while the wave vectors, λ, of the quasi particle and of the quasi hole in superconducting sector are set to be the same.

The BdG equation which is given by

$$\begin{pmatrix} H_0 - \xi & \Delta(z) \\ \Delta^*(z) & -H_0^* + \xi \end{pmatrix} \begin{pmatrix} u(z) \\ v(z) \end{pmatrix} = E \begin{pmatrix} u(z) \\ v(z) \end{pmatrix}, \tag{32.1}$$

where $H_0 = -(\hbar^2/2m_e)(\partial/\partial z)^2$ and ξ is the chemical potential, can be solved with the phase matching conditions at $z = -d/2$ and $z = d/2$. However consider another boundary condition, that is, the conservation of total current represented by the flux, $j = -(\hbar/2im_e)[\Psi\partial\Psi^*/\partial z - \Psi^*\partial\Psi/\partial z]$, at the boundaries. For normal segment we obtain $j = (\hbar/2m_e)(k_0 - k_1)$, but for superconducting sector $j = 0$, which shows that the current is not conserved. Hence we introduce a correction in the wave function that the wave vectors of quasi particle, λ_e, and of quasi hole, λ_h, have generally different values.

For $E < \Delta$, the wave function $\Psi(z)$ can be written as follows,

$$\begin{pmatrix} u(z) \\ v(z) \end{pmatrix} = \begin{cases} \begin{pmatrix} Ae^{ik_0 z} \\ Be^{ik_1 z} \end{pmatrix} & (-d/2 < z < d/2) \\ \begin{pmatrix} Ce^{i\lambda_{+e}(z-d/2)+i\eta+i\chi_2/2} \\ Ce^{i\lambda_{+h}(z-d/2)-i\eta-i\chi_2/2} \end{pmatrix} & (d/2 \le z) \\ \begin{pmatrix} De^{i\lambda_{-e}(z+d/2)-i\eta+i\chi_1/2} \\ De^{i\lambda_{-h}(z+d/2)+i\eta-i\chi_1/2} \end{pmatrix} & (z \le -d/2). \end{cases} \tag{32.2}$$

The additional phase factor η in quasi particle wave function in superconductor is introduced to describe the phase shift due to the Andreev reflections at the NS interfaces.

The BdG equation in Eq. (32.1) can be solved easily in normal sector to give relations for ξ and E such that

$$\xi = \frac{\hbar^2}{4m_e}(k_0^2 + k_1^2), \tag{32.3}$$

$$E = \frac{\hbar^2}{4m_e}(k_0^2 - k_1^2), \tag{32.4}$$

with the mass of electrons m_e. In above equations the total energy of a pair of electrons in normal sector has the fixed value of ξ corresponding to the chemical potential and E is the energy splitting due to the wave vector difference, $k_0 - k_1$, produced by the phase difference $\chi \equiv \chi_1 - \chi_2$.

When a uniform current of quasi particles flows with a wave vector $2q$ in a superconductor, the BdG equation can be solved with the wave functions $u(z) = Ce^{i\lambda_e z}$ and $v(z) = Ce^{i\lambda_h z}$ and the pair potential $\Delta(z) = \Delta e^{i(\lambda_e - \lambda_h)z}$, where $\lambda_e - \lambda_h = 2q$.[1] Therefore, if the phases of pair potential are considered at boundaries such as $\Delta(-d/2) = \Delta e^{i\chi_1}$ and $\Delta(d/2) = \Delta e^{i\chi_2}$, the pair potential can be written by $\Delta(z) = \Delta e^{i(\lambda_{+e} - \lambda_{+h})(z-d/2)+i\chi_2}$ $(z > d/2)$ and $\Delta(z) = \Delta e^{i(\lambda_{-e} - \lambda_{-h})(z+d/2)+i\chi_1}$ $(z < -d/2)$.

In superconducting sector for $z \geq d/2$, the BdG equation becomes $[(\hbar^2/2m_e)\lambda_{+e}^2 - \xi] + \Delta e^{-2i\eta} = E$ and $\Delta e^{2i\eta} - [(\hbar^2/2m_e)\lambda_{+h}^2 - \xi] = E$. Representing λ_{+e} and λ_{+h} such as $\lambda_{+e} = \lambda_0 + i\lambda_0'$ and $\lambda_{+h} = \lambda_1 + i\lambda_1'$, we get

$$\xi = \frac{\hbar^2}{4m_e}(\lambda_0^2 + \lambda_1^2)(1 - \alpha) \tag{32.5}$$

and, by eliminating E in Eq. (32.4) and the above BdG equations, we obtain the equation

$$\frac{\hbar^2}{4m_e}(k_0^2 - k_1^2) = \frac{\hbar^2}{4m_e}(\lambda_0^2 - \lambda_1^2)(1 + \alpha) + \Delta \cos 2\eta \tag{32.6}$$

with $\alpha \equiv (m_e \Delta \sin 2\eta/\hbar^2 \lambda_0 \lambda_1)^2$. For $z \leq L$, we can also solve the BdG equation and find that $\lambda_{-e} = \lambda_0 - i\lambda_0'$ and $\lambda_{-h} = \lambda_1 - i\lambda_1'$, where $\lambda_0' = m_e \Delta \sin 2\eta/\hbar^2 \lambda_0$ and $\lambda_1' = m_e \Delta \sin 2\eta/\hbar^2 \lambda_1$.

The phase matching conditions for the wave functions in Eq. (32.2) at $z = d/2$ and $z = -d/2$ produce the equations,

$$A_+ e^{ik_0 d/2 - i\eta - i\chi_2/2} = B_+ e^{ik_1 d/2 + i\eta + i\chi_2/2} = C_+, \tag{32.7}$$

$$A_+ e^{-ik_0 d/2 + i\eta - i\chi_1/2} = B_+ e^{-ik_1 d/2 - i\eta + i\chi_1/2} = D_+. \tag{32.8}$$

The condition for there being a solution leads to the boundary condition

$$(k_0 - k_1)d - 4\eta + \chi = 2\pi n, \tag{32.9}$$

and the current conservation condition at the boundaries results the condition

$$k_0 - k_1 = \lambda_0 - \lambda_1. \tag{32.10}$$

Thus there is one more variable in λ's and one more equation, Eq. (32.10), compared to Ref. [2].

From the equations, Eqs. (32.3), (32.5), (32.6), (32.9) and (32.10), is numerically calculated as the current, $I = e\hbar(k_0 - k_1)n_e/m_e$, with the density of electrons n_e as a function of the phase difference χ as shown in Fig. 32.1. In Ref. [2]. the current is given by the relation, $E = \Delta \cos\phi \approx \hbar(k_0 - k_1)v/2$, where the phase factor ϕ represents the phase shift through the Andreev reflection process, $\phi = 2\eta$, and $v = \sqrt{2\xi/m_e}$. The current then can be written by $I = I_0(\Delta, \xi)\cos 2\eta$ with $I_0 \equiv en_e v$. The currents are plotted in Fig. 32.1, where the only difference between present calculation and Ref. [2]. is the amplitude of current; the deviation is prominent for small d and G. Since $(k_0 - k_1) \propto 1/d$ as can be seen from the boundary condition in Eq. (32.9) and thus $E \propto 1/d^2$, it is seen that the differences become negligible for low energy shift, $E \ll \Delta$.

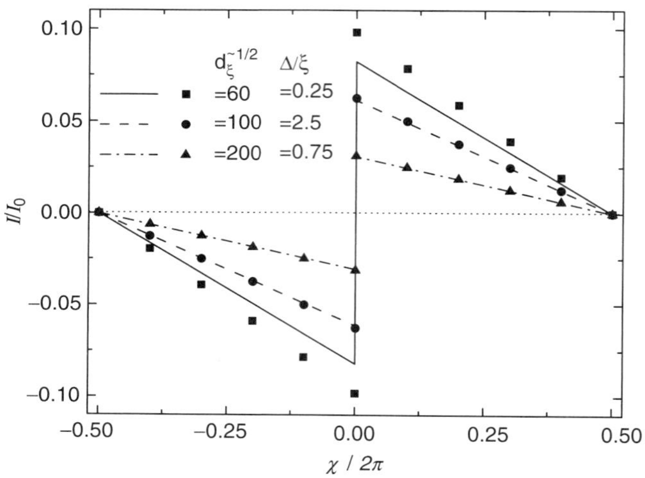

FIGURE 32.1. The calculated currents of NSN junction, where $\tilde{\xi} \equiv 2m\xi/\hbar^2$. The scattered marks denote the currents obtained from the equations in Ref. [2].

Until now we have the constraint that the total energy in Eq. (32.3) should have constant value corresponding to the chemical potential ξ independent of the phase difference χ. However in loop system it seems that this constraint is not natural as will be seen in followings.

32.3 NS Loop

For NS loop with piercing magnetic flux $\Phi_{\rm ext}$, H_0 in the BdG equation in Eq. (32.1) becomes $H_0 = (-i\hbar\partial/\partial z - eA/c)^2/2m_e$ with the vector potential $A = \Phi_{\rm ext}/L$ and the circumference of loop L. First consider a superconducting loop without junction penetrated by an AB flux, where the electron-like wave function $u(z)$ and the hole-like wave function $v(z)$ can be represented with the gauge factor such as $u(z) = Ce^{i(\lambda_e+2\pi f/L)z}$ and $v(z) = Ce^{i(\lambda_h-2\pi f/L)z}$ with $f \equiv \Phi_{\rm ext}/\Phi_0 = eAL/h$. The uniform flow of the persistent current is described by the BdG equation with the pair potential, $\Delta(z) = \Delta e^{i(\lambda_e-\lambda_h+2\pi f/L)z}$.[1]

For a superconducting loop with a normal segment inserted, the wave function Ψ of the electrons in normal sector ($0 < z < d$) and the quasi particles in superconducting sector ($d < z < L$) are given by

$$\begin{pmatrix} u(z) \\ v(z) \end{pmatrix} = \begin{cases} \begin{pmatrix} Ae^{i(k_0+\frac{2\pi}{L}f)z} \\ Be^{i(k_1-\frac{2\pi}{L}f)z} \end{pmatrix} & (0 < z < d) \\ \begin{pmatrix} Ce^{i(\lambda_{+e}+\frac{2\pi}{L}f)z+i\eta} \\ Ce^{i(\lambda_{+h}-\frac{2\pi}{L}f)z-i\eta} \end{pmatrix} & (d \le z) \\ \begin{pmatrix} De^{i(\lambda_{-e}+\frac{2\pi}{L}f)z-i\eta} \\ De^{i(\lambda_{-h}-\frac{2\pi}{L}f)z+i\eta} \end{pmatrix} & (z \le L). \end{cases} \tag{32.11}$$

The BdG equation with the wave function in Eq. (32.11) results the same equations as Eqs. (32.3) and (32.4) for normal sector. In superconducting sector for $z \geq d$ the BdG equation can be solved with the pair potential represented as $\Delta(z) = \Delta e^{i(\lambda_{+e}-\lambda_{+h})z+2\pi f/L}$. For SNS junction the total energy of electrons is constrained to a constant chemical potential as in Eq. (32.3). But in the loop system it is well known that the total free energy, $F(\Phi_{\rm ext})$, of the system varies with the external flux, $\Phi_{\rm ext}$. Hence we cannot constrain the total energy in Eq. (32.3) a constant ξ any more and would rather treat it as a function of f. Eliminating ξ from Eqs. (32.3) and (32.5) we obtain the equation,

$$\frac{1}{2}(k_0^2 + k_1^2) = \frac{1}{2}(\lambda_0^2 + \lambda_1^2)(1 - \alpha). \tag{32.12}$$

The phase matching conditions for the wave functions in Eq. (32.11) at $z = d$ and $z = L$ are obtained similarly,

$$\hat{A}e^{ik_0d - i\lambda_0 d - i\eta} = \hat{B}e^{ik_1d - i\lambda_1 d + i\eta} = \hat{C}, \tag{32.13}$$

$$\hat{A}e^{-i(\lambda_0 + \frac{2\pi}{L}f)L + i\eta} = \hat{B}e^{-i(\lambda_1 - \frac{2\pi}{L}f)L - i\eta} = \hat{D}, \tag{32.14}$$

where $\hat{A}$, $\hat{B}$, $\hat{C}$ and $\hat{D}$ represent the phase part of the coefficients, A, B, C and D, respectively. Then the boundary condition becomes

$$\left(k_0 - k_1 + \frac{2\pi}{L}2f\right)d + \left(\lambda_0 - \lambda_1 + \frac{2\pi}{L}2f\right)(L - d) - 4\eta = 2\pi n. \tag{32.15}$$

The total current is represented by the flux, $j = -(1/2m)[\Psi(-i\hbar\partial/\partial z - eA/c)\Psi^* - \Psi^*(-i\hbar\partial/\partial z - eA/c)\Psi]$, and the current conservation condition at the boundaries becomes the same as the equation in Eq. (32.10), $k_0 - k_1 = \lambda_0 - \lambda_1 \equiv 2q$. Thus the boundary condition in Eq. (32.14) can be rewritten as $(k_0 - k_1)L - 4\eta + 4\pi f = 2\pi n$, where the external flux $4\pi f$ plays the role of the phase difference χ in Eq. (32.9) of the SNS junction.

By minimizing the free energy another condition can be obtained to determine the five dynamic variables, $k_0, k_1, \lambda_0, \lambda_1$, and η.[9] In superconducting sector the Cooper pairs carry the persistent current corresponding to the Cooper pair wave vector $2q = \lambda_0 - \lambda_1$ and the energy of a Cooper pair can be written as $(\hbar^2/2m_c)(2q)^2$ with $m_c = 2m_e$. Therefore the total free energy per particle can be written as $U_{\rm tot} = (\hbar^2/2m_e)(1/2)(k_0^2+k_1^2)(d/L)+(1/2)(\hbar^2/2m_c)(\lambda_0-\lambda_1)^2(1-d/L)-U_0$, where the constant U_0 is the minimum total energy. ¿From the condition $\mathrm{d}U_{\rm tot}/\mathrm{d}\eta = 0$, we obtained the equation,

$$\left[(\lambda_0 + \lambda_1) + (\lambda_0 - \lambda_1)\frac{L}{d} + \frac{(2m\Delta\sin 2\eta)^2}{2\lambda_0^3\hbar^4}\right]\frac{\mathrm{d}\lambda_0}{\mathrm{d}\eta}$$
$$+ \left[(\lambda_0 + \lambda_1) - (\lambda_0 - \lambda_1)\frac{L}{d} + \frac{(2m\Delta\sin 2\eta)^2}{2\lambda_1^3\hbar^4}\right]\frac{\mathrm{d}\lambda_1}{\mathrm{d}\eta}$$
$$= \frac{\lambda_0^2 + \lambda_1^2}{2\lambda_0^2\lambda_1^2}\left(\frac{2m\Delta}{\hbar^2}\right)^2 \sin 4\eta, \tag{32.16}$$

where the expressions for $\mathrm{d}\lambda_0/\mathrm{d}\eta$ and $\mathrm{d}\lambda_1/\mathrm{d}\eta$ can be obtained by differentiating Eqs. (32.6), (32.10), (32.12) and (32.14). This equation corresponds to the current relation in Ref. [9] which is the relation between the momentum of particles and the phase difference across the junction at free energy minimum.

Solving the coupled equations Eqs. (32.5), (32.10), (32.12), (32.14) and (32.15) numerically, the excitation spectrum and the persistent current of the loop as a function of external flux f are plotted in Fig. 32.2 (a) and (b). The persistent current, $I = 2e\hbar 2qn_c/m_c$, with the density of Cooper pairs n_c can be written by $I = I_0 2q/\sqrt{\tilde{\Delta}_0}$, where $I_0 \equiv 2en_c\hbar\sqrt{\tilde{\Delta}_0}/m_c$ and $\tilde{\Delta}_0 \equiv 2m\Delta_0/\hbar^2$ is a gap potential chosen arbitrary. In Fig. 32.3 the persistent current for ground state are plotted for several values of L and Δ, which shows that the amplitude of the persistent current depends on the value of L and Δ while the period has the fixed value of $0.5\Phi_0$. In the previous studies another periodicity becomes developed as the length of the normal segment, d, increases.[7,8] However, to our knowledge, there has been no experimental evidence for the sub-periodicity. Actually in the experiments for the Andreev interferometer[10] the oscillation period is exactly the superconducting flux quantum regardless of the system parameters, which is consistent with present results.

Considering the sign of amplitude changes at the point where 2η is $\pi/2$ as can be seen in Fig. 32.2 (b) and (c) and the Δ dependence in Fig. 32.3, we can fit the data with a simple formula

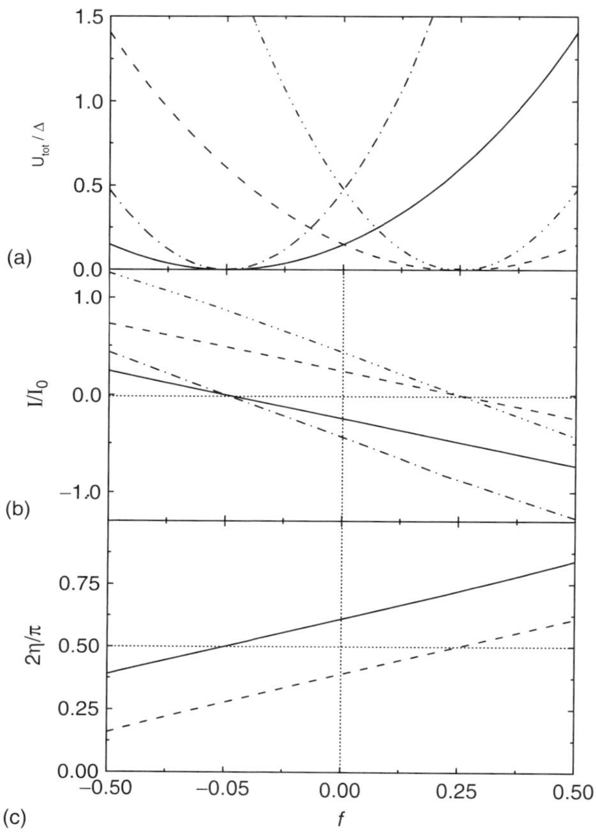

FIGURE 32.2. (a) Excitation spectrum of the NS loop. The solid (dashed) line for $f < 0$ ($f > 0$) shows the ground state energy. (b) Persistent currents corresponding to the energy levels in (a). (c) Phase shifts at SN boundaries. The solid (dashed) line corresponds to the energy level in (a) and the persistent current in (b) represented by the solid (dashed) lines.

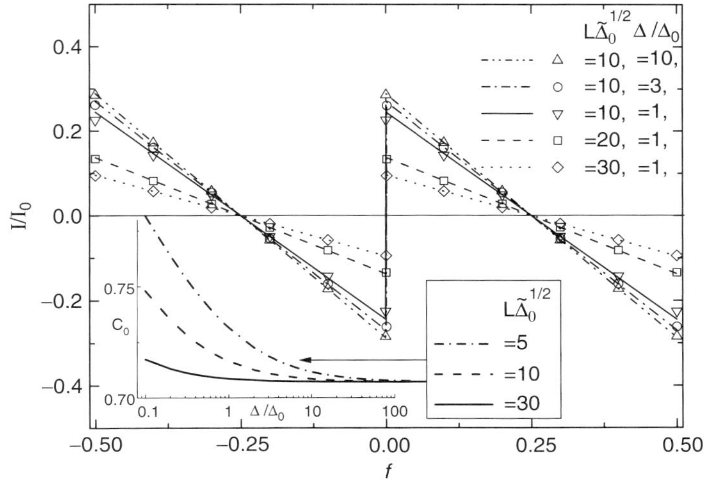

FIGURE 32.3. Ground state persistent currents, which correspond to solid (dashed) line for $f < 0$ ($f > 0$) in Fig. 32.2 (b). The scattered marks are given by the current relation in Eq. (32.17). Inset shows the constant C_0 converges to about 0.707.

between the current and the phase shift such that

$$I = C_0 I_0 \sqrt{\frac{\Delta}{\Delta_0}} \sin(\pi/2 - 2\eta(L, \Delta, f)), \tag{32.17}$$

where C_0 is a constant and the dependence of amplitude on L is included through $\eta(L, \Delta, f)$. In the inset of Fig. 32.3 the amplitude of persistent currents for several values of L as a function of Δ/Δ_0 at $f = 0.45$ is plotted, where we can determine the value of constant, $C_0 \approx 0.707 \approx 1/\sqrt{2}$. The persistent currents in Eq. (32.17) with the constant C_0 are represented as scattered graphs in Fig. 32.3. Here it is seen that the relation in Eq. (32.17) is well satisfied when $1/L\sqrt{\tilde{\Delta}_0} \ll \Delta/\Delta_0$, that is, for low energy splitting, $E \ll \Delta$, since $E \propto 1/L^2$. The main difference between the present SNS junction and the Josephson tunnel junction is in the dependence on gap function; $I/\sin(\pi/2 - 2\eta) \propto \sqrt{\Delta}$ in the SNS junction while $I/\sin\phi \propto \Delta$ for Josephson tunnel junctions.[1]

Recently quantum computing with the persistent current qubit using the two current states of the loop with Josephson junctions operating at $f \approx 0.5$ has been actively studied.[11] But, due to the severe decoherence problem, the two qubit coherent oscillation has not yet been achieved experimentally. One of the causes for decoherence is the noise due to the external magnetic field source. As can be seen in Fig. 32.2 (a) there are two degenerate ground states at $f = 0$. Thus the two current states in NS loop can be a qubit with a much low decoherence from the external magnetic field source, since it can operate at very small magnetic flux, $f \approx 0$.

In summary, we study the periodicity and the amplitude of the currents of the NSN junction and the NS loop. The currents are determined by the Andreev reflection phase shift 2η to give the periodicity of $\Phi_0/2$ regardless of system parameters and the dependence on the gap potential, $\sqrt{\Delta}$. It can be said that the periodicity comes from the current conservation condition put together with the phase matching condition. Taking the total energy as a variable and minimizing the free energy is not necessary for obtaining the periodicity, but it is indispensable for the amplitude dependence on gap potential. The results show that it can be used as a qubit with low decoherence.

References

1. de Gennes P. G. (1966). *Superconductivity of Metals and Alloys* New York:Benjamin.
2. Kulik I. O. (1969). Zh. Eksp. Teor. Fiz. **57**, 1745 [(1970). Sov. Phys. JETP **30**, 944].
3. Blonder G. E., Tinkham M., and Klapwijk T. M. (1982). Phys. Rev. B **25**, 4515.
4. Andreev A. F. (1964). Zh. Eksp. Teor. Fiz. **46**, 1823 [(1964). Sov. Phys. JETP **19**, 1228]; (1965). **49**, 655 [(1966). **22**, 455].
5. Tinkham M. (1996). *Introduction to Superconductivity* New York:McGraw-Hill.
6. Ambegaokar V. and Baratoff A. (1963). Phys. Rev. Lett. **10**, 486.
7. Büttiker M. and Klapwijk T. M. (1986). Phys. Rev. B **33**, 5114.
8. Cayssol J., Kontos T., and Montambaux G. (2003). Phys. Rev. B **67**, 184508.
9. Kim M. D., Shin D., and Hong J. (2003). Phys. Rev. B **68**, 134513.
10. Reulet B., Kozhevnikov A. A., Prober D. E., Belzig W., and Nazarov Yu. V. (2003). Phys. Rev. Lett. **90**, 066601;Charlat P., Courtois H., Gandit Ph., Mailly D., Volkov A. F., and Pannetier B. (1996). *ibid.* **77**, 4950; Pothier H., Guéron S., Esteve D., and Devoret M. H. (1994). *ibid.* **73**, 2488.
11. van der Wal Caspar H. *et al.* (2000). Science **290**, 773; Chiorescu I., Nakamura Y., Harmans C. J. P. M., and Mooij J. E. (2003). *ibid.* **299**, 1869.

33

Josephson junction ladders: a realization of topological order

GERARDO CRISTOFANO[1], VINCENZO MAROTTA[1], ADELE NADDEO[2], AND PASQUALE SODANO[3]

[1] Dipartimento di Scienze Fisiche, Universitá di Napoli "Federico II"
and INFN, Sezione di Napoli, Via Cintia - Compl. universitario M. Sant'Angelo - 80126 Napoli, Italy

[2] Dipartimento di Scienze Fisiche, Universitá di Napoli "Federico II"
and Coherentia-INFM, Unità di Napoli,Via Cintia - Compl. universitario M. Sant'Angelo - 80126 Napoli, Italy

[3] Dipartimento di Fisica, Universitá di Perugia and INFN, Sezione di Perugia, Via A. Pascoli - 06123 Perugia, Italy

Abstract

How a one-dimensional fully frustrated ladder of quantum Josephson junctions may develop quantum order is shown here. Such a property is crucial for its implementation as a "protected" solid state qubit.

33.1 Introduction

The concept of topological order was first introduced to describe the ground state of a quantum Hall fluid.[1] Although todays interest in topological order mainly derives from the quest for exotic non-Fermi liquid states relevant for high T_c superconductors,[2] the concept is of much more general interest.[3]

Two features of topological order are very striking: fractionally charged quasiparticles and a ground state degeneracy depending on the topology of the underlying space, which is lifted by quasiparticles tunnelling processes. For Laughlin Fractional Quantum Hall (FQH) states both these properties are well understood,[4] but for superconducting devices the situation is less clear.

Josephson junction networks may be regarded as good candidates for exhibiting topological order; in fact some studies evidence this possibility[5,6] within a Chern–Simons gauge field approach. Such a property may allow for their use as devices for quantum computation. In this contribution it is shown that fully frustrated Josephson Junction Ladders (JJL) may support topological order making use of conformal field theory techniques.[7] In particular the ground state is shown to be degenerate, the different states being accessible by adiabatic flux change

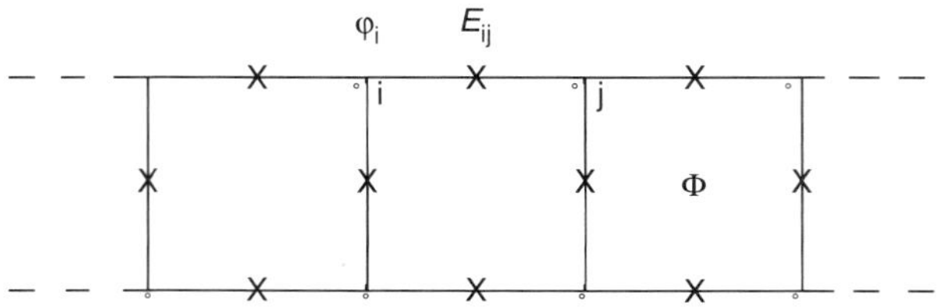

FIGURE 33.1. Josephson junction ladder

techniques. Such a degeneracy appears to be strictly related to the presence in the spectrum of quasiparticles with non abelian statistics and can be lifted non perturbatively through vortices tunnelling.

33.2 Josephson Junction Ladders

In this Section the system under study is briefly described in the following, that is a ladder of Josephson junctions (see Fig. 33.1) in the presence of a defect. With each side i we associate a phase φ_i and a charge $2en_i$, representing a superconducting grain coupled to its neighbours by Josephson couplings; n_i and φ_i are conjugate variables satisfying the usual phase-number commutation relation. The Hamiltonian describing the system is given by the Quantum Phase Model (QPM):

$$H = -\frac{E_C}{2}\sum_i \left(\frac{\partial}{\partial\varphi_i}\right)^2 - \sum_{\langle ij\rangle} E_{ij}\cos\left(\varphi_i - \varphi_j - A_{ij}\right), \tag{33.1}$$

where $E_C = \frac{(2e)^2}{C}$ (C being the capacitance) is the charging energy at site i, while the second term is the Josephson coupling energy between sites i and j and the sum is over nearest neighbours. $A_{ij} = \frac{2\pi}{\Phi_0}\int_i^j A\cdot dl$ is the line integral of the vector potential associated to an external magnetic field B and $\Phi_0 = \frac{hc}{2e}$ is the magnetic flux quantum. The gauge invariant sum around a plaquette is $\sum_p A_{ij} = 2\pi f$ with $f = \frac{\Phi}{\Phi_0}$, where Φ is the flux threading each plaquette of the ladder.

The phase fields are Labeled on the two legs with $\varphi_i^{(a)}$, $a = 1, 2$ and it is assumed that $E_{ij} = E_x$ for horizontal links and $E_{ij} = E_y$ for vertical ones. Also let the gauge choice $A_{ij} = +\pi f$ for the upper links, $A_{ij} = -\pi f$ for the lower ones and $A_{ij} = 0$ for the vertical ones, which corresponds to a vector potential parallel to the ladder and taking opposite values on upper and lower branches. So the phase fields $\varphi_i^{(a)}$, $a = 1, 2$, have opposite chiralities.

The correspondence between the effective quantum Hamiltonian Eq. (33.1) and the Twisted Model (TM) can be best shown performing the change of variables:[8] $\varphi_i^{(1)} = X_i + \phi_i$, $\varphi_i^{(2)} = X_i - \phi_i$, so getting:

$$H = -\frac{E_C}{2}\sum_i \left[\left(\frac{\partial}{\partial X_i}\right)^2 + \left(\frac{\partial}{\partial\phi_i}\right)^2\right] - \sum_i \left[2E_x\cos\left(X_{i+1} - X_i\right)\cos\left(\phi_{i+1} - \phi_i - \pi f\right) + E_y\cos\left(2\phi_i\right)\right] \tag{33.2}$$

where X_i, ϕ_i (i.e., $\varphi_i^{(1)}$, $\varphi_i^{(2)}$) are only phase deviations of each order parameter from the commensurate phase and should not be identified with the phases of the superconducting grains.[8]

When $f = \frac{1}{2}$ and $E_C = 0$ (classical limit) the ground state of the 1D frustrated quantum XY (FQXY) model displays — in addition to the continuous $U(1)$ symmetry of the phase variables — a discrete Z_2 symmetry associated with an antiferromagnetic pattern of plaquette chiralities $\chi_P = \pm 1$, measuring the two opposite directions of the supercurrent circulating in each plaquette. Thus it has two symmetric, energy degenerate, ground states characterized by currents circulating in the opposite directions in alternating plaquettes.

Performing the continuum limit of the Hamiltonian (33.2):

$$-H = \frac{E_C}{2}\int dx\left[\left(\frac{\partial}{\partial X}\right)^2 + \left(\frac{\partial}{\partial\phi}\right)^2\right]$$
$$+\int dx\left[E_x\left(\frac{\partial X}{\partial x}\right)^2 + E_x\left(\frac{\partial\phi}{\partial x} - \frac{\pi}{2}\right)^2 + E_y\cos(2\phi)\right] \tag{33.3}$$

we see that the X and ϕ fields are decoupled. In fact the X term of the above Hamiltonian is that of a free quantum field theory while the ϕ one coincides with the quantum sine-Gordon model. Through an imaginary-time path-integral formulation of such a model it can be shown that the 1D quantum problem maps into a 2D classical statistical mechanics system, the 2D fully frustrated XY model, where the parameter $\alpha = (\frac{E_x}{E_C})^{\frac{1}{2}}$ plays the role of an inverse temperature.[8] For small E_C there is a gap for creation of kinks in the antiferromagnetic pattern of χ_P and the ground state has quasi long range chiral order.

The modified ladder is now ready to be introduce[7] (see Fig. 33.2). In order to do so it is first required that the $\varphi^{(a)}$, $a = 1, 2$ variables recover the angular nature by compactification of both the up and down fields. In such a way the XY-vortices, causing the Kosterlitz–Thouless transition, are recovered. Also let us indicate the compactified phases $\varphi^{(1)}$, $\varphi^{(2)}$ as φ_{L}, φ_{R} respectively (where L, R stay for left, right components). As a second step at a point $x = 0$ a magnetic impurity is introduced, which couples the up and down edges through its interaction with the Cooper pairs of the two edges. In the limit of strong coupling such an interaction gives rise to non trivial boundary conditions for the fields:

$$\varphi_{\mathrm{L}}^{(1)}(x=0) = \mp\varphi_{\mathrm{R}}^{(2)}(x=0) - \varphi_0. \tag{33.4}$$

It is interesting to notice that such a condition is naturally satisfied by the twisted field $\phi(z)$ of the TM (see Eq. (33.7)). In the following further details are given on such an issue, in particular

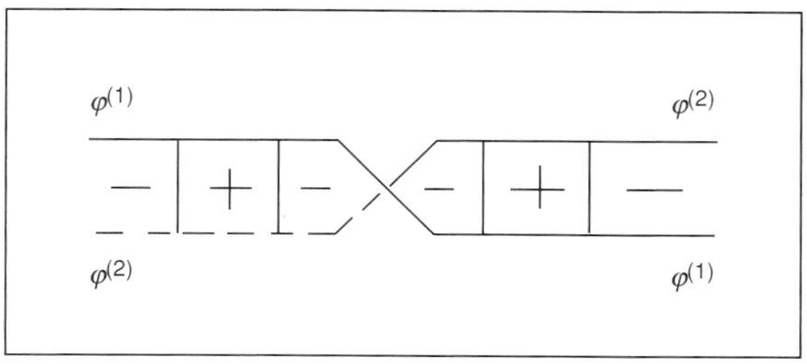

FIGURE 33.2. Josephson junction ladder with an impurity

the m-reduction technique[9,10] is adopted, which accounts for non trivial boundary conditions[11] for the Josephson ladder in the presence of an impurity.

33.3 m-reduction Technique

In this Section we focus on the m-reduction technique for the special $m = 2$ case and apply it to the system described by the Hamiltonian (33.3). In the previous Section a chirality was assigned to each of the two legs of the ladder, making a correspondence between up-down leg and left-right chirality states. Now we identify in the continuum the corresponding phase fields $\varphi^{(a)}$, $a = 1, 2$, each defined on the corresponding leg, with the two chiral fields $Q^{(a)}$, $a = 1, 2$ of the CFT, the TM, with central charge $c = 2$.

In order to construct such fields we start from a CFT with $c = 1$ described in terms of a scalar chiral field Q compactified on a circle with radius $R^2 = 2$, explicitly given by:

$$Q(z) = q - i\, p\, lnz + i \sum_{n\neq 0} \frac{a_n}{n} z^{-n} \tag{33.5}$$

with a_n, q and p satisfying the commutation relations $[a_n, a_{n'}] = n\delta_{n,n'}$ and $[q, p] = i$; its primary fields are the vertex operators $U^{\alpha}(z) =: e^{i\alpha Q(z)} :$. It is possible to give a plasma description through the relation $|\psi|^2 = e^{-\beta H_{\text{eff}}}$ where $\psi(z_1, ..., z_N) = \langle N\alpha| \prod_{i=1}^{N} U^{\alpha}(z_i)|0\rangle = \prod_{i<j=1}^{N}(z_i - z_j)^2$ is the ground state wave function. It can be immediately seen that $H_{\text{eff}} = -\sum_{i<j=1}^{N} \ln|z_i - z_j|$ and $\beta = 4$, that is only vorticity $v = 1$ vortices are present in the plasma.

Starting from such a CFT mother theory one can use the m-reduction procedure, which consists in considering the subalgebra generated only by the modes in eq. (33.5) which are a multiple of an integer m, so getting a $c = m$ orbifold CFT (daughter theory, i.e., the TM).[10] With respect to the special $m = 2$ case, the fields in the mother CFT can be organized into components that have well defined transformation properties under the discrete Z_2 (twist) group, which is a symmetry of the TM. By using the mapping $z \to z^{1/2}$ and by making the identifications $a_{2n+l} \longrightarrow \sqrt{2}a_{n+l/2}$, $q \longrightarrow \frac{1}{\sqrt{2}}q$ the $c = 2$ daughter CFT is obtained. It is interesting to notice that such a daughter CFT gives rise to a vortices plasma of half integer vorticity, that is to a fully frustrated XY model, as it will appear in the following.

Its primary fields content can be expressed in terms of a Z_2-invariant scalar field $X(z)$, given by

$$X(z) = \frac{1}{2}\left(Q^{(1)}(z) + Q^{(2)}(z)\right), \tag{33.6}$$

describing the continuous phase sector of the new theory, and a twisted field

$$\phi(z) = \frac{1}{2}\left(Q^{(1)}(z) - Q^{(2)}(z)\right), \tag{33.7}$$

which satisfies the twisted boundary conditions $\phi(e^{i\pi}z) = -\phi(z)$.[10] Such fields coincide with the ones introduced in Eq. (33.3).

The whole TM theory decomposes into a tensor product of two CFTs, a twisted invariant one with $c = \frac{3}{2}$ and the remaining $c = \frac{1}{2}$ one realized by a Majorana fermion in the twisted sector. In the $c = \frac{3}{2}$ subtheory the primary fields are composite vertex operators $V(z) = U_X(z)\psi(z)$ or $V_{qh}(z) = U_X(z)\sigma(z)$, where $U_X(z) = \frac{1}{\sqrt{z}} : e^{i\alpha X(z)} :$ is the vertex of the continuous sector with $\alpha^2 = 2$ for the $SU(2)$ Cooper pairing symmetry used here.

Regarding the other component, the highest weight state in the neutral sector can be classified by the two chiral operators:

$$\psi(z) = \frac{1}{2\sqrt{z}} \left(: e^{i\alpha\phi(z)} : + : e^{i\alpha\phi(-z)} : \right),$$

$$\overline{\psi}(z) = \frac{1}{2\sqrt{z}} \left(: e^{i\alpha\phi(z)} : - : e^{i\alpha\phi(-z)} : \right); \tag{33.8}$$

which correspond to two $c = \frac{1}{2}$ Majorana fermions with Ramond (invariant under the Z_2 twist) or Neveu–Schwartz (Z_2 twisted) boundary conditions[10] in a fermionized version of the theory. Let us point out that the energy-momentum tensor of the Ramond part of the neutral sector develops a cosine term:

$$T_\psi(z) = -\frac{1}{4}(\partial\phi)^2 - \frac{1}{16z^2}\cos\left(2\sqrt{2}\phi\right), \tag{33.9}$$

a clear signature of a tunneling phenomenon, which selects a new stable vacuum, is the linear superposition of the two ground states. The Ramond fields are the degrees of freedom which survive after the tunneling and the Z_2 (orbifold) symmetry, which exchanges the two Ising fermions, is broken.

So the whole energy-momentum tensor within the $c = \frac{3}{2}$ subtheory is:

$$T = T_X(z) + T_\psi(z) = -\frac{1}{2}(\partial X)^2 - \frac{1}{4}(\partial\phi)^2 - \frac{1}{16z^2}\cos\left(2\sqrt{2}\phi\right). \tag{33.10}$$

The correspondence with the Hamiltonian in Eq. (33.3) is more evident once it is observed that the neutral current $\partial\phi$ appearing above coincides with the term $\left(\partial\phi - \frac{\pi}{2}\right)$ of Eq. (33.3), since the $\frac{\pi}{2}$-term coming there from the frustration condition, appears here in $\partial\phi$ as a zero mode, i.e., a classical mode. Besides the fields appearing in Eq. (33.8) there are the $\sigma(z)$ fields, also called the twist fields, which appear in the primary fields $V_{qh}(z)$ combined to a vertex with charge $\frac{e}{4}$. The twist fields have non local properties and decide also for the non trivial properties of the vacuum state, which in fact can be twisted or not in our formalism. Such a property for the vacuum is more evident for the torus topology, where the σ-field is described by the conformal block $\chi_{\frac{1}{16}}$ (see Section 4).

Within this framework the $c = \frac{3}{2}$ ground state wavefunction is described as a correlator of N_{2e} Cooper pairs:

$$< N_{2e}\alpha| \prod_{i=1}^{N_{2e}} V^{\sqrt{2}}(z_i)|0> = \prod_{i<i'=1}^{N_{2e}} (z_i - z_{i'}) Pf\left(\frac{1}{z_i - z_{i'}}\right) \tag{33.11}$$

where $Pf(\frac{1}{z_i - z_{i'}}) = \mathcal{A}(\frac{1}{z_1 - z_2}\frac{1}{z_3 - z_4}\ldots)$ is the antisymmetrized product over pairs of Cooper pairs, so reproducing well known results.[12] In a similar way it is also possible to evaluate correlators of N_{2e} Cooper pairs in the presence of (quasi-hole) excitation[12,10] with non Abelian statistics.[13] It is now easy to check that the charged contribution appearing in eq. (33.11) gives

FIGURE 33.3. Boundary conditions: (a) untwisted; (b) twisted.

rise to a vortices plasma with $H_{\text{eff}} = -\frac{1}{4}\sum_{i<j=1}^{N} \ln|z_i - z_j|$ at the same "temperature" $\beta = 4$, that is it describes vortices with vorticity $\upsilon = \frac{1}{2}$!

On closed annulus geometries, as it is the discretized analogue of a torus, we must properly account for boundary conditions at the ends of the finite lattice since they determine in the continuum the pertinent conformal blocks yielding the statistics of quasiparticles as well as the ground state degeneracy. There are two possible boundary conditions which correspond to two different ways to close the double lattice, i.e., $\varphi_N^{(a)} \to \varphi_1^{(a)}$ or $\varphi_N^{(a+1)} \to \varphi_1^{(a)}$, $a = 1, 2$. It is not difficult to work out that, for the ladder case, the twisted boundary conditions can be implemented only on the odd lattice.

Indeed for $f = \frac{1}{2}$ the ladder is invariant under the shift of two sites, so that there are two topologically inequivalent boundary conditions for even or odd number of sites. In the even case the end sites are of the same kind of the starting one, while in the odd case a ferromagnetic line corresponds to an antiferromagnetic one. The even (odd) case corresponds to untwisted (twisted) boundary conditions, as depicted in Fig. 33.3. The odd case selects out two degenerate ground states which are in different topological sectors, so the system may develop quantum order in the twisted sector of our theory.

33.4 Topological Order and "Protected" Qubits

The aim of this Section is to fully exploit the issue of topological order in a quantum JJL. In order to meet such a request let us develop the two-reduction technique on the torus topology with an impurity.

For closed geometries (the torus) the JJL with an impurity gives rise to a line defect in the bulk. In order to describe it we resort to the folding procedure. Such a procedure is used in the literature to map a problem with a defect line (as a bulk property) into a boundary one, where the defect line appears as a boundary state of a theory which is not anymore chiral and its fields are defined in a reduced region which is one half of the original one. The approach, TM, is a chiral description of that, where the chiral ϕ field defined in $(-L/2, L/2)$ describes both the left moving component and the right moving one defined in $(-L/2,\ 0)$, $(0,\ L/2)$ respectively, in the folded description.[11] Furthermore to make a connection with the TM we consider more general gluing conditions:

$$\phi_{\mathrm{L}}(x=0) = \mp\phi_{\mathrm{R}}(x=0) - \varphi_0$$

the $-(+)$ sign staying for the twisted (untwisted) sector. The X field, which is even under the folding procedure, does not suffer any change in boundary conditions.[11]

Let us now recall briefly the TM Boundary States (BS) constructed in Ref.[11] The most convenient representation of such BS is the one in which they appear as a product of Ising and $c = \frac{3}{2}$ BS. These last ones are given in terms of the BS $|\alpha>$ for the charged boson and the Ising ones

$|f>, |\uparrow>, |\downarrow>$ according to (see Ref. [14] for details):

$$|\chi_{(0)}^{c=3/2}>=|0>\otimes|\uparrow> +|2>\otimes|\downarrow> \quad (33.12)$$

$$|\chi_{(1)}^{c=3/2}>=\frac{1}{2^{1/4}}(|1>+|3>)\otimes|f> \quad (33.13)$$

$$|\chi_{(2)}^{c=3/2}>=|0>\otimes|\downarrow> +|2>\otimes|\uparrow> . \quad (33.14)$$

Such a factorization naturally arises already for the TM characters.[10]

Now we turn to the whole $c = 2$ theory. The vacuum state for the TM model corresponds to the $\tilde{\chi}_{(0)}$ character, which is the product of the vacuum state for the $c = \frac{3}{2}$ subtheory and that of the Ising one. It appears in two TM characters, so a linear combination of them must be taken in order to define a unique vacuum state:

$$|\tilde{\chi}_{((0,0),0)}>=\frac{1}{\sqrt{2}}(|\tilde{\chi}_{(0)}^{+}>+|\tilde{\chi}_{(0)}^{-}>)=\sqrt{2}(|0>\otimes|\uparrow\bar{\uparrow}>+|2>\otimes|\downarrow\bar{\uparrow}>). \quad (33.15)$$

All the correct BS in the untwisted sector of our theory are given in Ref.[11] For the twisted sector we have:

$$|\chi_{(0)}>=(|0>+|2>)\otimes(|\uparrow\bar{f}>+|\downarrow\bar{f}>) \quad (33.16)$$

$$|\chi_{(1)}>=\frac{1}{2^{1/4}}(|1>+|3>)\otimes(|f\bar{\uparrow}>+|f\bar{\downarrow}>). \quad (33.17)$$

Now, by using as reference $|A>$ the vacuum state $|\tilde{\chi}_{((0,0),0)}>$, the chiral partition functions Z_{AB} are computed, where $|B>$ are the two BS given above. We obtain:

$$\chi_{(0)}=Z_{<\tilde{\chi}_{((0,0),0)}||\chi_{(0)}>}=\bar{\chi}_{\frac{1}{16}}\left(\chi_0+\chi_{\frac{1}{2}}\right)(K_0+K_2) \quad (33.18)$$

$$\chi_{(1)}=Z_{<\tilde{\chi}_{((0,0),0)}||\chi_{(1)}>}=\left(\bar{\chi}_0+\bar{\chi}_{\frac{1}{2}}\right)\chi_{\frac{1}{16}}(K_1+K_3). \quad (33.19)$$

In order to understand the physical significance of the $c = 2$ conformal blocks in terms of the charged low energy excitations of the system, let us evidence their electric charge and magnetic flux contents in the dual theory (it is obtained by exchanging the compactification radius $R_e^2 \to R_m^2$ in the charged sector of the CFT). Hence let us consider the "charged" sector conformal blocks appearing in Eqs. (33.18–33.19):

$$K_{2l+i}(w|\tau)=\frac{1}{\eta(\tau)}\Theta\left[\begin{array}{c}\frac{2l+i}{4}\\0\end{array}\right](2w|4\tau), \qquad \forall(l,i)\in(0,1)^2, \quad (33.20)$$

corresponding to primary fields with conformal dimensions $h_{2l+i}=\frac{1}{2}\alpha_{(l,i)}^2=\frac{1}{2}(\frac{2l+i}{2}+2\delta_{(l+i),0})^2$ and electric charges $2e(\frac{\alpha_{(l,i)}}{R_X})$ (magnetic charges in the dual theory $\frac{hc}{2e}(\alpha_{(l,i)}R_X)$), $R_X = 1$ being the compactification radius.

The topological order referring to the two degenerate ground states can how be discussed corresponding to the twisted characters in Eqs. (33.18–33.19):

$$|0\rangle \to \chi_{(0)}(w|\tau), |1\rangle \to \chi_{(1)}(w|\tau). \quad (33.21)$$

Such two states can be distinguished through the presence or absence respectively of a half flux quantum trapped inside the central hole which, in turn, is related to the presence or absence of the Ising character $\chi_{\frac{1}{16}}$ in the $c = \frac{3}{2}$ subtheory. The trapped half flux quantum can be experimentally detected, so giving a way to read out the state of the system. Upon performing an adiabatic change of local magnetic fields which drags one half vortex across the system, i.e., through the transport of a half flux quantum around the B-cycle of the torus, it was possible to flip the state of the system, so lifting the degeneracy. Furthermore it should be noticed that the presence of the twist operators σ, described here by the Ising character $\chi_{\frac{1}{16}}$, gives rise to non Abelian statistics, which can be evidenced by their fusion rules $\sigma\sigma = \mathbf{1} + \psi$.[10,12]

In conclusion, Josephson junction ladders with non trivial geometry may develop topological order allowing for the implementation of "protected" qubits, a first step toward the realization of an ideal solid state quantum computer.

References

1. X. G. Wen and Q. Niu, Ground-state degeneracy of the fractional quantum Hall states in the presence of a random potential and on high-genus Riemann surfaces, *Phys. Rev. B* **41**, 9377–9396 (1990).
2. T. Senthil and M. P. A. Fisher, Fractionalization, topological order and cuprate superconductivity, *Phys. Rev. B* **63**, 134521–134536 (2001).
3. X. G. Wen, Quantum order: A quantum entanglement of many particles *Phys. Let. A* **300**, 175–181 (2002); X. G. Wen, Quantum order and symmetric spin liquids, *Phys. Rev. B* **65**, 165113–165149 (2002).
4. X. G. Wen, Topological orders and edge excitations in fractional quantum Hall states, *Adv. Phys.* **44**, 405-473 (1995).
5. B. Doucot, M. V. Feigel'man and L. B. Ioffe, Topological order in the insulating Josephson junction array, *Phys. Rev. Lett.* **90**, 107003–107006 (2003); B. Doucout, M. V. Feigel'man, L. B. Ioffe, A. S. Ioselevich, Protected qubits and Chern Simons theories for Josephson junction arrays, cond-mat/0403712.
6. M. C. Diamantini, P. Sodano, C. A. Trugenberger, Gauge theories of Josephson junction arrays, *Nucl. Phys. B* **474**, 641–677 (1996).
7. G. Cristofano, V. Marotta, A. Naddeo, G. Niccoli, A conformal field theory description of magnetic flux fractionalization in Josephson junction ladders, preprint Napoli DSF-T-05/2004, INFN-NA-05/2004, ISAS/24/2004/FM.
8. M. Kardar, Josephson-junction ladders and quantum fluctuations, *Phys. Rev. B* **33**, 3125–3128 (1986); E. Granato, Phase transitions in Josephson-junction ladders in a magnetic field, *Phys. Rev. B* **42**, 4797–4799 (1990); E. Granato, Phase transitions in the one-dimensional frustrated quantum XY model and Josephson-junction ladders, *J. Appl. Phys.* **75**, 6960–6962 (1994).
9. V. Marotta, Relation between covariant and parafermionic construction of affine Kac–Moody algebras, *J. Phys. A* **26**, 3481–3487 (1993); V. Marotta, Stress-tensor for parafermions from winding subalgebras of affine algebras, *Mod. Phys. Lett. A* **13**, 853–860 (1998); V. Marotta, W_k structure of $SU(2)_k$ from covariant construction of Kac–Moody algebras, *Nucl. Phys. B* **527**, 717–737 (1998).
10. G. Cristofano, G. Maiella and V. Marotta, A conformal field theory description of the paired and parafermionic states in the quantum Hall effect, *Mod. Phys. Lett. A* **15**, 1679–1688 (2000); G. Cristofano, G. Maiella, V. Marotta and G. Niccoli, Paired states on a torus, *Nucl. Phys. B* **641**, 547–567 (2002).
11. G. Cristofano, V. Marotta and A. Naddeo, Twisted CFT and bilayer quantum Hall systems in the presence of an impurity, *Phys. Lett. B* **571**, 250–259 (2003); G. Cristofano, V. Marotta and A. Naddeo, A twisted conformal field theory description of dissipative quantum mechanics, *Nucl. Phys. B* **679**, 621–631 (2004).
12. G. Moore and N. Read, Nonabelions in the fractional quantum Hall effect *Nucl. Phys. B* **360**, 362–396 (1991).
13. C. Nayak and F. Wilczek, $2n$-quasihole states realize 2^{n-1}-dimensional spinor braiding statistics in paired quantum Hall states, *Nucl. Phys. B* **479**, 529–553 (1996).
14. P. Di Francesco, P. Mathieu and D. Senechal, Conformal Field Theories, (Springer–Verlag, 1996).

34

Single-electron charge qubit in a double quantum dot

TOSHIMASA FUJISAWA[1,2], TOSHIAKI HAYASHI[1], SUNG WOO JUNG[1,3], YOON-HA JEONG[3], AND YOSHIRO HIRAYAMA[1,4]

[1]NTT Basic Research Laboratories, NTT Corporation, 3-1 Morinosato-Wakamiya, Atsugi, 243-0198, Japan.
[2]Tokyo Institute of Technology, 2-12-1, O-okayama, Meguro, Tokyo, 152-8550, Japan
[3]Pohang University of Science and Technology, Pohang, Kyungpook, Korea.
[4]CREST, 4-1-8, Honmachi, Kawaguchi, Saitama, 331-0012, Japan.

Abstract

A semiconductor double quantum dot provides a simple artificial two-level system (qubit) that can be manipulated by electronic signals. Full one-qubit operation is demonstrated with a high-speed voltage pulse. In addition, strength of decoherence can be controlled to allow efficient initialization of the qubit. Remaining uncontrolled decoherence is discussed with background charge fluctuations, cotunneling, and electron–phonon coupling. Moreover, charge detection of a double dot is demonstrated with a quantum-point-contact charge detector.

34.1 Introduction

In the quantum computing scheme, quantum state of each particle (qubit) is sequentially manipulated in a programmable manner so that some sorts of information processing can be performed efficiently in a parallel way.[1,2] The advanced quantum information processing has been led by various experiments on natural atoms, such as pulsed nuclear magnetic resonance of molecules and optically manipulated atomic states. Integrating sufficiently large number of qubits is essential for realizing practical quantum computing hardware. In this sense, artificial quantum systems with nanofabrication capability has attracted much attention for future quantum information processing. Recent nanofabrication technology allows us to design artificial atoms (quantum dots) and molecules (double quantum dots), in which atomic (molecular)-like electronic states can be controlled with external voltages.[3–5] Quantum states of interacting electrons in quantum dots are well understood in terms of atomic physics language, but now can be controlled with external voltages. High controllability of the artificial quantum system is strongly desired for quantum information technology. Moreover, similar characteristics have been obtained in various kinds of nanostructures, including metallic particles and synthesized molecules. Nanofabrication capability is particularly meant for large-scale integration as well as designing an artificially controlled environment surrounding a quantum system.

Here is described experimental studies on coherent manipulation, dissipation, and decoherence is described in single-electron charge qubit in a double quantum dot. All the qubit parameters and strength of decoherence can be controlled by external voltages. Rotation gate is demonstrated, in which population of an electron is manipulated, and phase-shift gate, in which phase difference between the qubit states is accumulated, by tailoring a pulse waveform.

34.2 Charge Qubit in a Double Quantum Dot

A Double Quantum Dot (DQD) provides a simple and realistic two-level system, which can be discussed with an electron in a double dot potential. In a classical picture, where the two quantum dots are well isolated, the system can be described by two charge states, in which an electron occupies one of the states in the left dot ($|L>$) or the right dot ($|R>$). When the tunneling between the two dots is allowed, however, the electron state can be written as a linear superposition of the two localized states,

$$|\psi>= \cos(\theta/2)|L> + \sin(\theta/2)\exp(i\phi)|R>, \tag{34.1}$$

which can be referred to as charge qubit.[6]

Charge qubit can be realized in a double quantum dot fabricated in a GaAs/AlGaAs heterostructure with a two-dimensional electron gas. As shown in the scanning electron micrograph in Fig. 34.1(a), a narrow conductive channel is formed between the etched grooves (the upper and lower dark regions). Three tunneling barriers are formed by applying negative voltages to the gate electrodes (the bright vertical lines), leaving the left and right QDs (white circles) between the source and drain electrodes. Each QD contains tens of electrons, which form many-body ground and excited states. If we consider only one specific energy state in each QD, effective two-level system (qubit) can be defined. Other states can be neglected, when characteristic energies (addition energy, single-particle energy spacing, and electrostatic coupling energy) are greater than thermal energy or other excitation energies.

Fig. 1(b) shows a simplified energy diagram of the double dot device. E_L and E_R are the energy of the localized state $|L>$ and $|R>$, respectively. The important parameters of the qubit are the energy offset between the localized state, $\varepsilon = E_R - E_L$, and tunneling coupling between the two states, T_c. The qubit state can be detected by single electron tunneling current through the left and right barriers, whose tunnel rates are Γ_L and Γ_R, respectively. Advantage of the double dot device is that all the parameters can be adjusted by gate voltages. These parameters can be determined from single-electron tunneling current profile.[7]

Importantly, the strength of decoherence of the qubit can also be controlled by an external voltage. When the dot is in the Coulomb blockade region, tunneling in and out of the double dot is blocked. If other decoherence is neglected, the system Hamiltonian can be written in a simple form

$$H = \frac{1}{2}\varepsilon\sigma_z + T_c\sigma_x, \tag{34.2}$$

where σ_x and σ_z are Pauli matrices. For instance, when gate voltages are swept so as to cross E_L and E_R, the eigenstates of the system become bonding and antibonding states, whose energies are E_b and E_a, respectively, as shown in Fig. 34.1(c).

In contrast, when a large source-drain voltage is applied, single-electron tunneling current flows through a double dot. Corresponding energy diagram is shown in Fig. 34 1(d). In this regime, an electron in the left dot can escape to the left electrode and another electron can enter the right dot. These tunneling processes given rise to significant decoherence of the system, which is often discussed with density matrix calculations.[7,8]. Under the condition that Γ_L and Γ_R are much greater than T_c, stationary solution of the rate equations becomes a density matrix element $\rho_{LL} = 1$, that is the localized state in the left dot.[9]

In the following experiments, first a localized state $|L>$ in the transport regime is prepared, and the system is switched into the Coulomb blockade regime by applying a rectangular pulse to the source-drain electrode. Since $|L>$ is no longer an eigenstate in the Coulomb blockade regime, coherent time evolution is expected. After adjustable period of the pulse, the double dot is again set in the transport regime, so that the electron, if it exists in the right dot, contributes to the current. One electron tunneling per pulse at most is expected. By repeating the pulse, reasonable current is obtained.

34.2.1 Rotation gate and phase-shift gate operation

The dynamics of the charge qubit is determined by the Hamiltonian (Eq. (34.2)), which includes two spin matrices. In the Bloch sphere representation of the qubit, the state rotates about total fictitious magnetic field $B = (T_c, 0, \varepsilon)$. In our case, ε can be varied from negative through zero to positive, while T_c is always positive. Full one-qubit operation can be realized with a combination of rotation gate and phase shift gate.

Rotation gate can be performed by applying $B = (T_c, 0, 0)$ at $\varepsilon = 0$, where the state rotates about x-direction. Figure 34.2(a) shows the pulse induced current observed in this situation.[6]

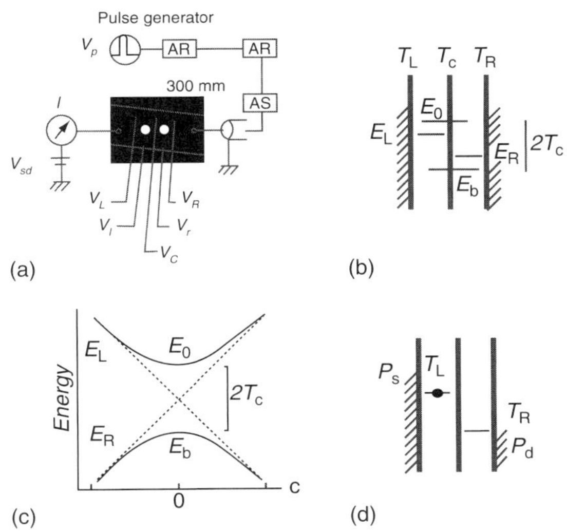

FIGURE 34.1. (a) Schematic of the measurement setup with a scanning electron microscope image of the sample. (b) Schematic energy diagram of the double quantum dot. (c) Energy of localized states (E_L and E_R for $|L>$ and $|R>$, respectively) and delocalized states (E_b and E_a for bonding and anti-bonding states). (d) Schematic energy diagram in the transport regime.

The current reflects the probability of finding an electron in the right dot, and clear oscillation observed by changing the pulse length is a signature of coherent rotation gate. The oscillation frequency should be given by $2T_c/h$, and can be adjusted by changing the central gate voltage, as shown in Fig. 34.2(b). A rectangular pulse that gives a half (quarter) period of the oscillation corresponds to a logical NOT (square-root-NOT) gate. The amplitude of the oscillation is approximately half of the ideal case (16 pA with a repetition frequency of 100 MHz), which might be due to populating other excited states during initialization. The decay of the oscillation is attributed to decoherence, which will be discussed in Section 34.2.2.

Phase-shift gate may be performed by applying $B = (0, 0, \varepsilon)$ in principle. Since T_c cannot be set exactly zero in this experiment, the phase shift is approximated by applying $B = (T_c, 0, \varepsilon)$ with $\varepsilon \gg T_c$. In order to demonstrate phase shift operation, a sharp tipping pulse is added at the center of the NOT gate operation (π pulse). The first $\pi/2$ pulse prepare the state in a superposition ($|L> +i|R>$). The sharp tipping pulse gives a phase difference ϕ, and the second $\pi/2$ pulse brings the system to the North Pole or South Pole on the Bloch sphere depending on ϕ, as illustrated in the insets of Fig. 34.3. The height of the tipping pulse, V_{tip}, is changed with a constant pulse width of about $t_{tip} \sim 40$ps. The phase shift is approximately given by the area of the pulse,

$$\phi \sim eV_{tip}t_{tip}/h. \tag{34.3}$$

Figure 34.3 shows the pulse induced current obtained in this way, indicating proper phase shift operation. Since the oscillation is obtained with a fixed pulse width, the amplitude decay with V_{tip} implies increase of decoherence rate.

It is known that any one-qubit unitary operation can be prepared by combining rotation and phase-shift gates. Quantum operation may be constructed from arbitrary field $B = (T_c, 0, \varepsilon)$. Hadamard gate, which is often used to prepared a superposition $|L> +|R>$ from a localized state $|L>$, can be performed with the pulse width $t_H = \sqrt{2}\pi h/\varepsilon$ at $\varepsilon = T_c$.

34.2.2 Decoherence of the system

Decoherence is a common problem for any quantum system. Some possible decoherence mechanisms of our device are discussed here. Firstly, background charge fluctuations in the sample and noise in the gate voltage fluctuate the parameters ε and T_c, which give rise to decoherence of the

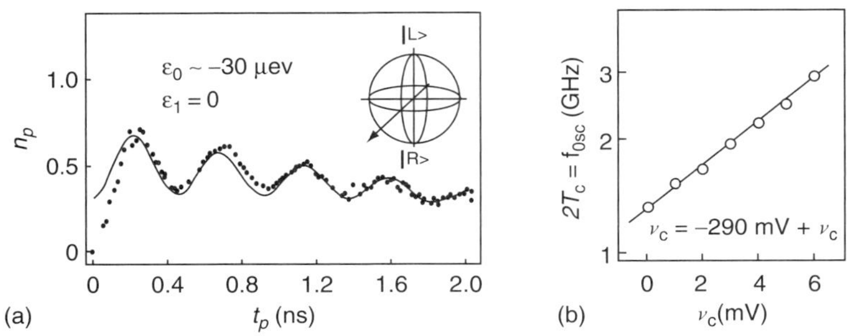

FIGURE 34.2. (a) Coherent charge oscillation observed at resonant condition, $\varepsilon \sim 0$. Inset shows Bloch sphare representation of the qubit dynamics. (b) The central gate voltage dependence of the oscillation frequency, f_{osc}.

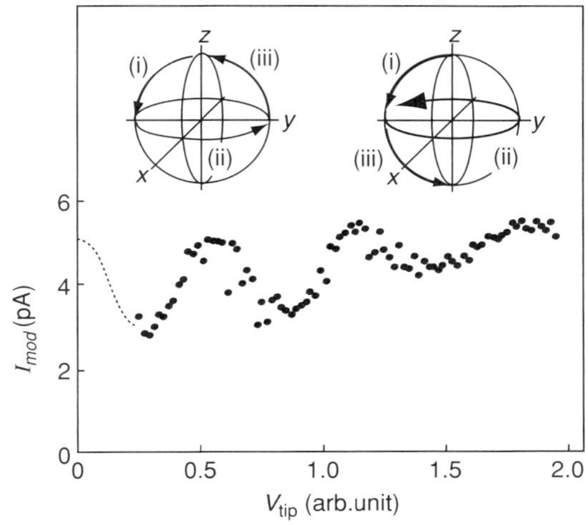

FIGURE 34.3. Phase-shift operation performed by applying a tipping pulse. Inset shows schematic dynamics in the Bloch share.

system.[11,12] Low-frequency fluctuation of ε is estimated to be about 1.6μeV, which is obtained from current fluctuation in the single-electron current spectrum. This fluctuation explains the decoherence rate observed at off-resonant condition, while the decoherence at on-resonance condition may be dominated by other mechanisms. Microscopic origin of the background charge fluctuation is not known well, and the magnitude differs from sample to sample, even which are fabricated in the same batch. Understanding the fluctuation is important for developing quantum information devices. It seems from recent noise measurement that the fluctuation can be reduced by decreasing temperature.[13]

Although the first-order tunneling processes are suppressed in the Coulomb blockade regime, higher-order tunneling (cotunneling) processes can take place and decohere the system.[14] Actually the cotunneling rate estimated from tunneling rates is close to the observed decoherence rate. However, cotunneling is not an intrinsic problem because the cotunneling effect can be reduced by making tunneling barrier less transparent.

Electron–phonon interactions is an intrinsic decoherence mechanism in semiconductor QDs.[15] Spontaneous emission of acoustic phonon remains even at zero temperature and causes an inelastic transition between the two states. Actually, the negative background slope shown by dashed line in Fig. 34.2(a) comes from the inelastic tunneling process, and its rate is also comparable to the observed decoherence rate. Moreover, a single quantum dot also shows phonon emission rate of the order of a few ns.[16] Strong electron–phonon coupling is related to the fact that wavelength of the corresponding phonon is comparable to the size of the quantum dot.[17] Electron–phonon coupling may be reduced by redesigning quantum dot structure.

Other mechanisms, such as electromagnetic environment, have to be considered to fully understand the decoherence. It should be noted that the quality of the coherent oscillation was improved by reducing high-frequency noise from the gate voltages and the coaxial cable. The remaining noise may still contribute to the decoherence. Some decoherence effects can be reduced by further studies.

34.3 Charge Detection of a Double Quantum Dot

Quantum state in a DQD is measured with a current through the DQD. It would be nice if the charge state of the qubit can be measured with a sensitive electrometer placed in the vicinity of the charge qubit. Single-electron transistors and Quantum Point Contact (QPC) structures have been demonstrated as a sensitive electrometer, and radio frequency circuit provides high speed detection for high-impedance devices.[18–20] These techniques are interesting especially for realizing single-shot readout scheme, which is useful for studying correlation between two or more qubit system.

Figure 34.4(a) shows a device structure containing a DQD in the upper electrical channel and a QPC charge detector in the lower channel. For testing feasibility of detecting charge states of the DQD, dc conductance measurement is carried out with the QPC electrometer. The QPC conductance is adjusted at the maximal sensitivity condition (about half of the quantized conductance). Gate voltages (V_L and V_R) of the DQD are swept to change the potential of the two dots, independently. Since these voltages also affect the QPC conductance, the QPC gate voltage is simultaneously changed so as to compensate for the crosstalk. In addition, in order to improve the signal to noise ratio, the charge state of the DQD was modulated by applying low-frequency (100 Hz) sinusoidal voltage to the DQD, and corresponding modulated current through the QPC was measured with a lock-in amplifier.

Figure 34.4(b) shows the intensity plot of the modulated QPC current (dI_{det}/dV_{mod}), when V_L and V_R were swept. Positive or negative signal appears when an electron is added or removed from the DQD. The sign of the signal depends on the symmetry of the left and right tunneling rates. When the left barrier is more transparent (upper-left region of the figure), an electron is injected into the DQD from the left electrode by making the modulation voltage positive. In contrast, when the right barrier is more transparent (lower-right region), an electron is extracted to the drain electrode from either dot.

The signal pattern in Fig. 34.4 can be understood well as charging on the two quantum dots. As shown in the magnified sweep of the inset of Fig. 34.5, one can identify stable charge states of the DQD. Here $(\boldsymbol{n}, \boldsymbol{m})$ corresponds to $\boldsymbol{n}$ and $\boldsymbol{m}$ excess electrons occupied by the left and right dots, respectively. Signal should appear along the boundary of the charging diagram, when two charge states are degenerated.

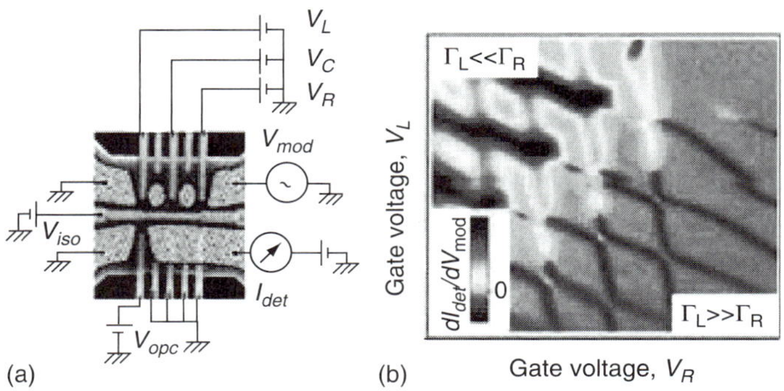

FIGURE 34.4. (a) Schematic of the charge detection measurement of a double quantum dot. (b) Intensity (color) plot of the charge detection measurement, dI_{det}/dV_{mod}.

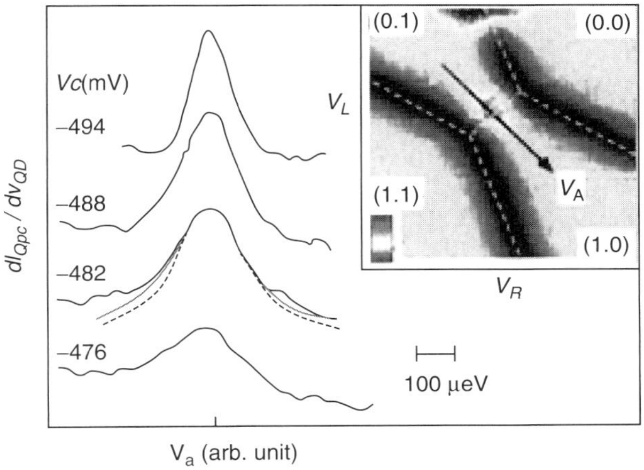

FIGURE 34.5. dI_{det}/dV_{mod} traces across the charge degeneracy condition between $(n+1, m)$ and $(n, m+1)$ states. Dashed and dotted lines are fitting curves when the localized charge states are thermally distributed and coherently coupled, respectively. Inset shows a charging diagram of a double dot.

Now, the focus is on the charge qubit measurement, in which an electron occupies one of the two dots. The qubit state is defined around the degeneracy condition between $(n + 1, \boldsymbol{m})$ and $(\boldsymbol{n}, m + 1)$ charge states. When an electron jumps from the right dot to the left dot, the QPC conductance should always increase in our geometry. This change is clearly observed as positive signal in the inset of Fig. 34.5.

Figure 34.5 shows the traces of dI_{det}/dV_{mod} around the degeneracy condition. The energy scale in the figure was determined from photon assisted tunneling spectroscopy with microwave irradiation (not shown). As the central gate voltage V_C is made less negative to increase the coupling, the peak broadens. This suggests that the two dots can be coherently coupled with the tunneling coupling. On the other hand, the peak in the topmost trace at $V_C = -494$mV is close to the narrowest peak observed in this charge measurement, indicating significant broadening (decoherence) probably from external noise or ac modulation voltage.

In order to test the coherency of the qubit, the traces are fit by assuming two extreme cases. One is based on coherent system described by Eq. 34.2, but it is assumed that the ground state (bonding state) is always occupied. By assuming that the charge detection signal is proportional to the probability of finding an electron in the left dot, the peak profile of the signal for detecting an electron in the ground state is given by

$$dI_{det}/dV_{mod} \propto T_c^2/(\varepsilon^2 + T_c^2)^{\frac{3}{2}} \tag{34.4}$$

The peaks for the lowest two traces in Fig. 34.5 can be fitted well with this formula (See dotted line fitted with $T_c \sim 60\,\mu$eV for $V_C = -482$mV). Disagreement on the left hand side might come from the broadened positive signal from another peak.

The other extreme case is based on the statistical mixture of localized states, which is distributed by an effective temperature, T_{eff}. Tunneling coupling is thus neglected. In this case, the peak profile is given by

$$dI_{det}/dV_{mod} \propto \cosh^{-2}(\varepsilon/k_B T_{eff}) \tag{34.5}$$

Actually, the topmost trace in Fig. 34.5 can be fitted well with this model as shown by a dashed line with $k_B T_{\rm eff} \sim 30$ μeV. However, fitting to the trace at $V_C = -482$ mV is not perfect even with an elevated (unrealistic) effective temperature (dashed line with $k_B T_{\rm eff} \sim 50$ μeV).

The experiment demonstrates that the charge detection can determine the charge distribution is superposition or statistical mixture of the localized states. More accurate parameters may be obtained by taking into account both tunneling coupling and thermal distribution.[21,22] High-frequency technique would allow us to determine instantaneous charge state, and provide single-shot charge readout scheme.[23]

34.4 Summary

Charge qubit in a double quantum dot has been discussed. While various kinds of qubit are realized in superconducting system, it was shown that the qubit can be realized with semiconductor electronics. It can be expected that quality of a qubit can be improved by utilizing various semiconductor nanostructure technologies.

References

1. M. A. Nielsen and I. L. Chuang, Quantum Computation and Quantum Information (Cambridge University Press, Cambridge, 2000).
2. G. Burkard and D. Loss, *Semiconductor Spintronics and Quantum Computation*, ed. by D. D. Awschalom, D. Loss, and N. Samarth (2002), p.229.
3. L. P. Kouwenhoven, C. M. Marcus, P. L. McEuen, S. Tarucha, R. M. Westervelt, and N. S. Wingreen, in *Mesoscopic Electron Transport* ed. by L. L. Sohn, L. P. Kouwenhoven, and G. Sch'on, NATO ASI series E **345** (Kluwer Academic, Dordrecht 1997), p. 105.
4. S. Tarucha, D. G. Austing, T. Honda, R. J. van der Hage, and L. P. Kouwenhoven, *Phys. Rev. Lett.* **77**, 3613 (1996).
5. T. H. Oosterkamp, T. Fujisawa, W. G. van der Wiel, K. Ishibashi, R. V. Hijman, S. Tarucha and L. P. Kouwenhoven, *Nature* **395**, 873 (1998).
6. T. Hayashi, T. Fujisawa, H. D. Cheong, Y. H. Jeong and Y. Hirayama, *Phys. Rev. Lett.* **91**, 226804 (2003).
7. N. C. van der Vaart, S. F. Godijn, Yu. V. Nazarov, C. J. P. M. Harmans, and J. E. Mooij, *Phys. Rev. Lett.* **74**, 4702 (1995).
8. Yu. V. Nazarov, *Physica B* **189**, 57 (1993).
9. T. Fujisawa, T. Hayashi, and Y. Hirayama, *J. Vac. Sci. Technol.* (in press).
10. T. Fujisawa, T. Hayashi, H. D. Cheong, Y. H. Jeong, and Y. Hirayama, *Physica E* **21**, 1046 (2004).
11. Y. Nakamura, Yu, A. Pashkin, T. Yamamoto, and J. S. Tsai, *Phys. Rev. Lett.* **88**, 047901 (2002).
12. E. Paladino, L. Faoro, G. Falci, and R. Fazio, *Phys. Rev. Lett.* **88**, 228304 (2002).
13. S. W. Jung, T. Fujisawa, Y. Hirayama, Y. H. Jeong, *Appl. Phys. Lett.* (in press).
14. U. Hartmann and F. K. Wilhelm, *Physica Status Solidi B*. **233**, 385 (2002).
15. T. Brandes and T. Vorrath, *Phys. Rev. B* **66**, 075341 (2002).
16. T. Fujisawa, D. G. Austing, Y. Tokura, Y. Hirayama, and S. Tarucha, *Nature* **419**, 278 (2002).
17. T. Fujisawa, T. H. Oosterkamp, W. G. van der Wiel, B. W. Broer, R. Aguado, S. Tarucha and L. P. Kouwenhoven, *Science* **282**, 932 (1998).
18. M. Field, C. G. Smith, M. Pepper, D. A. Ritchie, J. E. F. Frost, G. A. C. Jones, and D. G. Hasko, *Phys. Rev. Lett.* **70**, 1311 (1993).
19. D. S. Duncan, C. Livermore, R. M. Westervelt, K. D. Maranowski, and A. C. Gossard, *Appl. Phys. Lett.* **74**, 1045 (1999).

20. R. J. Schoelkopf, P. Wahlgren, A. A. Kozhevnikov, P. Delsing, D. E. Prober, *Science*, **280**, 1238 (1998).
21. J. M. Elzerman, R. Hanson, J. S. Greidanus, L. H. W. van Beveren, S. De-Franceschi, L. M. K. Vandersypen, S. Tarucha, L. P. Kouwenhoven, *Phys. Rev. B* **67**, 161308 (2003).
22. L. DiCarlo, H. J. Lynch, A. C. Johnson, L. I. Childress, K. Crockett, C. M. Marcus, M. P. Hanson, and A. C. Gossard, cond-mat/0311308v4.
23. A. Aassime, G. Johansson, G. Wendin, R. J. Schoelkopf, and P. Delsing, *Phys. Rev. Lett.* **86**, 3376 (2001).

35

Quantum dots for single photon and photon pair technology

R. M. STEVENSON[1], R. J. YOUNG[1,2], R. M. THOMPSON[1,2], Z. YUAN[1], B. E. KARDYNAL[1], I. FARRER[2], M. B. WARD[1], D. A. RITCHIE[2], AND A. J. SHIELDS[1]

[1] Toshiba Research Europe Limited, 260 Cambridge Science Park, Milton Road, Cambridge, CB4 0WE, UK.

[2] Cavendish Laboratory, University of Cambridge, Madingley Road, Cambridge, CB3 0HE, UK

Abstract

The excitation of a quantum dot from its ground state depends on the probabilistic nature of carrier capture, and thus radiative decay can occur via a multi photon cascade. It is also possible that non-radiative initial exciton states can be formed when the spins of the component electron and heavy hole have the same sign. Time integrated and time resolved micro-photoluminescence experiments are presented that reveal the broader properties of single photon and multi-photon emission from InAs quantum dots, including single photon emission, polarisation correlated photon emission, and dark state blocking, and their implications in terms of quantum information technology are discussed.

Applications in quantum optics, such as quantum communication[1] and computing,[2,3] require the development of a new kind of light source, which ideally would emit a fixed number of photons in a given time interval. Such device has very different properties to typical light sources, for which the number of photons emitted in a given time interval obeys Poissonian statistics. One approach to the realization of fixed photon number sources utilizes the radiative decay of excitations in zero dimensional systems.[4] In general, optical transitions between discrete states in these systems gives rise to photons with unique energies. Thus, by preparing the system in an excited state, and filtering the emission corresponding to one or more chosen transition energies, the emission of a fixed number of photons per excitation cycle can be achieved. One of the most appealing systems in terms of ease of integration into standard optoelectronic fabrication techniques, are semiconductor quantum dots.[5–10] Authors' progress on the study of quantum dot single photon, and photon pair sources are summarized here.

Self assembled quantum dot samples were grown by molecular beam epitaxy on a (100)-oriented undoped GaAs substrate. Following a GaAs buffer, an InAs layer was grown, followed by a 300 nm GaAs cap. The thickness of the InAs layer was close to the critical thickness for 3D island formation, around 1.7 monolayers. The resulting quantum dots are rather small, resulting in emission around 900 nm, allowing photon detection using silicon CCDs and Avalanche Photodiodes (APD). The low dot density ($< 1\mu m^{-2}$) allowed the isolation of single dots by the fabrication of small mesas, or by etching small holes in a metal mask evaporated on the surface.

A single quantum dot was isolated by etching 0.8 μm diameter pillars and studied in a micro-photoluminescence experiment[9] at a temperature of 5 K. Excitation was provided by a titanium sapphire laser, tuned above the GaAs band gap, producing 1ps pulses at 77 MHz repetition rate. A microscope objective was used both to focus the laser onto the sample, and collect the emission. Fig 35.1(a) plots the PL spectrum recorded on a CCD for average excitation powers between 0.13 and 40 nW. The sharp lines observed in the spectra derive from the recombination of different exciton configurations confined in the quantum dot, and the dominance of the lines marked X and X* at low power indicates the capture of a smaller number of carriers relative to X_2 and X_2*. The observed linewidths are unlikely to be resolution limited, and are thought to be broadened by fluctuating charge distribution close to the dot, a consequence of non-resonant excitation. No additional lines are observed at energies between those shown, and that of the wetting layer, even for the higher powers. This is in contrast to other reports that show excited state emission 30–60 meV to higher energy than the lines observed at lowest powers.[11–13] It is believed that this type of quantum dot cannot confine triexcitons due to the lack of a second confined electron level as a result of their small size. The single lines X, X_2, and X* and the doublet X_2* are attributed to emission from exciton, biexciton, charged exciton and charged biexciton states respectively. These assignments are made on the basis of the power dependence, temporal characteristics, and correlation of the lines, and are described in detail in Ref. [9] The corresponding exciton configurations are shown in Fig. 35.1(b).

The time response of the four exciton complexes by time correlated single photon counting is measured, using a silicon APD, that limits the temporal resolution to ~400 ps. Figure. 35.1(c) shows the PL intensity as a function of time for all 4 lines. Remarkably each of the excitons is observed to have a different temporal characteristic. All show an approximately exponential decay with fitted lifetimes for X: 1.36 ns, X*: 1.07±0.02 ns, X_2: 0.59±0.02 ns and X_2*: 0.52±0.05 ns. Notice that the radiative lifetimes of X and X* are more than a factor of two

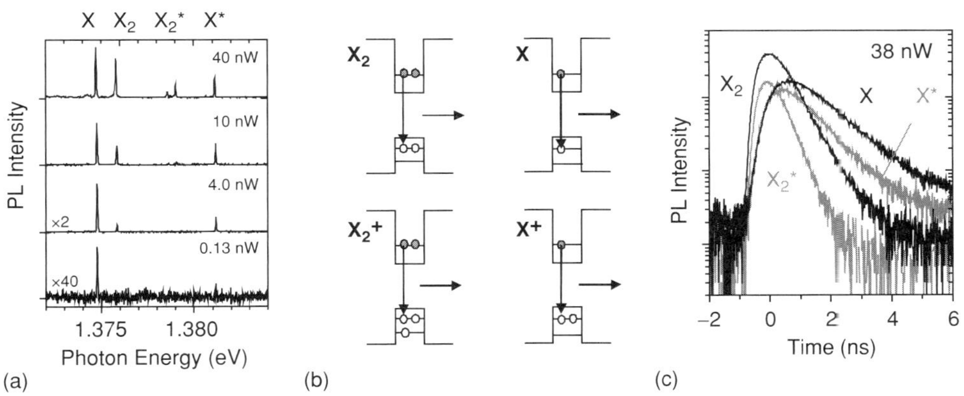

FIGURE 35.1. (a) Photoluminescence from a single quantum dot as a function of laser excitation power. (b) Schematic showing different electron and charge configurations within a quantum dot. (c) Time resolved photoluminescence from a single dot for different emission lines.

longer than the corresponding biexciton. This is attributed to the two possible recombination paths for the two electron-hole pairs in the biexciton. It is also apparent that there is a delay in the emission of X relative to X_2. This is explained by the fact that the biexciton state decays radiatively into the single exciton ($X_2 \rightarrow$X+photon). Similarly, the peak of the emission due to X* follows that of X_2*, as expected from the radiative decay of a charged biexciton state into a charged single exciton (X_2*$\rightarrow$X*+photon).

The second order correlation function, $g^{(2)}(\tau)$, between photons emitted from the X, X* and X_2 states was measured using a Hanbury–Brown and Twiss intensity interferometer. PL excited with a relatively high laser power so as to saturate the emission intensities, was spectrally filtered by the spectrometer so as to contain just one emission line. Figure 35.2(a) plots the second order correlation function recorded for the X, X*, and X_2 lines of the dot, as well as that measured for the laser itself. Each correlation trace consists of a series of peaks separated by the laser period of 13.0 ns. For the laser, the correlation peaks have roughly equal height, as expected for a coherent light source. In contrast the emission from the quantum dot displays a strong suppression of the peak around zero time delay for each of the complexes. This is clear evidence for single photon emission from each of the exciton complexes.

The area of the zero time delay peak observed for the X transition, compared to those at finite delay, suggests the fraction of multi-photon pulses emitted to be just 6% of that of a coherent source of the same intensity. These multi-photon pulses derive from stray emission from the buffer, substrate and wetting layer of the structure. Their number could be reduced by redesign of the sample structure or better spectral rejection.

The smooth grey lines in Fig. 35.2(a) are calculated from a statistical description of the correlation measurement.[14] The parameters for the calculation are all determined from the experiment. The excitation rate is determined by comparing the ratio of the X and X_2 peaks in the calculated and measured spectra. A finite time resolution of the measuring system of 0.85 ns was measured from the correlation of a ps laser pulses. The background PL level was measured directly by tuning the spectrometer to a nearby wavelength where there is no excitonic transition. There is close agreement between the measured and calculated curves. This demonstrates that

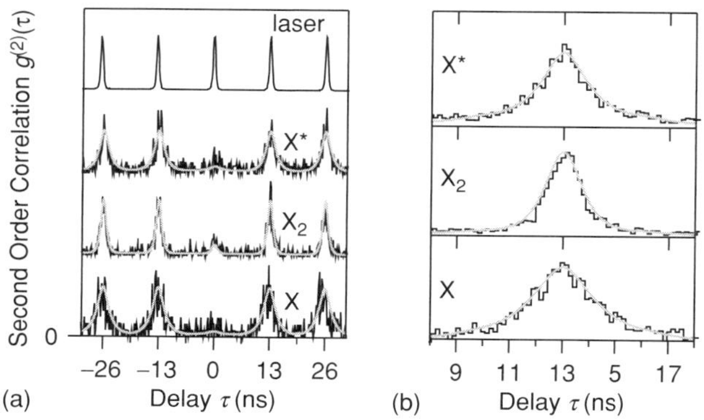

FIGURE 35.2. (a) Second order correlation function measured under pulsed optical excitation for emission from different exciton complexes. Smooth lines are calculated for the same experimental conditions. The top trace shows the measured correlation of an attenuated laser is for comparison. (b) Uncertainty in time (jitter) between subsequent photon emission for different exciton complexes.

after taking account of the level of background PL, the device can be modeled as an ideal single photon emitter.

Notice that the correlation peaks observed for X_2 appear to be significantly narrower than those for X. This can be seen more clearly in Fig. 35.2(b) that plots the average peak shape for exciton, biexciton, and charged exciton emission. The FWHM of these peaks demonstrates a reduction in the jitter between single photon emission events for the biexciton (1.5 ns), and to a lesser extent the charged exciton (2.1 ns) compared to the simple exciton (2.8 ns). This reduction in jitter derives from the shorter radiative decay time of the biexciton (charged exciton) state, measured above to be 0.59 ns (1.07 ns) compared to 1.36 ns for X.

There are potential advantageous in designing single-photon emitting devices around biexcitonic emission from quantum dots rather than the single exciton transition. Since the radiative lifetime of the biexciton is significantly shorter than that of the exciton, by at least a factor of two in these measurements, the maximum possible emission rate from the biexciton state can be higher. Another advantage is a reduction in the timing jitter associated with the uncertainty in the time between photons. This would allow the photon detector used in an application to be gated 'on' for a shorter time, thus reducing its dark count probability.

Finally, we briefly note that it is also possible to realise electrically driven single photon emission, described in detail in Ref. [15]. The device shown in Fig. 35.3(a) consists of a layer of InAs quantum dots grown within the intrinsic region of a p-i-n diode. Electroluminescence spectra are shown as a function of current in Fig. 35.3(b), and are reminiscent of the PL described above from optically excited devices. Second order correlation measurements shown in Fig. 35.3(c) reveal $g^{(2)}(\tau)$ to be 0.11 for applied Voltage pulses of 150 mV high and 400 ps wide.

The complete decay of the X_2 state to the ground state emits 2 photons. It is intriguing to consider the relationship between the properties of the exciton photon, and the biexciton photon, and to look for correlations.

PL was measured from a quantum dot isolated by a 2 μm hole in a metal mask. The spectra from quantum dots in such samples tend to have narrower linewidths, and less dominant

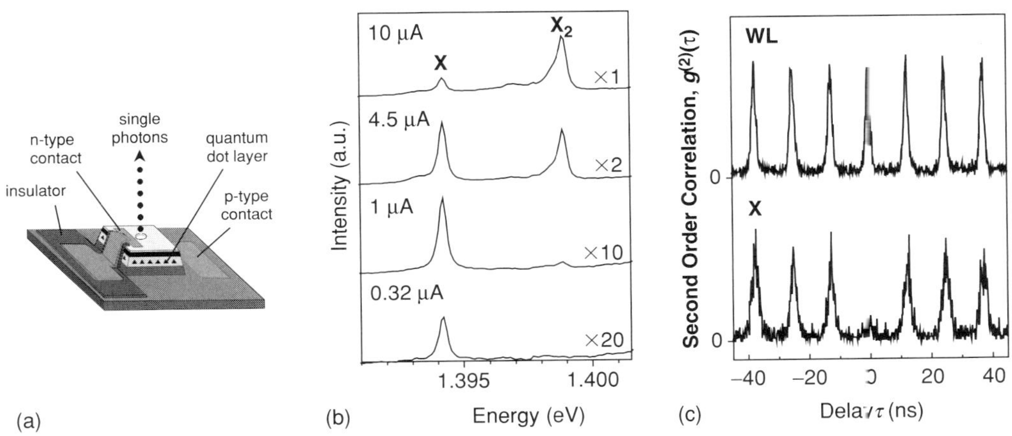

FIGURE 35.3. (a) Schematic of single photon emitting diode. (b) Electro-luminescence spectra from the device for different injection currents. (c) Second order correlation function of the electro-luminescence of the single exciton line. For comparison, the correlation trace of the wetting layer electro-luminescence is also shown.

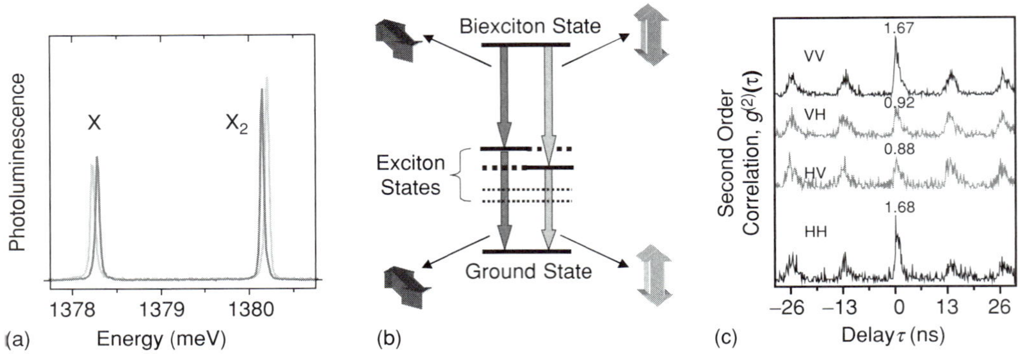

FIGURE 35.4. (a) Linear polarization dependent spectra from a quantum dot. (b) Sketch showing polarization dependent energy level structure of a quantum dot. (c) Second order correlation of biexciton photons with exciton photons, for different combinations of linear polarision.

contribution from charged states, both attributed to lower probability of transient charge trapping, a feature inherent of etched side walls. The detection polarization dependent PL spectra are shown in Fig. 35.4(a) for X and X_2 emission. Clear polarization dependence is observed, and both X and X_2 show doublet structure with orthogonally linear polarised components. The polarization splitting is equal for the X and X_2 doublets, 56 μeV in this example, although a wide range of splittings have been observed for other dots. The total energy of the X and X_2 photons is equal for the horizontal and vertical polarization, which implies two alternative decay paths for the biexciton state as shown in Fig. 35.4(b). The parallel decay paths each produce a pair of colinearly polarised photons, the polarization of which is opposite for each path.

This behaviour is due to the non equivalent properties of the exciton wavefunction in the x and y planar directions, which lifts the degeneracy of the optically active exciton level and results in linearly polarised transitions[16,17]. Elongation of the quantum dot,[18] and built in strain fields,[19] are both likely to contribute this splitting of the optically active exciton level, and both would lead to emission linearly polarised along the [110] and [1-10] directions of the crystal, which is the same as measured here.

The correlation between the different polarised components of the X and X_2 lines were also measured. This experiment was performed using a modification of the Hanbury–Brown and Twiss set up.[10] A beamsplitter directed PL emission to two spectrometers, set to pass emission from the first (X_2) and second (X) photon respectively. The polarization and arrival time of the photons transmitted through each spectrometer was then measured by a polarising beam splitter, and pair of APDs. The correlation between X_2 and X photons for all possible polarization combinations are shown in Fig. 35.4(b). In contrast to the single line correlation measurements of Fig. 35.2, the peaks at finite delay have an asymmetric lineshape, with the decay on the longer time delay side of the peak slower than the rise at shorter delays. This derives from the difference in the lifetime of the X_2 and X states used for the start and stop channels. The second order correlation of identically linear polarised pairs are very similar to each other, and show a relatively strong peak at $\tau = 0$. The shape of this peak demonstrates suppression in the probability of detecting the exciton photon before the biexciton, characterised by the sharp rising edge.[19,20] The second order correlation of photons of opposite linear polarization are also similar to each other, and have zero delay peaks that are similar in size to the other peaks, in strong contrast to the pairs of the same polarization. The area of the central peaks is normalized to the average integrated area of the other peaks arising from pairs of photons in different laser periods. The resulting areas are 1.67 ± 0.18 and 1.68 ± 0.33 for the photons of the same polarization, and 0.92 ± 0.14 and 0.88 ± 0.23 for photons of opposite polarization. This shows that there is strong correlation in the polarization of the first and second photon emitted from the quantum dot, where 65% of the photon pairs contain photons of the same linear polarization, while only 35% have opposite polarizations. This result is found to be independent of the polarization of the laser excitation, and the time integrated total emission is found to be un-polarized within experimental error.

By the insertion of an appropriately oriented quarter wave plate in the path of the emission we analyze the relationship between circular polarised photons emitted from the sample was analyzed, and found no polarization dependence, showing that the circular component of the detected photons, including those components due to potential phase shifts in the measurement system, is small.

It is noted that the degree of polarization correlation is strongly dependent on the rotation of the linear polarization relative to the polarization beam splitters. The correlation in the polarization of the photon pair reduces from a maximum close to no rotation of linear polarized light, to the case when no polarization dependence is resolvable beyond a rotation of the polarization by 20°. This is consistent with the expected behavior for correlated photon pairs, as the path selection by the polarization splitter becomes random as the orientation of the photon polarization approaches 45° to the vertical axis. It is also in contrast to the expected behaviour of polarization entangled photon pairs,[21] where the correlation is expected to be independent of the wave plate rotation. It is the splitting of the exciton level in the formation of two distinct decay paths, which destroys the entanglement of the system.

The generally unpolarised emission of the quantum dot here suggests that there is no preferential selection of a particular decay path, and the fact that a significant number of pairs are emitted in the opposite polarization suggests scattering of the exciton between the non-degenerate levels, or imperfect selection of the polarization in our measurement system. The scattering of linearly polarized laser light reflected off the sample surface into the orthogonal polarization is found to be less than 1%, and the very strong polarization of the X and X_2 emission doublets suggests that exciton scattering may be the cause, although from these measurements a scattering time 1.2 ns can be determined, of a similar order to the exciton lifetime.[22] This is at least an order of magnitude faster than the spin scattering lifetime measured elsewhere.[23]

The results presented above have only probed the optically active states of quantum dots. However, there are two additional states that are unable to couple to external radiation due to their total spin angular momentum of $m = \pm 2$. These so called dark states have been relatively little studied, as they cannot be observed directly in common experiments such as PL. However, by using magnetic fields to break in-plane symmetry, it is possible to at least partially mix the dark states with optically active bright states, making them accessible via PL[24–25].

One question is what role do dark states play in the dynamics of emission at zero field. For example, the formation of optically inactive states is clearly not desirable in single photon sources, and although the radiative decay of biexciton states cannot populate dark states, it is possible for weakly driven sources.

Magnetic fields were generated using a superconducting magnet. In the Faraday geometry, where the field is perpendicular to the sample surface, the four polarised lines observed at 0T shown in Fig. 35.5(b) are also seen in the unpolarised PL in the Faraday geometry, shown at 5T in Figure 35.5(c). There is a diamagnetic shift to higher energy of all lines, and Zeeman splitting

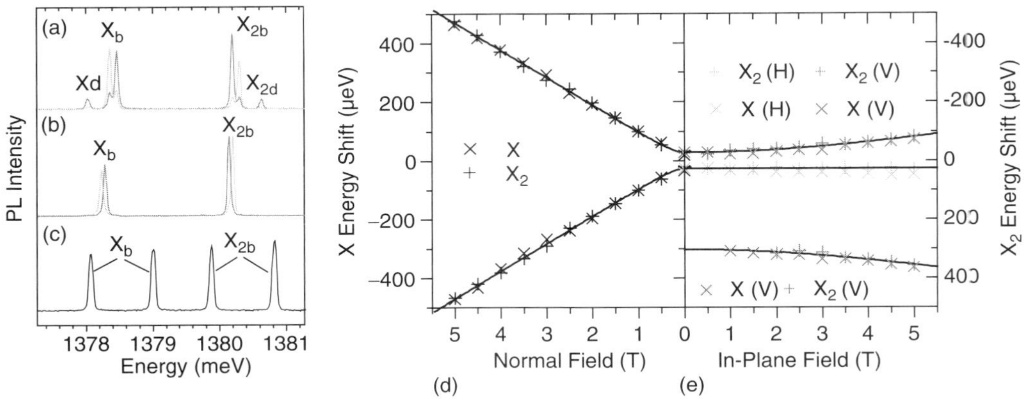

FIGURE 35.5. Photoluminescence spectra from a single quantum dot for (a) 5T applied in the plane of the sample, (b) zero magnetic field, and (c) 5T applied normal to the sample. Dark and light lines represent vertically and horizontally polarised detection for (a) and (b). In the Faraday geometry all features were measured from unpolarised spectra. (d) and (e) Zeeman splitting of exciton (left axis) and biexciton (right inverted axis) for the (d) Faraday and (e) Voigt geometries. Diamagnetic shifts have been subtracted for clarity. Solid lines show fitted theoretical behavior.

of the X and X_2 doublets. In the Voigt geometry, the field is applied in the plane of the sample, and the corresponding PL spectra is shown on Figure 35.5(a). The energies of all lines have changed to a much lesser extent from 0T, and in addition two new lines are observed in the nominally horizontal polarization. These new lines are ascribed to the mixing of the dark and bright states that exist at 0T. In the range of fields accessible in these experiments, the emission lines that evolve from the bright states at 0T remain dominant. Thus for the purposes of identification these states are denoted as 'brighter' excitons, and the weaker states 'darker' excitons.

The magnetic field dependence of the exciton and biexciton lines was measured for both the Faraday and Voigt geometries. The influence of the diamagnetic shift was then isolated from the dispersion of the average energies of pairs of exciton and biexciton lines. For the Voigt geometry, the Zeeman splitting effectively acts between darker and brighter states rather than pairs of bright states as in the Faraday geometry, so we have taken the average of the brighter and predominantly darker state of the same polarization. The diamagnetic shifts are all found to be proportional to the square of the field, and are measured to be 10.7 ± 0.3 and 7.4 ± 0.3 μeVT^{-2} respectively for the exciton and biexciton in the Faraday geometry, and 5.4 ± 0.2 and 3.9 ± 0.1 μeVT^{-2} respectively for the exciton and biexciton in the Voigt geometry. These values are not unlike those quoted for the single exciton by others.[26] The larger diamagnetic coefficients for the Faraday geometry are attributed to stronger quantum confinement in the vertical direction[27] as the diamagnetic shift is determined by the exciton's extent normal to the magnetic field. The relatively small diamagnetic coefficient for biexcitons could reflect the interplay between magnetic confinement and coulomb repulsion, the latter of which gives rise to the negative biexciton binding energies in small quantum dots such as those studied here.[28]

The diamagnetic shifts are subtracted from the magnetic field dependence, and the energies are plotted in Fig. 35.5(d) and (e) relative to the average exciton energy at zero field. It should be noted that the biexciton energies are similarly plotted on a relative scale, but that the scale is inverted. The apparent coincidence of points from exciton and biexciton measurements in this highlights the fact that Zeeman shifts in biexciton emission energies are equal and opposite to those for exciton emission. This arises because the biexciton state and ground state are spin singlet states, so Zeeman shifts of all transitions are determined only by the single exciton state. The g factors describing the Zeeman splitting of the exciton and biexciton lines in the Faraday geometry are 3.23 ± 0.01 and 3.25 ± 0.01 respectively. This is several times larger than the Zeeman splitting between horizontally polarized exciton and biexciton lines in the Voigt geometry, which we measure to be 1.00 ± 0.01 and 0.98 ± 0.03. The vertically polarized photons for this geometry have an extremely weak dependence on magnetic field. These lines may be expected to repel with a second dark state, which we do not observe in the range of energies studied in these measurements. A large energy separation to the lowest energy dark state could mask any Zeeman splitting, and would also result in only weak mixing with optically active states, making it difficult to observe.

Time resolved photoluminescence was measured by switching the laser to pulsed mode, filtering the emission using the spectrometer so that the band pass of ~60 μeV contains only one emission line, and measuring the arrival time of photons detected by a silicon avalanche photodiode. The laser power was adjusted so that the biexciton and exciton intensities were close to saturation. Example time resolved photoluminescence spectra for the horizontally polarized exciton states in the Voigt geometry are shown in Fig 35.6(a). A small time independent background, from dark counts and stray light, has been removed from all traces. The lifetime for the brighter of the two horizontally polarized excitons increases by more than 20% for a 4T increase in the field. In contrast, the darker of the two excitons, not detectable at 0T, shows an almost 30% reduction in lifetime over only 1 T. Moreover, the lifetime for the darker exciton is many times that of the brighter exciton, even at 4.5 T. We note that the long lifetime of the darker exciton results in a low count rate per time bin, which allows the time resolved background peak from wetting layer and substrate to appear as a bump close to time $\tau = 0$. As this emission is only spread across ~0.4 ns, it does not affect the lifetime measurements.

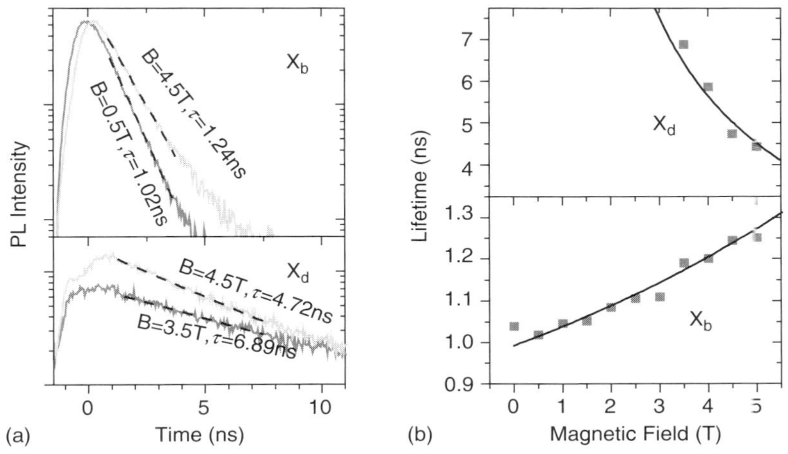

FIGURE 35.6. Time resolved photoluminescence for the brighter (X_b) and darker (X_d) horizontally polarised exciton for several in-plane magnetic fields as shown. (b) Measured lifetimes of as a function of in-plane magnetic field. Lines represent fitted behavior according to a weighted average decay rate model.

The measured lifetimes of these two states are plotted in Fig. 35.6(b) as a function of in-plane magnetic field. For the brighter horizontally polarized exciton, a clear trend of increasing lifetime with magnetic field is observed. In contrast, for the darker horizontally polarised exciton, we see a trend of decreasing lifetime with magnetic field is seen.

The lifetimes of the mixed exciton states are thus found to be strongly dependent on the applied field. To model the change in lifetimes, the effective radiative decay rate of the mixed state as the weighted average of the radiative decay rates of the pure bright and dark states are described. Fits reveal a pure dark state decay rate insignificant compared to the pure exciton decay rate. An accurate determination of the dark exciton lifetime is not possible due to the strong gradient of the fitting function close to 0 T. An estimated lower limit of 20 ns can be placed, although we stress that the real value is much likely to be longer.

In conclusion, these experiments have shown that single quantum dots show great promise as single photon emitters for applications in quantum information. By using biexciton emission, the jitter of the photon emission time is halved, allowing higher frequency operation, and detectors to be gated on for shorter times, reducing detector dark counts and increasing the maximum distance over which a quantum cryptographic key can be formed. This mode of operation also prevents the formation of dark exciton states, which we have determined from these experiments would block photon emission for at least 20 ns, which corresponds to several excitation periods. Photon emission following dark state formation would otherwise proceed via excitation into the biexciton state and subsequent decay, a process that these experiments show delays exciton photon emission and thus increases the jitter of the device. A final use of a such a device is to rely upon the randomness of the polarization of each photon pair emitted, and employ polarization dependent path selection to prepare biexciton photons into states compatible with the B92 quantum cryptography protocol. Provided the splitting of the exciton level is not larger than the linewidth, a quantum key can be generated, passively encoded and transmitted, and its value determined independently by the sender by measurement of the polarization correlated exciton photons.

References

1. See, for instance, D. Bouwmeester, A. Ekert and A. Zeilinger, (Eds)., The Physics of Quantum Information (Springer, Berlin, 2000).
2. E Knill, R Laflamme and G J Milburn (2000), *Nature* **409**, 46 (2001).
3. A J Shields, *Science* **297**, 1821 (2002).
4. F. Walls and G. J. Milburn, *Quantum Optics* (Springer, Berlin, 1994)
5. P. Michler, A. Kiraz, C. Becher, W. V. Scoenfeld, P. M. Petroff, Lidong Zhang, E. Hu, A. Imamoglu, *Science* **290**, 2282 (2000).
6. C. Santori, M. Pelton, G. Solomon, Y. Dale, and Y. Yamamoto, *Phys. Rev. Lett.* 86, 1502 (2001).
7. V. Zwiller, H. Blom, P. Jonsson, N. Panev, S. Jeppesen, T. Tsegaye, E. Goobar, M.-E. Pistol, L. Samuelson, and G. Björk, *Appl. Phys. Lett.* **78**, 2476 (2001).
8. E. Moreau, I. Robert, J. M. Gérard, I. Abram, L. Manin, and V. Thierry-Mieg et al, *Appl. Phys. Lett.* **79**, 2865 (2001).
9. R. M. Thompson, R. M. Stevenson, A. J. Shields, I. Farrer, C. J. Lobo, D. A. Ritchie, M. L. Leadbeater, and M. Pepper, *Phys. Rev. B* **64**, 201302R (2001).
10. R. M. Stevenson, R. M. Thompson, A. J. Shields, I. Farrer, B. E. Kardynal, D. A. Ritchie and M. Pepper, *Phys. Rev. B* **66**, 081302R (2002)
11. K. Hinzer, P. Hawrylak, M. Korkusinski, S. Fafard, M. Bayer, O. Stern, A. Gorbunov, and A. Forchel, *Phys. Rev. B* **63**, 075314 (2001).

12. J. J. Finley, A. D. Ashmore, A. Lemaître, D. J. Mowbray, M. S. Skolnick, I. E. Itskevich, P. A. Maksym, M. Hopkinson, and T. F. Krauss, *Phys. Rev. B* **63**, 073307 (2001).
13. E. Dekel, D.V. Regelman, D. Gershoni, E. Ehrenfreund, W.V. Schoenfeld, and P.M. Petroff, *Solid State Communications*, **117**, 395 (2001).
14. A J Shields, R.M. Stevenson, R.M. Thompson, Z. Yuan and B.E. Kardynal, *Nano-Physics and Bio-Electronics: A New Odyssey* (Elsevier, Amsterdam, 2002).
15. Z. Yuan, B. E. Kardynal, R. M. Stevenson, A. J. Shields, C. J. Lobo, K. Cooper, N. S. Beattie, D. A. Ritchie, and M. Pepper, *Science* **295**, 102 (2002).
16. V. D. Kulakovskii, G. Bacher, R. Weigand, T. Kümmell, A. Forchel, E. Borovitskaya, K. Leonardi and D. Hommel, *Phys. Rev. Lett.* **82**, 1780 (1999).
17. M. Paillard, X. Marie, P. Renucci, T. Amand, A. Jbeli, and J. M. Gérard, *Phys. Rev. Lett.* **86**, 1634 (2001)
18. D. Bimberg, M. Grundmann, N. N. Ledenstov *Quantum Dot Heterostructures*, (Wiley, Chichester, 1999)
19. M. Grundmann, O. Stier and D. Bimberg, Phys. Rev. B 52, 11969 (1995).
20. E. Moreau, I. Robert, L. Manin, V. Thierry-Mieg, J. M. Gérard, and I. Abram, *Phys. Rev. Lett.* **87**, 183601 (2001).
21. O. Benson, C. Santori, M. Pelton and Y. Yamamoto, *Phys. Rev. Lett.* **84**, 2513 (2000)
22. C. Santori, D. Fattal, M. Pelton, G. S. Solomon, and Y. Yamamoto, *Phys. Rev. B* **66**, 045308 (2002).
23. T. Flissikowski, A. Hundt, M. Lowisch, M. Rabe, and F. Henneberger, *Phys. Rev. Lett. 86* 3172 (2001)
24. M. Bayer, O. Stern, A. Kuther and A. Forchel, *Phys. Rev. B* **61**, 7273 (2000).
25. J. G. Tischler, A. S. Bracker, D. Gammon, and D. Park, *Phys. Rev. B* **66**, 081310(R) (2002).
26. J. J. Finley, D. J. Mowbray, M. S. Skolnick, A. D. Ashmore, C. Baker, A. F. G. Monte, M. Hopkinson, *Phys. Rev. B* **66**, 153316 (2002).
27. M. Sugisaki, H-W, Ren, S. V. Nair, K. Nishi, Y. Masumoto, *Phys. Rev. B* **66**, 235309 (2002).
28. S. Rodt, R. Heitz, A. Schliwa, R. L. Sellin, F. Guffarth, and D. Bimberg, *Phys. Rev. B* **68**, 035331 (2003).

36

Semiconductor few-electron quantum dots as spin qubits

J. M. ELZERMAN, R. HANSON, L. H. WILLEMS VAN BEVEREN, L. M. K. VANDERSYPEN, AND L. P. KOUWENHOVEN

Kavli Institute of Nanoscience Delft and ERATO Mesoscopic Correlation Project, Delft University of Technology, PO Box 5046, 2600 GA Delft, The Netherlands

Abstract

The experimental steps taken towards using the spin of a single electron trapped in a semiconductor quantum dot as a spin qubit[1] is described. Fabrication and characterization of a double quantum dot containing two coupled spins has been achieved, as well as initialization and single-shot read-out of the spin state. The single-spin relaxation time was found to be of the order of a millisecond, but the decoherence time is still unknown. Concrete ideas – using a charge detection approach – on how to proceed towards demonstrating superposition and entanglement of spin states are presented. Single-spin manipulation relies on a microfabricated wire located close to the quantum dot, and two-spin interactions are controlled via the tunnel barrier connecting the respective quantum dots.

36.1 Qubit

Any implementation of a quantum bit has to satisfy the five DiVincenzo requirements.[2] The first requirement is to have a scalable physical system with well-characterized qubits. Double quantum dot devices have been fabricated in which a single electron can be confined in each of the two dots.[3] The spin states $\uparrow$ and $\downarrow$ of the electron, subject to a large magnetic field B, correspond to the two states of the proposed qubit twolevel system. The Zeeman splitting, ΔE_Z, between the two states can be tuned with the magnetic field, according to $\Delta E_Z = g\mu_B B$, with $g \approx -0.44$ the electron g-factor in GaAs,[4] and μ_B the Bohr magneton.

These one-electron dots can be fully characterized using a QPC as a charge detector.[3,4] First of all, the QPC can be used to monitor the charge configuration of the double dot, in order to reach the regime where both dots contain just a single electron. Then the tunnel rate from each dot to the reservoir can be evaluated and tuned using a lock-in technique. The same technique can be employed to determine the energy spectrum of each of the two dots, i.e., the Zeeman splitting between the two qubit states, as well as the energy of orbital excited states. Furthermore, the QPC can be used to monitor the inter-dot tunnel barrier, both qualitatively (from the curvature of lines in the honeycomb diagram) and quantitatively (by performing photon-assisted tunneling spectroscopy to measure the tunnel splitting between the one-electron bonding and antibonding state[5]). In principle, it is even possible to use the lock-in technique to measure the exchange splitting J between the delocalized two-electron singlet and triplet spin states. However, in practical situations the splitting might be too small ($< 20\,\mu$eV) to be resolved using tunneling spectroscopy.

All relevant parameters of the two-spin system can thus be determined without performing transport measurements. The essential advantage of the QPC technique is that it works even for a dot that is very weakly coupled to just a *single* reservoir, with a tunnel rate between zero and ~ 100 kHz (limited by the bandwidth of the current measurement setup). This gives us more freedom to design simpler dots with fewer gates, which could therefore be easier to operate.

36.2 Read-out

Single-shot read-out of the spin orientation of an individual electron in a quantum dot[6] has been achieved. The approach used here utilizes the Zeeman splitting, induced by a large magnetic field parallel to the 2DEG, to create spin-to-charge conversion (Fig. 36.1a). This is followed by real-time detection of single-electron tunneling events using the QPC. The total visibility of the spin measurement is $\sim 65\%$, limited mostly by the ~ 40 kHz bandwidth of our current measurement setup, and also by thermal excitation of electrons out of the quantum dot, due to the high effective electron temperature of ~ 300 mK.

It was estimated that the visibility of the spin read-out technique could be improved to more than 90% by lowering the electron temperature below 100 mK, and especially by using a faster way to measure the charge on the dot. This could be possible with a 'Radio-Frequency QPC' (RF-QPC), similar to the well-known RF-SET.[7] In this approach, the QPC is embedded in an LC circuit with a resonant frequency of ~ 1 GHz. By measuring the reflection or transmission of a resonant carrier wave, it is estimated that it should be possible to read out the charge state of the nearby quantum dot in $\sim 1\,\mu$s, an order of magnitude faster than is currently attainable.

A disadvantage of the read-out technique based on the Zeeman splitting is that it relies on accurate positioning of the dot-levels with respect to the Fermi energy of the reservoir, E_F (see Fig. 36.1(a)). This makes the spin read-out very sensitive to charge switches, which can easily push the $\uparrow$ level above E_F, or pull $\downarrow$ below E_F,resulting in a measurement error. To counteract this effect, a large enough Zeeman splitting is required (in Ref. [6] a magnetic field of more than 8 Tesla was used, although with a more stable sample a lower field might be sufficient). On the

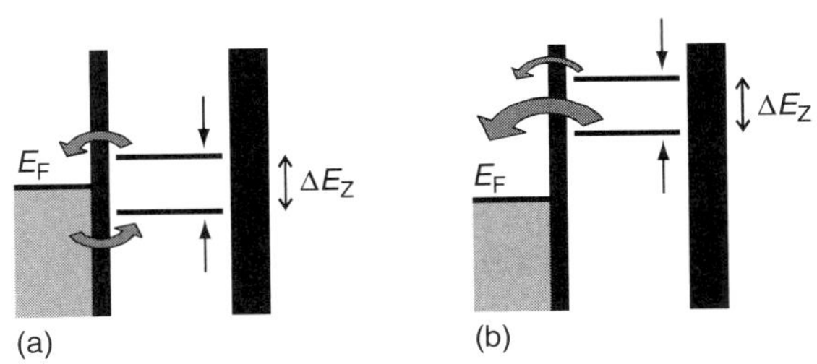

FIGURE 36.1. Schematic energy diagrams depicting (**a**) spin-to-charge conversion based on a difference in *energy* on a difference in *tunnel rate* (**b**) between $\uparrow$ and $\downarrow$.

other hand, a smaller Zeeman splitting is desirable because it implies a lower and therefore more convenient resonance frequency for coherent spin manipulation. In addition, the spin relaxation time is expected to be longer at smaller ΔE_Z. Therefore, a different spin read-out mechanism that is less sensitive to charge switches and can function at lower fields would be very useful.

A particularly convenient way to perform spin-to-charge conversion could be provided by utilizing not a difference in *energy* between spin-up and spin-down, but a difference in *tunnel rate* (Fig. 36.1(b)). To read out the spin orientation of an electron on the dot, we simply raise both dot levels above E_F, so that the electron can leave the dot. If the tunnel rate for spin-up electrons, $\Gamma_\uparrow$ is much larger than that for spin-down electrons, $\Gamma_\downarrow$, then at a suitably chosen time the dot will have a large probability to be already empty if the spin was up, but a large probability to be still occupied if the spin is down. Measuring the charge on the dot within the spin relaxation time can then reveal the spin state.

This scheme is very robust against charge switches, since no precise positioning of the dot levels with respect to the leads is required: both levels simply have to be above E_F. Also, switches have a small influence on the tunnel rates themselves, as they tend to shift the whole potential landscape up or down, which does not change the tunnel barrier for electrons in the dot.[8] Of course, the visibility of this spin measurement scheme depends on the difference in tunnel rate tha can be achieved .

A difference in tunnel rate for spin-up and spin-down electrons is provided by the magnetic field. From large-bias transport measurements in a magnetic field parallel to the 2DEG,[9] it was found that the spin-selectivity ($\Gamma_\uparrow/\Gamma_\downarrow$) grows roughly linearly from ~ 1.5 at 5 T to ~ 5 at 14 T. This is in good agreement with the spin-selectivity of about 3 that was found at 10 T using the single-shot spin measurement technique of Ref. [6].

In a magnetic field parallel to the 2DEG, the effect only leads to a modest spin-selectivity that does not allow a single-shot measurement. However, a much larger spin-selectivity is possible in a perpendicular magnetic field, i.e., in the quantum Hall regime. Magnetotransport measurements in 2DEGs with odd filling factor have shown that the effective g-factor can be enhanced by as much as a factor of ten, depending on the field strength.[10] In this case, $\downarrow$ electrons tunneling into or out of the dot experience a thicker tunnel barrier than $\uparrow$ electrons, resulting in a difference in tunnel rates.[11] A convenient perpendicular field of ~ 4 T is anticipated to give enough spin-selectivity to allow high-fidelity spin read-out.

36.3 Initialization

Initialization[6] of the spin to the pure state $\uparrow$ – the desired initial state for most quantum algorithms have been demonstrated. By waiting long enough, energy relaxation will cause the the spin on the dot to relax to the $\uparrow$ ground state (Fig. 36.2(a)). This is a very simple and robust initialization approach, which can be used for any magnetic field orientation (provided that $g\mu_B B > 5k_B T$). However, as it takes about $5T_1$ to reach equilibrium, it is also a very slow procedure (~ 10 ms), especially at lower magnetic fields, where the spin relaxation time T_1 might be very long.

A faster initialization method is to place the dot in the read-out configuration (Fig. 36.2(b)), where a spin-up electron will stay on the dot, whereas a spin-down electron will be replaced by a spin-up.[6] After waiting a few times the sum of the typical tunnel times for spin-up and spin-down ($\sim 1/\Gamma_\uparrow + 1/\Gamma_\downarrow$), the spin will be with large probability in the $\uparrow$ state. This initialization procedure can therefore be quite fast (< 1 ms), depending on the tunnel rates.

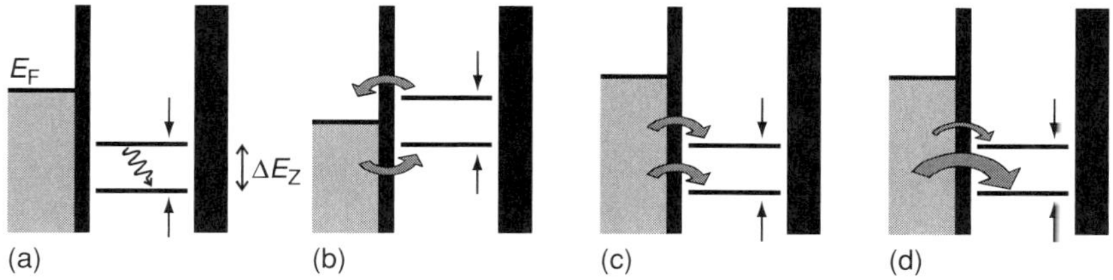

FIGURE 36.2. Schematic energy diagrams depicting initialization procedures in a large parallel or perpendicular magneticfield. (**a**) Spin relaxation to pure state ↑. (**b**) The 'read-out' configuration can result in ↑ faster. (**c**) Random spin injection gives a statistical mixture of ↑ and ↓. (**d**) In a large perpendicular field providing a strong spin-selectivity, injection results mostly in ↑.

To initialize the dot to a mixed state is also possible, where the spin is probabilistically in ↑ or ↓. In Ref. [6], mixed-state initialization was demonstrated in a parallel field by first emptying the dot, followed by placing both spin levels below E_F during the 'injection stage' (Fig.36.2(c)). The dot is then randomly filled with either a spin-up or a spin-down electron. This is very useful, e.g., to test two-spin operations.

In a large perpendicular field providing a strong spin-selectivity, initialization to the ↑ state is possible via spin relaxation (Fig.36.2(a)) or via direct injection (Fig. 36.2(d)). Initialization to a mixed state (or in fact to any state other than ↓) is very difficult due to the spin-selectivity. It probably requires the ability to coherently rotate the spin from ↑ to ↓ (see Section 36.5).

36.4 Coherence Times

The long-term potential of GaAs quantum dots as electron spin qubits clearly depends crucially on the spin coherence times T_1 and T_2. It has been shown that the single-spin relaxation time, T_1, can be very long – on the order of 1 ms at 8 T.[6] This implies that the spin is only very weakly disturbed by the environment. The dominant relaxation mechanism at large magnetic field is believed to be the coupling of the spin to phonons, mediated by the spin-orbit interaction.[12]

The fundamental quantity of interest for spin qubits is the decoherence time of a single electron spin in a quantum dot, T_2, which has never been measured. Experiments with electrons in 2DEGs have established an ensemble-averaged decoherence time, T_2^*, of ~ 100 ns.[13] A similar lower bound on T_2 has been claimed for a single trapped electron spin, based on the linewidth of the observed electron spin resonance.[14] Theoretically, it has been suggested that the real value of T_2 can be much longer,[12] and under certain circumstances could even be given by $T_2 = 2T_1$, limited by the same spin-orbit interactions that limit T_1.

36.5 Coherent Single-spin Manipulation: ESR

The key requirement for an actual spin qubit is yet to be satisfied: coherent manipulation of one- and two-spin states. To create controllable superpositions of ↑ and ↓, the well-known Electron Spin Resonance (ESR) effect can be used. A microwave magnetic field B_{ac} oscillating in the plane perpendicular to B, at a frequency $f = g\,\mu_B B/h$ (in resonance with the spin precession about B) causes the spin to make transitions between $\uparrow$ and $\downarrow$. The choice of B strength is a trade-off between reliable initialization and read-out (strong B is better) and experimental convenience (low f is easier). It is expected that a perpendicular field of 4 T should be sufficient to provide high-fidelity read-out and initialization, with $f \approx 25$ GHz (for $g = -0.44$). Alternatively, in a parallel field we may have to go up to 8 T, corresponding to $f \approx 45$ GHz,[15] for high-fidelity spin measurement. However, since single-shot read-out is not strictly required, a somewhat lower field could also be enough.

Properly timed bursts of microwave power tip the spin state over a controlled angle, e.g., 90° or 180°. In order to observe Rabi oscillations, the Rabi period must be at most of the order of the single-spin decoherence time T_2. For a Rabi period of 150 ns, we need a microwave field strength B_{ac} of ~ 1 mT. If T_2 is much longer, there is more time to coherently rotate the spin, so a smaller oscillating field is sufficient.

We intend to generate the oscillating magnetic field by sending an alternating current through an on-chip wire running close by the dot (Fig. 36.3(a)). If the wire is placed well within one wavelength (which is a few mm at 30 GHz near the surface of a GaAs substrate) from the quantum dot, the dot is in the near-field region and the electric and magnetic field distribution produced by the AC current should be the same as for a DC current. With a wire 200 nm from the dot, a current of ~ 1 mA should generate a magnetic field of about 1 mT and no electric field at the position of the dot. To minimize reflection and radiation losses, the wire is designed to be a shorted coplanar stripline (Fig. 36.3(b)) with a 50Ω impedance.

To detect the ESR and obtain a lower bound on T_2 from the linewidth of the resonance signal, various methods have been proposed, either using transport measurements[16] or relying on charge detection.[17] In both cases, the required spin-to-charge conversion is achieved by positioning the dot levels around the Fermi energy of the reservoir (Figs. 36.4(a) and (b)). The ESR field induces spin flips, exciting $\uparrow$ electrons to $\downarrow$, which can then tunnel out of the dot. This leads to an average current (Fig. 36.4(a)) or to a change in the average occupation of the dot (Fig. 36.4(b)). However, in this configuration the dot is particularly sensitive to spurious effects induced by the microwaves, such as $\uparrow$ electrons being excited out of the dot via thermal excitation or photon-assisted tunneling. These processes can completely obscure the spin resonance.

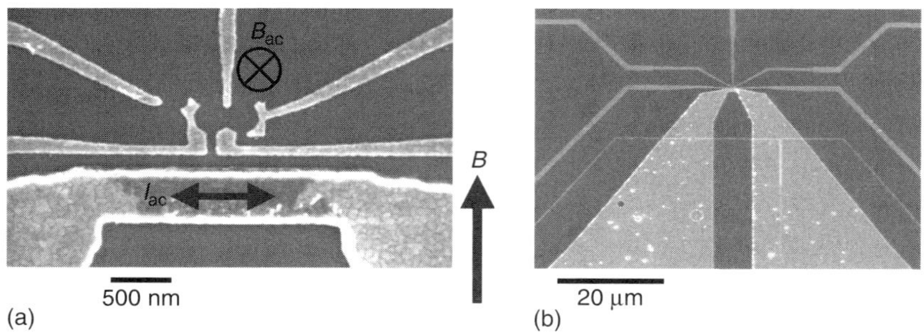

FIGURE 36.3. On-chip wire to apply microwaves to a nearby quantum dot. (**a**) Scanning electron microscope image of a device consisting of a double quantum dot in close proximity to a gold wire. An AC current through the wire, I_{ac}, generates an oscillating magnetic field, B_{ac}, perpendicular to the plane. If the AC frequency is resonant with the Zeeman splitting due to a large static in-plane magnetic field, B, a spin on the dot will rotate. (**b**) Large-scale view of the wire, designed to be a 50 Ω coplanar stripline.

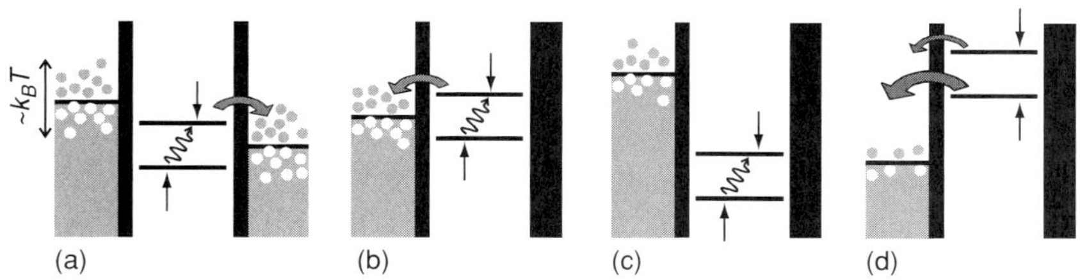

FIGURE 36.4. Detecting ESR. (**a**) To detect ESR in a transport measurement,[16] the dot is placed in Coulomb blockade, so that electron spins that are flipped by the ESR field can contribute to a current. (**b**) A similar configuration is used to detect ESR via changes in the occupation of the dot,[17] measured using a charge detector. (**c**) If the dot is deep in Coulomb blockade during the spin-flip stage, the electron is not easily excited to the reservoir via thermal excitation or photon-assisted tunneling. (**d**) The microwaves are off during the spin read-out stage to enhance the measurement fidelity.

Such problems can be avoided by combining (pulsed) electron spin resonance with *single-shot* spin measurement. This allows us to separate the spin manipulation stage (during which the microwaves are on) from the spin read-out stage (without microwaves). In this way, excitation out of the dot is prevented by Coulomb blockade (Fig. 36.4(c)), until spin read-out is initiated (Fig. 36.4(d)). In contrast to the techniques described above – which require a large spin flip rate to generate a *measurable* current or disturbance of the dot occupation – this approach only requires the spin flip rate to be faster than the decoherence rate. Therefore, a longer T_2 allows us to use a smaller B_{ac}, corresponding to (quadratically) smaller microwave power This should help to suppress heating and photon-assisted tunneling.

In principle, an ESR experiment can be performed in a parallel or a perpendicular magnetic field. The read-out in a perpendicular field is particularly suitable for ESR detection, as the dot levels are far above E_F (so are not affected by photon-assisted tunneling or heating). If B is perpendicular to the surface, B_{ac} must run through the dot in a direction parallel to the surface, so the wire must be placed above the dot rather than to its side. The wire could be located on top of an insulating dielectric layer that covers the gate electrodes.

36.6 Coherent Spin Interactions: $\sqrt{}$SWAP

Two electron spins S_1 and S_2 in neighboring quantum dots are coupled to each other by the exchange interaction, which takes the form $J(t) = S_1 \cdot S_2$. If the double dot is filled with two identical spins, the interaction does not change their orientation. However, if the left electron spin starts out being $\uparrow$ and the right one $\downarrow$, then the states of the two spins will be swapped after a certain time. An interaction active for half this time performs the $\sqrt{}$SWAP gate, which has been shown to be universal for quantum computation when combined with single qubit rotations.[18] In fact, the exchange interaction is even universal by itself when the state of each qubit is encoded in the state of three electron spins.[19]

The strength $J(t)$ of the exchange interaction depends on the overlap of the two electron wavefunctions, which varies exponentially with the voltage applied to the gate controlling the inter-dot tunnel barrier. By applying a (positive) voltage pulse with a certain amplitude and duration, the exchange interaction can be temporarily turned on, thereby performing a $\sqrt{}$SWAP gate.

It is expected that J may correspond to a frequency of $\sim 10\,\text{GHz}$, so two-qubit gates could be performed in $\sim 100\,\text{ps}$. A much larger value would not be convenient experimentally, as the exact amplitude and duration of the pulse have to be controlled very precisely. On the other hand, a very slow exchange operation would be more sensitive to decoherence resulting from fluctuations in the tunnel rate, due to charge noise. The value of J can in principle be determined in a transport measurement,[20] or alternatively by using a QPC tunneling spectroscopy technique.[4] However, in practical situations J might be too small to be resolved.

To explore the operation of the SWAP gate, reliable initialization and read-out is only needed, without requiring ESR. Imagine qubit 1 is prepared in a pure state $\uparrow$ and qubit 2 is prepared in a statistical mixture of $\uparrow$ and $\downarrow$. Measurement of qubit 1 should then always give $\uparrow$, while measurement of qubit 2 should give probabilistically $\uparrow$ or $\downarrow$. After application of the SWAP gate, in contrast, measurement of qubit 2 should always give $\uparrow$, while measurement of qubit 1 should give a probabilistic outcome. This and other spin-interaction experiments are probably easiest in a parallel magnetic field, where initialization to a statistical mixture is convenient. In addition, a large perpendicular field shrinks the electron wavefunctions, lowering the tunnel coupling and thus the exchange interaction between the two dots.

36.7 Conclusion

It has been demonstrated that single electrons trapped in GaAs lateral quantum dots are promising candidates for implementing a spin qubit. The 'hardware' for such a system is: a device consisting of two coupled quantum dots that can be filled with one electron spin each, flanked by two quantum point contacts. Using these QPCs as charge detectors, we can determine all relevant parameters of the double dot can be determined. In addition, a technique has been developed to measure the spin orientation of an *individual* electron. Now all these ingredients can be combined with the ability to generate strong microwave magnetic fields close to the dot, and gate voltage pulses to control the inter-dot coupling, in order to demonstrate superposition and entanglement of spin states.

Acknowledgments. We thank C. J. P. M. Harmans, W. G. van der Wiel, M. Blaauboer, S. Tarucha, and D. Loss for useful discussions. We acknowledge financial support from the Specially Promoted Research Grant-in-Aid for Scientific Research from the Japanese Ministry of Science; DARPA grant DAAD19-01-1-0659 of the QuIST program; the Dutch Organization for Fundamental Research on Matter (FOM); and the European Union through a TMR Program Network.

References

1. D. Loss and D. P. DiVincenzo, *Phys. Rev. A* **57**, 120 (1998).
2. D. P. DiVincenzo, *Fortschr. Phys.* **48**, 771 (2000).
3. J. M. Elzerman, R. Hanson, J. S. Greidanus, L. H. Willems van Beveren, S. De Franceschi, L. M. K. Vandersypen, S. Tarucha, and L. P. Kouwenhoven *Phys. Rev. B* **67**, 161308 (2003).
4. J. M. Elzerman, R. Hanson, L. H. Willems van Beveren, L. M. K. Vandersypen, and L. P. Kouwenhoven, *Appl. Phys. Lett.* **84**, 4617 (2004).
5. J. R. Petta, A. C. Johnson, C. M. Marcus, M. P. Hanson, and A. C. Gossard, cond-mat/0408139 (2004).

6. J. M. Elzerman, R. Hanson, L. H. Willems van Beveren, B. Witkamp, L. M. K. Vandersypen, and L. P. Kouwenhoven, *Nature* **430**, 431 (2004).
7. R. J. Schoelkopf, P. Wahlgren, A. A. Kozhevnikov, P. Delsing, and D. E. Prober, *Science* **280**, 1238 (1998).
8. S .W. Jung, T. Fujisawa, Y. H. Jeong and Y. Hirayama, *Appl. Phys. Lett.* **85**, 768 (2004).
9. R. Hanson, B. Witkamp, L. M. K. Vandersypen, L. H. Willems van Beveren, J. M. Elzerman, and L. P. Kouwenhoven, *Phys. Rev. Lett.* **91**, 196802 (2003).
10. T. Englert, D. C. Tsui, A. C. Gossard, and C. Uihlein, *Surf. Sci.* **113**, 295 (1982).
11. M. Ciorga *et al.*, *Phys. Rev. B* **61**, R16315 (2000).
12. V. N. Golovach, A. Khaetskii, and D. Loss, *Phys. Rev. Lett.* **93**, 016601 (2004).
13. J. M. Kikkawa and D. D. Awschalom, *Nature* **397**, 139 (1999).
14. M. Xiao, I. Martin, E. Yablonovitch, and H. W. Jiang, *Nature* **430**, 435 (2004).
15. R. Hanson, L. M. K. Vandersypen, L. H. Willems van Beveren, J. M. Elzerman, I. T. Vink, and L. P. Kouwenhoven, cond-mat/0311414 (2003).
16. H. A. Engel and D. Loss, *Phys. Rev. Lett.* **86**, 4648 (2001).
17. I. Martin, D. Mozyrsky, and H. W. Jiang, *Phys. Rev. Lett.* **90**, 018301 (2003).
18. G. Burkard, D. Loss, and D. P. DiVincenzo, *Phys. Rev. B* **59**, 2070 (1999).
19. D. P. DiVincenzo, D. P. Bacon, D. A. Lidar, and K. B. Whaley, *Nature* **408**, 339 (2000).
20. V. N. Golovach and D. Loss, *Phys. Rev. B* **69**, 245327 (2004).

37

Spin amplifier for single spin measurement

PAOLA CAPPELLARO[1], JOSEPH EMERSON[2], NICOLAS BOULANT[1], CHANDRASEKHAR RAMANATHAN[1], SETH LLOYD[3], DAVID G. CORY[1]

[1] Department of Nuclear Engineering, Massachusetts Institute of Technology - Cambridge, MA 02139, USA.
[2] Perimeter Institute of Theoretical Physics, Waterloo, ON N2J 2W9, Canada.
[3] Department of Mechanical Engineering, Massachusetts Institute of Technology - Cambridge, MA 02139, USA.

Abstract

A new approach to the measurement of the state of a collapsed single spin is described by using many entangled spins as an amplifier. A single target spin is coupled via the natural dipolar Hamiltonian to a large collection of spins. Applying external radio frequency (r.f.) pulses, we can control the evolution of the system so that the ensemble spins reach one of two orthogonal states whose collective properties differ depending on the state of the target spin and are easily measured. The result of an experiment simulating this method on an ensemble liquid state NMR quantum processor is reported. That entanglement assisted metrology is compatible with the real control we have over physical spins is suggested, since the measurement process can actually be described in terms of the physical Hamiltonian of the spin system. By building on this work, and with the needed technical advances, it should be possible to detect a single nuclear spin.

37.1 Introduction

The measurement of a single nuclear spin is an experimentally challenging task that has stimulated a wide interest since it could bring useful applications as well as valuable physical insights. Single spin measurement, for example, could find applications in spintronics devices[1], a fast evolving field that promises to enhance modern classical computer. It could lead to biomolecular microscopy, which would be a powerful instrument in medical research and structural biology. Finally, it is a critical enabling technology for Quantum Information Processing[2] (QIP), where nuclear spins in solids are candidates to be used as qubits in a quantum computer.

Different techniques are used to attack the problem: force detection[3] and near field optics[4], for example, have reached the limit of single electron spin detection, but they have failed to prove the measurement of the quantum state of a single nuclear spin. Instead of exploring new experimental techniques, a quantum measurement device is built based on NMR methods, which relies on coherent collective properties of a quantum system. The proposed device, called *Spin Amplifier*, could represent an important step in the road towards single spin detection.

The Spin Amplifier is composed of an ensemble of nuclear spins (about 10^6 spins, which is the current limit for low temperature conventional NMR detection), possibly of a different chemical species than the single spin of interest. The device is put in contact with the spin to be measures (target spin), to which it is coupled by the usual spin-spin interaction (for example the magnetic dipolar interaction). Because of this coupling, the dynamics imposed to the Amplifier spins can be modulated conditionally on the state of the target spin and the final state of the Amplifier will provide information about the quantum state of the target spin. In its simplest implementation, the interaction of the spin of interest with the ensemble of spins transfers the magnetization properties of the target to the ensemble spins, resulting in the amplification of the target spin signal, up to the point where it is detectable by the conventional inductive means.

This method requires the ability to control a large number of spins by inducing a coherent dynamics, dictated by the interaction with a single spin. Although this is experimentally challenging, it is shown that the desired dynamics can actually be described in terms of the physical Hamiltonian of the spin system, and therefore it can be implemented.

37.2 Measurement Schemes

To illustrate how a collective measurement of a macroscopic observable can provide the knowledge of the target spin state, we present a simple first scheme (Scheme 1) is presented. The ensemble of spins forming the Amplifier is first prepared in the fiducial state: $|00...0\rangle$ (or more generally in a highly polarized state). For simplicity it is assumed that the target spin has already collapsed into one of the two states $|0\rangle$ or $|1\rangle$. This is not a limitation in the context of QIP, since the measurement of only two orthogonal states is sufficient for the read-out stage in a quantum computation[5].

The desired evolution of the Spin Amplifier is described by a simple quantum circuit consisting of a train of Controlled Not (C-NOT) gates between the target spin and each one of the Amplifier spins.

The C-NOT gate inverts the state of the controlled qubit if the controlling qubit is in the state $|1\rangle$ and does nothing otherwise. If the controlled spin is in the (known) state $|0\rangle$, this amounts to copying the controlling spin state on the controlled qubit. The "no-cloning" theorem[6], which forbids the creation of identical copies of an arbitrary unknown quantum state, does not apply

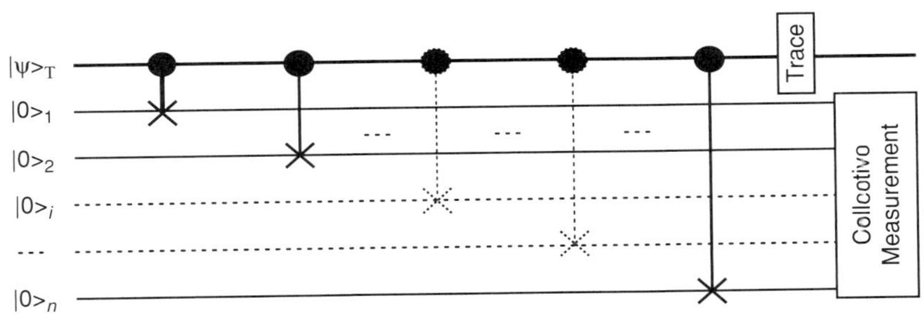

SCHEME 1. Series of C-NOT gates between the target and Amplifier spins. A collective measurement is sufficient to detect the state of the target spin.

here, since the target spin is already in one of two orthogonal states. Since the C-NOTs act on the fiducial state $|00...0\rangle_A$, the final state of the Amplifier is $|00...0\rangle_{gA}$ or $|11...1\rangle_{gA}$ if the target spin is in the state $|0\rangle_T$ or $|1\rangle_{gT}$ respectively. At the end of the circuit, the measurement of the Amplifier magnetization along the z-direction, $M_z \propto \langle\psi_A|\Sigma_i\sigma_i^z|\psi_A\rangle g = gg_z(0)$, will indicate the state of the target spin. Even if the amplifier are addressed spins individually during the scheme, the measurement, on the contrary, involves only collective properties of the Amplifier; the states of the individual spins are not measured.

The C-NOT gate can be easily implemented on a NMR system, with r.f. pulses and periods of evolution under the spin-spin coupling. This first scheme, however, demands a direct coupling between the target spin and each one of the Amplifier spins. Since any spin-spin interaction is local, this requirement is difficult to meet in practice. To avoid invoking direct interactions between the target spin and each ensemble spin, entanglement among the Amplifier spins can be used to develop equivalent schemes relying only on the collective behavior of the system. These schemes are therefore more susceptible to be realizable in the near term because they connect better to the control we have on the physical system.

If the Amplifier spins remain in a factorable state as in Scheme 1, the interaction with the target spin can produce only local changes on individual spin states. On the other hand, when the Amplifier is in a macroscopic entangled state (cat-state)[7], a single interaction with the target spin can have a global effect, even if this operation acts on just one spin. For example, a C-NOT gate can be applied between the target spin and its closest neighbor, which is most strongly coupled to it, after creating a cat-state in the Amplifier. Upon refocusing of the entanglement, two different states are reached depending on the state of the target spin. The propagator creating the cat-state can be looked upon as if it performed an effective change of basis to a reference frame, where the local C-NOT gate is now a global operator on the Amplifier.

37.3 Experimental Demonstration

A scheme has been implemented experimentally that illustrates the effectiveness of entanglement to propagate a local perturbation (Scheme 2) on a small QIP NMR liquid system. In the experiment, the target spin is the single proton spin of a ^{13}C labeled Alanine molecule ($^2CH_3{-}^1C^T\mathbf{H}(NH_2){-}^3COOH$), while the 3 carbons compose the Amplifier. The target spin is therefore represented experimentally by a macroscopic ensemble of spins (the ratio target spins/Amplifier spins is 1:3). Although their state is detectable, it is measured only indirectly, following the scheme proposed.

In a strong external magnetic field, this spin system exhibits a weakly coupled spectrum corresponding to the Hamiltonian:

$$\begin{aligned} H_{\text{int}} = \omega_1\sigma_1^z + \omega_2\sigma_2^z + \omega_3\sigma_3^z) + \omega_T\sigma_T^z + \pi/2(J_{12}\sigma_1^z\sigma_2^z + J_{13}\sigma_1^z\sigma_3^z + J_{23}\sigma_2^z\sigma_3^z) \\ +\pi J_{1T}\sigma_1^z\sigma_T^z + J_{2T}\sigma_2^z\sigma_T^z + J_{3T}\sigma_3^z\sigma_T^z), \end{aligned} \quad (37.1)$$

where the ω's are Larmor frequencies and the J's the spin-spin coupling constants in Hertz. The experiments were carried out on a Bruker AVANCE-300 spectrometer in a field of 7.2 Tesla. The first carbon has a large coupling to the proton, J_{T1}=145.2. Therefore, as the spectrum of the first Carbon at thermal equilibrium shows (Fig. 37.1), the coupling with the proton are completely resolved and the signal arising from carbons coupled to protons in the $|1\rangle$ state (left) can be separated from the signal due to carbons coupled with $|0\rangle$ state protons (right).

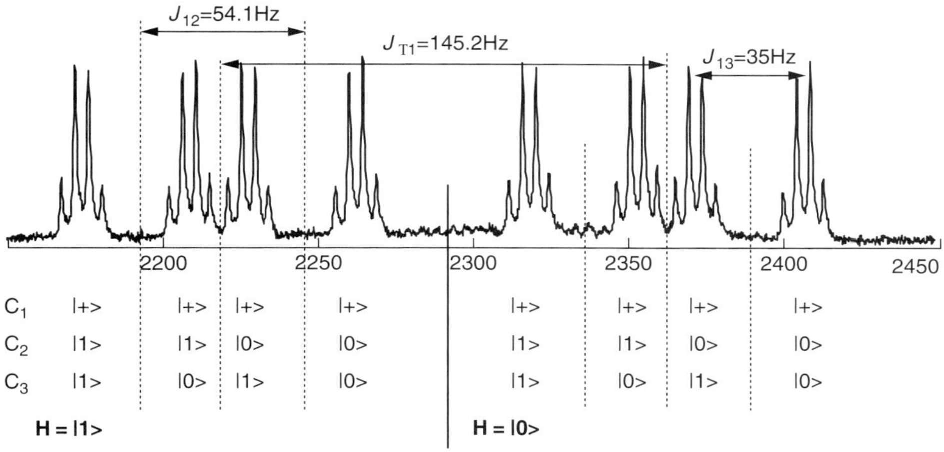

FIGURE 37.1. Spectrum of the first carbon at thermal equilibrium, showing the coupling with the proton and the other 2 carbons (The methyl group produces the multiplet splitting).

Before applying the circuit, the proton spin is first put into the identity state and the carbons are prepared in the pseudo-pure ground state $|000\rangle$[8,9]. The state of the proton is a mixture of the two decohered states $|0\rangle\langle 0|$ and $|1\rangle\langle 1|$, obtained by simulating a strong measurement with a magnetic field gradient. Two experiments can therefore effectively be performed in parallel, and the outcomes can be readout from just one spectrum, where the lines are separated by the carbon-proton coupling splitting. The pseudo-pure ground state is prepared from the thermal equilibrium state by spatial averaging, which is obtained by creating (by means of magnetic-field gradients) a spatial distribution of states across the ensemble whose mean density matrix is pseudo-pure[10].

Starting from the fiducial state (the pseudo-pure state as in the experiment or more generally the fully polarized state $|00...0\rangle$) the Amplifier is first transformed into the cat-state $(|00...0\rangle - i|11...1\rangle)/\sqrt{2}$ by a $\pi/2$ rotation about σ_x and a series of C-NOTs (see Scheme 2). Then, the state of the first Amplifier spin is inverted, conditionally on the state of the target spin, obtaining either

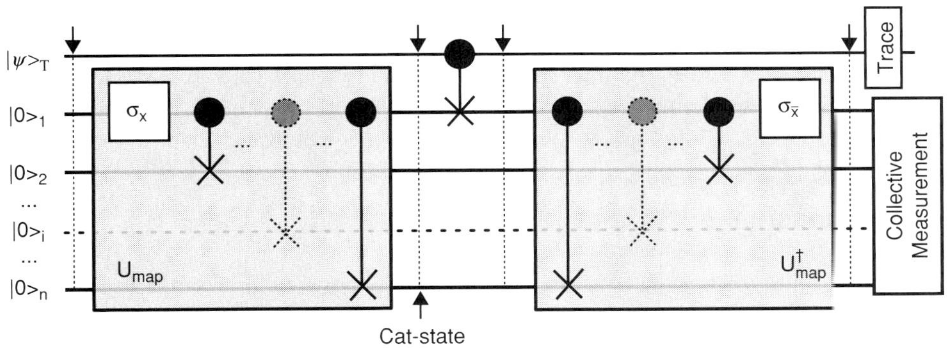

SCHEME 2. The creation of an entangled cat-state permits to use only one local action of the target spin. The arrows indicate where diagnostic measurements were performed in the experiment (see Fig. 37.2)

$|1\rangle_T(|10...0\rangle - i|01...1\rangle)/\sqrt{2}$ or $|0\rangle_T(|00...0\rangle - i|11...1\rangle)/\sqrt{2}$. When we next apply the inverse transformation to undo the entanglement, what was a local perturbation is propagated through the entire Amplifier. If the target spin is in the $|0\rangle_T$ state, the following C-NOT and σ_x gates bring back the Amplifier to the initial state. Otherwise the C-NOTs produce the state: $|1\rangle_T(|11...1\rangle$-i$|01...1\rangle)/\sqrt{2}$, which upon the σ_x rotation becomes: $|1\rangle_T|111...1\rangle$. As in the previous scheme, measuring the Amplifier magnetization provides information about the target spin state.

Diagnostic measurements were performed at four steps in the algorithm illustrated in Scheme 2 (as indicated by arrows). The measurement of the first carbon magnetization is sufficient to determine the state of the 3 Amplifier spins, since it is assumed that only population elements in the density matrix are nonzero (we would need full tomography to determine the whole density matrix).

The experimental results are shown in Fig. 37.2, where, as said, the left side of the spectrum corresponds to the case in which the target spin (the proton) is in the $|0\rangle_T$ state and the right part corresponds to the target spin in the $|1\rangle_T$ state. The four spectra illustrate clearly the evolution of the Amplifier spins under the measurement algorithm. The spectrum of the final state of the Amplifier, showing that the polarization of the three carbons has been inverted conditionally on the state of the proton, gives an indirect measurement of its state.

The superiority of Scheme 2 on the previous scheme is that only one interaction is invoked between the target spin and a privileged spin in the Amplifier, a more practical requirement, given the locality of any spin-spin interaction. Extending this scheme to larger systems with a much smaller target to Amplifier spins ratio, to obtain a sensitivity enhancement, is experimentally challenging. Even if was the interaction with the target spin was invoked only once, still needed to act on each Amplifier spin a kind of control that is available only for a limited number of spins, separately.

Other schemes can be developed that rely only on collective dynamics and control, if final states are allowed to be just distinguishable, based on their magnetizations, instead of requiring the maximum and minimum magnetization as before. Techniques for introducing entanglement, using only the natural Hamiltonian and r.f. pulses, have been developed in the context of spin counting experiments[11]. The aim of these experiments is to calculate the size of a cluster of interacting spins by measuring its entanglement or more precisely the coherence order of the system (which is the difference in the Hamming weight between two states). The entanglement of a state is related to its coherence order: for example, a cat-state corresponds to an n-spin, n-quantum coherence. To obtain an entangled state therefore, the well established techniques of Multiple Quantum Coherence (MQC)[12,13] can be used, and selective MQC[14] to create the n-quantum coherence operator, $\mathrm{H}^{(\mathrm{n})} = \Pi_{k=1}^{n}\sigma_k^{+} + \Pi_{k=1}^{n}\sigma_k^{-}$, which rotates the fully polarized state into the cat-state. In this case, the final magnetization will be either zero or equal to the initial one, but we will have used only the control that is available in conventional NMR will be used.

37.4 Conclusions

In conclusion, it was shown that it is possible to transfer polarization from a target spin to an ensemble of spins, through the creation of a highly entangled state with r.f. pulses and the coupling among the target spin and its closest neighbors. This in turn permits us to measure the state of a collapsed single spin, by inferring its state from the collective measurement of the magnetization of a large spin ensemble.

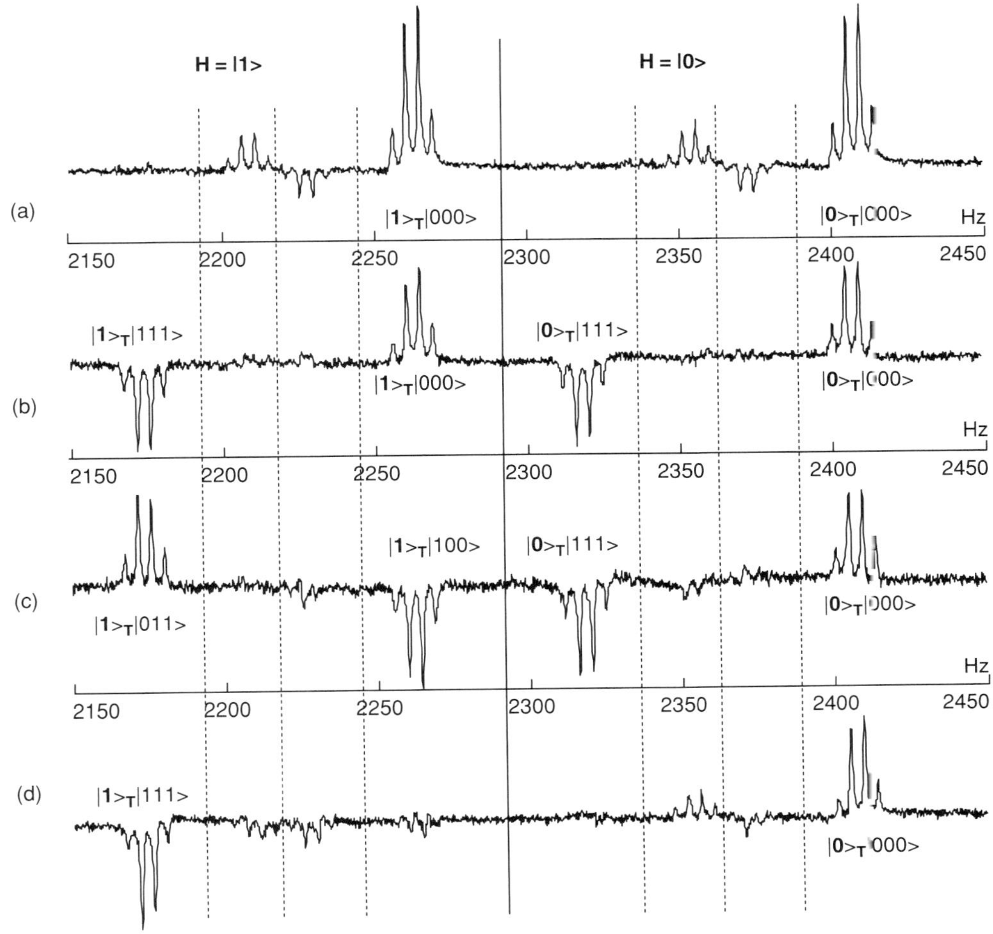

FIGURE 37.2. Experimental Results: **A**. Initial State: Pseudo-pure state of the 3 carbons. **B**. Cat-state. **C**. Cat-state, after inverting the state of the 1st carbon conditionally on the state of the proton. **D**. Final state: The polarization of the Carbons coupled to the target spin in state $|1\rangle_T$ (left) has been inverted, while it is unchanged for spins coupled to target spin in the $|0\rangle_T$ state (right).

The use of entanglement and multiple pulse sequences allows the scheme to be implemented using only the natural Hamiltonian and the control available in conventional NMR. The methods and physical systems proposed open the possibility to a new class of devices, where quantum effects, such as entanglement, are used to make a transition from microscopic to macroscopic properties.

Acknowledgments. This work was supported by NSF grant EEC 0085557, DARPA grant MDA972-01-1-0003, and by ARDA/ARO grants DAAD19-01-1-0519, DAAD19-03-1-0125, DAAD19-01-1-0654.

References

1. G. Burkard, H. Engel, D. Loss, *Fortschr. Phys.* **48**, 965 (2000).
2. B. E. Kane, *Nature* **393**, 133 (1998).
3. D. Rugar *et al.*, *Science* **264**, 1560 (1994).
4. J. Wrachtrup, A. Gruber, L. Fleury, C. von Borczyskowski, *Chem. Phys. Lett.* **267**, 179 (1997).
5. D. P. DiVincenzo, *Fortschr. Phys.* **48**, 771 (2000).
6. W. K. Wootters, W. H Zurek, *Nature* **299**, 802 (1982).
7. A. J. Legget, A. Garg, *Phys. Rev. Lett.* **54**, (9) 857 (1985).
8. D. G. Cory, A. F. Fahmy, T. F. Havel, *Proc. Natl. Acad. Sci.* **94**, 1634 (1997).
9. N. A. Gershenfeld, I. L. Chuang, *Science* **275**, 350 (1997).
10. M. D. Price, T. F. Havel, and D. G. Cory, *Physica D* **120**, 82 (1998).
11. M. Munowitz, A. Pines, M. Mehring, *J. Chem. Phys.* **86**, 3172 (1987).
12. J. Baum, M. Munowitz, A. N. Garroway, A. Pines, *J. Chem. Phys.* **83**, 2015 (1985).
13. Y. S. Yen, A. Pines, *J. Chem. Phys.* **78**, 3579 (1983).
14. W. S. Warren, D. P. Weitekamp, A. Pines, *J. Chem. Phys.* **73**, 2084 (1980).

38

Entanglement in quantum-critical spin systems

TOMMASO ROSCILDE[1], PAOLA VERRUCCHI[2], ANDREA FUBINI[3], STEPHAN HAAS[1], AND VALERIO TOGNETTI[3,4]

[1] Department of Physics and Astronomy, University of Southern California, Los Angeles, CA 90089-0484.
[2] Istituto Nazionale per la Fisica della Materia, UdR Firenze, Via G. Sansone 1, I-50019 Sesto Fiorentino (FI), Italy.
[3] Dipartimento di Fisica dell'Università di Firenze and Istituto Nazionale per la Fisica della Materia, UdR Firenze, Via G. Sansone 1, I-50019 Sesto Fiorentino (FI), Italy.
[4] Istituto Nazionale per la Fisica della Materia, UdR Firenze, and Istituto Nazionale di Fisica Nucleare, Sez. di Firenze, Via G. Sansone 1, I-50019 Sesto Fiorentino (FI), Italy.

Abstract

We present an extensive study of entanglement properties in easy-plane quantum spin systems with a quantum phase transition driven by a magnetic field applied in the easy plane. Making use of quantum Monte Carlo simulations, we are able to monitor the behavior of englement in the ground state and at finite temperature as a function of the applied field for different lattice geometries of the system. Our calculations focus on the entanglement of formation, quantified by the *one-tangle* for the global entanglementof the system and by the *concurrence* for the pair-wise englement. For the case of a one-dimensional spin chain, we observe that the entanglement estimators are able to single out with high precision both the quantum critical point, through a minimum of the paiwise-to-global entanglement ratio, and the known occurrence of a factorized state below the critical field, through the vanishing of all entanglement estimators. The vanishing of entanglement at a given field also persists at finite-temperature, so that the system displays a temperature-resistant *entanglement* switch effect. We then extend the same analysis to the case of a spin ladder and of a square lattice, and, thanks to the entanglement estimators, we find that the existence of a factorized state close to the quantum critical point is a general feature independent of the geometry.

38.1 Introduction

One of the most striking aspects of quantum coherence in a quantum many-body system is the occurrence of *entanglement*, namely the realization of a superposition of many-body states that cannot be factorized into a product of single-particle wave functions. An entangled state possesses correlations that cannot be accounted for by classical-like quantities; for instance an entangled state might not show any form of classical order, and nonetheless be at the same time strongly correlated. The possibility of a local description of such state is partially or completely lost, depending on the degree of entanglement contained in the state. In particular, the non-local nature of special collective quantum states is the fundamental ingredient that allows quantum communication protocols and quantum computation algorithms[1] to outperform their classical counterparts.

The idea of entanglement as a resource naturally demands a systematic investigation of which quantum many-body systems are able to display sizable entanglement in a controllable way. An intriguing perspective is that pure-quantum correlations are strongly enhanced when a system undergoes a Quantum Phase Transition (QPT)[2] analogously to what classical correlations do at a thermal phase transition. Indeed quantum fluctuations show up at all length scales at a quantum critical point. In what sense quantum correlations, and thus entanglement, 'diverge' at a QPT has been the subject of investigation in several studies,[3–6] although the resulting picture is still controversial. In this study a quite general $S = 1/2$ quantum spin system displaying a field-induced QPT in arbitrary dimensions is considered. In particular, how the behavior of entanglement at and around the quantum critical point shows strong features providing new insight in the drastic change of the system's ground state under the effect of strong quantum fluctuations is discussed.

38.2 The Model

A very general example of a quantum phase transition in spin models is offered by the the antiferromagnetic *XYZ model* in a field. The model Hamiltonian reads:

$$\hat{\mathcal{H}} = J \sum_{\langle ij \rangle} \left[\hat{S}_i^x \hat{S}_j^x + \Delta_y \hat{S}_i^y \hat{S}_j^y + \Delta_z \hat{S}_i^z \hat{S}_j^z - \frac{2h}{\mathrm{z}} \hat{S}_i^z \right] \tag{38.1}$$

where $J > 0$ is the exchange coupling, the sum $\langle ij \rangle$ runs over the nearest neighboring sites of a bipartite lattice with coordination number z, and $h \equiv g\mu_{\mathrm{B}}H/J$ is the reduced magnetic field. In the following we perform the canonical transformation $\hat{S}_i^{x,y} \to (-1)^i \hat{S}_i^{x,y}$ on Eq. (38.1), so that the relevant correlations along the x and y axes are ferromagnetic. The parameters $0 \leq \Delta_y, \Delta_z \leq 1$ control the anisotropy of the system. In the most general case of $\Delta_y \neq 1$ the Zeeman term in Eq. (38.1) does not commute with the rest of the hamiltonian. This property is at the core of the field-driven quantum phase transition occurring at a critical field $h_{\mathrm{c}}(\Delta)$, which separates a Néel-ordered phase ($h \leq h_{\mathrm{c}}$) from a partially-polarized disordered phase ($h > h_{\mathrm{c}}$). When $h \leq h_{\mathrm{c}}$ the field favors long range magnetic order along the x-axis. This order disappears at the critical field h_{c}, where magnetic correlations along x become short-ranged, while quantum fluctuations prevent the spins from being fully polarized along the field.[7–11]

The case $\Delta_z = 0$ reproduces the XY model in a transverse field, which is exactly solvable in one dimension,[7] and whose entanglement properties have been the subject of several recent investigations.[3–6] When $\Delta_z \neq 0$, and/or in higher dimensions, the model is no longer exactly solvable. The one-dimensional case has been indeed investigated within approximate analytical and numerical approaches.[9–11] A renewed interest in the model stems from the experimental in-field studies on the quantum spin chain compound Cs_2CoCl_4,[12] displaying a strong planar anisotropy, $\Delta_y \approx 0.25$, $\Delta_z \approx 1$, and $J \approx 0.23$ meV.

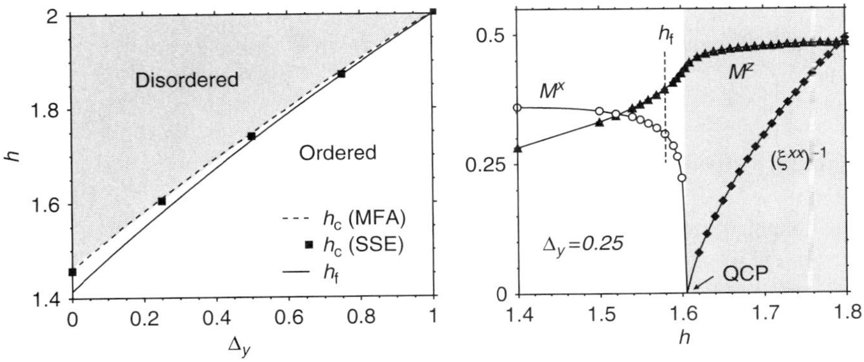

FIGURE 38.1. Upper panel: ground state phase diagram of the XYX model in a field. Mean-field (MFA) results are taken from Ref.,[11] the factorizing field h_f from Ref.[8] Lower panel: quantum critical behavior of $x-$ and $z-$magnetizations and correlation length for the model with $\Delta_y = 0.25$, $L = 100$, $\beta = 200$. The factorizing field is indicated by a dashed line. The arrow indicates the quantum critical point (QCP).

38.3 Linear Chain: Critical Point and Factorized State

Motivated by the existing theoretical and experimental results, we first concentrate on the case of a linear chain, and on the parameter range $0 \leq \Delta_y \leq 1$, $\Delta_z = 1$, which defines the *XYX model* in a field.[13] The qualitative behavior of the more general XYZ model is expected to be close to that of the XYX model in a field, since the two models share the same symmetries.

The analysis of the theoretical model is performed via Stochastic Series Expansion (SSE) Quantum Monte Carlo (QMC) simulations,[14] based on a modified version of the directed-loop algorithm to account for the low symmetry of the Hamiltonian.[15] Chains of various lengths, $L = 40, ..., 120$, have been considered, at an inverse temperature $\beta = 2L$ high enough to mimic the $T = 0$ behaviour for each lattice size.

The left panel of Fig. 38.1 shows the ground-state phase diagram of the one-dimensional XYX model in the $\Delta_y - h$ plane. The quantum Monte Carlo data confirm that the transition belongs to the universality class of the 1D transverse-field Ising model (or of the 2D Ising model),[9] and are in very good agreement with predictions from a mean-field treatment of the Hamiltonian.[9] On the right panel of Fig. 38.1 the signatures of the transition in the critical behavior of the magnetizations along x (estimated through the asymptotic value of the spin-spin correlator as $M^x = |\langle \hat{S}_i^x \hat{S}_{i+L/2}^x \rangle|^{1/2}$) and along z, and the divergence of the correlation length along x is clearly seen.

Remarkably, none of these standard magnetic observables bears signatures of the second striking feature of the model, namely the occurrence of an exactly factorized state for a field $h_f(\Delta_y)$ lower than the critical field h_c. The *factorizing field* reads $h_f = \sqrt{2(1 + \Delta_y)}$ in the case of the XYX model. The factorized state has the form $|\Psi\rangle = \bigotimes_{j=1}^{N} |\psi_j\rangle$ with $|\psi_j\rangle = \cos\theta| \uparrow\rangle + \sin\theta e^{i\phi_j}| \downarrow\rangle$ and $\phi_j = (1 + (-1)^j)\pi/2$, $\theta = \cos^{-1}\sqrt{(1 + \Delta_y)/2}$. This corresponds to a configuration in which the spins have perfectly staggered x components but also cant out of the xy plane by an angle θ. The occurrence of such a factorized ground state is particularly surprising if one considers that we are dealing with the $S = 1/2$ case, characterized by the most pronounced effects of quantum fluctuations. However, in the class of models here considered, such fluctuations are fully uncorrelated[8] at $h = h_{\mathrm{f}}$, thus leading to a classical-like ground state.

38.4 Entanglement Estimators

The study of entanglement properties in the XYX model turns out to be very insightful, and it has the unique feature of unambiguously detecting both the factorized state and the quantum critical point. This study has focussed on the *entanglement of formation*[16] which is quantified through the *one-tangle* and the *concurrence*. The one-tangle[17,18] is an estimate of the $T = 0$ entanglement between a single site and the remainder of the system. It is defined as $\tau_1 = 4 \det \rho^{(1)}$, where $\rho^{(1)} = (I + \sum_\alpha M^\alpha \sigma^\alpha)/2$ is the one-site reduced density matrix, $M^\alpha = \langle \hat{S}^\alpha \rangle$, σ^α are the Pauli matrices, and $\alpha = x, y, z$. In terms of the spin expectation values M^α, τ_1 takes the simple form:

$$\tau_1 = 1 - 4 \sum_\alpha (M^\alpha)^2. \tag{38.2}$$

The concurrence[19] quantifies instead the pairwise entanglement between two spins at sites i, j both at zero and finite temperature. For the model of interest, in absence of spontaneous symmetry breaking ($M^x = 0$) the concurrence takes the form[18]

$$C_{ij} = 2 \max\{0, C_{ij}^{(1)}, C_{ij}^{(2)}\}\,, \tag{38.3}$$

where

$$C_{ij}^{(1)} = g_{ij}^{zz} - \frac{1}{4} + |g_{ij}^{xx} - g_{ij}^{yy}|\,, \tag{38.4}$$

$$C_{ij}^{(2)} = |g_{ij}^{xx} + g_{ij}^{yy}| - \sqrt{\left(\frac{1}{4} + g_{ij}^{zz}\right)^2 - (M^z)^2}\,, \tag{38.5}$$

with $g_{ij}^{\alpha\alpha} = \langle \hat{S}_i^\alpha \hat{S}_j^\alpha \rangle$.

38.5 Results

The QMC results for the model Eq. (38.1) with $\Delta_y = 0.25$ are shown in Fig. 38.2, where we plot τ_1, the sum of squared concurrences

$$\tau_2 = \sum_{j \neq i} C_{ij}^2\,, \tag{38.6}$$

and, in the inset, $C_{i,i+n}$ for $n = 1, 2, 3$. The following discussion, although directly referred to the results for $\Delta_y = 0.25$, is actually quite general and applies to all the other studied values of Δ_y.

Unlike the standard magnetic observables plotted in Fig. 38.1, the entanglement estimators display a marked anomaly at the factorizing field, where they clearly vanish as expected for a factorized state. In particular, the power of the entanglement estimators is to rigorously unveiling the factorized nature of the ground state. It can indeed be easily shown that the ground state is factorized *if and only if* the one-tangle τ_1 vanishes for each spin. When the field is increased

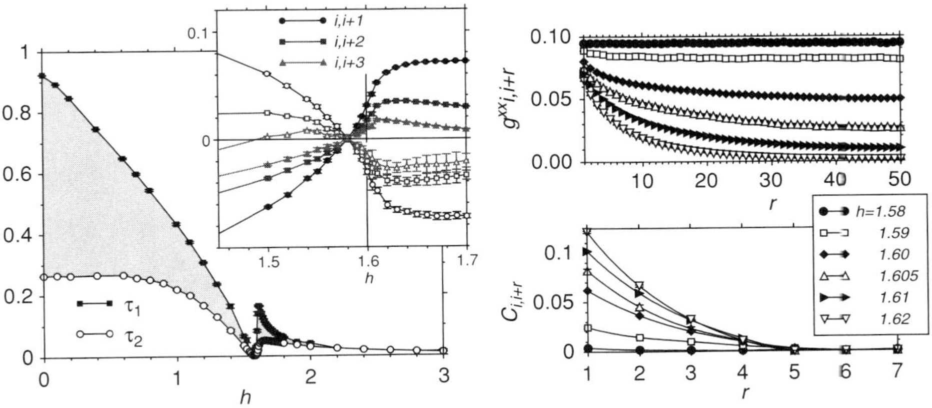

FIGURE 38.2. *Left*. One-tangle τ_1 and sum of squared concurrences τ_2 as a function of the applied field for the $S = 1/2$ XYX model with $\Delta_y = 0.25$, $L = 100$ and $\beta = 200$. Inset: contributions to the concurrence between *j-th* neighbors; full symbols stand for $C^{(1)}_{i,i+j}$, open symbols for $C^{(2)}_{i,i+j}$. The dashed line marks the critical field h_c. *Right*. Spin-spin correlator $g^{xx}_{i,i+r}$ (upper panel) compared with the concurrence $C_{i,i+r}$ (lower panel) for various field values around the critical value $h_c = 1.605(2)$. Other simulation parameters as in the left panel.

above h_f, the ground-state entanglement has a very steep recovery, accompanied by the quantum phase transition at $h_c > h_f$. The system realizes therefore an interesting *entanglement switch* effect controlled by the magnetic field.

As for the concurrence terms Eqs. (38.4) and (38.5), the factorizing field divides two field regions with different expressions for the concurrence:

$$C^{(1)}_{ij} < 0 < C^{(2)}_{ij} \quad \text{for} \quad h < h_f\,, \tag{38.7}$$

$$C^{(2)}_{ij} < 0 < C^{(1)}_{ij} \quad \text{for} \quad h > h_f\,, \tag{38.8}$$

whereas $C^{(1)}_{ij} = C^{(2)}_{ij} = 0$ at $h = h_f$. In presence of spontaneous symmetry breaking occurring for $h < h_c$, the expression of the concurrence is generally expected to change with respect to Eqs. (38.4) and (38.5), as extensively discussed in Ref.[20] . For the model under investigation, the expression of the concurrence stays unchanged when the condition[20] $C^{(2)}_{ij} < C^{(1)}_{ij}$ is satisfied, i.e. for $h > h_f$. This means that our estimated concurrence is accurate even in the ordered phase above the factorizing field; in the region $0 < h < h_f$ it represents instead a lower bound to the actual $T = 0$ concurrence. Alternatively it can be regarded as the concurrence for infinitesimally small but finite temperature.

A naive analogy between classical and quantum correlations would lead to the expectation that the range of pairwise entanglement, expressed through the concurrence, is critically enhanced at a quantum phase transition, reflecting the divergence of the length scale for quantum effects. Indeed this is clearly *not* the case as shown in the right panel of Fig. 38.2, where the behavior of the spin-spin correlator g^{xx} as a function of the distance is contrasted with that of the concurrence. We observe that, while the correlator becomes long-ranged below the critical field and even completely flat at the factorizing field, the concurrence remains short-ranged when passing through the transition, and it basically vanishes after four lattice spacings.

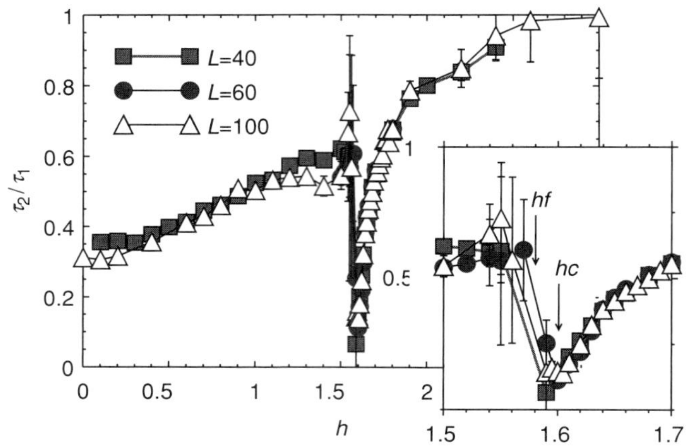

FIGURE 38.3. Entanglement ratio τ_2/τ_1 as a function of the field for $\Delta_y = 0.25$ and $\beta = 2L$. Inset: zoom on the critical region.

To clarify this issue, one has to consider the special nature of quantum correlations. At variance with classical correlations, entanglement has basically to satisfy a sum rule, or, as it is often phrased, is subject to a constraint of *monogamy*. This means that the more partners a spin is entangled with, the less entanglement will be shared with each partner. Moreover, entanglement is present not only in the form of pairwise correlations as in the case of the concurrence, but also in the form of n-spin correlations with $n > 2$, which, unlike classical correlations, can be completely independent of pairwise correlations. This is exemplified for instance by the maximally entangled Greenberger–Horne–Zeilinger (GHZ) state for n spins $|\text{GHZ}\rangle = (|\uparrow\uparrow \dots \uparrow\rangle + |\downarrow\downarrow \dots \downarrow\rangle)/\sqrt{2}$, on which the concurrence between any two spins is vanishing.[17] Therefore, according to monogamy, pairwise entanglement and n-wise entanglement with $n > 2$ are mutually exclusive, a condition which is absurd from the point of view of classical correlations.

The mathematical expression of the monogamy constraint for the entanglement is provided by the Coffman–Kundu–Wootters (CKW) conjecture,[17] which indeed represents an approximate sum rule for entanglement correlations. Such conjecture states that $\tau_2 \leq \tau_1$, and, although it can be rigorously proved only in the case of three spins, it has always been verified so far in the case of an arbitrary number of spins.[18] Indeed Fig. 38.2 shows that the CKW conjecture is verified also in the case of the present model.

In particular, the *entanglement ratio* $R = \tau_2/\tau_1$ is interpreted as a measure of the fraction of the total entanglement stored in pairwise correlations. This ratio is plotted as a function of the field in Fig. 38.3. As the field increases, the general trend of pairwise entanglement saturating the whole entanglement content of the system is observed. But a striking anomaly occurs at the quantum critical field h_c, where R displays a very narrow dip. According to our interpretation, this result shows that the weight of pairwise entanglement decreases dramatically at the quantum critical point in favor of multi-spin entanglement. Indeed, due to the monogamy constraint, multi-spin entanglement appears as the only possible quantum counterpart to long-range spin-spin correlations occurring at a quantum phase transition. The divergence of entanglement range has to be interpreted in the sense that, at a quantum phase transition, finite n-spin entanglement appears with $n \to \infty$ at the expense of the pairwise one. A unique estimator for the

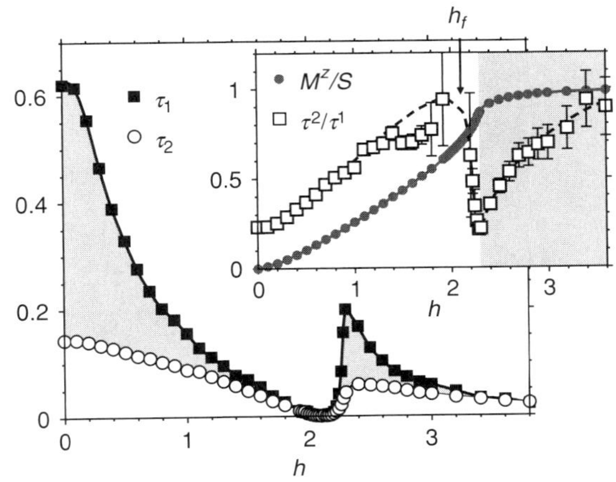

FIGURE 38.4. One-tangle and sum of squared concurrences as a function of the field for the XYX model on the two-leg ladder with $L = 40$ on each leg and $\Delta_y = 0$. Inset: entanglement ratio and z-magnetization as a function of the field. The shaded area marks the quantum disordered region. All lines are guides to the eye.

n-spin entanglement with $n > 2$ has not been found yet, so that justifying the above statement on a more quantitative level is problematic. Nonetheless strong indications of the relevance of n-spin entanglement with large n at a quantum critical point are given in Ref.[5] Finally, the above results evidence the serious limitations of concurrence as a reliable estimate of entanglement at a quantum critical point. In turn, the minimum of the entanglement ratio R is proposed as a novel estimator of the quantum critical point, fully based on entanglement quantifiers.

38.6 Two-leg Ladder

The above analysis of entanglement is extended to the less investigated case of the XYX model on a two-leg ladder, thanks to the fact that the SSE quantum Monte Carlo approach is easily generalizable to any bipartite lattice. In this case the Hamiltonian of Eq.(38.1) is still expected to display a field-driven quantum phase transition, whose existence is independent of the lattice geometry and it is instead uniquely due to the non-commutativity between the exchange term and the Zeeman term. There is instead no specific reason to expect that the model still displays a perfectly factorized state for geometries other than that of the linear chain.

Fig. 38.4 shows the one-tangle and the sum of squared concurrences for the XYX model on the two-leg ladder in the case $\Delta_y = 0$. We notice that, for such a strong anisotropy, the two-leg ladder does not display a gapped Haldane phase, as shown by the finite value of the magnetization for low fields (inset of Fig. 38.4). Remarkably, the qualitative behavior of both τ_1 and τ_2 as a function of the field is the same as in the case of a single chain, and in particular the vanishing of the one-tangle signals the rigorous existence of a factorized state for a lower field than the critical one. The CKW conjecture is verified for all the field values considered, and therefore the entanglement ratio τ_2/τ_1 can be still interpreted as a measure of the fraction of entanglement stored in pairwise correlations. This quantity, plotted in the inset of Fig. 38.4, again displays a deep minimum corresponding to the inflection point of the uniform magnetization and therefore marking the quantum critical point.

38.7 Conclusions

In this study it is we have shown that entanglement estimators give a precious and novel insight in the ground state properties of lattice $S = 1/2$ spin systems. In the case of anisotropic spin chains with a field-driven quantum phase transition, it has been shown that the quantum critical point can be detected through a narrow dip in the pairwise-to-global entanglement ratio. Moreover, unlike the more conventional magnetic observables, entanglement estimators are able to single out the occurrence of an exactly factorized state in these systems. The use of quantum Monte Carlo techniques naturally allows to extend this analysis to different lattice geometries. In particular we have discussed here the case of a two-leg ladder is discussed here, in which the calculation of entanglement estimators remarkably shows the existence of a factorized state below the quantum critical point. The entanglement ratio displays again a minimum at the critical field, confirming the generality of this feature of the entanglement behavior as a signature of a quantum phase transition. Finally, the proximity of a quantum critical point to the factorized state of the system gives rise to an interesting field-driven entanglement-switch effect. This suggests that many-body effects, driven by a macroscopic parameter as an applied field, are a powerful tool for the control of the microscopic entanglement in a multi-qubit system. The application of this kind of concepts in the design of quantum computing devices therefore looks appealing.

Acknowledgments. Fruitful discussions with L. Amico, T. Brun, P. Delsing, G. Falci, R. Fazio, A. Osterloh, and G. Vidal are gratefully acknowledged. We acknowledge support by DOE under grant DE-FG03-01ER45908 (T.R. and S.H.), by INFN, INFM, and MIUR-COFIN2002 (A.F., P.V., and V.T.).

References

1. M. A. Nielsen and I. L Chuang, *Quantum Computation and Quantum Information*, (Cambridge University Press, 2000).
2. S. Sachdev, *Quantum Phase Transitions*, (Cambridge University Press, Cambridge 1999).
3. T. J. Osborne and M. A. Nielsen, *Phys. Rev. A* **66**, 032110 (2002).
4. A. Osterloh, L. Amico, G. Falci, and R. Fazio, *Nature* (London) **416**, 608 (2002).
5. G. Vidal, J. I. Latorre, E. Rico, and A. Kitaev, *Phys. Rev. Lett.* **90**, 227902 (2003).
6. F. Verstraete, M. Popp, and J. I. Cirac, *Phys. Rev. Lett.* **92**, 027901 (2004); F. Verstraete, M. A. Martin-Delgado, and J. I. Cirac, *ibid.* **92**, 087201 (2004).
7. E. Barouch, B. M. McCoy, and M. Dresden, *Phys. Rev. A* **2**, 1075 (1970); *ibid.* **3**, 786 (1971).
8. J. Kurmann, H. Thomas, and G. Mueller, *Physica A*, **112**, 235 (1982).
9. D. V. Dmitriev, V. Ya. Krivnon, A. A. Ovchinnikov, and A. Langari, *J. Exp. Th. Phys.* **95**, 538 (2002).
10. F. Capraro and C. Gros, *Eur. Phys. J B* **29**, 35 (2002).
11. J.-S. Caux, F. H. L. Essler, and U. Loew, *Phys. Rev. B* **68**, 134431 (2003).
12. M. Kenzelmann *et al.*, *Phys. Rev. B* **65**, 144432 (2002).
13. T. Delica and H. Leschke, *Physica A* **168**, 736 (1990).

14. O. F. Syljuåsen and A. W. Sandvik, *Phys. Rev. E* **66**, 046701 (2002).
15. T. Roscilde, P. Verrucchi, A. Fubini, S. Haas, and V. Tognetti, cond-mat/0404403 (unpublished).
16. C. H. Bennett, D. P. DiVincenzo, J. A. Smolin, and W. K. Wootters, *Phys. Rev. A* **54**, 3824 (1996).
17. V. Coffman, J. Kundu, and W. K. Wootters, Phys. Rev. A **61**, 052306 (2000).
18. L. Amico, A. Osterloh, F. Plastina, R. Fazio, and G. M. Palma, *Phys. Rev. A* **69**, 022304 (2004).
19. W. K. Wootters, *Phys. Rev. Lett.* **80**, 2245 (1998).
20. O. F. Syljuåsen, *Phys. Rev. A* **68**, 060301 (2003).

39

Control of nuclear spins by quantum Hall edge channels

S. KOMIYAMA[1], T.MACHIDA[2], K.IKUSHIMA[1], AND Y.YAMAZAKI[1]

[1] Department of Basic Sciences, University of Tokyo, Komaba 3-8-1, Meguro-ku, Tokyo 153-8902, Japan.
[2] PRESTO, Japan Science and Technology Agency, Building 16-622, Komaba 3-8-1, Meguro-ku, Tokyo 153-8902, Japan. Present address; Institute of Industrial Science, University of Tokyo, Komaba 4-6-1,Meguro-ku,Tokyo 153-85-5, Japan.

39.1 Introduction

Weak interaction of nuclear spins with their physical environment makes them a unique object. It lends them exceedingly long decoherence time (T_2), but on the other hand, makes them difficult objects to control and detect. Though a quantum algorithm has been experimentally demonstrated by liquid-state Nuclear Magnetic Resonance (NMR),[1–3] contriving ingenious means of controlling/detecting nuclear spins in solid state devices may in general contribute to strengthen the basis for exploring implementation of spintronics, quantum information processing or quantum control of nuclear-spin quantum states has not been achieved in earlier works. [4–9] Here, we show that quantum-Hall edge channels provide a unique and powerful tool to initialize, coherently control, and sensitively detect nuclear spin polarization.[10,11]First, selectively populated spin-resolved edge channels dynamically polarize nuclear spins via the spin-flip scattering, where only those nuclei along the edge channels that are within the reach of the edge-state electron wave-functions are affected. Secondly, a pulsed Radio-Frequency (RF) magnetic field at the NMR frequency, generated by a micro-metal wire lithographically fabricated on the substrate surface right above the edge channels, causes the nuclear-spin state to evolve coherently. Finally, the nuclear-spin state is read out through the Hall resistance: The conductance oscillates as a function of the pulse duration, demonstrating all-electrical detection of Rabi oscillation in the solid-state nuclear spin system.

39.2 Samples

Fig. 39.1(a) illustrates a Hall-bar device fabricated in a GaAs/AlGaAs heterostructure crystal with a Two-Dimensional Electron Gas (2DEG) layer located 100 nm beneath the crystal surface.[10,11] The electron mobility and the sheet electron density are μ = 30 m^2/Vs and ns = [illegible].6 × $10^{15}/m^2$ at 4.2 K, respectively. Micrographs of its central region is shown in Fig. 39.1(b).

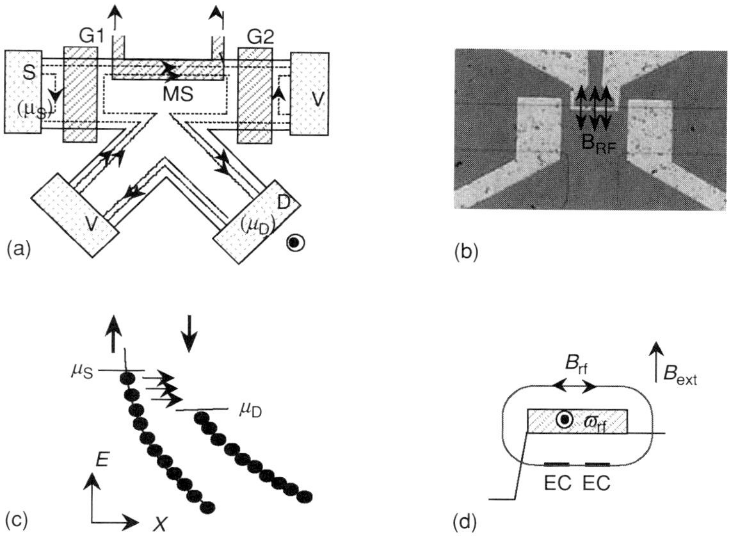

FIGURE 39.1. (a) Schematic representation of a Hall-bar device, (b) a micrograph, (c) unequally occu pied spin-split edge channels, and (d) cross sectional view of the RF magnetic field generated by the metal strip.

Ohmic contacts serve as source and drain reservoirs (S and D) and voltage probes (V). Two Schottky cross gates (G1 and G2) are prepared. A metal strip (MS) with a 2μm width and 20μm length is deposited along a boundary of the 2DEG region sandwiched by the two gates. The device is mounted in a ^{3}He/^{4}He dilution refrigerator at a base temperature of 50 mK. The four-terminal differential Hall resistance, $R'_H \equiv \partial V_H/\partial I$, is studied by superposing an AC-current I_{ac} = 1 nA (18 Hz) on a background DC-current I_{dc}. The Landau-level filling factor in the region without the gates (G1 and G2) is adjusted to $\nu = 2$ in an external static magnetic field B_{ext} (about 7T). The gates (G1 and G2) are negatively biased so that the outer spin-up edge channel of the lowest orbital Landau level is transmitted while the inner spin-down edge channel is totally reflected at the potential barriers underneath the gates as schematically shown in Fig. 39.1(a). In this condition, the spin-resolved edge channels along the upper boundary of the regionbetween the two gates are unequally populated, that is, the outer spin-up states at the left corner of the boundary are fed with the electrons from the source reservoir of the electrochemical potential μ_S, while the inner spin-down edge states with those from the drain reservoir of the electrochemical potential μ_D. This implies that the spin-up states are overpopulated compared to the spin-down states by $\Delta\mu = \mu_S - \mu_D$ at the left corner of the boundary as shown in Fig. 39.1(c), if $\mu_S > \mu_D$, which we define as the positive polarity of current, $I_{SD} > 0$. It follows that the spin-up to spin-down inter-edge-state scattering ($\uparrow\rightarrow\downarrow$) takes place along the boundary. Due to the inter-edge-channel scattering, the magnitude of unequal population is reduced to a certain extent when the electrons in the edge channels reach the right corner of the boundary. If the polarity of current is negative, $I_{SD} < 0$ ($\mu_S < \mu_D$), the spin-down states are overpopulated to cause the opposite spin-flip scattering ($\downarrow\rightarrow\uparrow$). In either condition, it is important that the amplitude of the Hall resistance R'_H is determined by the magnitude of unequal population at the right corner of the boundary, or the rate of inter-edge-channel scattering rate along the boundary, i.e., the resistance is largest, $R'_H = h/e^2$, if the unequal population, $\Delta\mu = \mu_S - \mu_D$, is retained in the absence of scattering, while it is reduced by the factor of two, $R'_H = h/2e^2$, in the perfect equilibration, $\Delta\mu = 0$, with strong scattering.[12,13]

39.3 Dynamical Nuclear Polarization

For the spin-flip inter-edge-channel scattering discussed in the above, spin-orbit interaction is the dominant mechanism.[13] However, there is a finite contribution of the contact hyperfine interaction between the nuclear spin $\boldsymbol{I}$ of GaAs nuclei and the electron spin $\boldsymbol{S}$ of conduction electrons in edge channels,[14]

$$H_{\text{Hyperfine}} = A\boldsymbol{I}\cdot\boldsymbol{S} = \frac{A}{2}\left(I^+S^- + I^-S^+\right) + AI_zS_z, \tag{39.1}$$

with $A(>0)$ the hyperfine constant. The term $(I^+S^- + I^-S^+)$ yields the probability of electron-spin flip scattering that is accompanied by the nuclear spin flop. The second term is the hyperfine splitting, which reduces or increases the effective Zeeman splitting for the electron spin. It follows from the term $(I^+S^- + I^-S^+)$ in Eq. (39.1), that electron

spin-flip scattering between unequally populated edge channels induces strong Dynamical Nuclear Polarization (DNP) along the edge channels due to the simultaneous flip-flop process.[5–8] When I_{SD} is positive ($\mu_S > \mu_D$), up-to-down electron-spin-flip scattering (Fig. 39.1(c)) induces a positive DNP ($I_z > 0$) through the hyperfine interaction. The positive nuclear-spin polarization, in turn, reduces the effective Zeeman energy through the term AI_zS_z in Eq. (39.1), which promotes the inter-edge-channel scattering. Thus the up-to-down electron-spin-flip scattering is self-accelerating, and reduces R'_H as positive DNP develops. With the negative polarity of I_{SD} ($\mu_S < \mu_D$), a negative DNP ($I_z < 0$) develops, which increases R'_H through the increase in effective Zeeman splitting.

When I_{DC} is scanned over a few minutes, the DNP manifests itself as a hysteretic structure in the curves of R'_H vs. I_{DC} as shown in Fig. 39.2(a)[9,11] When one switches I_{DC} to – 4 nA after having maintained I_{DC} at + 4 nA ($t < 0$) for several minutes, negative DNP slowly develops increasing R'_H (Fig. 39.2(b)). When I_{DC} = – 4 nA ($t < 0$) is switched to I_{DC} = + 4 nA ($t > 0$), the opposite feature is seen as positive DNP develops (Fig. 39.2(c)).

39.4 Local and Coherent Control

To carry out NMR,[10] I_{DC} = + 4 nA is initially maintained for several minutes to fully develop the positive DNP and R'_H is fully suppressed. We then pass RF current, I_{RF}, through the metal strip MS. The RF current generates RF magnetic field, B_{RF}, in the region of DNP along the edge channels in the direction parallel to the 2DEG layer, or normal to the orientation of DNP (Figs. 39.1(b) and (d)). Figure 39.3(a) shows a gray-scale plot of the parallel component of B_{RF} at the 2DEG layer, 100 nm below MS, calculated by the finite element method. Figure 39.3(b) shows that the distribution of B_{RF} is uniform along the MS but rapidly decays outside the region beneath MS.[15] When the frequency, f, of B_{RF} is scanned, R'_H sharply increases at the NMR frequencies for ^{75}As, ^{69}Ga, and ^{71}Ga as shown in Figs. 39.4(a)–(c).[10] The NMR is detected by R'_H because the nuclear spins, positively polarized in the initial condition, is depolarized by the

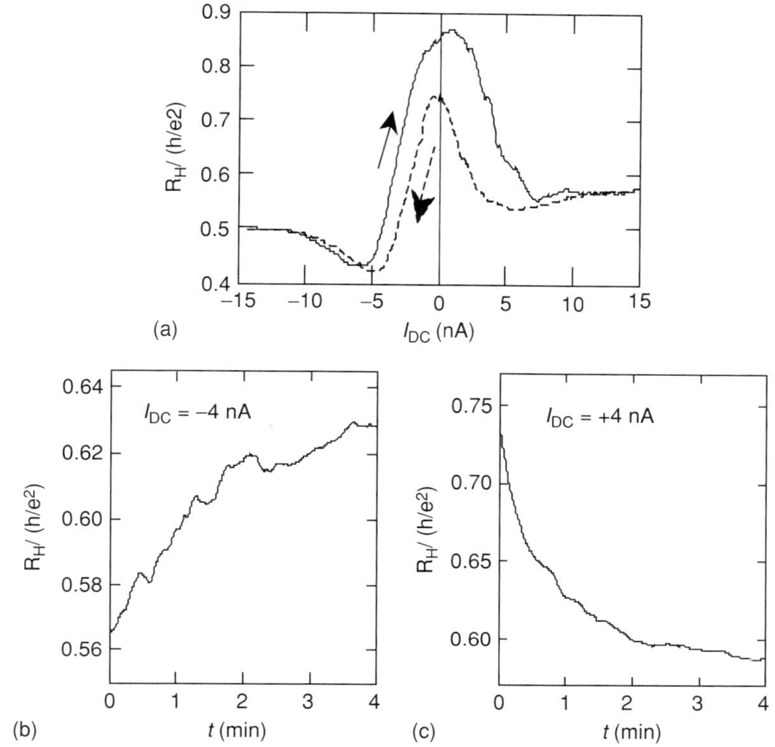

FIGURE 39.2. (a) Hall resistance in the scan of I_{DC}. (b) and (c) Real-time traces of the Hall resistance obtained after I_{dc} is switched to – 4nA and + 4nA shown in (b) and (c) respectively.

NMR. Additional experiments carried out with $I_{DC} < 0$ provides similar NMR signals with the opposite polarity for all the nuclei of GaAs.

To carry out pulsed mode NMR,[11] the initial nuclear-spin state is again prepared by maintaining $I_{DC} = +4$ nA. The nuclear spins in the narrow region along the edge channels are initialized in the up-spin state |0>. Transmitting a pulse of RF current I_{RF} through the metal strip MS generates superposition of the up-spin and the down-spin states, |0> and |1>. The nuclear-spin state evolves with time during the pulse application. The final state reached at the end of the pulse, $|t\rangle = \alpha(t)|0\rangle + \beta(t)|1\rangle$, is read out by the change in $R\prime_H$ which gives a measure of $|\beta|^2$.

Figure 39.5(a) displays a change in $R\prime_H$ caused by the application of a pulse with a duration of $\tau_{pulse} = 9\ \mu$s and a peak-to-peak amplitude of $I_{RF} = 62$ mA at the resonance frequency for ^{71}Ga ($f = 94.8$ MHz). The increase of $R\prime_H$is caused by an increase in $|\beta^{\ 2}$. By carrying out similar measurements with different pulse durations ($0 \leq \tau_{pulse} \leq 600\mu$s) at a 1μs- step, the amplitude of the resistance was found to increase, $\Delta R\prime_H$, oscillates as a function of τ_{pulse} as shown by the top curve in Fig. 39.5(b). Reducing the amplitude of I_{RF} decreases the oscillation frequency as displayed by the lower curves in Fig. 39.5(b). The oscillation frequency is

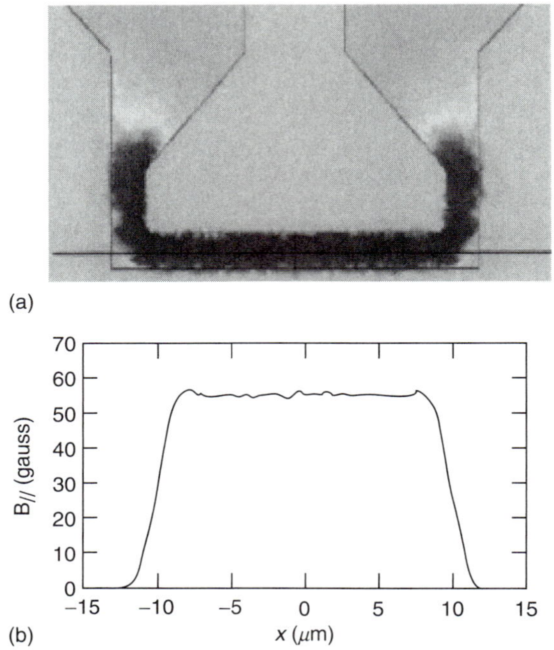

FIGURE 39.3. Spatial distribution of RFmagnetic field parallel to the 2DEG plane (B_{RF}). I_{RF} = 44 mA (100 MHz). (a) Two-dimensional gray-scale plot, and (b) the distribution of along the thick solid line in (a).

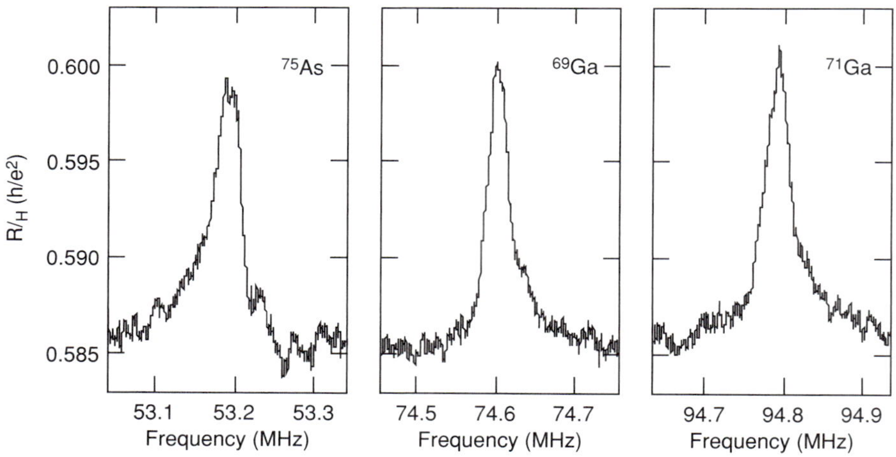

FIGURE 39.4. NMR signals detected by the Hall resistance.

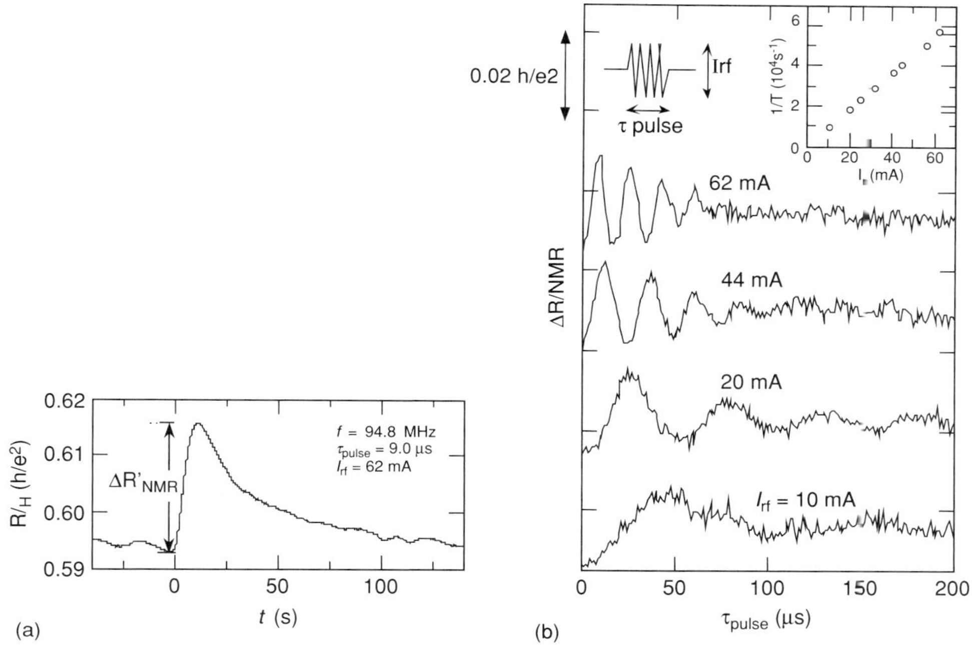

FIGURE 39.5. Pulsed-mode NMR signal (71Ga) detected by the Hall resistance. (a) Time trace of the Hall resistance after p-pulse is applied. (b) Coherent evolution of nuclear-spin quantum state: The Rabi oscillation. The inset shows that the Rabi frequency linearly increases with the amplitude of RF magnetic fields.

found to be proportional to the amplitude of I_{RF} as shown in the inset. The series of curves in Fig. 39.5(b) thus demonstrates coherent nuclear-spin evolution in the time domain, i.e., the Rabi oscillation.

Spin-echo experiments, carried out also by studying the Hall resistance, shows that the dephasing time is $T_2 \approx 80\ \mu s$.[15] We expect that the phase breaking process is dominated by indirect dipole interaction mediated by the 2DEG or edge channels.

39.5 Discussion

The nuclear spins controlled in this work are limited to those in a narrow region along the edge channels where the edge channel electrons have finite probability amplitudes. The depth-wise spatial extent of the 2DEG wave function (in the direction normal to the heterostructure interface) is about 15 nm. The width-wise extent of individual wave function of edge states formed by the lowest orbital Landau level is roughly given by the magnetic length, $l_B = (h/eB)^{1/2} \approx 10$ nm. The location of the edge channels can be scanned width-wise by negatively biasing a metal side gate.[16] Preliminary experiments carried out to probe the cross sectional profile of DNP by scanning the edge channels suggest that the width of the DNP region is about 20 nm for a short-time initialization (< 20 s) but expands to about 45 nm as the initialization time increases to

180 s. The expansion of the DNP region suggests nuclear spin diffusion, but clarifying detailed dynamics is left for studies in the future.

The number of relevant nuclei is estimated to be $N \approx 10^9$ by assuming the width of the DNP region to be 20 nm and noting the thickness of the 2DEG wave function to be 15 nm and the length of edge channels to be 20 μm. This value is much smaller than that of the optical manipulation of nuclear spins, $N \gg 10^9$,[17] where the spot size of the circularly polarized light (diameter $\approx$ 50 μm) defines the active region. More importantly, coherent evolution of nuclear-spin state is not probed by the optical method.[17] Standard pulsed-NMR experiments on bulk quantum Hall systems have been reported,[18] where an external coil detects nuclear magnetic moments from the entire regions of 2DEG layers in multiple quantum wells. The number of relevant nuclear spins is hence larger than that in the present experiments by a factor more than five orders of magnitude.

39.6 Summary

Coherent control of local nuclear spins has been demonstrated based on a pulsed NMR in a quantum-Hall device. Compared to standard NMR measurements, the present work is distinguished by (39.1) its efficient initialization of nuclear-spin state via edge channels, (2) the local controllability of nuclear spins in a well-defined narrow region along the edge channels, and (3) the high sensitive electrical detection of nuclear polarization via conductance through the edge channels. Another potential strength of the present scheme may be its flexible device-design capability and the controllability. The location of polarized/detected nuclear spins is primarily defined by the incompressible region between spin-resolved edge channels. The location and the spatial pattern of edge channels can be manipulated during measurements with a simple gate biasing technique: The location of the incompressible strip can be controlled on nanometer-scale accuracy. Potentially, indirect interaction mediated by edge channels may serve as a means to control scalar coupling between different groups of nuclear spins.

Acknowledgments. This work is supported by Grant-in-Aid for Specially Promoted Research from the Japanese Ministry of Education, Culture, Sports, Science and Technology, by Solution Oriented Research for Science and Technology (SORST), Japan Science and Technology Corporation (JST), and by Precursory Research for Embryonic Science and Technology (PRESTO), Japan Science and Technology Corporation (JST).

References

1. L. M. K. Vandersypen et al., *Nature* (London) **414**, 883 (2001).
2. B. E. Kane, *Nature* (London) **393**, 133 (1998).
3. T. D. Ladd et al., E. Abe, and K. M. Itoh, *Phys. Rev. Lett.* **89**, 017901 (2002).
4. S. Krommüller et al., *Phys. Rev. Lett.* **81**, 2526 (1998).
5. B. E. Kane, L. N. Pfeiffer, and K. W. West, *Phys. Rev. B* **46**, 7264 (1992).
6. K. R. Wald et al., *Phys. Rev. Lett.* **73**, 1011 (1994).
7. D. C. Dixon, K. R. Wald, P. L. McEuen, and M. R. Melloch, *Phys.Rev. B* **56,** 4743 (1997).
8. T. Machida et al., *Phys. Rev. B* **65**, 233304 (2002).

9. K. Hashimoto, K. Muraki, T. Saku, and Y. Hirayama, *Phys. Rev. Lett.* **88**, 176601 (2002).
10. T. Machida, T. Yamazaki, and S. Komiyama, *Appl. Phys. Lett.* **80,** 4178 (2002).
11. T. Machida, T. Yamazaki, K. Ikushima, and S. Komiyama, *Appl. Phys. Lett.* **82,** 409 (2003).
12. H. Hirai, S. Komiyama, S. Sasa, and T. Fujii, *J. Phys. Soc. Jpn.* **58,** 4086 (1989).
13. G.Müller et al., *Phys. Rev. B* **45,** 3932 (1992).
14. M. Dobers et al., *Phys. Rev. Lett.* **61,** 1650 (1988).
15. T. Machida, T. Yamazaki, K. Ikushima, and S. Komiyama, *Physica E* **21,** 921–927 (2004).
16. Y. Acremann et al., *Physical Review B* **59,** 2116–2119 (1991).
17. G. Salis et al., *Phys. Rev. Lett.* **86,** 2677 (2001).
18. N. Freytag et al., *Phys. Rev. Lett.* **87,** 136801 (2001).

40

Cloning of single photon by high gain amplifier

F. SCIARRINO AND F. DE MARTINI

Dipartimento di Fisica, Universita' di Roma "La Sapienza" Istituto Nazionale di Fisica della Materia

Abstract

In the last years the simultaneous realization of the optimal cloning and optimal flipping of an input qubit has been reported adopting a quantum injected optical parametric amplifier. The previous process spreads the quantum information contained in the initial qubit into a three qubits state in an universal and optimal way. In the present chapter the previous results are extended to a large number of qubits. The successful generation of a multiparticle quantum superposition of photon states by a *universal* cloning of a single photon qubit amplified by a high gain parametric amplifier is reported. The *information preserving* property of the process suggests for these states the name of "*multi-particle qubits*". They are ideal objects for investigating the emergence of the classical world in quantum systems with increasing complexity, the decoherence processes and may allow the practical implementation of the universal two-qubit logic gates.

A fundamental aspect of the field of quantum information is what physical transformations to the state of a quantum system are allowed. Since manipulations of qubits, the basic amounts of quantum information, are constrained to the quantum mechanical rules, several classical information tasks cannot be perfectly extended to the quantum world. A well known and relevant QI limitation consists of the impossibility of perfectly cloning (copying) any unknown qubit $|\phi\rangle$.[1] In other words, the map $|\phi\rangle \overset{U}{\rightarrow} |\phi\rangle|\phi\rangle$ cannot be realized by Nature because it does not belong to the set of Completely Positive (CP) maps. Another forbidden operation is the NOT gate that maps any $|\phi\rangle$ in its orthogonal state $|\phi^{\perp}\rangle$.[2]

Even if these two processes are unrealizable in their exact forms, they can be optimally approximated by the so-called *universal quantum machines*, i.e., which exhibit the minimum possible noise. The *efficiency* of a gate, that measures how close its action is to the desired one, is generally quantified by the *fidelity* $\mathcal{F}.\mathcal{F} = 1$ implies a perfect implementation, while noisy processes correspond to: $\mathcal{F} > 1$. The Universal NOT (UNOT) gate, the optimal approximation of the NOT gate, maps N identical input qubits $|\phi\rangle$ into M optimal flipped ones in the state σ_{out}. It achieves the fidelity: $\mathcal{F}^{*}_{N\rightarrow M}(|\phi^{\perp}\rangle, \sigma_{out}) = \langle\phi^{\perp}|\sigma_{out}|\phi^{\perp}\rangle = (N+1)/(N+2)$ that depends only on the number of the input qubits.[3] Indeed the fidelity of the UNOT gate is exactly the same as the optimal quantum estimation fidelity.[4] This means that such process may be modeled as a "classical", i.e., exact, preparation of M identical flipped qubits following the quantum, i.e., inexact, estimation of N input states. Only this last operation is affected by noise. Differently from the UNOT gate, the Universal Optimal Quantum Cloning Machine (UOQCM) that transforms N identical qubits $|\phi\rangle$ into M identical copies ρ_{out}, achieves as optimal fidelity: $\mathcal{F}^*{}_{N\to M}(|\phi\rangle, \rho_{out}) = \langle\phi|\rho_{out}|\phi\rangle = (N+1+\beta)/(N+2)$ with $\beta \equiv N/M \leq 1$ [5] As we can see $\mathcal{F}_{N\to M}(|\phi\rangle, \rho_{out})$ is larger than the one obtained by the N estimation approach and reduces to that result for $\beta \to 0$, i.e., for an infinite number of copies. The extra positive term β in the above expression accounts for the excess of quantum information which is originally stored in N states and is optimally redistributed by entanglement among the $M-N$ remaining blank qubits encoded by UOQCM.

The UNOT gate and the UOQCM can be implemented following two different approaches. The first one has been based on finding a suitable unitary operator U_{NM}, acting on N input qubits and on $2(M-N)$ ancillary qubits.[5] At the output of this device M and $(M-N)$ qubits are obtained, which are, respectively, the optimal clones and the best flipped qubits of the input ones (Fig. 40.1(a)). The $N \to M$ cloning and the $N \to (M-N)$ flipping can also be implemented adopting a probabilistic approach that exploits a symmetrization process.[6,7] The initial state of the overall system consists of the N input qubits and of $(M-N)$ pairs of entangled qubits. The two optimal quantum machines are implemented by applying a projective operation on the symmetric subspace over the N input qubits and over $(M-N)$ ancillary qubits, each one belonging to a different entangled pair (Fig. 40.1(b)). This transformation assures the uniform distribution of the initial information into the overall system and guarantees that all output clone qubits are indistinguishable.

The present work reports on a nearly decoherence-free all optical cloning scheme based on the Quantum-Injected Optical Parametric Amplification (QI-OPA) of a single photon in a

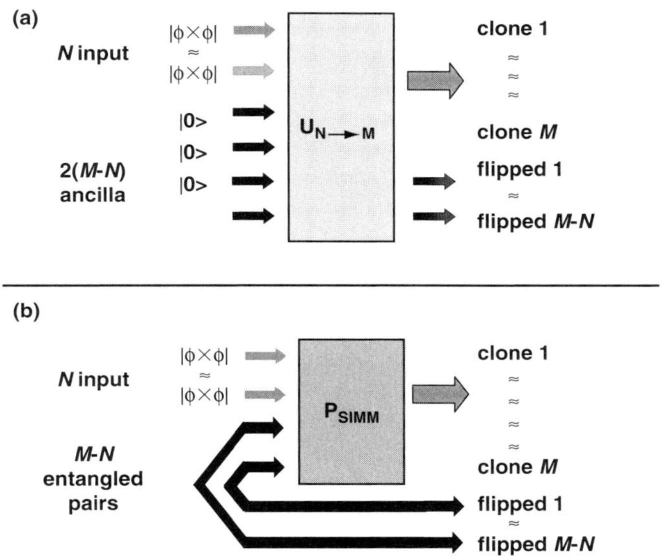

FIGURE 40.1. General scheme for the simultaneous realization of the UNOT gate and of the Universal Quantum Cloning Machine. (a) Unitary transformation acting on the N input qubits and $2(M-N)$ ancilla qubits initially in the state $|0\rangle$. (b) Symmetrization process acting on the input qubits and $(M-N)$ entangled pairs of qubits.

quantum superposition state of polarization (π), i.e., a π *–encoded qubit*.[8–10] In addition, a similar method was successfully adopted recently to realize the first $1 \rightarrow 2$ UOQCM and the first universal quantum NOT-gate.[11–13] Conceptually, the method consists of transferring the well accessible condition of quantum superposition characterizing a single-photon qubit, $N = 1$, to a *mesoscopic*, i.e., multi-photons amplified state $M \gg 1$, here referred to as a *multi-particle qubit* (*M-qubit*). In quantum optics this can be done by injecting in the QI-OPA the single-photon *qubit*, $\alpha|H\rangle + \beta|V\rangle$, here expressed in terms of two mutually orthogonal π–states, e.g., horizontal and vertical linear π's: $|H\rangle, |V\rangle$. In virtue of the general *information preserving* property of the OPA, the generated multi-particle state is found to keep the *same* superposition character and the interfering capabilities of the injected qubit, thus realizing the most relevant and striking property of the *M-qubit* condition.[10] Since the present scheme basically realizes the *deterministic* $1 \rightarrow M$ *universal optimal quantum cloning machine* (UOQCM), i.e., able to copy *optimally* any unknown input qubit into $M \gg 1$ copies with the same fidelity, the output state will be necessarily affected by *Squeezed-Vacuum* (SV) noise arising from the input vacuum field. Furthermore, since any UOQCM can be designed to redistribute *optimally* the initial information into many output channels, the present scheme is expected to find useful QI applications, e.g., as it can be used as a optimal eavesdropping in quantum cryptography[2,14,15] and in error correction schemes.[16]

Let's refer to the apparatus: Fig. 40.2. The OPA active element was a nonlinear (NL) crystal slab (BBO: β–barium borate), 1.5 mm thick cut for Type II phase-matching, able to generate by spontaneous parametric down conversion (SPDC) $\vec{\pi}$*–entangled* pairs of photons. Precisely, the OPA *intrinsic phase* was set as to generate by SPDC *singlet* entangled states on the output modes, a condition assuring the *university* of the cloning transformation by the QI-OPA scheme.[8,11,13] The photons of each pair were emitted with equal wavelengths (wls) $\lambda = 795$ nm over two spatial modes $-\mathbf{k}_1$ and $-\mathbf{k}_2$ owing to a SPDC process excited by a coherent pump UV field provided by a Ti:Sa Coherent MIRA modelocked pulsed laser coupled to a Second Harmonic Generator

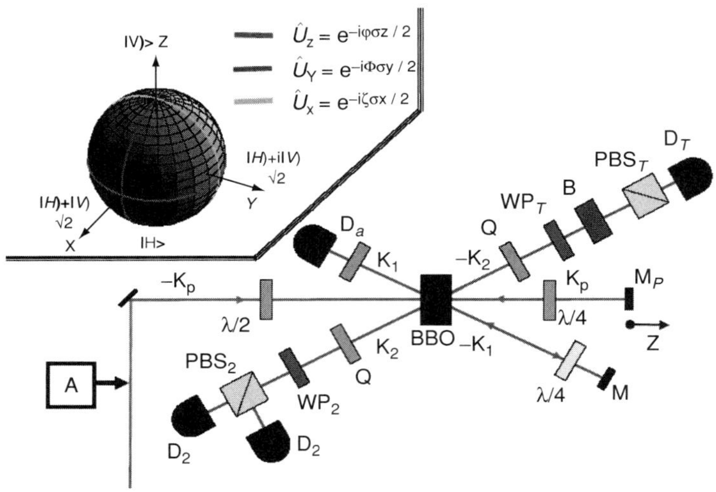

FIGURE 40.2. Layout of the *quantum-injected OPA* apparatus. The device **A** represents the Regenerative Ti:Sa laser Amplifier to attain the High Gain (HG) dynamical condition. INSET: Bloch sphere representation of the SU (2) unitary transformations applied to the input qubit.

(SHG) and associated with a mode with wavevector (wv) $-\mathbf{k}_p$ and wl $\lambda_p = 397.5$ nm. The average UV power was 0.25 W, the *pulse repetition rate* (*rep-r*), $r_{UV} = 7.6{*}10^7 s^{-1}$, and the *coherence time* of each UV pulse as well of the generated single photon pulses were $\tau_{coh} =$ 140 fs. The SPDC process implied a three-wave NL parametric interaction taking place towards the right hand side (r.h.s.) of Fig. 40.2. The UV pump was back-reflected over the mode $\mathbf{k}_p$ onto the NL crystal by a spherical mirror $\mathbf{M}_p$, with μ-metrically adjustable position $\mathbf{Z}$, thus exciting the main OPA "cloning" process, towards the left hand side (l.h.s.) of Fig. 40.2 By the combined effect of two adjustable optical UV waveplates (wp) $(\lambda/2+\lambda/4)$ acting on the projections of the linear polarization π_p of the UV field on the fixed optical axis of the BBO crystal for the $-\mathbf{k}_p$ and $\mathbf{k}_p$ counter propagating excitation processes, the SPDC excitation was always kept at a very low level while the main OPA amplification could reach *any large intensity*, as will be seen later in this chapter. Precisely, by smartly unbalancing the orientation angles $\vartheta_{\frac{1}{2}}$ and $\vartheta_{\frac{1}{4}}$ of the UV (wp's), the probability ratio of SPDC emission towards the r.h.s. of Fig. 40.2 two simultaneous correlated photon pairs and of a single pair was always kept below 3×10^{-2} in any *High Gain* condition. One of the photons of the SPDC emitted pair, back-reflected by a fixed mirror $\mathbf{M}$, was *reinjected* onto the amplifying NL crystal by the input mode $\mathbf{k}_1$, while the other photon emitted over mode $(-\mathbf{k}_2)$ excited the detector D_T, the *trigger* of the overall *conditional* experiment. The detectors (D) were single-photon SPCM-AQR14. A proper setting of $\mathbf{Z}$ secured the space-time overlapping into the NL crystal of the interacting re-injected pulses with wl's λ_p and λ, and then determined the optimal QI-OPA condition. The time optical walk-off effects due to the crystal birefringence were compensated by inserting in the modes $\mathbf{k}_1, \mathbf{k}_2$ and $-\mathbf{k}_2$ three fixed X-cut quartz plates Q and one $\lambda/4$ wp. Before re-injection into the NL crystal, the *pure* input qubit on mode $\mathbf{k}_1$, $|\Psi\rangle_{in} = (\tilde{\alpha}|\Psi\rangle^{\alpha}_{in} + \tilde{\beta}|\Psi\rangle^{\beta}_{in})$, $|\tilde{\alpha}|^2+|\tilde{\beta}|^2 = 1$, represented by the Bloch sphere shown in Fig. 40.2, underwent unitary *SU* (2) *rotation* Φ transformations: $\hat{U}_i \equiv \exp(-i\sigma_i \Phi/2)$ around the three Cartesian axes $i = x, y, z$ by the combined action of the $\lambda/2$ wp WP_T, of the adjustable *Babinet Compensator* B and of the polarizing beam-splitter PBS_T acting on mode $(-\mathbf{k}_2)$ in virtue of the nonlocality correlating the modes $-\mathbf{k}_1$ and $-\mathbf{k}_2$. These SU (2) transformations are represented in Fig. 40.2 (inset) by circles drawn on the surface of a Blochsphere. The main OPA process, i.e., acting towards the l.h.s. of Fig. 40.2 on the injected qubit $|\Psi\rangle_{in}$, was characterized by two different excitation regimes, establishing two corresponding *sizes* of the output *M-qubit*:

A) Low Gain (LG) regime, characterized by a low excitation UV energy (3.5 nJ) per pulse, leading to a small value of the NL parametric *gain*: $g = 0.07$.

B) High Gain (HG) regime, characterized by a larger value (by a factor ≈ 16) of the *gain*: $g = 1.13$. This condition was attained by a further amplification of the UV "pump" beam by a Ti-Sa regenerative amplifier Coherent-REGA operating at a pulse *rep-r*: $r_{UV} = 2.5 * 10^5$ s^{-1}, with pulse duration: 180 fs: Fig. 40.1.

Let us rewrite the state $|\Psi\rangle_{in}$ by expressing the interfering states as Fock product states: $|\Psi\rangle^{\alpha}_{in} = |1\rangle_{1h}|0\rangle_{1v}|0\rangle_{2h}|0\rangle_{2v} \equiv |1,0,0,0\rangle$; $|\Psi\rangle^{\beta}_{in} = |0,1,0,0\rangle$, accounting respectively for 1 photon with horizontal (h) polarization on the input $\mathbf{k}_1$, vacuum state on the input $\mathbf{k}_2$, and 1 photon with vertical $(v)\pi$ on $\mathbf{k}_1$, vacuum state on the mode $\mathbf{k}_2$. The solution of the QI-OPA dynamical equations is found to be expressed by the *M-qubit*: $|\Psi\rangle_{out} \equiv (\tilde{\alpha}|\Psi\rangle^{\alpha} + \beta|\Psi\rangle^{\beta})$, with

$$|\Psi\rangle^{\alpha} \equiv \gamma \sum_{i,j=0}^{\infty} (-\Gamma)^i \Gamma^j \sqrt{i+1}|i+1, j, j, i\rangle \tag{40.1}$$

$$|\Psi\rangle^{\beta} \equiv \gamma \sum_{i,j=0}^{\infty} (-\Gamma)^{i} \Gamma^{j} \sqrt{j+1} |i, j+1, j, i\rangle \tag{40.2}$$

where $\gamma \equiv C^{-3}, C \equiv \cosh g$ and $\Gamma \equiv \tanh g$.[11] These interfering *entangled*, multi-particle states are *orthonormal*, i.e., $|^{i}\langle\Psi||\Psi\rangle^{j}|^{2} = \delta_{ij}\{i, j = \alpha, \beta\}$ and *pure*, i.e., fully represented by the operators $\rho^{\alpha} = (|\Psi\rangle\langle\Psi|)^{\alpha}$, $\rho^{\beta} = (|\Psi\rangle\langle\Psi|)^{\beta}$. Hence the *pure* state $|\Psi\rangle_{\text{out}}$ is an entangled quantum superposition of two multi-photon *pure states* and bears the *same* superposition properties of the injected single-photon qubit. The genuine quantum signature of this superposition, namely the *non-definite positive* character of the Wigner function $W(\alpha, \beta)$ of the output $|\Psi\rangle_{\text{out}}$ in the eight-dimensional complex phase-space $\alpha_{j}, \beta_{j}, j = 1, 2$, was confirmed by a previous theoretical analysis.[8] For the sake of completeness, consider the overall output density operator $\rho \equiv |\Psi\rangle\langle\Psi|$ and his *mixed-state* reductions over the $\vec{\pi}$*–vector* spaces relative to the modes $\mathbf{k}_1$ and $\mathbf{k}_2$: $\rho_1 = Tr_2\rho$; $\rho_2 = Tr_1\rho$. These ones may be expanded as a weighted superpositions of p*–square* matrices of order $p = (n+2)$, the relative weight Γ^2 of each two successive matrices being determined by the parametric *gain*. Note that Γ^2 approaches asymptotically the unit value for large g. In turn, the p*–square* matrices may be expressed as sum of 2×2 matrices as shown by the following expressions:

$$\rho_1 = \gamma^2 \sum_{n=0}^{\infty} \Gamma^{2n} \times \sum_{i=0}^{n} \begin{bmatrix} |\tilde{\beta}|^2(n-i+1) & \tilde{\alpha}^*\tilde{\beta}\sqrt{(i+1)(n-i+1)} \\ \tilde{\alpha}\tilde{\beta}^*\sqrt{(i+1)(n-i+1)} & |\tilde{\alpha}|^2(i+1) \end{bmatrix} \tag{40.3}$$

written in terms of the Fock basis: $\{|i\rangle_{1h}|n-i+1\rangle_{1v}; |i+1\rangle_{1h}|n-i\rangle_{1v}\}$. Correspondingly

$$\rho_2 = \gamma^2 \sum_{n=0}^{\infty} \Gamma^{2n} \times \sum_{i=0}^{n+1} \begin{bmatrix} |\tilde{\beta}|^2(n-i+1) & -\tilde{\alpha}^*\tilde{\beta}\sqrt{(n-i+1)i} \\ -\tilde{\alpha}\tilde{\beta}^*\sqrt{(n-i+1)i} & |\tilde{\alpha}|^2 i \end{bmatrix} \tag{40.4}$$

in terms of the Fock basis: $\{|n-i\rangle_{2h}|i\rangle_{2v}; \ |n-i+1\rangle_{2h}|i-1\rangle_{2v}\}$. Interestingly, the value n appearing in Eqs. (40.3) and (40.4) coincides with the number of photon pairs generated by the QI-OPA amplification. Note the

non-diagonal character of these matrices implying the quantum superposition property of the overall state. Furthermore the interfering states possess the *nonlocal* property of all entangled quantum systems.

The actual $|\Psi\rangle_{\text{out}}$ experimental detection of the output states could be undertaken by measurements taken either (a) on the injection, i.e., cloning mode $\mathbf{k}_1$, or (b) on the *anticloning* mode $\mathbf{k}_2$, or (c) on both output modes. The option (b) was selected since there the field on mode $\mathbf{k}_2$ is not affected by the input qubit in absence of the QI-OPA action and then any registered *interference* effect is by itself an unambiguous demonstration of the *M-qubit* condition. This condition was carefully verified experimentally. In addition, it was checked that only *non interfering* SV-noise was detected on the output mode $\mathbf{k}_2$, in absence of the injection qubit $|\Psi\rangle_{\text{in}}$.

Since the 1st-order interference property of any quantum object is generally expressed by the 1st-*order correlation-function* $G^{(1)}$,[17] this quantity was measured by two detectors D_2 and D_{2^*} coupled to the mutually orthogonal fields $\hat{c}_j(t)$, $j = 1, 2$ emerging from the polarizing beam-splitter PBS_2 inserted in the output $\mathbf{k}_2$: $G^{(1)}_{2j} \equiv_{\text{out}} \langle\Psi|\hat{N}_j(t)|\Psi\rangle_{\text{out}}$. In the present context the $G^{(1)}$ are expressed by ensemble averages of the number operators $\hat{N}_j(t) \equiv \hat{c}^{\dagger}_j(t)\hat{c}_j(t)$, written in terms of the detected fields: $\hat{c}_1(t) \equiv 2^{-\frac{1}{2}}[\hat{b}_h(t) + \hat{b}_v(t)]$, $\hat{c}_2(t) \equiv 2^{-\frac{1}{2}}[\hat{b}_h(t) - \hat{b}_v(t)]$. There the $\hat{b}_i(t)$, $i = h, v$, defined on the QI-OPA output $\mathbf{k}_2$, underwent a $45°\pi$-rotation by the $\lambda/2$-wp WP_2 before injection into PBS_2: Fig. 40.2. The same operations transformed the basis $\{h, v\}$ into H,V. By expressing the injected qubit as: $|\Psi\rangle_{in} \equiv (\alpha|\Psi\rangle^{\alpha}_{in} + \beta e^{i\varphi}|\Psi\rangle^{\beta}_{in})$ with α and β real numbers, the $G^{(1)}_{2j}$ could be given as: $G^{(1)}_{2H} = \bar{n} + \frac{1}{2}\bar{n}[12\alpha\beta\cos\varphi]$; $G^{(1)}_{2V} = \bar{n} + \frac{1}{2}\bar{n}[1 - 2\alpha\beta\cos\varphi]$, showing the superposition character of the output average field per mode with respect to α, β and φ, where $\bar{n} \equiv \sinh^2 g$. By comparison of this result with the corresponding *non interfering* averages $G^{(1)}_{21,\text{vac}} = G^{(1)}_{22,\text{vac}} = \bar{n}$ taken in absence of the injected qubit, i.e., over the *input vacuum state*, the signal-to-noise ratio for the *M-qubit* detection: was found: $S/N = 2$, for $\varphi = 0$ and $\alpha = \beta = 2^{-\frac{1}{2}}$.

The above result immediately suggests, as a demonstration of the 1st-*order* π-interference of the *M-qubit*, to draw the *fringe* patterns expressing the quantity: $\Delta G^{(1)}_2 \equiv (G^{(1)}_{2H}) - G^{(1)}_{2V}$ and bearing the fringe "visibility": $\mathcal{V} = (G^{(1)}_{2\max} - G^{(1)}_{2\min})/(G^{(1)}_{2\max} + G^{(1)}_{2\min}) = 2\alpha\beta/3$. The experimental values of $G^{(1)}$ were obtained by coincidence measurements involving the couples of detectors: $[D_2D_T]$ and $[D^*_2D_T]$. Note that the fringe patterns, reported in Fig. 40.3 and the corresponding values of $\mathcal{V}$ were strongly affected by the superposition character of the injection qubit and by the corresponding SU(2) transformations expressed by the closed paths drawn on the Bloch sphere: Fig. 40.2 (inset). In order to show more clearly the quantum efficiency of the *M-qubit* for the two "gain" regimes, the *relative coincidence rates*, $\xi = r_{SC}/r_{UV}$ were reported in Fig. 40.3, where r_{SC} expressed the values of $\Delta G^{(1)}_2 \equiv (G^{(1)}_{2H} - G^{(1)}_{2V})$. The experimental best-fit curves drawn in Fig. 40.3 were found to reproduce very closely the quantum theoretical results.

In the LG excitation condition the *average* number of quantum interfering photon pairs was $\overline{N} \approx 0.009$ while the HG condition it was $\overline{N} \approx 4$, corresponding to the realization of a much "fatter" *M-qubit*. The value of *average* number of generated pairs was $\overline{N} = 3\bar{n}$ in virtue of the stimulated emission process. They were estimated by accounting for the overall *quantum efficiency* of the D detectors: QE $\approx$ 18%. Calibrated attenuation filters placed in front of the D's assured the condition of single-photon detection in the HG regime. By the Eqs. (40.3) and (40.4) straightforward evaluation of the output photon-pair distribution $p(n)$ could be carried out. By this one, the probability of generating a number of pairs $N \geq 2\overline{N} = 8$, i.e., a quantum superposition in excess of 16 photons, was found to be $P(\overline{N}) \equiv \sum_{k=2\overline{N}}^{\infty} p(k) \approx 14\%$. By moving the injected state with $\alpha = \beta$ over the Bloch sphere, a visibility $V_{th} = 33\%$ of the fringe pattern was expected from theory. Due to experimental imperfections, mostly due to the mode selection before detection, we have measured $V \approx 4\%(4\%)$ with a two-coincidence detection measurement scheme and $V \approx 25\%(7\%)$ with a four-coincidence scheme respectively for the LG and HG regimes. The results above are expected to be *linearly scaled* by adoption of a more efficient NL crystal and of a more powerful UV source.

Unlike other systems involving atoms or ions a limitation of this method consists of the impossibility of controlling the "distance", d on the *phase-space* of the interfering multi-particle states by means of the QI-OPA parameters. This condition, implied by the results of the Wigner function analysis[8] can be expressed here by the Hilbert–Schmidt (d) of the interfering states: $d(\rho^{\alpha};\rho^{\beta}) = Tr[(\rho^{\alpha} - \rho^{\beta})^2]$.[18–20] This quantity, not affected by the amplification process, is expressed in our case by: $d(\rho^{\alpha};\rho^{\beta})_{in} = d(\rho^{\alpha};\rho^{\beta})_{out} = 2$. However, as a lucky counterpart of this condition, the decoherence of our system can be deemed generally irrelevant as it is mostly determined by the stray reflection losses on the surfaces of the optical components of the apparatus. Accordingly, a number of photons in the range $(10^2 \div 10^3)$ could be made simultaneously excited in quantum superposition without sizable decoherence effects: this is but one of the commendable qualities of the device. Of course, the injection into the OPA of a many particle state will result in a better *S/N* ratio.

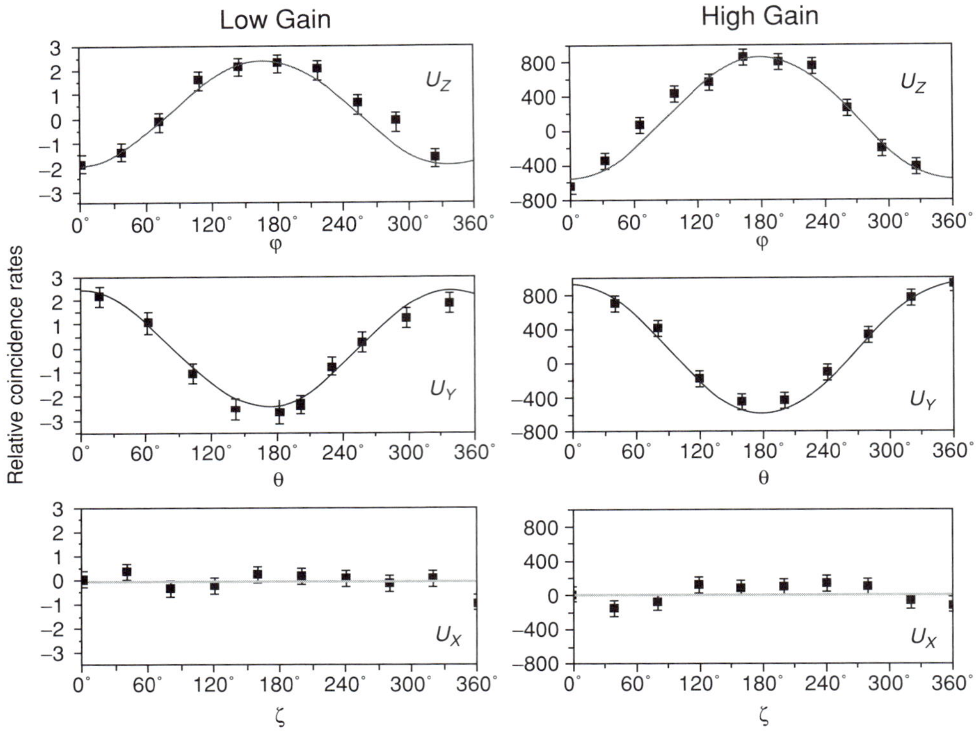

FIGURE 40.3. Multi-photon interference fringe patterns expressed by the *relative coincidence rate*, $\xi = \eta(r_{SC}/r_{UV})$ proportional to the ratio of the two-detector coincidence rates expressing the quantity: $\Delta G_2^{(1)} \equiv (G_{2H}^{(1)} - G_{2V}^{(1)})$ and of the repetition rate of the UV pump excitation in LG and HG regimes ($\eta = 7.6 \times 10^7$). The patterns correspond to the SU(2) transformations shown in Fig. 40.1 (inset) and affecting the injected qubit. $G^{(1)}$ are the 1st *-order correlation functions.*

In summary, the interference of classically distinguishable *multi-particle, orthonormal, pure* quantum states have been realized. The adoption of the OPA to *clone universally* a single qubit in a large gain regime witnesses the scalability of the information preserving property of this device in the quantum domain. In a more conceptual perspective, the present realization could open a new trend of studies on the persistence of the validity of several crucial laws of quantum mechanics for entangled mixed-state systems of increasing complexity and on the violation of Bell-type inequalities in the multi-particles regime.[21]

Acknowledgments. This work has been supported by the FET European Network on Quantum Information and Communication (Contract IST-2000-29681: ATESIT), by Istituto Nazionale per la Fisica della Materia (PRA "CLON") and by Ministero dell'Istruzione, dell'Università e della Ricerca (COFIN 2002).

References

1. W. K. Wootters, and W. H. Zurek, *Nature* (London) **299**, 802 (1982); D. Dieks, *Phys. Lett.* **92**, 271 (1982).
2. H. Bechmann-Pasquinucci, and N. Gisin, *Phys. Rev. A* **59**, 4238 (1999).
3. V. Buzek, M. Hillery, and R. F. Werner, *Phys. Rev. A* **60**, R2626 (1999); N. Gisin, and S. Popescu, *Phys. Rev. Lett.* **83**, 432 (1999).
4. S. Massar, and S. Popescu, *Phys. Rev. Lett.* **74**, 1259 (1995).
5. V. Buzek, and M. Hillery, *Phys. Rev. A* **54**, 1844 (1996); N. Gisin, and S. Massar, *Phys. Rev. Lett.* **79**, 2153 (1997); D. Bruss, A. Ekert, and C. Macchiavello, *Phys. Rev. Lett.* **81**, 2598 (1998); V. Buzek, and M. Hillery, *Phys. Rev. Lett.* **81**, 5003 (1998); R. Derka, V. Buzek and A. Ekert, *Phys. Rev. Lett.* **80**, 1571 (1998).
6. M. Ricci, F. Sciarrino, C. Sias, and F. De Martini, *Phys. Rev. Lett.* **92**, 047901 (2004); F. Sciarrino, C. Sias, M. Ricci, and F. De Martini, *Phys. Lett. A* **323**, 34 (2004); F. Sciarrino, C. Sias, M. Ricci, and F. De Martini, *Phys. Rev. A* (in press), quant-ph/0407176.
7. R. F. Werner, *Phys. Rev. A* **58**, 1827 (1998).
8. F. De Martini, *Phys. Rev. Lett.* **81**, 2842 (1998).
9. F. De Martini, G. Di Giuseppe and S. Padua, *Phys. Rev. Lett.* **87**, 150401 (2001). There the double-injection did not allow a direct detection of the interference.
10. F. De Martini, and F. Sciarrino, *Phys. Rev. Lett.* (in press).
11. F. De Martini, V. Buzek, F. Sciarrino, and C. Sias, *Nature* (London) **419**, 815 (2002).
12. D. Pelliccia, V. Schettini, F. Sciarrino, C. Sias, and F. De Martini, *Phys. Rev. A* **68**, 042306 (2003); F. De Martini, D. Pelliccia, and F. Sciarrino, *Phys. Rev. Lett.* **92**, 067901 (2004).
13. A. Lamas-Linares, C. Simon, J.C. Howell, and D. Bouwmeester, *Science* **296**, 712 (2002).
14. W. Dur, C. Simon, and J.I. Cirac, *Phys. Rev. Lett.* **89**, 210402 (2002).
15. E. F. Galvão, and L. Hardy, *Phys. Rev. A* **62**, 022301 (2000).
16. D. Bruβ, J. Calsamiglia, and N. Lütkenhaus, *Phys. Rev. A* **63**, 042308 (2001).
17. D. Walls, G. Milburn, *Quantum Optics* (Springer, N.Y. 1994), Chs 5, 16.
18. M. Nielsen and I. Chuang, *Quantum Computation and Information* (Cambridge, 2000) Ch 9.
19. M. Brune, *et al.*, *Phys. Rev. Lett.* **77** 4887 (1996), A. Auffeves, *et al. Phys. Rev. Lett.* **91**, 230405 (2003), M.W. Noel, *et al.*, *Phys. Rev. Lett.* **77**, 1913 (1996).
20. C. Monroe, *et al*, *Science* **272**, 1131 (1996); D. Leibfried, *et al.*, *Science* **304**, 1476 (2004), C. F. Roos, *et al.*, *Science* **304**, 1478 (2004).
21. M. D. Reid, W. L. Munro, and F. De Martini, *Phys. Rev. A* **66**, 033801 (2002).